POWER ELECTRONICS
Principles and Applications

Joseph Vithayathil
California Polytechnic
State University

McGraw-Hill, Inc.

New York   St. Louis   San Francisco   Auckland   Bogotá   Caracas   Lisbon
London   Madrid   Mexico City   Milan   Montreal   New Delhi
San Juan   Singapore   Sydney   Tokyo   Toronto

7-95   # 31970379

This book was set in Times Roman.
The editors were George T. Hoffman and John M. Morriss;
the production supervisor was Paula Keller.
The cover was designed by Edward Smith Design.
Project supervision was done by The Universities Press (Belfast) Ltd.
R. R. Donnelley & Sons Company was printer and binder.

POWER ELECTRONICS
Principles and Applications

This book is printed on acid-free paper.

1 2 3 4 5 6 7 8 9 0 DOC DOC 9 0 9 8 7 6 5

ISBN 0-07-067555-4

Library of Congress Cataloging-in-Publication Data

Vithayathil, Joseph.
    Power electronics: principles and applications / Joseph
Vithayathil.—International ed.
        p.   cm.—(McGraw-Hill series in electrical and computer
engineering. Power and energy)
    Includes bibliographical references and index.
    ISBN 0-07-067555-4
    1. Power electronics.   I. Title.   II. Series: McGraw-Hill series
in electrical and computer engineering.   III. Series: McGraw-Hill
series in electrical and computer engineering. Power and energy.
TK7881.15.V58   1995
621.31'7—dc20   95-2753

ABOUT THE AUTHOR

Joseph Vithayathil received his Ph.D. degree in Electrical Engineering from Imperial College, London University, where he was a research scholar from the Indian Institute of Science, Bangalore. While on the faculty at IISc, Bangalore, Dr. Vithayathil conducted several research projects in the area of Power Electronics, and was recognized for his work by the Bimal Bose Award from the IETE in India. From 1984 until his retirement in 1992, Dr. Vithayathil taught at the California Polytechnic State University in San Luis Obispo.

The author dedicates this work
to the memory of his father,
Joseph Vithayathil (Senior),
who passed away while this work was in progress

CONTENTS

3 AC Switching Controllers

4 Choppers

PREFACE

Power electronics is the technology that links the two major traditional divisions of electrical engineering, namely, electric power and electronics. It has shown rapid development in recent times, primarily because of the development of semiconductor power devices that can efficiently switch large currents at high voltages, and so can be used for the conversion and control of electrical energy at high power levels. The parallel development of functional integrated circuits for the controlled switching operation of power electronic converters for specific applications has also contributed to this development. Power electronic techniques are progressively replacing traditional methods of power conversion and control, causing what may be described as a technological revolution, in power areas such as regulated power supply systems, adjustable speed DC and AC electric motor drives, high voltage DC links between AC power networks, etc. The need to include power electronics in the undergraduate curriculum for electrical engineers is now well accepted.

But presumably because this is a developing technology, the selection of topics and the boundaries of treatment for the teaching of power electronics do not seem to have emerged very clearly. As a consequence, a well-accepted pattern of teaching does not seem to exist. The choice and treatment of topics seem to be dependent, to a more than legitimate extent, on the background and areas of interest of individual instructors. This book embodies the experience of the author in teaching power electronics over several years, and presents what he considers as a logical organization of the presentation of the subject. The author has attempted to present the topics in a manner that keeps in focus an underlying unity, and proper interrelationships. The title of the book "Power Electronics, Principles and Applications" reflects its coverage. Although it is not appropriate to totally exclude the "applications" aspects from a treatment of "principles," the priority in the first five chapters is on the latter. Similarly, the remaining four chapters deal primarily with the applications aspects. A brief overview of the coverage may be appropriate here.

The practical semiconductor power devices that are used in power electronic converters are presented in the first chapter. Power electronics is essentially the technology of power conversion and control, by using semiconductor power devices as static switches. Therefore a background knowledge of the power switching devices available should serve to provide a proper perspective. The treatment of power converters in subsequent chapters is general, and—to the extent possible—independent of the actual switching element employed. The first chapter can be quickly covered if the reader has some background of semiconductor behavior and the properties of the pn junction.

Line commutated converters constitute the subject matter of the second and third chapters. A large number of industrial power electronic converters belong to this category. Line commutated converters have, historically, evolved from earlier converters that used the mercury arc valve, which was the forerunner of the present-day thyristor. Therefore the treatment of this class of converters follows traditional lines, since it has evolved from mercury arc converter circuits, and became adapted to suit modern thyristor circuits. Chapter 3 deals with power control in AC circuits. This is basically the use of line commutated circuits, using inverse parallel thyristor pairs, or triacs, to control bi-directional currents, and is therefore an extension of Chapter 2.

Chapters 4 and 5 present respectively DC/DC converters, commonly known as choppers, and DC/AC inverters. The technique of pulse width modulated switching, widely used in power electronic converters, is introduced here. DC/AC inverter configurations are basically extensions of the DC/DC chopper converter. The concept of sinusoidal pulse width modulation is introduced from basics in the chapter on inverters. Inverter circuit configurations are introduced sequentially, starting from the half-bridge and proceeding to the full three-phase bridge. The treatments in Chapters 4 and 5 are, by and large, independent of the previous two chapters on line commutated converters. Therefore the instructor has a choice of starting with Chapter 4 directly after the introductory treatment of devices covered in Chapter 1.

Chapters 2–5 may generally be described as the "principles" part of the coverage. The remaining chapters deal essentially with the "applications" aspects. It is, however, difficult to isolate these two aspects of the treatment, and the difference is actually in the emphasis. Chapter 6 deals with switching power supplies. These are now widely used for the power supply block of personal computers and various types of medium and large size electronic equipment. These in essence are applications of the pulse width modulated switching technique, introduced earlier in Chapter 4. This chapter includes a treatment of the technique of using a high power factor preregulator to make the current drawn from the power lines close to a sinusoid, to avoid the pollution of the power mains through current harmonics. Present-day concerns regarding the adverse effects on the power supply quality resulting from the excessive creation of harmonic currents by the large scale use of electronic equipment is making this technique popular. A treatment of uninterruptible power supplies (UPS) is also included in Chapter 6. This is basically a practical application of the inverters dealt with in Chapter 5.

DC motor drives are treated in Chapter 7. The focus is on the separately excited DC motor drive, which is the type most widely used in industrial adjustable speed drive systems, and provides the best dynamic performance. A more detailed treatment of the phase controlled dual converter, which is commonly used for the four-quadrant operation of industrial DC drives, is included in this chapter. The decoupled control of the armature current and the field flux, which is the basis of vector control of AC motor drives, is introduced here, in the simpler context of the separately excited DC motor drive. Also introduced here is the technique, also usable in vector controlled AC drives, of nested control loops, with the closed-loop control loops organized as an inner torque loop, and outer speed and position loops sequentially introduced for speed and position controls where necessary. The analysis and simulation of controller performance is not included in the scope of the treatment.

Chapters 8 and 9 are devoted to the treatment of AC motor drives, which are rapidly growing in popularity at present. The treatment is limited to drive systems using conventional induction and synchronous motors. The essential background theory of these two types of motors is also developed. This may be redundant for a student who has taken an earlier course on electrical machines. But it will still be useful for review, and it is in a summary form with only the essentials necessary to make the treatment independent. Chapter 8 outlines the different converter topologies and control schemes that have found practical use in induction and synchronous motors. Chapter 9 is devoted to an introductory treatment of the vector control technique. This is in view of the fact that vector controlled AC motor drives provide superior dynamic performance, and are beginning to replace DC motor drive systems. The concepts necessary for a mathematical description of vector control, such as the concept of space vectors, transformations between two-phase and three-phase equivalent machines, and the transformation of space vectors from one rotating reference frame to another are developed from first principles, followed by a typical implementation of this control technique using a field oriented reference frame.

The book is primarily intended for use at the senior undergraduate level in an engineering degree program. The complete coverage of the book is most conveniently done in two sequential courses, extending over two semesters. It can also be done in two quarters by eliminating certain topics. It could also be used for a one-semester introductory course on the principles of power electronics, if the bulk of the material from Chapters 3 and 6–9 is omitted. The actual topics to be skipped for a condensed course will be evident to an instructor who skims through the pages of the book.

The book is written in a manner suitable for self study also, without aid from an instructor. The basic background of solid state electronics and electric circuit theory that is generally available to a student entering the senior undergraduate year or earlier will be sufficient for this purpose. Possible questions that may arise are anticipated to the extent possible, and explained. There are illustrative examples worked out throughout the text. A student who works through these should not experience difficulty with the end-of-chapter problems, which are chosen to aid understanding of the text. It is the author's hope that this book will help towards setting an efficient pattern for the teaching of power electronics, and thereby contribute towards the training of engineers in this expanding area of practical applications.

The author is conscious of his debt of gratitude to the many students who have taken his courses in power electronics, both at the Indian Institute of Science, Bangalore, and at the California Polytechnic State University, San Luis Obispo, whose responses to his teaching have contributed to the shaping of many of the presentations in this book. He also wishes to express his thanks to all his colleagues and co-workers for their cooperation and encouragement during the times when he served on the faculty at these institutions. In particular, he wishes to thank Dr. Samir Datta and Dr. James Harris of Cal-Poly, and Dr. Ranganathan, Dr. Ramanarayanan, Dr. S. K. Sinha and Dr. Gopakumar of IISc.

He is very grateful to Ashoka, K. S. Bhat, University of Victoria; L. L. Brigsby, Auburn University; Prasad Enjeti, Texas A&M University; A. A. El Keib, University of Alabama; Ali Keyhani, The Ohio State University; Phil Krein, University of Illinois at Urbana-Champaign; V. Rajagopalan, Directeur GREI–Université du Quebec à Trois-Rivières; Satish Ranade, New Mexico State University; Andrzej Trzynadlowski, University of Nevada; and Longya Xu, The Ohio State University for reviewing the manuscript and offering constructive suggestions. Several others have helped directly or indirectly in this work and the author wishes to thank Dr. G. K. Dubey of IIT, Kanpur, Dr. S. K. Biswas of Jadavpur University, Dr. S. Sathiakumar of Sydney University, Dr. K. S. Rajashekara, Dr. R. Palaniappan, Dr. H. V. Manjunath and Dr. A. K. Kargekar.

This work would not have been possible without the patient support of the author's wife Gracy, to whom he wishes to express his deep gratitude. He also wishes to thank Jose, Rosemarie, Varghese, Maymol, Kurian and Mini, for their encouragement and support.

The author wishes to express his thanks to Anne Brown, the former editor at McGraw-Hill, who initiated this project, to the present editor, George Hoffman, and to the editing supervisor, John Morriss, all of whom and whose associates have been very helpful and understanding during the progress of this project. It has been a pleasure to work with Wallace McKee and the Staff of The Universities Press, Belfast, Northern Ireland, during the production of this work.

Joseph Vithayathil

CHAPTER

1

POWER
SEMICONDUCTOR
DEVICES

1.1 INTRODUCTION

Power Electronics is the technology of converting electric power from one form to another using electronic power devices. Several types of solid state power semiconductor devices have been developed in recent years, making it possible to build efficient power converters with excellent facility for control of output parameters, such as voltage, current or frequency. In a static power converter, the power semiconductor devices function as switches, which operate statically, that is, without moving contacts. The time durations, as well as the turn ON and turn OFF operations of these switches, are controlled in such a way that an electrical power source at the input terminals of the converter appears in a different form at its output terminals. In most types of converters, the individual switches in the converter are operated in a particular sequence in one time period, and this sequence is repeated at the switching frequency of the converter. Figure 1.1 shows two simple conversion schemes to illustrate this statement.

Figure 1.1(a) shows a scheme for converting DC to AC. This type of power conversion is called inversion, and the circuit itself is called an inverter. Our inverter power circuit consists of four switches labeled S_1, S_2, S_3 and S_4 connected in the manner shown in Fig. 1.1(a). The input is a DC voltage source of magnitude V (in V) connected to the input terminals of the inverter,

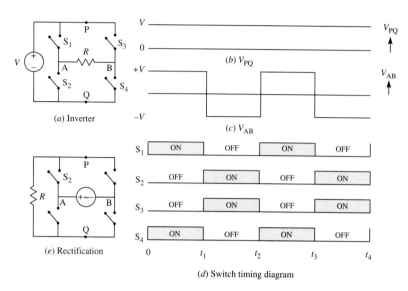

FIGURE 1.1
Power conversion by switching.

which are P (positive) and Q (negative). The timing of the switches is shown in Fig. 1.1(d). For example, from instant $t = 0$ to instant $t = t_1$, switches S_1 and S_4 are kept ON, the other two being kept OFF. Therefore the input DC voltage appears at the output terminals with terminal A positive. During the next interval from t_1 to t_2, S_1 and S_4 are kept OFF, but S_2 and S_3 are kept ON. Therefore, during this interval, the input DC voltage appears at the output terminals with reversed polarity (A negative). This sequence of switching is repeated, and in this way the input voltage V of fixed polarity shown in Fig. 1.1(b) is presented at the output terminals PQ as an AC square wave voltage as shown in Fig. 1.1(c).

Figure 1.1(e) shows how the same circuit configuration of four switches can be used to convert an AC voltage source connected at the terminals AB into a DC voltage at the terminals PQ. This type of power conversion is called rectification, and the circuit itself a rectifier. In our rectifier of Fig. 1.1(e), the switches S_1, S_2, S_3 and S_4 are operated according to the same timing as indicated by Fig. 1.1(d). The input terminals of the rectifier are now A and B, to which the AC square wave voltage such as that shown in Fig. 1.1(c) is connected. The output terminals of our rectifier are P (positive) and Q (negative). The input AC voltage is now presented as a unidirectional voltage at the output terminals. Notice that the directions of current through the switches will be different in Figs 1.1(a) and (e).

The above examples are only intended to highlight the role of static switches in the power conversion process. The objective of this book is to present the principles underlying power conversion by the use of static switches and the techniques employed for controlling output parameters such as voltage, current, power, frequency and waveform. We shall, in a progressive sequence, present all the important types of power converters that have proved useful in the application areas of electric power. We shall also present important application areas, and this will bring out how converter schemes and control strategies can be tailored to meet specific needs. We begin this by describing the static semiconductor power switches themselves in the present chapter.

1.2 TYPES OF STATIC SWITCHES

1.2.1 Uncontrolled Static Switch—The Power Diode

The simplest static switch is the diode. A power diode is a two-terminal device whose circuit symbol is shown in Fig. 1.2(*a*). If a diode is present in an electrical circuit in such a way that its anode (terminal A) has a positive potential with respect to its cathode (terminal K), it is said to be forward-biased. An ideal diode conducts when forward-biased, with negligible voltage drop across it and a forward current, shown as I_F in Fig. 1.2(*a*), flows through it. If, however, it is reverse-biased, ideally, it does not conduct. A real diode will have a small forward voltage drop across it when it conducts and a small reverse leakage current when it is reverse-biased. For the purposes of our present description, we shall treat the power diode as ideal. Therefore an ideal diode, considered as a static switch, turns ON automatically whenever the external circuit can send a forward current through it. It turns OFF automatically whenever the external circuit attempts to send a reverse current though it by impressing a reverse voltage. The power terminals of the switch are A and K. It has no control terminal through which we can control its ON and OFF

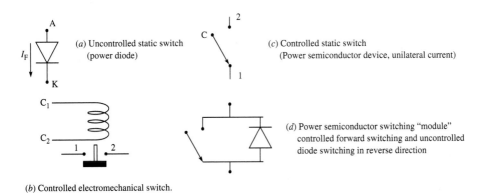

(*a*) Uncontrolled static switch
  (power diode)

(*c*) Controlled static switch
  (Power semiconductor device, unilateral current)

(*b*) Controlled electromechanical switch.
  Has moving contacts. Nonstatic operation

(*d*) Power semiconductor switching "module"
  controlled forward switching and uncontrolled
  diode switching in reverse direction

FIGURE 1.2
Types of switches.

switching operations. The switch blocks reverse voltages, but has no capability to block forward voltages. We can describe the ideal power diode as an uncontrolled static switch that turns ON and turns OFF by itself, depending on the polarity of the voltage. For a switch of this kind, we shall use the diode circuit symbol itself, shown in Fig. 1.2(a).

1.2.2 Controlled Switch

A controlled switch is one that could be turned ON and OFF by activating and deactivating a control circuit. Figure 1.2(b) shows a nonstatic (with moving contacts) switch of this kind. It has a control coil whose terminals are labeled C_1 and C_2. The power terminals of the switch are labeled 1 and 2.

To turn ON this switch, we send a current through the control coil, which will cause the plunger to move and connect the power terminals. The following aspects should be noted here.

1. The switch has four terminals—two for the power circuit and two for the control circuit.
2. The control circuit is electrically isolated from the power circuit.

The electrical isolation between the power terminals and the control circuit is very often a requirement in static power converters. The control circuit block in a static converter provides the electrical signals to perform the switching operations of the static switches. The control circuit often consists of low voltage electronic components such as analog and digital integrated circuits, working from a low voltage power supply with respect to ground. The control circuit will be damaged if large voltages with respect to ground are impressed on it. In a static power converter, the power terminals of a static switch may reach high voltages with respect to ground. Also, during the switching, these potentials change by large magnitudes in each switching cycle. It is important to ensure that these large voltages, as well as the fast changes that occur in them, do not disturb the control circuit. Unfortunately, power semiconductor switching devices presently available do not provide any isolation at all between the power and control terminals. In fact, in the case of every power semiconductor switch we shall present in this chapter, one of the control terminals is common with one of the power terminals. All are therefore three-terminal devices. A typical example is the power transistor. The power terminals are the collector and the emitter. The third terminal, namely the base, and the emitter are the control terminals. We turn ON the switch by sending the control current between the base and emitter. When electrical isolation between power and control circuits is a requirement, we have to use an external isolating device, typically a pulse transformer, to couple the control circuit to the control terminals of the switching device.

The circuit symbol that we shall use to represent an ideal three-terminal controlled unidirectional static switch is shown in Fig. 1.2(c). The power terminals of the switch are labeled 1 and 2. The arrow shows the direction of ON state current through it. The contol input is across terminals C and 1.

1.2.3 Directional Properties of Static Switches—Current Direction

We have described the power diode as a static uncontrolled switch with only one direction for current flow. The general circuit symbol that we have chosen and shown in Fig. 1.2(c) is for a controlled switch, which is also unidirectional. Of the power semiconductor switches described in this chapter, the unidirectional ones, besides the diode, are (1) the bipolar power transistor, (2) the insulated gate bipolar transistor (IGBT), (3) the thyristor, also known as the silicon controlled rectifier (SCR), (4) the asymmetrical silicon controlled rectifier (ASCR), (5) the gate turn off thyristor (GTO) and (6) the MOS controlled Thyristor (MCT). The switches with bidirectional current capability are (1) the power MOSFET, (2) the reverse conducting thyristor and (3) the triac. Of these, the power MOSFET and the reverse conducting thyristor function as controlled switches in the forward direction and as uncontrolled switches in the reverse direction. The triac works as a controlled switch in both directions.

In several types of static power converter circuits that we shall study in subsequent chapters, the static switches have to be bidirectional, with control only for one direction. To represent such a switch, or a combination of two switches that will achieve the same function, we shall combine our circuit symbols of the uncontrolled switch and the controlled switch in the manner shown in Fig. 1.2(d). We can use a power MOSFET or a reverse conducting thyristor if such a function is to be achieved using a single device. Alternatively, we can use a switching "block" or "module" consisting of one of the unidirectional controlled switches and a power diode connected in "antiparallel" with it. Such switching blocks or "modules", consisting of a unilateral power semiconductor device (which may typically be a bipolar transistor or an IGBT), and an "antiparallel" diode, are currently readily available from manufacturers of power semiconductor devices.

1.2.4 Directional Voltage Capabilities of Static Switches

A distinction must be drawn between directional current flow capability and directional voltage blocking capability for a static power semiconductor switch. For example, the bipolar power transistor is a static switch that can only switch current in the "forward" direction. This does not mean that it has ability to block reverse voltages. In fact, it has no significant capability to withstand reverse voltages, and will be permanently damaged if it is subjected to an appreciable magnitude of reverse voltage. It can only block forward voltages, and should be used in such a way that no significant reverse voltage ever comes across it. In contrast, the thyristor, which is also a unidirectional switch, has a symmetrical voltage blocking capability in that it can block approximately the same voltages in both forward and reverse directions. The asymmetrical SCR, which is otherwise similar to the thyristor, has only a very low reverse voltage blocking capability. Certain GTO types have symmetrical voltage blocking capability, whereas others do not have this feature and can only block forward voltages. In DC-to-DC converters and in DC/AC inverters generally, reverse voltage blocking capability is seldom required. In such converters, the controlled switching element is typically used with an anti-parallel diode. Therefore the maximum reverse voltage such a device will be called upon to withstand will be the forward voltage drop of its antiparallel diode when this diode conducts current. The directional voltage and current capabilities of commonly used semiconductor power switches are summarized in Table 1.1.

TABLE 1.1
Properties of power semiconductor switching devices

	Capability to block forward voltages	Significant capability to block reverse voltages	Reverse conduction	Type of forward ON switching control		Is control available for switching OFF forward current If so, type		Is control available for reverse conduction? If so, type		Is control available for switching OFF reverse current? If so, type	
				C†	L†	C	L	C	L	C	L
Power diode	No	Yes	No	No control							
Bipolar power transistor	Yes	No	No	Yes		Yes		Not applicable			
Power MOSFET	Yes	No	Yes	Yes		Yes		No		No	
Insulated gate bipolar transistor (IGBT)	Yes	No	No	Yes		Yes		Not applicable			
Thyristor (SCR)	Yes	Yes	No		Yes	No		Not applicable			
Asymmetrical SCR	Yes	No	No		Yes	No		Not applicable			
Reverse conducting thyristor	Yes	No	Yes		Yes	No		No		No	
Gate turn off thyristor (GTO), reverse blocking type	Yes	Yes	No		Yes		Yes	Not applicable			
GTO without reverse voltage blocking capability	Yes	No	No		Yes		Yes	Not applicable			
Triac	Yes	Yes	Yes		Yes	No			Yes	No	
MOS controlled thyristor (MCT)	Yes	No	No		Yes		Yes	Not applicable			

† C = continuous; L = latching.

1.2.5 Types of Switching Control—Continuous or Latching

There is a major difference between static switches in the manner in which the control terminal performs the switching operation. In some devices, such as the bipolar power transistor and the power MOSFET, after the turn ON switching is implemented by an input to the control terminal, this input should continue to be present, to keep the switch in the ON state. If the control input ceases, the switch will turn OFF. With such a switch, both the turn ON and the turn OFF operations can be implemented by the same control circuit. Turn ON is implemented by giving a voltage or current input to the control terminal. Turn OFF is achieved by terminating this input. This type of control may be described as "continuous". In devices like the thyristor, the control input to implement turn ON need be only a pulse of very short duration. Once the switch has turned ON, there is no further need for the turn ON control pulse to be present. Another example of this type of control is the gate turn OFF thyristor (GTO). The GTO is turned ON by a short positive pulse and turned OFF by a short negative pulse on its control terminal. This type of control may be described as "latching", because the device is latched into the required state by a pulse of short duration.

The thyristor, which is a latching device, has a serious limitation. Its control terminal (gate) has the ability only to control the ON switching operation. Once the device has been latched into the ON state, the gate loses control and the device behaves like a diode. Its OFF switching has to take place by reverse bias of the main terminals like a diode. The directional and control features of the commonly used power semiconductor devices are summarized in Table 1.1.

> **Illustrative example 1.1.** Figure 1.3(*a*) shows four switching blocks linking the DC voltage source *V*, which is connected across P and Q to a series *R–L* load connected across A and B. Each switching block consists of a controlled unidirectional switch S, in antiparallel with an uncontrolled (diode) switch D. $V = 100$ V, $R = 20\,\Omega$, $L = 1$ H.
>
> This is the circuit configuration of an inverter that converts a DC voltage at terminals PQ into an AC voltage at terminals AB. In this example, we use this circuit configuration to illustrate certain aspects of switching, by finding answers to the questions posed in (*a*), (*b*) and (*c*).
>
> (*a*) Prior to instant $t = 0$, all switches are OFF. At $t = t_0$, S_1 and S_4 are both turned ON.
>
> (1) Express *i* as a function of time.
> (2) State which switches are blocking (in the OFF state), and indicate, in each case, whether it is blocking a "forward" voltage or a "reverse" voltage. Give the blocking voltage magnitudes.
>
> (*b*) After S_1 and S_4 had been ON for a long time, they are both turned OFF simultaneously.
>
> (1) Take a new reference zero of time *t* at the instant of opening of S_1 and

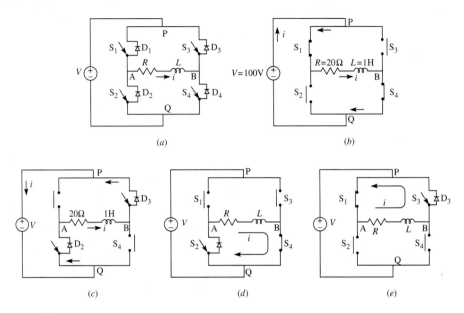

FIGURE 1.3
Circuit changes according to states of the switches in Example 1.1.

S_4. Show the new circuit configuration immediately after $t = 0$. List the switches that are now in the OFF state and give the magnitude and polarity of the blocking voltage of each. Give the expression for i for $t > 0$.

(2) Determine t at which the next switching, and the consequent change in the circuit configuration, will take place without any input from the control circuit.

(c) If the statement in (b) is modified to the effect that only S_1 is turned OFF and not S_4 at $t = 0$, show the path of current flow. Obtain an expression for this current. If instead of S_1 it is S_4 that is turned OFF, what will be the difference in the current path?

Solution

(a)(1) With S_1 and S_4 alone ON, the circuit will be shown by Fig. 1.3(b). The loop equation may be written as

$$V = L\frac{di}{dt} + Ri, \quad i = 0 \quad \text{at } t = 0$$

The solution with zero initial current will be:

$$i = \frac{V}{R}(1 - e^{-t/\tau}), \quad \text{where } \tau = \frac{L}{R}$$

Substituting numerical values,

$$i = 5(1 - e^{-20t}) \quad \text{A}$$

(2) By reference to Fig. 1.3(b) for the switches that are in the OFF state, the blocking voltage polarity and magnitude are as follows:

S_2	forward blocking	100 V
D_2	reverse blocking	100 V
S_3	forward blocking	100 V
D_3	reverse blocking	100 V

D_1 and D_4 are both OFF with zero blocking voltage

(b)(1) When S_1 and S_4 have been ON for a long time $i = 5$ A. This follows from the equation for i obtained in (a). When S_1 and S_4 are opened at $t = 0$ (the new reference zero for t), the current cannot fall instantly to zero, because of the inductance L. The decay of the current will cause an induced e.m.f. to appear across the inductance, which will forward-bias the diodes D_2 and D_3 and turn them ON. Some of the energy stored in the inductance will be returned to the voltage source V and the rest will be dissipated in R. The new circuit configuration will be as shown in Fig. 1.3(c).

For the switches that are now blocking, the polarity and magnitudes of the blocking voltages are as follows:

S_1	forward blocking	100 V
D_1	reverse blocking	100 V
S_4	forward blocking	100 V
D_4	reverse blocking	100 V

S_2 and S_3 are OFF, with zero voltage.

The loop equation now becomes

$$-V = L\frac{di}{dt} + Ri, \quad \text{with } i = 5 \text{ A at } t = 0$$

This gives, for the given values of R and L and the above stated initial conditions,

$$i = -5 + 10e^{-20t} \quad \text{A}$$

We notice that a negative current flows in the source, feeding power into it. The power fed back into the source comes from the stored energy in the inductor. Such a flow of current, due solely to stored energy in an inductor, is termed "freewheeling".

(2) In the above equation, i will decay to zero at $t = t_1$ given by

$$t_1 = 34.7 \text{ ms}$$

For $t > t_1$, the equation shows that i reverses in sign. This will be prevented by D_2 and D_3, which will turn off at $t = t_1$. Therefore the next switching takes place at $t = 34.7$ ms without any input from the control circuit. This happens because of the decay of stored energy in the inductor, resulting in the termination of freewheeling.

(c)   If S_1 alone is turned off and not both S_1 and S_4, the circuit configuration will be as shown in Fig. 1.3(d). Freewheeling will now be through S_4 and D_2. Terminals A and B will be short-circuited through these two switches, and both these output terminals will be at the potential of the negative terminal Q of the DC source. In this case, the freewheeling current will not flow through the source. The loop equation for the freewheeling current will now be

$$0 = L\frac{di}{dt} + Ri, \quad \text{with } i = 5\,A \text{ at } t = 0$$

For the given values of R and L and the initial condition stated above, the freewheeling current will be given by

$$i = 5e^{-20t}\,\text{A}$$

Similar consideration as stated above shows that if the switch turned OFF is S_4 and not S_1, the freewheeling path will be through S_1 and D_3. The circuit configuration will be as shown in Fig. 1.3(e). The expression for the freewheeling current will be the same. Now the switches that short-circuit the terminals A and B will be D_3 and S_1. These two terminals will now be at the potential of the positive terminal of the DC source.

1.3   Ideal and Real Switches

To assess the performance of a switch, we look at two aspects of its behavior—static and dynamic. If the switch is either in its ON or OFF state, we call this a static condition. The dynamic condition is the transition from one static state to the other. We shall look at the limitations of power semiconductor switches from these two aspects.

1.3.1   Static Performance

An ideal switch should have zero voltage across it in the ON state and zero current through it in the OFF state when it is blocking a voltage. The product of current and voltage, which gives the power dissipated in the switch, is zero in both conditions. This is the basic reason why a power conversion scheme based on switching is more efficient than other methods, because ideally, there is no internal power loss. Power semiconductor switches depart to some extent from the ideal—there is a small but finite voltage drop in the ON state and a small but finite "leakage current" in the OFF state. The power dissipated in a switch during its ON state is given (in W) by

$$p = v_f i_f$$

For an ON period duration t_1 to t_2, during which v_f and i_f may vary, the total energy (in J) dissipated in it will be

$$J = \int_{t_1}^{t_2} v_f i_f \, dt$$

Energy dissipated in the switch causes its temperature to rise. For satisfactory operation, the maximum temperature of the silicon pellet that constitutes the switching element has to be limited below the specified safe limit. The efficiency of power conversion is also lowered, because of the power dissipation in the switches.

The relationship between i_f and v_f for a semiconductor switching element is typically nonlinear. This relationship, when graphically plotted, is called the output characteristic or forward characteristic of the device. This characteristic, or relevant parameters from it, are generally available from the manufacturer's data sheet for the device.

The leakage current that flows in the OFF state may be a "forward" leakage or a "reverse" leakage, depending on whether the switch is blocking a forward or a reverse voltage. In both cases, there will be power dissipation in the device. However, in present-day devices, the leakage current is quite small and does not vary significantly with voltage. Under normal load current conditions, the power dissipation due to this leakage current is small in comparison with the power dissipation in the ON state. We shall therefore generally neglect the power dissipation due to OFF state leakage and the resulting loss of energy.

Illustrative example 1.2. Figure 1.4 shows an inverter configuration using four static switches, for converting a DC voltage input across PQ into an AC voltage across AB. Each switch is to be treated as nonideal, with a forward characteristic given by

$$v_f = 1.6 + 0.1i_f \quad V$$

The repetitive switching frequency is 500 Hz. In each switching cycle, the sequence is as follows. S_1 and S_4 are both turned ON at $t = 0$. Both are turned OFF at $t = 0.6$ ms. S_2 and S_3 are next turned ON at $t = 1.0$ ms and turned OFF at $t = 1.6$ ms. Neglect all power losses in the switch other than that due to forward voltage drop in the ON state.
(a) Determine (1) the maximum instantaneous power loss in any one switch; (2) the average power loss in it.

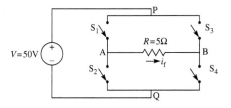

FIGURE 1.4
Circuit for Example 1.2.

(*b*) Determine the maximum instantaneous power output from the converter and its average efficiency.

Solution

(*a*) When S_1 and S_4 are on, the current is given by Kirchhoff's law as follows:

$$2(1.6 + 0.1i_f) + 5i_f = 50 \text{ V}$$

This gives $i_f = 9$ A. The voltage drop across a switch is $1.6 + (0.1)(9) = 2.5$ V.
(1) The instantaneous maximum power dissipation within a switch is $(2.5)(9) = 22.5$ W. The energy dissipated in a switch in one period is

$$J = 22.5 \times 0.6 \times 10^{-3} = 13.5 \times 10^{-3} J$$

$$T = \text{time period} = \tfrac{1}{500} = 2 \text{ ms}$$

(2) The average power dissipation in one switch is

$$J/T = 13.5 \times 10^{-3}/2 \times 10^{-3} = 6.75 \text{ W}$$

(*b*) When two switches are ON, the output voltage v_{AB} is

$$v_{AB} = i_f R = 45 \text{ V}$$

The power output is

$$i_f v_{AB} = 45 \times 9 = 405 \text{ W}$$

The total energy output in one period is given by

$$2 \times \text{energy in a half-period} = 2 \times 405 \times 0.6 \times 10^{-3} = 0.486 \text{ J}$$

The average output power is given by

$$\text{output energy in one second} = 0.486 \times 500 = 243 \text{ W}$$

The input power from source when switches are ON is $50 \times 9 = 450$ W.
The input energy in one period is $2 \times 450 \times 0.6 \times 10^{-3} = 0.54$ J.
The average input power is $0.54 \times 500 = 270$ W.
The average efficiency is $243/270 = 0.9 = 90\%$.

1.3.2 Dynamic Performance

An ideal switch should change from the OFF state to the ON state, *instantaneously*, when the required switching control signal is applied to its control terminal. Similarly, the transition time for the turn OFF switching

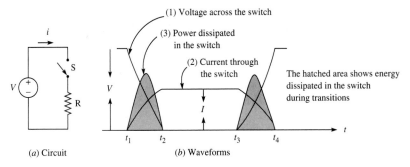

FIGURE 1.5
Instantaneous voltages, currents and power in a static switch during the switching transitions.

should also be zero. A real switch needs a finite t_{on} for ON switching and t_{off} for OFF switching. These finite switching times have two major consequences.

1. They limit the highest repetitive switching frequencies possible.
2. They introduce additional power dissipation in the switches themselves.

Of these, the second phenomenon needs further explanation.

Referring to Fig. 1.5(a), we assume that the switch shown there has ideal static characteristics, but nonideal dynamic performance. Figure 1.5(b) shows the waveforms of (1) the voltage drop across the switch and (2) the current through it, when a turn ON switching is implemented at $t = t_1$ and a turn OFF switching at $t = t_3$. Let us first look at the turn ON switching.

Prior to $t = t_1$, the switch is in the OFF state, and the blocking forward voltage across it is equal to the source voltage V. At $t = t_1$, the turn ON control signal arrives at its control terminal. During the turn ON transition, which takes a finite time, the instantaneous value of the voltage across the switch is assumed to be shown by the waveform in Fig. 1.5(b). During the same time, the current through the switch rises from zero to the static ON state magnitude I. The instantaneous value during the transition is assumed to be as shown by the current waveform.

We notice that during the transition there is power dissipation taking place inside the switch, whose instantaneous value is given by the product of voltage and current. The instantaneous power is also plotted (3) as a function of time in Fig. 1.5(b). Depending on the nature of the current and voltage waveforms during the transition, the peak power can reach a relatively large magnitude. The energy dissipation in the switch is equal to the area shown hatched under the power waveform.

Similar considerations are valid for the turn OFF switching operation, which takes place from t_3 to t_4 in Fig. 1.5(b). During this transition, the voltage rises from zero to V and the current falls from I to zero. The transition times t_{on} and t_{off} are generally not equal in power semiconductor switches, t_{off} usually

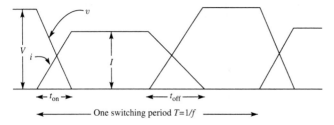

FIGURE 1.6
Switch voltage and current waveforms in Example 1.3.

being larger. The total energy J (in J) dissipated in the switch in one switching cycle is given by the sum of the areas under the power waveform during t_{on} and t_{off}. The average switching power loss is therefore proportional to the switching frequency, and is given (in W) by

$$P_{switching} = fJ$$

f being the converter switching frequency (in Hz).

Illustrative example 1.3. Assume linear variation with respect to t for both voltage and current, both during t_{on} and t_{off}. The OFF state voltage is V and the ON state current is I. The switching frequency is f. Assume that the switch is ideal as regards its static performance. Obtain expressions for (1) the average switching power loss and (2) the instantaneous peak power loss in the switch.

Solution. Figure 1.6 shows the current through the switch and the voltage across it as functions of time. We shall first consider the turn ON switching transition.

Taking the reference zero of t at the commencement of the turn ON transition, we may express i and v during t_{on} by the following expressions based on linear variation:

$$i = It/t_{on}$$
$$v = V(1 - t/t_{on})$$

Therefore the instantaneous power will be

$$p = VI\left(\frac{t}{t_{on}} - \frac{t^2}{t_{on}^2}\right)$$

The energy loss during t_{on} is obtained by integrating p from $t = 0$ to $t = t_{on}$. This gives

$$J_{on} = \int_0^{t_{on}} p\, dt = \tfrac{1}{6}VIt_{on}$$

To determine the instantaneous maximum power dissipation, we equate $dp/dt = 0$. This gives $t = \tfrac{1}{2}t_{on}$. Therefore the instantaneous maximum power dissipation will be given by putting $t = \frac{1}{2}t_{\text{on}}$ in the expression for p. This gives

$$p_{\text{max}} = \frac{1}{4}VI$$

A similar analysis gives the energy loss during t_{off} as

$$J_{\text{off}} = \frac{1}{6}VIt_{\text{off}}$$

The total energy loss due to switching in one cycle will be

$$J = J_{\text{on}} + J_{\text{off}} = \frac{1}{6}VI(t_{\text{on}} + t_{\text{off}})$$

For a switching frequency f, the average power dissipation in the switch will be

$$p_{\text{switching}} = \frac{1}{6}VI(t_{\text{on}} + t_{\text{off}})f$$

The instantaneous maximum power dissipation during turn OFF is found in the same way as during turn ON and will have the same value.

Illustrative example 1.4. In the previous example, assume that the static performance of the switch is also nonideal. Assume that the forward voltage drop during the ON state is constant at 2 V. Neglect leakage during the OFF state. The transition times needed by the switch are $t_{\text{on}} = 2\ \mu s$ and $t_{\text{off}} = 4\ \mu s$. The OFF state voltage is 120 V and the ON state current is 10 A. The switch has a repetitive switching frequency f (in Hz). The switching duty cycle is $D = 0.6$, defined as the ON time of the switch/switching period.

Determine the frequency f at which the switching power loss begins to exceed the static power loss. What is the total power dissipation in the switch at this frequency?

Solution

Switching power loss calculations. Figure 1.7 shows the waveforms of voltage across the switch and the current through it. Consider the turn ON transition. During this interval, the voltage falls from $V = 120$ V to $v_f = 2$ V, assumed linearly. On this basis, the expressions for v and i during transition will be:

$$v = V - (V - v_f)t/t_{\text{on}}$$

$$i = It/t_{\text{on}}$$

This gives the instantaneous power during turn ON as

$$p = \frac{VIt}{t_{\text{on}}} - \frac{(V - v_f)It^2}{t_{\text{on}}^2}$$

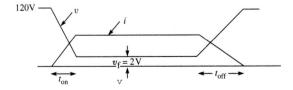

FIGURE 1.7
Switch current and voltage waveforms in Example 1.4.

The energy dissipated during t_{on} will be

$$J_{on} = \int_0^{t_{on}} p \, dt = \tfrac{1}{6} VI t_{on} + \tfrac{1}{3} v_f I t_{on}$$

A similar analysis will give, during t_{off},

$$J_{off} = \tfrac{1}{6} VI t_{off} + \tfrac{1}{3} v_f I t_{off}$$

The total energy dissipation in the switch during switching transitions in one cycle will be

$$J_{switching} = J_{on} + J_{off} = (\tfrac{1}{6} VI + \tfrac{1}{3} v_f I)(t_{on} + t_{off})$$

Therefore the average switching power dissipation at a switching frequency f (in Hz) will be

$$p_{switching} = (\tfrac{1}{6} VI + \tfrac{1}{3} v_f I)(t_{on} + r_{off})f$$

for the given values of V, I, v_f, t_{on} and t_{off}, this becomes

$$p_{switching} = 1.24 \times 10^{-3} f \quad \text{W}$$

Static power loss calculations. We shall assume that in defining the switching duty cycle D, the effective ON time used is

$$T_{on} = \text{static ON time} + \tfrac{1}{2}(t_{on} + t_{off})$$

Therefore the static ON state duration of the switch will be

$$T_s = \frac{D}{f} - \tfrac{1}{2}(t_{on} + t_{off})$$

The energy dissipated in the switch during its static ON state in one period will be

$$J_{static} = v_f I \left[\frac{D}{f} - \tfrac{1}{2}(t_{on} + t_{off}) \right]$$

Therefore the average static power dissipation at a frequency f will be

$$p_{static} = v_f I \left[\frac{D}{f} - \tfrac{1}{2}(t_{on} + t_{off}) \right] f$$

For the given values of v_f, I, D, t_{on} and t_{off}.

$$p_{static} = 12 - 60 \times 10^{-6} f \quad \text{W}$$

The switching power loss increases with frequency. The frequency at which the two are equal is obtained by equating $p_{switching} = p_{static}$. This gives

$$f = 9.23 \text{ kHz}$$

At this frequency, the switching power loss is

$$p_{\text{switching}} = 11.45 \text{ W}$$

The static loss being the same at this frequency, the total power dissipation in the switch becomes

$$p = 22.9 \text{ W}$$

1.3.3 Temperature Rise—Use of Heat Sinks

The immediate consequence of energy dissipation in a static switch is its temperature rise. The need for cooling arises to keep the switch temperature below its safe limit. The exact location where the heat is generated is in the silicon pellet that actually is the switch. A power semiconductor device is basically a very thin wafer of silicon in which the necessary impurity profiles are created during manufacture. Electrical contacts are made to appropriate areas of this thin wafer, and these constitute the terminals of the device, which are brought outside the casing in which the silicon pellet is housed. The power losses of the switch raises the temperature of the pellet. The temperature gradient thus created causes the heat power to flow outwards to the casing surface. To facilitate the easier flow of heat power to the ambient atmosphere outside, it is a common practice to mount the casing on a "heat sink." Heat sinks are made of metal, and provide a large surface area from which the heat power can pass by convection and radiation to the ambient atmosphere. The convection flow can be further enhanced, if needed, by using a fan to provide forced air cooling.

With the device mounted on a heat sink, the path of heat flow can be viewed as a series combination of the following individual paths: (1) from the junction (J) to the surface of the casing (C); (2) from the surface of the casing (C) to the outer surface of the heat sink (S); (3) from S to the ambient atmosphere (A), which we shall assume to be an external region sufficiently distant from the heat sink, at which the thermal gradient is negligible.

With constant power dissipation inside the pellet, thermal equilibrium conditions will be attained after a period of time. After this has happened, the temperatures and temperature gradients stay constant in such a way that all the power dissipated inside the device continually flows out into the ambient. Taking thermal power flow as proportional to temperature difference, we can represent this condition by an electrical analog in which electric current (in A) represents thermal power (in W) and potential difference (in V) represents temperature difference (in °C). Such an analogous electric circuit will consist of resistances ($\Omega = \text{V/A}$) as the analog of "thermal resistance" (expressed as °C/W). On this basis, Fig. 1.8 shows the thermal power flow model for a static switch mounted on a heat sink. In this model, J represents the inside of the silicon pellet. Θ_{JC} is the thermal resistance (in °C/W) between J and the outer surface of the casing C. The other thermal resistances are labeled by

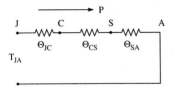

FIGURE 1.8
Model of heat flow under thermal equilibrium conditions.

appropriate subscripts. The model of Fig. 1.8 can be used to make estimates of the junction temperature rise when the power dissipation is known, or to estimate the maximum power dissipation possible for a specified junction temperature. The following example illustrates this.

Illustrative example 1.5. A power semiconductor switch has a specified thermal resistance of 0.6°C/W from junction to casing. It is mounted on heat sink whose thermal resistance is 0.25°C/W. The thermal resistance from heat sink surface to ambient is 0.15°C/W.

(*a*) The switch is operating with a total internal power dissipation of 60 W. The ambient temperature inside the cabinet in which the equipment is housed is 45°C. Determine (1) the junction temperature; (2) the temperatures of the casing and heat sink surfaces.

(*b*) The power dissipation in the switch consists of a static power loss, which may be expressed as $p_{static} = 5I$ W, where I is the ON state current (in A) and a switching power loss, which may be expressed as $p_{switching} = (2 \times 10^{-3})fI$ W, where f is the switching frequency (in Hz). Determine, from thermal considerations, the limiting ON state current at a frequency of 500 Hz, if the temperature of the junction is not to exceed the value determined in (*a*).

(*c*) Repeat (*b*) for $f = 5$ kHz.

Solution

(*a*)(1) Referring to Fig. 1.8, the total thermal resistance from junction to ambient will be

$$\Theta_{JA} = \Theta_{JC} + \Theta_{CS} + \Theta_{SA} = 0.6 + 0.25 + 0.15 = 1°C/W$$

The temperature rise from ambient to junction will therefore be

$$T_{JA} = \Theta_{JA} p = 1 \times 60 = 60°C$$

The junction temperature $T_J = 60 + T_A = 105°C$.

(2) The temperature difference between junction and casing will be

$$T_{JC} = \Theta_{JC} p = 0.6 \times 60 = 36°C$$

The casing temperature is $105 - 36 = 69°C$.

The temperature drop from casing to heat sink surface is

$$T_{CS} = \Theta_{CS} p = 0.25 \times 60 = 15°C$$

The surface temperature of the heat sink is $69 - 15 = 54°C$.

(b)   For the stated conditions of switching, the power dissipated in the switch will be

$$p = p_{static} + p_{switching} = 5I + 2 \times 10^{-3} \times 500 \times I = 6I \quad W$$

If the temperature of the junction is not to exceed the value calculated in (a), p should be limited to 60 W. This gives the maximum ON state current as $I = 10$ A.

(c)   If the switching frequency is increased to 5 kHz, I should be reduced so that the total power dissipation is again limited to 60 W. Therefore

$$60 = 5I + 2 \times 10^{-3} \times 5 \times 10^3 \times I = 15I \quad W$$

This gives $I = 4$ A.

The above example illustrates the need to derate the current rating of the switch as the switching frequency is increased. When using the current rating of a static switching element listed by the manufacturer in its data sheet, it is important to remember that this rating is valid only if an adequate heat sink is used in mounting the device, to limit the junction temperature within the specified value.

We have now completed the general treatment of static switches from the point of view of their use in power converters. We shall now take up the study of individual types of power semiconductor switching devices in the remaining sections of this chapter.

1.4 POWER DIODES

1.4.1 Available Ratings, Types

Diodes employed in static converters are essentially high power versions of conventional low power devices used in analog and digital signal processing circuits. Power diodes are manufactured in a wide range of current and voltage ratings. Current ratings vary from a few amperes to several hundred amperes for a single device. Similarly, voltage ratings extend from tens of volts to several thousand volts.

In addition to current and voltage ratings, an important parameter that determines the selection of a power diode for a static power converter application, is its "reverse recovery time." A diode that has been in the conducting state needs a short, but finite time after the forward conduction has stopped before it is able to recover its ability to block reverse voltages. Based on this time, diodes are classified as "fast recovery" and "slow recovery" types. Certain types of converters require the use of fast recovery power diodes. There are other types of converters in which short recovery time is not

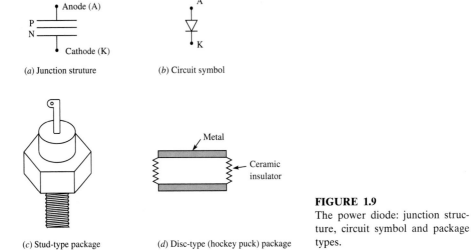

FIGURE 1.9

The power diode: junction structure, circuit symbol and package types.

(a) Junction struture

(b) Circuit symbol

(c) Stud-type package

(d) Disc-type (hockey puck) package

important and may be traded off for other benefits, such as lower ON state voltage drop and lower cost.

1.4.2 Junction Structure, Packaging

The diode consists of a silicon pellet with a single internal pn junction [Fig. 1.9(a)]. Typically, it is fabricated by diffusing p-type impurity atoms into one side of a n-type crystal wafer of silicon. It has two terminals, of which the anode (A) makes contact with the p side surface of the pellet and the cathode (K) with the n side. The forward direction of current flow is from A to K, as shown by the diode circuit symbol [Fig. 1.9(b)].

The area of the pellet, by and large, determines the current rating. For example, a large area device using a silicon wafer of 4 in. diameter may have a current rating in the neighbourhood of 1000 A. The resistivity of the starting material and its thickness are mainly responsible for the voltage blocking capability of the device. For example, if the diode is fabricated starting from n-type wafer on which the p layer is formed by diffusion of p-type impurities, its voltage rating will go up with the purity level of the starting wafer and its thickness.

The impurity profile in the silicon pellet has a major effect on both its ON state forward voltage drop and also on its reverse recovery time. During forward current flow in the ON state of the diode, holes cross from the p side of the junction to the n side and disappear by recombining with the electrons there. Also, electrons cross the junction from the n side to the p side, and disappear by recombining with the holes there. Since holes and electrons have opposite charge polarities and are crossing in opposite directions, the total electric current is obtained by adding up the current contribution due to the flow of holes and that due to the flow of electrons. The obvious question that arises in this context is how the total current is shared between electron current flow and hole current flow across the junction. The answer to this question is important for the understanding of power diode behavior during switching transitions and also of the operation of power transistors.

The relative proportions of the electron and hole currents will be decided by the relative concentrations of electrons on the n side and of holes on the p side. For example if the n-side electron concentration is equal to the p-side hole concentration, each current will be half the total current. If on the other hand, the p side has a very much larger density of holes in comparison with the density of electrons on the n side, as is often the case, a very large part of the total diode current at the junction will be due to the flow of holes. As a result, in such a diode, there will be an excess concentration of injected holes in the junction zone on the n side, while the diode is carrying current in its ON state. These excess holes, which are minority carriers on the n side, need a finite time to disappear before the diode is able to block reverse current flow. If the diode is subjected to reverse voltage before this, a reverse current will flow through it. During this reverse current flow, holes will be pushed back to the p side. The reverse current flow will last until the excess hole concentration has disappeared. This will be due to some of the holes moving across to the p side and the rest recombining with electrons. From this, it may be expected that a shorter "lifetime" for minority carriers (which implies a faster recombination rate) will shorten the reverse recovery time of the diode, because this will result in faster disappearance of excess minority carriers. The presence of gold atoms as an impurity in the silicon has been found to shorten the minority carrier lifetime. Therefore diffusion of gold into the silicon is a technique used to shorten the reverse recovery time.

A trade off exists between blocking voltage capability and reverse recovery time. High resistivity silicon has a long minority carrier lifetime. Therefore, if a high resistivity silicon wafer is used as the starting material to achieve a high blocking voltage capability, the reverse recovery time is likely to go up. A trade off also exists between blocking voltage capability and ON state forward voltage drop. If a thicker and higher resistivity wafer is employed to achieve higher voltage rating, this will result in a higher forward voltage drop in the ON state. In some power diodes, a lower resistivity n region, called an n^+ region, is formed on the outer side of the high resistivity n region. This helps to reduce the ON state voltage drop and also to make better electrical contact to the cathode terminal.

There are two types of packages commonly used for housing the diode pellet. In the stud type package shown in Fig. 1.9(c), the casing, which is of metal, has a threaded stud for easy mounting on a heat sink. The metal casing and the stud constitute one terminal of the diode. The other terminal is brought out on the side opposite to the stud, with electrical insulation from the casing. Stud-type diodes are available with either polarity for the stud—anode or cathode. This choice enables a designer of equipment to use fewer heat sinks. For example, of the converter circuit configuration has several diodes with all their anodes connected together then all of these can be mounted on the same heat sink if the devices selected are of the stud anode polarity. Also, in a "switching module," which we shall later encounter in power converter circuits, a diode has to be connected across the terminals of a controlled switching device. By choosing the stud polarity of the diode to match the casing terminal of the controlled switching device, both devices can be mounted on the same heat sink.

Diodes with very large current and voltage ratings employ the disc package, also known as "hockey puck," illustrated in Fig. 1.9(d). Here, the two terminals are flat metallic surfaces separated by a ceramic insulator. Disc-type diodes are mounted with at least one side in contact with a flat heat sink surface. For better cooling, two separate heat sinks may be used—one on each side. Special mounting hardware, which provide for adequate contact pressure and insulation between the two sides is usually available from the manufacturer.

1.4.3   Reverse Recovery Characterstics

For most practical purposes, we can consider the turn ON switching of a diode as instantaneous. But the turn OFF switching behavior requires special consideration, because this has a significant effect on the performance of most types of static converters.

In Fig. 1.10, D is the power diode whose switching behavior we shall

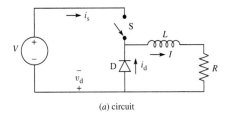

(a) circuit

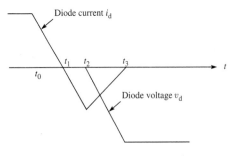

(b) Voltage and current waveforms

FIGURE 1.10
Power diode voltage and current waveforms during a turn OFF switching transition.

consider. In the circuit shown, we have a DC source in series with a controlled static switch labeled S. Let us assume that S has been ON for a long time, as a result of which a current I has been established in the $R–L$ circuit. After this, S is turned OFF. But the current in the $R–L$ circuit cannot instantly fall to zero, because of the presence of the inductance L. When the current tends to fall, the induced e.m.f. in the inductance will forward-bias the diode, and the current, will "freewheel," that is, continue to flow through it due to the stored energy in the inductance, even though there is no voltage source present to drive this current. We shall assume that L is large, so that the decay of current is negligible during the short period of our next sequence of observations. In other words, I will be assumed to be constant.

Our study of the diode behavior commences at the instant marked $t = t_0$ in Fig. 1.10(b), when S is turned ON again. This will cause a current i_s to build up in the DC source and through switch S. We will assume that the rate of increase of i_s is constant. This rate di_s/dt is determined by the residual inductance in the circuit mesh consisting of the source, the diode D and S. Since I is constant, the increase in i_s results in an equal decrease in the diode current i_d:

$$i_s + i_d = I$$

Therefore

$$\frac{di_d}{dt} = -\frac{di_s}{dt}$$

This linear fall of current in the diode is shown by the diode current waveform in Fig. 1.10(b). We now come to the instant $t = t_1$ at which the diode current has decreased to zero.

So far, the diode had been conducting in the forward direction, with a small forward voltage drop, which we shall consider negligible. If the diode is ideal in its dynamic behavior, it should stop conducting at $t = t_1$ and recover its reverse voltage blocking ability. But, because of the excess minority carriers, as explained earlier, it continues to conduct after $t = t_1$, now in the reverse direction. During this reverse current flow, the excess minority carriers are being pushed back across the pn junction. Therefore the diode continues to be a conducting switch and the current waveform continues unaffected, until the excess carriers are removed. This happens at $t = t_2$. After $t = t_2$ the reverse current falls towards zero and the diode recovers its reverse voltage blocking capability. In the reverse blocking condition, the diode junction is like a charged capacitor, charged to the reverse voltage. The current flowing from t_2 to t_3 may be viewed as the charging current of this capacitance. Neglecting the small reverse leakage that continues to flow, the diode current may be considered to have fallen to zero at $t = t_3$, when this junction capacitance has been charged to the full reverse voltage. At this instant, the turn OFF switching transition of the diode is complete. The voltage and current waveforms of the diode are shown in Fig. 1.10(b). We have approximated the actual waveforms by straight lines, primarily to focus attention on the essential aspects of the diode switching behavior, which are of interest in the operation of static power converters. There is also a reverse over-voltage spike, which is omitted in Fig. 1.10(b). This is described later.

The "reverse recovery time" t_{rr} is defined by reference to the waveforms of Fig. 1.10(b). It is the time interval measured from the instant at which the forward current has fallen to zero [t_1 in Fig. 1.10(b)] to the instant at which the reverse voltage recovery is completed [t_3 in Fig. 1.10(b)]. Therefore t_{rr} is in effect, the duration of reverse current flow during turn OFF switching. To be exact, the specified reverse recovery time in the data sheet of a power diode is correct only for the specified test conditions. It does vary to some extent, depending on the test conditions, and is higher for a larger ON state current before the commencement of turn OFF.

The nature of the reverse current waveform, during the decay period of the reverse current (t_2 to t_3), depends primarily on the impurity profile in the junction transition zone of the diode. If this time is very short, the diode is described as belonging to the "hard recovery" category. If this interval is relatively long, it is a "soft recovery" type. It follows that a soft recovery diode has a lower di/dt during reverse recovery. Soft recovery diodes are generally preferable in static power converters, because inductive voltage spikes resulting from the diodes' reverse recovery transients, are smaller.

1.4.4 Effects of Reverse Recovery Transient

The major consequences of the reverse recovery characteristic of a power diode on power converter operation are listed and explained below.

1.4.4.1 HIGH CURRENTS IN THE CONTROLLED SWITCHES. The circuit situation considered in Fig. 1.10 occur in DC/DC converters and DC/AC inverters, which we shall be studying in subsequent chapters. In Fig. 1.10, when S is repetitively turned ON and turned OFF in each switching cycle, there is a current transfer taking place at each switching, between S and the diode. If the diode were ideal, with no reverse recovery current, S would have to carry only the output current in the $R–L$ circuit. In the real situation shown in Fig. 1.10, we find that S has to handle the sum of the output current and the reverse recovery current immediately after it is turned ON. This means that it should be capable of carrying a pulse current of very large amplitude and of dissipating the additional power due to this current spike. Alternatively, the amplitude of the current pulse itself should be limited by inserting additional inductance in series with the source. In some types of controlled static switches, where the peak current magnitude possible through the switch in the ON state is determined by the input on its control terminal, it may be necessary to provide adequate control terminal inputs to handle the large current spike. An example is the power transistor, in which the maximum current possible in the collector circuit depends on the base current input.

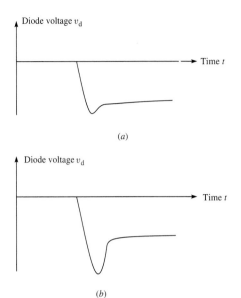

FIGURE 1.11
Reverse over-voltage spike for (*a*) a typical freewheeling duty and (*b*) a typical rectifier duty.

1.4.4.2 OVER-VOLTAGE SPIKE ACROSS DIODE. In the circuit of Fig. 1.10, there is usually a small, but unavoidable residual inductance, in the mesh consisting of the source, the diode and S. Additionally, there could be inductance that we intentionally put in, to limit the reverse current peak. Any inductance present in the loop causes a voltage spike to appear across the diode, during the decay of the reverse current. The magnitude of this reverse over-voltage spike depends on the loop inductance value and the slope di/dt of the reverse current. In a real diode, the di/dt value during the decay of reverse current is not constant, as we have assumed by the approximated straight line from t_2 to t_3 in Fig. 1.10(*b*). Therefore, the reverse over-voltage varies and has a peak value during this interval.

Typical reverse recovery voltage waveforms for power diodes are shown in Fig. 1.11. In Fig. 1.11(*a*), the loop inductance is assumed to be small, and therefore the over-voltage spike has a smaller amplitude. This is typical of DC/DC and DC/AC converters. In "rectifier" circuits used for converting from AC to DC, the loop inductances are generally larger, and therefore the over-voltage spikes are also larger, as shown in Fig. 1.11(*b*). If the reverse over-voltage exceeds the reverse voltage rating of the diode, it could break down, resulting in its permanent damage. Therefore it is usual in such situations to use external circuit elements to limit the magnitude of the reverse over-voltage spike. Such circuits are called snubber circuits or switching aid circuits. Snubber circuits for diodes are described later in this chapter.

1.4.4.3 LIMITATION OF CONVERTER SWITCHING FREQUENCY. In a static converter that has power diode in its circuit configuration, if the switching frequency is increased, the recovery time t_{rr} of the diodes will constitute an increasing fraction of the total cycle time. This will generally decrease the efficiency of the converter. Therefore the reverse recovery time sets an upper limit to the converter switching frequency. Diodes with short recovery times are designated as "fast recovery" types. Fast recovery diodes are always necessary for high frequency applications.

Illustrative example 1.6. In the circuit of Fig. 1.12, immediately before $t = 0$, S is OFF and $I = 40$ A is freewheeling through the diode D. At $t = 0$, S is turned ON. The peak reverse recovery current is observed to be $I_{RN} = 25$ A. Subsequent decay of the reverse current is found to be at the rate of 20 A/μs. Neglect ON state voltage drop and OFF state leakage. Assume L_s is large.

(a)(1) Sketch the current waveform through the diode and determine t_{rr}.
(2) Determine the maximum reverse voltage to be expected across the diode.
(3) Estimate the excess charge recovered by flow of reverse current until the reverse current peak is reached.
(4) Sketch the current waveform through the static switch S and determine the peak instantaneous current through it.
(b) L_1 is changed to 20 μH. Assume that the charge recovered till the instant of reverse current peak is the same as before. Determine (1) the new value of t_{rr}; (2) the new amplitude of i_s.

Solution
(a)(1) When S is turned ON, the source loop equation will be

$$V = L_1 \frac{di_s}{dt}$$

This gives

$$\frac{di_s}{dt} = \frac{V}{L_1} = 40 \text{ A}/\mu\text{s}$$

Since $i_s + i_d = I$ is assumed constant,

$$\frac{di_d}{dt} = -\frac{di_s}{dt} = -40 \text{ A}/\mu\text{s}$$

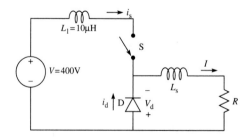

FIGURE 1.12
Circuit of Example 1.6.

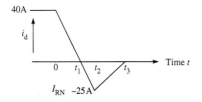

FIGURE 1.13
Diode current waveform in Example 1.6.

The diode current falls at $40\,\text{A}/\mu\text{s}$, and will therefore hit zero at $t_1 = 40/40 = 1\,\mu\text{s}$. This is shown in the current waveform sketched in Fig. 1.13.

Referring to the waveform, the reverse current commences at $t = t_1$ and reaches a peak magnitude of 25 A at $t = t_2$ with the same rate of change. This gives

$$t_2 - t_1 = 25/40 = 0.625\,\mu\text{s}$$

From the instant t_2, the reverse current decreases at $di_\text{d}/dt = 20\,\text{A}/\mu\text{s}$. Therefore the reverse current becomes zero at $t = t_3$, and $t_3 - t_2 = 25/20 = 1.25\,\mu\text{s}$. This gives

$$t_3 = 1 + 0.625 + 1.25 = 2.875\,\mu\text{s}$$

$$t_\text{rr} = t_3 - t_1 = 2.875 - 1 = 1.875\,\mu\text{s}$$

(2) To estimate approximately the peak reverse voltage across the diode, we assume that the diode voltage during the interval $t_2 - t_3$ is determined by the current waveform of Fig. 1.13. Therefore

$$V_\text{d} = -V + L_1 \frac{di_\text{s}}{dt} = -400 + (10 \times 10^{-6})(-20 \times 10^6) = -600\,\text{V}$$

(3) The charge recovered during the interval t_1 to t_2 is given by the area enclosed by the current waveform during this interval. This gives

$$Q_\text{rr} = (t_2 - t_1)(25)/2 = 7.81\,\mu\text{C}$$

(4) $i_\text{s} = I - i_\text{d}$. From this relationship the waveform of i_s is sketched in Fig. 1.14. From this waveform, we get the peak current through S as 65 A.

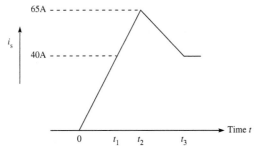

FIGURE 1.14
Current waveform in the switch S.

(b)(1) For $L_1 = 20\ \mu H$, the rate of decay of i_d will be $400/20 = 20\ A/\mu s$. Let I_{RN} be the amplitude of the reverse current peak, which occurs at $t = t_2$:

$$I_{RN} = 20(t_2 - t_1)$$

The charge recovered during $t_2 - t_1$ will be given by the area enclosed during this interval. Therefore

$$Q_{rr} = \tfrac{1}{2}I_{RN}(t_2 - t_1) = \tfrac{1}{40}I_{RN}^2 = 7.81\ \mu C \quad \text{[from } (a)(3)\text{]}$$

This gives $I_{RN} = 17.67\ A$.

The time interval t_1 to t_2 will be $I_{RN}/20 = 0.884\ \mu s$.

The rate of decay of reverse current being $20\ A/\mu s$, it will fall to zero at $t = t_3$ such that

$$t_3 - t_2 = 17.67/20 = 0.884\ \mu s$$

The new reverse recovery time becomes

$$t_{rr} = t_3 - t_1 = 1.768\ \mu s$$

(2) The peak current amplitude through S will now be $40 + 17.67 = 57.67\ A$.

1.4.5   Schottky Diodes

Schottky diodes do not have a pn junction. Instead, they employ a metal-to-silicon barrier. In a Schottky diode, the current flow is due to the flow of majority carriers. The barrier is not subject to recovery transients. Therefore they are suitable for use at very high frequencies. The barrier does have a capacitance, which accounts for a certain amount of reverse current, to charge this capacitance to the reverse voltage. At present, Schottky power diodes are not available for large voltages and currents. Current ratings of individual diodes available are well below 100 A. Reverse voltage ratings are well below 100 V. They are very widely used in "switch mode power supplies," which work at high switching frequencies.

1.4.6   Snubber Circuits (Switching Aid Circuits)

A snubber circuit for a diode serves to protect it from damage that can arise due to an over-voltage spike during reverse recovery. A typical snubber circuit for a power diode consists of a resistance in series with a capacitor connected across the diode, as shown in Fig. 1.15.

During the decay of the reverse recovery current, the capacitor serves to limit the voltage spike. The energy stored in the inductance of the reverse

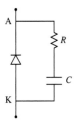

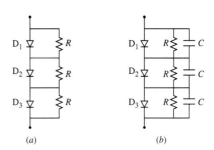

(a)   (b)

FIGURE 1.15
Snubber circuit for power diode.

FIGURE 1.16
Voltage division networks for series connected diodes.

recovery current loop serves to charge the capacitor, thereby reducing the
over-voltage spike. The resistance R dissipates some of this energy, and, if
suitably chosen, will damp out oscillations in the $L-C$ circuit.

1.4.7   Series and Parallel Operation of Diodes

When the voltage rating of an available diode type is not high enough for a
high voltage application, several such diodes connected in series can be used to
achieve the required voltage capability. If this is done, it is necesssary to ensure
that each diode in the series chain shares the total reverse voltage nearly
equally. There can be differences in the reverse current characteristics between
diodes of the same type due to manufacturing tolerances and/or temperature
differences. This can be the reason for large imbalances in the reverse voltage
distribution between the diodes. It is always advisable to use matched diodes.
Since perfect matching may be impossible, the difficulty can be overcome by
connecting a resistance in parallel with each diode, as shown in Fig. 1.16(a). If
the value of the resistance is such that the current through it is significantly
higher than the current through the parallel diode, the voltage division
between the individual diode sections will be nearly decided by the resistances.

The use of resistances only, as shown in Fig. 1.16(a), will help only during
the static reverse blocking conditions. There can still be excessive reverse
voltage across a diode during reverse recovery, because of differences in the
reverse recovery times. The diode with the shortest recovery time will recover
first and will have to withstand the total reverse voltage, until the next one
recovers, and so on.

This difficulty can be overcome by additionally connecting a capacitance
in parallel with each diode, as shown in Fig. 1.16(b). If this is done, when the
first diode recovers, it will take a finite time for the voltage across it to build
up, because of the time required to charge the parallel capacitor. This delay
will give time for the other diodes to recover.

To achieve high current capability, several diodes of one type can be connected in parallel. This is often done when the required current rating is not available in a single diode of a chosen category. When diodes are used in parallel, it is important to ensure proper current sharing. In a parallel connected bank of diodes, the voltage across each diode is the same. Therefore the current sharing will be determined by the forward current characteristics. So the recommended practice is to choose diodes that are matched as regards forward voltage drop. Additionally, it is also necessary to ensure temperature equality, because forward characteristics change with temperature. This calls for careful attention to the mounting and cooling arrangement, so that equality of temperature for all the parallel connected diodes is assured under working conditions. Improvement in current sharing can also be achieved by the use of an additional circuit element in series with each diode. But it is best to avoid such schemes for current sharing, because of the additional voltage drop and power dissipation that will result.

1.4.8 Current and Voltage Ratings of Power Diodes

1.4.8.1 CURRENT RATINGS. Current ratings are, in general, based on temperature rise considerations. Therefore they are valid only if diodes are mounted on heat sinks of adequate size to ensure that the internal temperature does not exceed the specified limit. The data sheet of a power diode usually gives three separate types of current ratings. These are (1) the average current, (2) the r.m.s. current and (3) the peak current. It is important to ensure that none of these ratings is exceeded in actual operation. To explain the significance of each of these ratings, we shall use the current waveform shown in Fig. 1.17, which is typical in a static DC/DC converter to be studied in Chapter 4.

In this waveform, the current consists of pulses that have a duration of only 0.2 ms in a switching period of 1 ms. The amplitude of the pulse is 50 A, and therefore the different current values are

$$I_{average} = 50 \times 0.2/1 = 10 \text{ A}$$
$$I_{rms} = \sqrt{(50^2 \times 0.2/1)} = 22.4 \text{ A}$$
$$I_{peak} = 50 \text{ A}$$

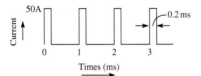

FIGURE 1.17
A diode current waveform for typical chopper duty.

To ensure safe operation of the diode, we have to look at the maximum rating specified separately for each of the above in the diode data sheet, and make sure that none is exceeded.

The peak current specified here is the "repetitive peak." A "surge current" rating is also frequently given which indicates the capability of the diode to handle an occasional transient, or a circuit fault. The permissible duration of the surge is also specified.

Very large power diodes are usually protected by fuses in the event of a circuit fault. Normal types of fuses used in power circuits are incapable of protecting power diodes. Special fast acting fuses are available for power semiconductors. To enable the selection of the proper fuse, the diode data sheet also gives a specification, called the i^2t rating, in $A^2 s$. To ensure protection, the fuse chosen must have an i^2t rating that is lower than that of the diode.

1.4.8.2 VOLTAGE RATINGS. For high power diodes, the data sheets usually specify two voltage ratings. These are (1) the repetitive peak reverse voltage and (2) the nonrepetitive peak reverse voltage. The non-repetitive peak is specified to indicate the capability of a diode to withstand an occasional over-voltage surge that may occur due to a circuit fault.

To choose the repetitive peak reverse voltage, we first look at the theoretical waveform of the voltage across the diode, that is to be expected in the circuit operation. From this, we get the peak repetitive voltage to which the diode is subjected. However, this theoretical waveform is usually on the assumption of ideal circuit elements. In an actual circuit, higher voltages can occur, due to reasons such as stray circuit inductances and transients as in the reverse recovery of diodes. Such over-voltages are difficult to estimate accurately. Therefore it is usual for circuit designers to use a safety factor, as regards the repetitive peak reverse voltage rating, in choosing a suitable diode. This is illustrated in the following example.

> **Illustrative example 1.7.** The AC voltages v_{AN} and v_{BN} in Fig. 1.18 are sinusoidal and equal in magnitude, but opposite in phase. The r.m.s. value of each is 120 V.
> (*a*) Sketch the theoretical voltage waveform across one of the diodes, and determine the repetitive peak.
> (*b*) Choose a factor of safety of 2.3, and determine the actual voltage rating of the diode to be selected on this basis.

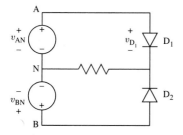

FIGURE 1.18
Circuit of Example 1.7.

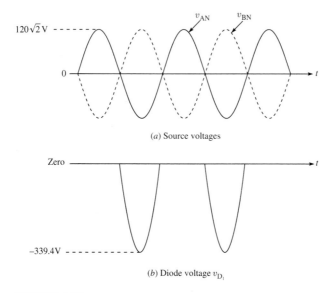

(a) Source voltages

(b) Diode voltage v_{D_1}

FIGURE 1.19
Voltage waveforms in Example 1.7.

Solution

(a) Figure 1.19(a) shows the two source voltage waveforms, which may be expressed as

$$v_{AN} = 120\sqrt{2} \sin \omega t$$

$$v_{BN} = 120\sqrt{2} \sin (\omega t - 180°)$$

When v_{AN} is positive, diode D_1 conducts and D_2 is OFF and vice versa. Therefore, when D_1 is OFF, the voltage v_{D_1} across it is $v_{AN} - v_{BN}$. On this basis, V_{D_1} is shown in Fig. 1.19(b). From this, the theoretical peak reverse voltage $= 2 \times \sqrt{2} \times 120 = 339.4$ V.

(b) With a safety of factor 2.3, the chosen diode should have a peak repetitive reverse voltage rating of $2.3 \times 339.4 = 780.6$ V.

1.5 POWER BIPOLAR JUNCTION TRANSISTORS AND POWER DARLINGTONS

1.5.1 Types, Ratings

These are high power versions of conventional small signal junction transistors, which are widely used for signal processing, both as discrete devices and in integrated circuits. High power discrete devices with individual current ratings of several hundred amperes and voltage ratings of several hundred volts are presently available, and such devices are widely used as static switches in power electronic converters. Power devices are available both in the npn and the pnp format. But the available current and voltage ratings are higher for npn devices. Because of the greater mobility of electrons compared with holes, npn devices can be fabricated on a smaller chip area to provide the same performance as an equivalent pnp device.

In static power converters, junction power transistors are invariably used as switches. These switches do not have any significant ability to block reverse voltages, and should be used in such a way that they are only required to block forward voltages. They are current controlled devices. This means that the operation of the switch is specified by the current input at its control terminal. There is a minimum threshold current input I_B necessary at the control terminal (which is labeled "base" in a junction transistor), for a given ON state current I_C through the main terminal (labeled the "collector terminal"). This minimum requirement to ensure the proper ON state is specified by the parameter h_{FE} for these devices, which is defined as

$$h_{FE} = I_C/I_B$$

The h_{FE} values for high power transistors are relatively low compared with low power devices, and may be as low as 20 or even less. This means that to switch 200 A using a transistor that has an h_{FE} of 20, we shall need to input at least $200/20 = 10$ A at its base terminal. This is still a large current requirement from a control circuit point of view.

This difficulty can be alleviated by using the "Darlington" arrangement. This scheme employs two transistors, one of which is the main transistor, the other being a smaller one. They are interconnected, in the manner to be explained later, so that the smaller "drive" transistor provides the base current to the main transistor. To turn the Darlington switch ON, it is only necessary to provide a very much smaller input at the base of the drive transistor, to enable it to provide a higher base current to the main transistor. A Darlington pair can be assembled from suitably chosen discrete transistors. However, integrated power Darlingtons are presently manufactured and are widely used. In these, the two interconnected transistors are fabricated on the same silicon pellet.

1.5.2 Junction Structure, Static Characteristics

Figure 1.20(a) shows the junction structure of a double diffused npn power transistor. In the fabrication of this device, the starting material is an n-type silicon wafer. First a p layer is formed, by diffusion of impurities, on one side. This is the base layer. A second diffusion, after masking the base terminal area, creates an n zone, which is the emitter layer. Electrical contacts are made by forming layers of metal, by vapor deposition, which are indicated by thickened lines in Fig. 1.20(a). The three external terminals of the device, which are collector (C), base (B) and emitter (E), make contact with the appropriate zones in the pellet through these metal deposition areas, as shown in Fig. 1.20(a). The circuit symbol for a power transistor is the same as for a low

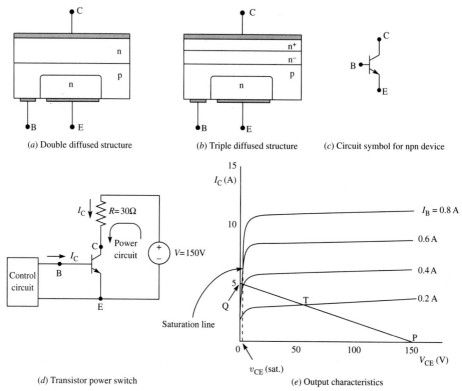

(a) Double diffused structure   (b) Triple diffused structure   (c) Circuit symbol for npn device

(d) Transistor power switch   (e) Output characteristics

FIGURE 1.20
npn bipolar junction power transistor.

power device, and is shown in Fig. 1.20(c) for an npn type. For a pnp, the arrow should be reversed. Figure 1.20(b) shows the "triple diffused" construction employed in some high power transistors. In this, an additional low resistivity n region labeled the n^+ layer is formed on top of the collector layer by means of a third diffusion. This is done primarily to provide a low resistance ohmic contact between the collector and the collector metal layer with good mechanical properties. It does not create any fundamental difference in the operating principle of the transistor.

When a transistor is used as a controlled switch, the control current input is provided at the base terminal. The control circuit is connected between the base and emitter. The power terminals of the switch are the collector and the emitter. This is shown in Fig. 1.20(d). In this circuit, by sending an appropriate current into the base terminal the "load" resistance R is connected across the DC supply voltage V. The manner in which the switching is achieved may be explained with reference to the "output" characteristics of the transistor shown in Fig. 1.20(e). The output characteristic is a plot of the current I_C through the switch versus the voltage V_{CE} across it for a fixed value of the control terminal current I_B. In Fig. 1.20(e), we have a family of output characteristics for different discrete values of I_B.

OFF **state.** If I_B is made zero, the value of I_C is negligibly small. This is the OFF state of the switch. This is also called the cut-off condition of the transistor. For negative values of V_{BE} also, there is no base current, and the transistor remains OFF. But the reverse voltage capability of the base–emitter junction is quite small, and it is important to ensure that this is not exceeded.

ON **state.** Let us assume that in Fig. 1.20(d), $V = 150$ V. Let $R = 30\,\Omega$ initially. Let us also decide to keep I_B at 0.6 A. The voltage V_{CE} across the switch and the current I_C through it must be given by a point on the characteristic for $I_B = 0.6$ A. To locate this point, we used a second relationship resulting from the application of Kirchhoff's law to the power circuit loop, which gives

$$V_{CE} = V - I_C R$$

This relationship is given by a straight line called the "load line." The load line corresponding to $R = 30\,\Omega$ is shown as PQ in Fig. 1.20(e). It is drawn by choosing two points on it that satisfy the above equation. These are usually chosen as the one for $I_C = 0$ for which $V_{CE} = V$ (point P) and the other for $V_{CE} = 0$ for which $I_C = V/R$ point Q). The intersection of the output characteristic and the load line gives us the current through the switch and the voltage across it.

We notice that the characteristics for all the different values of I_B are overlapping on the left-hand side, along a near-vertical line, indicated in the figure as the "saturation" line. The load resistance of 30 Ω gives us a load line that intersects the $I_B = 0.6$ A characteristic in the saturation region. The voltage across the transistor is seen to be very small, and is denoted by the symbol $V_{CE(sat)}$ in the figure. This is the ON state of the transistor switch, and $V_{CE(sat)}$ is the unavoidable small forward voltage drop acrosss the switch in its ON state.

Let us now see what will happen if we progressively reduce the base current I_B. We notice from the figure that until the base current falls to about 0.4 A, there is no noticeable fall in the load current. But further reduction in I_B causes the intersection point to move away from the saturation line, resulting in a large increase in the voltage across the transistor and a corresponding decrease in the voltage across the load. For example, for $I_B = 0.2$ A, the intersection point (labeled T in the figure) gives $v_{CE} = 60$ V and a current of 3 A. The transistor is no longer in the saturated ON state. Such a condition is to be avoided, because clearly the switch is very far from perfect. More importantly, there will be excessive power dissipation in the transistor, which can result in its damage. In our example, the collector power dissipation is seen to be 60 V × 3 A = 180 W. Compared with this, the power dissipation in the saturated ON state is very small. From the figure, the current is seen to be approximately 5 A and the voltage $V_{CE(sat)}$ approximately 2.5 V, resulting in a power dissipation of about 12.5 W.

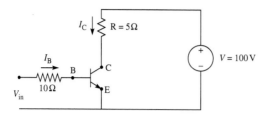

FIGURE 1.21
Circuit of Example 1.8.

The above consideration highlights the need to ensure a saturated ON state, by providing adequate base drive current, for the safe and satisfactory operation of the switch. To determine the minimum base current to ensure the saturated ON state, we use a parameter specified as h_{FE} in the data sheet of the transistor:

$$h_{FE} = I_C/I_B$$

From this, the minimum base current needed to ensure a saturated ON state will be given by $I_B = I_C/h_{FE}$. For values of the base current higher than this, the transistor will be saturated. Often it will be advisable to use a somewhat higher value of base current than that indicated by the above formula, as a safety feature, to take care of possible increases in I_C above the anticipated value. When this is done, we say that the circuit is designed to work normally with a "forced h_{FE}" that is less than the specified h_{FE} of the transistor. These statements are illustrated by the following example.

Illustrative example 1.8
(a) The transistor in the circuit of Fig. 1.21 has the following data: $V_{CE(sat)} = 1.5$; $h_{FE} = 50$; $V_{BE(sat)} = 1.8$ V. Determine the minimum value of V_{in} necessary to ensure a satisfactory ON state.
(b) Determine the total ON state power dissipation in the switch and its break up into collector dissipation and base dissipation.
(c) A transient over-voltage spike occurs in this circuit due to external causes, resulting in V going up from 100 V to 150 V for a short interval. Since the current is limited by I_B, assume that there is no significant change in I_C. What will be the power dissipation in the device under these conditions?
(d) If a "forced h_{FE}" is used in the original design of the base drive circuit [in (a)], what should its value be to ensure a saturated ON state under the above stated transient over voltage condition? What will be the power dissipation in that case?

Solution
(a) The ON state current is given by $[V - V_{CE(sat)}]/R = (100 - 1.5)/5 = 19.7$ A
The minimum I_B is given by

$$19.7/h_{FE} = 19.7/50 = 0.394 \text{ A}$$

$$V_{in} = 0.394 \times 10 + 1.8 = 5.74 \text{ V}$$

(b) The collector power dissipation

$$p_1 = I_C V_{CE(sat)} = 19.7 \times 1.5 = 29.55 \text{ W}$$

The base power dissipation

$$p_2 = V_{BE(sat)}I_B = 1.8 \times 0.394 = 0.71 \text{ W}$$

The total internal power dissipation is given by $p_1 + p_2 = 30.26$ W

Notice that the base power dissipation is very small compared with the collector power dissipation. For this reason, it is often ignored in power dissipation calculations.

(c) If I_C is limited to the same value of 19.7 A, because I_B is unchanged when the voltage goes up to 150 V, the new value of V_{CE} will be

$$V_{CE} = 150 - (19.7 \times 5) = 51.5 \text{ V}$$

The new value of collector power dissipation is given by

$$V_{CE}I_C = 51.5 \times 19.7 = 1014.55 \text{ W}$$

The new value of the total power dissipation in the transistor is $1014.55 + 0.71 = 1015.26$ W (for the duration of the transient over-voltage condition). This shows how the internal power dissipation can go up to potentially destructive levels, because the existing base drive is inadequate to handle the transient condition.

(d) We can ensure that the transistor stays in the saturated ON state while the transient is taking place, by using a "forced h_{FE}" lower than 50, in designing the base drive. If the transistor is not to come out of saturation, the collector current should be allowed to rise to $I_C = (150 - 1.5)/5 = 29.7$ A.

The base drive needed to maintain a saturated ON state will be

$$I_B = 29.7/h_{FE} = 29.7/50 = 0.594 \text{ A}$$

The forced h_{FE} under normal working conditions will therefore be

$$h_{FE} \text{ (forced)} = I_C/I_B = 19.7/0.594 = 33.2$$

If this had been done, the collector power dissipation during the occurrence of the transient would have been limited to

$$p_1 = 29.7 \times 1.5 = 44.6 \text{ W}$$

and the total power dissipation to

$$p = 44.6 + 0.71 = 45.31 \text{ W}$$

Proportional drive. The above example serves to highlight the fact that the minimum base current drive needed to ensure the saturated ON state of the transistor switch depends on the ON state current. In practical converters, the ON state current through the switch may vary according to load conditions. Therefore, if we employ a fixed base current drive, this should be sufficient for the highest ON state current to be expected. This implies that the base will be over-driven whenever the ON state current is less than the maximum value.

A major disadvantage of over-driving the base is the increase in the transition time for turn OFF switching. This happens because excesssive base current will cause excessive injection of minority carriers into the base region of the transistor, from the emitter side. Because of this, the collector current will persist for a longer time, until the excess minority carriers are removed, during turn OFF switching. To overcome this difficulty, circuit designers some times use "proportional drive." In such a scheme, the base current is automatically increased or decreased according to the magnitude of the collector current.

1.5.3 Safe Operating Area (SOA)

When a transistor functions in an electrical circuit, we can define its "operating point" at any given instant of time by means of the voltage V_{CE} across it and the current I_C through it. The operating point so defined can be graphically located by a point on the I_C versus V_{CE} plane such as in Fig. 1.20(e). Whenever there is a change of V_{CE} or I_C, or both, the operating point moves to a different location on this plane. The transition will be along a curve on the I_C versus V_{CE} plane, whose path will be determined by the instantaneous values of I_C and V_{CE} during the change. To ensure safe operation of the transistor without damage to it, all the operating points should be within finite boundaries on the I_C versus V_{CE} plane during transitions between operating points, which may occur during switching or for other reasons. This is called the Safe Operating Area (SOA). The boundaries of the SOA are usually specified by the manufacturer of the device, for stated conditions of working. Figure 1.22(a) shows a typical safe operating area. We shall examine the nature of the SOA and the parameters that determine each of the boundary lines.

1.5.3.1 MAXIMUM VOLTAGE—AVAILABLE BREAKDOWN LIMIT. A transistor has a maximum collector-to-emitter voltage V_{CE} that it can withstand, above which avalanche breakdown at the collector junction will occur. This determines the maximum voltage limit P in the SOA in Fig. 1.22 and the vertical boundary line PU. An indication of the maximum voltage capability is also provided in the data sheet of the transistor, by a parameter labeled as the "sustaining voltage" ($V_{CE(sus)}$). The significance of $V_{CE(sus)}$ may be explained as follows.

If the voltage V_{CE} across the transistor is increased progressively, the voltage at which avalanche breakdown commences will be determined by the manner in which the base terminal is connected. If the base terminal is left open (which means that the external impedance between the base and emitter is infinite), breakdown commences at a lower value of V_{CE} in comparison with the case where the base terminal is shorted to the emitter (zero external impedance between base and emitter). The breakdown voltage for the case of base terminal open is labeled V_{CEO} and that for the case of base terminal shorted to emitter is labeled V_{CEX}. For an intermediate value R of external base-to-emitter impedance, the breakdown voltage will be V_{CER}, intermediate between V_{CEO} and V_{CEX}. This is shown in Fig. 1.22(b). In all cases, after the avalanche breakdown has occurred, the voltage tends to remain constant at or near a particular value, as shown in Fig. 1.22(b). This value is the specified $V_{CE(sus)}$ of the transistor. $V_{CE(sus)}$ is usually about the same as V_{CEO}.

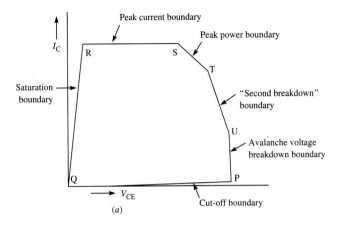

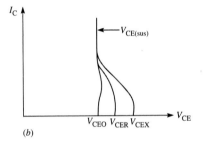

FIGURE 1.22
(*a*) Reverse-biased safe operating area for a junction power transistor. (*b*) Avalanche breakdown limits.

The value of $V_{CE(sus)}$ sets one of the operating limits of the transistor switch. The following example illustrates how a large V_{CE} can occur during turn OFF switching.

Illustrative example 1.9. In the circuit shown in Fig. 1.23, determine the forward blocking voltage of the transistor switch under the following condition. Assume the switch to be ideal.
(*a*) Initially, before switching ON.

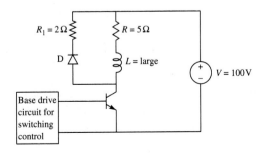

FIGURE 1.23
Circuit of Example 1.9.

(b) During turn OFF switching, after the switch was ON for a long enough time for the current to reach its steady state value.

Solution

(a) Initial blocking voltage $V = 100$ V.

(b) After the switch had been ON for a long time, the steady current through it will be $V/R = 100/5 = 20$ A, assuming $V_{CE(sat)}$ to be negligibly small.

When turn OFF switching is implemented, the current cannot fall to zero instantly, but will freewheel (continue to flow) through D and R_1.

Assuming that the turn OFF switching took place instantly, the voltage across R_1 immediately after turn OFF will be

$$V_{R_1} = 20 \times 2 = 40 \text{ V}$$

Therefore the blocking voltage that the transistor will have to withstand will be

$$V + V_{R_1} = 100 + 40 = 140 \text{ V}$$

1.5.3.2   CUT OFF AND SATURATION BOUNDARIES. Since normal operation is above the cut-off line PQ in Fig. 1.22(a) and to the right of the saturation line QR, these two lines constitute two other boundaries of the SOA.

1.5.3.3   PEAK CURRENT LIMIT. The lines RS in Fig. 1.22(a) corresponding to the maximum permissible collector current constitutes another boundary of the SOA.

1.5.3.4   MAXIMUM POWER. Neglecting the small base power dissipation, the total power dissipation in the transistor is equal to the collector power dissipation given by $p = V_{CE}I_C$. The maximum permissible value $p_{max} = V_{CE}I_C$ constitutes the bounary of the SOA indicated as ST in Fig. 1.22(a).

When used as a static switch, the peak current boundary RS and the maximum power boundary ST are very important during the turn OFF switching transition. The turn OFF switching of a power transistor is usually achieved by applying a reverse voltage to the base terminal. This reverse voltage is for the purpose of speeding up the turn OFF transition by "sucking out" the excess minority carriers from the base region. Therefore, this reverse voltage should last at least till the turn OFF switching transition is completed. So, in practice, these two boundaries of the SOA and the last boundary, which we shall consider next, are applicable under conditions of reverse-biased base voltage. As may be expected from Example 1.9, for a real transistor a large voltage and a large current can simulaneously occur during the turn OFF transition.

1.5.3.5   SECOND BREAKDOWN. In addition to the five boundaries of the SOA already described, there is another one, shown as TU in Fig. 1.22(a) and labeled "second breakdown." This is a phenomenon that can occur in a

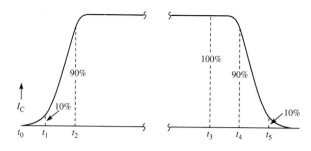

FIGURE 1.24
Power transistor switching times.

junction power transistor when voltage current and power dissipation are high, but still below the levels indicated by the limits discussed earlier. If we assume that during the turn OFF switching transition the collector current is uniformly distributed over the collector junction area, the power distribution will also be uniform over this area. If, on the other hand, the current distribution is nonuniform, local hot spots can occur due to excessive power dissipation, in locations in the junction area that experience high current densities. A failure of the device due to such occurrence of local hot spots is described as "second breakdown." On an oscillogram, the failure of a transistor can be identified as due to second breakdown by the collapse of V_{CE} to a low value. In the case of avalanche breakdown, the voltage across the device does not collapse, but stays at the $V_{CE(sus)}$ level. While avalanche breakdown need not necessarily result in permanent damage, second breakdown always does.

1.5.4 Switching times

Typical waveforms of collector current during turn ON and turn OFF transitions are shown in Fig. 1.24. The instants of time marked therein have the following significance:

t_0 the instant at which the turn ON switching is initiated by the arrival of the base current pulse

t_1 the instant at which the collector current has risen to 10% of its final value

t_2 the instant at which I_C reaches 90% of its final value

t_3 the instant at which turn OFF switching is initiated
(This is typically done by the application of a small reverse voltage on the base, to speed up the transition, resulting in a reverse base current pulse of short duration due to the excess minority carriers in the base region. This reverse base current lasts until the excess carriers are swept out of the base region. Notice that the collector current continues without any significant decrease, for a short time after t_3, because of these stored excess minority carriers.)

t_4 the instant at which the collector current has fallen to 90% of its ON state value

t_5 the instant at which the collector current has fallen to 10% of its ON value

The time delays stated in a typical data sheet of the device are defined as follows:

t_r "rise time" $= t_2 - t_1$
t_s "storage time" $= t_4 - t_3$
t_f "fall time" $= t_5 - t_4$

1.5.5 Base Drive Circuits for Power Transistor Switches

A well-designed base drive circuit should provide adequate base current to guarantee a saturated ON state under all conditions of collector current that can occur during operation. Also, a fast rising base current waveform will ensure fast turn ON switching. During turn OFF switching, a reverse base current of sufficient amplitude will result in the reduction of the storage time and therefore faster switching. For this purpose, it will be necessary to apply a reverse voltage pulse. It is not necessasry to maintain this reverse voltage after the reverse current has fallen to zero, when the excess minority carriers in the base zone have been removed. It is important to remember that the reverse voltage capability of the base emitter function is very small, and any reverse voltage should be well below the rated value to avoid damage to the transistor.

An example of a base drive circuit for a power transistor switch is given in Fig. 1.25. The base is driven from the secondary coil C_3 of a three-winding transformer. The transformer has two primary coils labeled C_1 and C_2. To

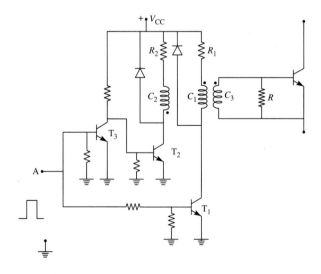

FIGURE 1.25
Example of a base drive circuit for a junction power transistor.

implement the turn ON switching, a positive voltage is applied at the terminal labeled A in Fig. 1.25. This causes the transistor T_1 to turn ON, energizing the primary coil C_1. At this time, the primary C_2 is turned OFF by transistor T_2. This happens because T_3 turns ON and connects the base of T_2 to ground. When the voltage pulse at A goes to the zero-voltage level, both T_1 and T_3 turn OFF. When T_1 turns OFF, the primary coil C_1 is disconnected from the power supply. The turning OFF T_3 causes the primary coil C_2 to be energized by the turning ON of T_2. Therefore a negative voltage appears on the secondary coil C_3. This negative voltage should be well below the rated reverse voltage of the base emitter junction. In this circuit, the reverse voltage on the base terminal may be maintained during the entire OFF period of the transistor depending on the pulse width capability of the pulse transformer. The pulse transformer provides isolation between the switching control circuit and the power circuit of the transistor. This pulse transformer must have the necessary pulse width capability to ensure the continued presence of the output pulse, during the entire duration of the ON time of the transistor.

1.5.6 Switching Aid Circuits (Snubber Circuits)

Switching aid circuits, also called snubber circuits, are used for the purpose of limiting the stress on static semiconductor switching devices during switching transitions. A typical switching aid circuit used with a junction power transistor is shown in Fig. 1.26. This circuit is for the purpose of limiting the operating point within the safe operating area (SOA) during turn OFF switching.

The circuit consists of a capacitor C, a diode D and a resistor R. When the transistor is in the ON state, the voltage across it, and therefore across the switching aid circuit, is nearly zero. The purpose of the capacitor–diode combination is to slow down the rate of rise of voltage across the switch during the turn OFF switching transition. This happens because during this time the diode turns ON and the capacitor starts charging. In the OFF state of the transistor, the capacitor remains charged to the full blocking voltage. It discharges during the next turn ON switching of the transistor. The resistor R is

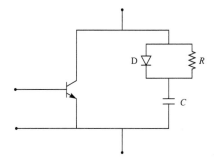

FIGURE 1.26
Turn OFF snubber circuit for junction transistor.

for the purpose of limiting the peak value of the discharge current through the transistor. Each time the transistor is turned ON, the total energy stored in the capacitor is dissipated in the resistor. Therefore the power dissipation in R is proportional to the switching frequency and to the square of the blocking voltage. The operation of the snubber circuit is illustrated by the following example.

Illustrative example 1.10. In the circuit of Fig. 1.27, the transistor switch has a storage time of 1.5 μs and a fall time of 1 μs. The current in the ON state is 10 A. Assume that the load inductance is large and therefore the load current is constant during the turn OFF switching transition.

(a) Choose suitable values for the snubber circuit capacitance and resistance to satisfy the following requirements:

(1) the maximum value of V_{CE} should not exceed 100 V before the turn OFF switching is completed;

(2) the peak discharge current of the snubber capacitor during turn ON should not exceed 8 A.

During the turn OFF switching transition, assume that the current through the transistor falls linearly.

(b) Determine the peak collector power dissipation during the turn OFF switching transition (1) with the snubber circuit and (2) without the snubber circuit.

(c) What is the power rating of the snubber resistance?

Solution

(a) During the storage time, there is no significant decrease in the collector current. We therefore assume that the transistor switch continues in its ON state during this interval. Since the current falls linearly during the fall time, its value at any instant t from the commencement of the fall may be expressed (in A) as

$$i = 10\left(1 - \frac{t}{t_f}\right)$$

The current (in A) through the capacitor is therefore,

$$i_c = 10t/t_f$$

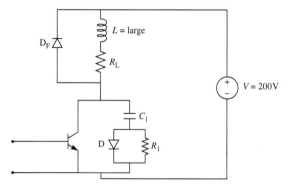

FIGURE 1.27
Circuit of Example 1.10.

The voltage across the capacitor at instant t will therefore be

$$v_{CE} = \frac{1}{C_1} \int i_c \, dt = \frac{1}{C_1} \frac{10}{2} \frac{t^2}{t_f}$$

This is equated to 100 V at $t = t_f = 10^{-6}$ s and we get $C_1 = 0.05 \, \mu F$

Before the next ON switching, the capacitor will be charged to 200 V. Therefore the peak capacitor discharge current at turn ON will be $200/R_1$. Equating this to 8 A, we get $R_1 = 25 \, \Omega$.

(b) The power dissipated in the transistor at any instant during the fall time will be

$$p = v_{CE} i_c$$

$$= \frac{1}{C_1} \frac{10}{2} \frac{t^2}{t_f} \times 10 \left(1 - \frac{t}{t_f} \right)$$

$$= \frac{100}{2C_1 t_f} \left(t^2 - \frac{t^3}{t_f} \right)$$

At maximum dissipation, $dp/dt = 0$. Therefore, differentiating the above expression and equating to zero, we get

$$t = \tfrac{2}{3} t_f = 0.67 \, \mu s$$

Therefore the maximum power dissipation occurs at $t = 0.67 \, \mu s$. At this instant the voltage

$$v_{CE} = \frac{1}{C_1} \frac{10}{2} \frac{t^2}{t_f} = 44.4 \text{ V}$$

The current

$$i_C = 10 \left(1 - \frac{t}{t_f} \right) = 3.33 \text{ A}$$

The maximum power dissipation is $v_{CE} i_C = 148$ W.

Note that in the above solution we have assumed that the current falls from its full value of 10 A to zero during the specified fall time of 1 μs. The exact definition of fall time is the time for fall from 90% to 10%.

Without the switching aid circuit, as soon as the current begins to decrease, the diode D_F will turn ON, because the total current cannot change instantaneously. Therefore, the full supply voltage of 200 volts will appear at the collector terminal. Under this condition, the maximum power dissipation in the transistor switch will reach a potentially destructive level given by

$$p = 200 \times 10 = 2000 \text{ W}$$

(c) At the end of every turn OFF switching, the snubber capacitor gets charged to the full DC voltage of 200 V. During the next turn ON switching, all the energy is dissipated in the snubber resistance. Therefore the power rating of the snubber resistor should not be less than the switching frequency times the energy stored in the snubber capacitor.

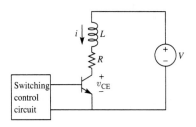

FIGURE 1.28
Turn ON switching of an inductive load.

The energy stored in the capacitor is $\frac{1}{2}C_1 V^2 = 10^{-3}$ J.

The minimum power rating of the snubber goes up proportionately for the switching frequency and also for the value chosen for the snubber capacitor. In the present case, at a switching frequency of 20 kHz, the resistance should dissipate 20 W.

1.5.6.1 TURN ON SWITCHING. Turn ON switching into a highly inductive load does not normally create a power dissipation problem. Referring to Fig. 1.28, during the turn ON transition, the voltage v_{CE} is given by

$$v_{\text{CE}} = V - \left(L\frac{di}{dt} + Ri \right)$$

The inductance slows down the rate of rise of current through the transistor during the transition time. For this reason, the voltage transition from V to zero will be completed well before the current rises to significant levels. Therefore the internal power dissipation during the transition will be small. For such a condition of working a "turn on" snubber may not be necessary.

There are, however, other situations, which occur very often in practice, in which a turn ON snubber will be necessary. This is illustrated by the circuit shown in Fig. 1.29. This is essentially the circuit considered in Example 1.6. We shall look at such a circuit again, to highlight the need to use a turn ON snubber.

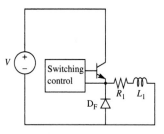

(a) Transistor switch without turn ON snubber

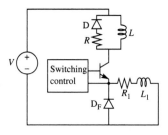

(b) Switch with turn ON snubber

FIGURE 1.29
Turn ON snubber for a junction transistor.

In the circuit of Fig. 1.29(a), the transistor switch is turned ON and turned OFF at a chosen repetitive frequency. This is for the purpose of controlling the current in the R_1–L_1 load circuit, which can be done by controlling the ON time in the switching period. During the ON time of the switch, the voltage V is applied to the load, and during the OFF time, the current freewheels (continues to flow due to the stored energy in L_1) through the diode D_F. Let us focus attention on the instant at which the transistor is turned ON, when the current is already freewheeling through D_F. Because D_F has a finite recovery time, initially, when the transistor switch is turned ON, the full voltage V comes across it, without any circuit element in the loop to limit the current. This is therefore a situation in which the voltage and current magnitudes can take the device outside its SOA and cause damage to it. This type of situation occurs in the majority of static DC/DC converters and DC/AC inverters. Unless there are external circuit elements to limit the growth of current, it will become necessary to use a "turn ON" snubber.

A typical turn ON snubber circuit is shown with the transistor switch in Fig. 1.29(b). This consists of a small inductance L, a diode D and a resistor R. During turn ON switching, the diode D is OFF and the inductance limits the rate of rise of current in the transistor. When the transistor is turned OFF, the current in L freewheels through the diode, and the energy stored in L is dissipated in R. Since L is needed only for a short time initially during the recovery time of D_F, some designers use a saturable core inductance, consisting of a few turns of wire on a toroidal magnetic core. The core gets saturated when the current rises to a certain level, and afterwards the inductance becomes negligible. Therefore such a coil will present the needed inductance at the beginning only of the turn ON switching. In Fig. 1.29(b), only the turn ON section of the snubber circuit is shown, in the interests of clarity. But the total switching aid circuit may actually consist of both the turn ON and the turn OFF snubbers. The use of the turn ON snubber is illustrated by the following example.

Illustrative example 1.11
(a) In the circuit of Fig. 1.29(b), $V = 100$ V. The load current may be assumed constant at 10 A. Choose the snubber inductance value L to limit the rate of rise of current through the switch to 10 A/μs.
(b) The reverse recovery peak current in D_F occurs 500 ns after the occurrence of zero current through it. Determine (1) the peak current pulse amplitude through the transistor; (2) the peak energy stored in the turn ON snubber in a switching cycle.

Solution
(a) Immediately after the transistor is turned ON and before D_F has recovered, the loop equation for the voltage source loop will be:

$$V = L\frac{di}{dt}$$

To limit di/dt to $10 \text{ A}/\mu\text{s} = 10^7 \text{ A/s}$, the value of L should be

$$L = V \Big/ \frac{di}{dt} = \frac{100}{10^7} = 10\,\mu\text{H}$$

(b) The current through the transistor, rising at the rate of $10 \text{ A}/\mu\text{s}$, reaches 10 A in 1 μs. At this instant, the current through D_F has fallen to zero. But D_F has not yet recovered, and is continuing to conduct. In the next 500 ns, the current continues to rise in the transistor at the same rate. Therefore the further increase in current in the transistor will be $0.5\,\mu\text{s} \times 10$ A. Therefore the peak current amplitude

$$I_{\text{peak}} = 10 + 5 = 15 \text{ A}$$

The peak energy stored in $L = \frac{1}{2}Li_{\text{peak}}^2 = 1.125 \times 10^{-3} \text{ J}$.

1.5.6.2 SNUBBER POWER LOSS. Examples 1.10 and 1.11 illustrate how the snubber circuits help to protect the transistor switch by limiting its operation within the SOA. But there is a price to be paid for this, in the form of additional power loss. In the turn OFF snubber, the energy stored in the snubber capacitor C and dissipated in each switching cycle is $\frac{1}{2}CV^2$. In the turn ON snubber, the energy stored in the snubber inductance and dissipated in each cycle is $\frac{1}{2}LI_{\text{peak}}^2$. Since these energy losses occur in every switching cycle, the average power losses in the snubber circuits increase directly in proportion to the converter switching frequency. Some snubber circuits that incorporate the recovery of stored energy from the snubber circuit elements, instead of dissipating it in resistances, have been developed and are described in the literature. The additional circuit complexity involved in using such schemes may be justified in the interests of improving converter efficiency when the switching frequencies are high.

1.5.7 Power Darlingtons

Power Darlingtons are used primarily for the purpose of reducing the control current requirement for turn ON switching. Figure 1.30 shows how the two junction transistors that constitute the Darlington switch are interconnected. T_M is the main power transistor. T_A is the auxiliary transistor, of lower power, which provides the base current to the main transistor. At the present time,

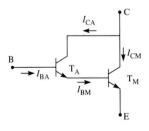

FIGURE 1.30
Power Darlington.

integrated power Darlingtons, in which both the transistors and their intercon-nections are fabricated on the same silicon chip, are available for large ratings. Externally, they have only three terminals, as in single transistors.

Referring to Fig. 1.30, the base current I_{BA} needed to maintain the saturated ON state for the auxiliary transistor with a collector current I_{CA} will be given by the relationship

$$I_{CA} = h_{FE(A)}I_{BA}$$

The corresponding emitter current, which is the base current of the main transistor, will be

$$I_{BM} = (1 + h_{FE(A)})I_{BA}$$

The corresponding collector current I_{CM} will be

$$I_{CM} = h_{FE(M)}I_{BM} = h_{FE(M)}(1 + h_{FE(A)})I_{BA}$$

Therefore the total load current in the external circuit will be

$$I = I_{CA} + I_{CM} = (h_{FE(A)} + h_{FE(M)} + h_{FE(A)}h_{FE(M)})I_{BA}$$

Since the h_{FE}s are relatively large numbers, their product will be very much larger, and therefore we may approximate the above expression as

$$I = h_{FE(A)}h_{FE(M)}I_{BA}$$

This means that the Darlington switch has an overall current gain approxim-ately equal to the product of the current gains of the individual transistors constituting the switch. This explains why the Darlington switch needs only a very much smaller control current for its operation, in comparison with a single power transistor, to switch the same load current. However, the switching times of power Darlingtons are somewhat longer than comparable single transistors, as may be expected from the fact that two transistors need to switch in a Darlington.

1.6 POWER MOSFETS

1.6.1 Types, Comparison with BJT

The power MOSFET is the high power version of the low power metal oxide semiconductor field effect transistor (MOSFET) that is widely used in analog and digital signal processing circuits, both in discrete and in integrated circuit (IC) forms. Relatively recent developments in power device technology have resulted in the production of high power devices, with large voltage and current capabilities. Individual devices with typical ratings of tens of amperes and hundreds of volts are presently available. Both "n-channel" and "p-channel" devices are being made, but the former are available in higher ratings.

Although the working principle of a power MOSFET is the same as that of its low power version, there are significant differences in the internal geometry. A major difference is that IC MOSFETs have a "planar" structure. This means that all the terminals of the device are on one side of the silicon pellet. Therefore the internal current flow paths are parallel to the surface of the pellet. In contrast, power MOSFETs have a vertical structure, meaning that the current flow is across the pellet, between its power terminals, which make contact on opposite sides of it. This results in lower internal voltage drop and higher current capability.

In the n-channel device, the current flow is due to electrons, whereas in the p-channel type, it is due to holes. The p-channel device works in exactly the same way as an n-channel device, with the voltage polarities and current directions reversed. Therefore it will be sufficient for us to describe the n-channel power MOSFET only, it being understood that our statements are applicable to p-channel devices also, with reversed signs for voltages and currents. Electrons have a higher mobility than holes inside the silicon crystal. Therefore it is possible to make devices with higher ratings in a pellet of the same size in the n-channel format.

A power MOSFET can be used either as a static switch or for analog operation. In analog working, the current magnitude through it is continuously adjustable by varying the voltage at its control terminal. In static power converters, power MOSFETs are invariably used as switches. We shall therefore confine our treatment to the switching mode of operation. The main considerations in the choice between power MOSFETs and power junction transistors are summarized below.

1. The power MOSFET is a voltage controlled device, which requires negligible current in its control terminal to maintain the ON state. In contrast, the junction power transistor is a current controlled device, which needs appreciable control current for keeping it in the ON state.

2. Power MOSFETs have relatively shorter switching times. Therefore they can be used at higher switching frequencies.

3. The internal junction structure of a power MOSFET is such that there exists a diode path in the reverse direction across the main terminals of the switch. Therefore it is, in effect, a parallel combination of two static switches—a controlled switch for forward current flow and an uncontrolled diode switch for reverse currents. Such a combination of switches is very frequently required in static converters in which the diode serves to provide a freewheeling current path. The integral reverse diode of the power MOSFET is called its "bodydiode." It has adequate current rating and

switching speed to meet most freewheeling needs. Therefore the need to use a separate antiparallel diode for free wheeling is rare. When using a junction power transistor for such an application, it is always necessary to use a separate fast recovery diode of adequate rating in antiparallel with it.

4. Junction power transistors generally have lower ON state voltage drop than power MOSFETs. Therefore they have less static power dissipation.

5. Junction transistors are available in much higher current and voltage ratings than power MOSFETs.

1.6.2 Junction Structure

Power MOSFETs are fabricated in the form of arrays. This means that a single power MOSFET is in reality a parallel combination of thousands of individual cells, each cell being a MOSFET in itself. The number of cells on a silicon pellet may be as high as 1000 on an area as small as 1 mm^2. The device has three external terminals, called drain, source and gate. The drain and the source are the power terminals of the switch. Gate is the control terminal. The control voltage to implement turn ON is applied between the gate and the source terminals. The direction of forward current flow in an n-channel device is from the drain to the source, through it. This results from the flow of electrons from the source to the drain.

The junction structure of an n-channel device is shown in Fig. 1.31(a),

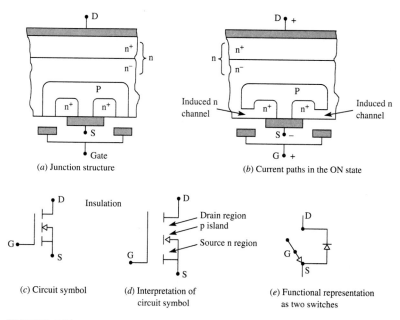

(a) Junction structure

(b) Current paths in the ON state

(c) Circuit symbol

(d) Interpretation of circuit symbol

(e) Functional representation as two switches

FIGURE 1.31
n-Channel power MOSFET.

which is drawn with the drain surface on the top. Only one cell, out of the tens of thousands that constitute the device, is shown in the figure. All the cells have a common drain surface. The source metal depositions for all the cells are connected in parallel and constitute the source terminal of the device. A similar statement applies to the gate terminal.

The n layer on the top constitutes the drain. This layer is actually made up of an outer n^+ layer of low resistivity and an inner high resistivity (low impurity concentration) n^- region. The inner high resistivity region serves to give high voltage capability, while the outer low resistivity region serves to make a strong low resistance electrical contact with the drain surface metal deposition. Adjacent to the n^- zone is a relatively large p island, as shown in Fig. 1.31(a). Inside the p island are again n^+ islands. The source metal deposition covers a good part of the n^+ islands and also the middle part of the p island between the n^+ islands, as shown. The gate terminal does not make any electrical contact with the silicon pellet, because of the presence of a layer of silicon dioxide, which is an insulator, between the silicon surface under the gate. The gate zone is over the p island between the n^- drain region and the n^+ source region. A conducting polycrystalline silicon layer, deposited over the gate zone, over the silicon dioxide layer, serves as the gate layer. This is found to give better performance than deposited metal layer.

1.6.3 Principle of Operation

In normal operation, the drain terminal is positive with respect to the source. But if there is no input on the gate terminal, no current can flow from drain to source, because the junction between the n^- drain region and the p island is reverse biased. The only current that flows is the reverse leakage current of this junction, which is negligibly small. This is therefore the OFF state of the switch.

If a positive voltage with respect to the source is now applied to the gate, the electric field so created pulls electrons from the n^+ zone into the p zone immediately near the gate. In this way an n "channel" is created linking the source n^+ region and the drain n^- region, as shown in Fig. 1.31(b). This n channel now provides the path for flow of current from drain to source. If the positive potential on the gate is not of sufficient magnitude to create a channel, no current will flow. Therefore there is a threshold value for the gate-to-source voltage V_{GS}, below which the switch will be completely OFF. Above this threshold, the cross-sectional area of the channel will increase with increasing V_{GS}.

For a given value of V_{GS}, however, there is a limit to the maximum current that can flow through the channel, without appreciable voltage drop. If we keep increasing the drain-to-source voltage V_{DS} in an attempt to increase the current, there will initially be a steep increase in current. Afterwards, the

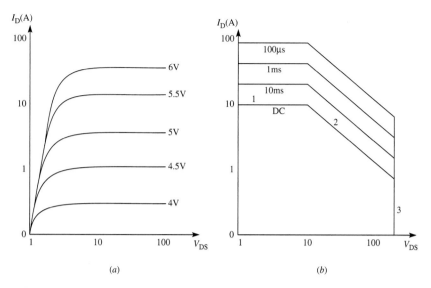

FIGURE 1.32
(*a*) Output characteristics and (*b*) safe operating area.

current will reach a saturation value I_{DS}, which is limited by the size of the channel, that is, by V_{GS}. There will be no further significant increase in current for that particular value of V_{GS}. If the power MOSFET is to be used as a switch, with the lowest possible value of voltage drop in the ON state, the current should be limited below the saturation level. Once the saturation value is reached, further increase in V_{DS} will only cause increased voltage drop across the device and increased power dissipation in it, without increase in current. These statements are evident from the output characteristics shown in Fig. 1.32(*a*). These characteristics show the relationship between the drain current I_D and V_{DS} for different values of V_{GS}.

A power MOSFET, when used as a switch, should be in the "unsaturated" region of the output characteristic. This region is understood for a MOSFET as the region on the left side of the characteristics where the current rises linearly with respect to V_{DS}. The ratio of voltage to current in this region is described as the ON state resistance of the device $R_{DS(ON)}$. This is an important parameter listed in the data sheet of a power MOSFET. A device with lower $R_{DS(ON)}$ has lower internal power dissipation and lower voltage drop in the ON state, but generally costs more.

In practice, a V_{GS} value of $+12$ to $+15\,\text{V}$ will be adequate to turn the switch fully ON, in the case of most power MOSFETs. Power MOSFETs are also being manufactured that can be turned fully ON by lower positive voltage levels such as $5\,\text{V}$ used in TTL logic ICs. These are called logic level MOSFETs.

The circuit symbol for an n-channel MOSFET is shown in Fig. 1.31(c). The basis of this circuit symbol can be understood by reference to the junction structure shown in Fig. 1.31(a) and the explanatory labels given in Fig. 1.31(d).

The integral reverse diode or the "body diode" of the power MOSFET. Reference to the junction structure in Fig. 1.31(a) shows that if the source is made positive with respect to the drain, there is a direct path for current flow across the junction between the p region and the drain n$^-$ region, which becomes forward-biased under this condition. Therefore the device functions like a power diode in this direction. This integral antiparallel diode is an advantageous feature for most switching applications of the power MOSFET. The functional representation of a power MOSFET switch will therefore consist of a controlled static switch in the forward direction in antiparallel with an uncontrolled diode switch. This is shown in Fig. 1.31(e).

1.6.4 Output Characteristics

From the output characteristics in Fig. 1.32(a), we notice that for any value of V_{GS} above the threshold level, initially for low values of V_{DS}, the device behaves like a resistance, the current increasing linearly with voltage. This is to be expected, because, once the channel has been created, there is no pn junction in the current path. The current can be looked upon as flowing through a series of combination of resistances, consisting of the bulk resistance to vertical current flow in the drain n$^+$ and n$^-$ regions, the resistance of the channel and the resistance to transverse current flow in the source n$^+$ region from the source terminal to channel opening. The ratio V_{DS}/I_D is the total resistance in the ON state, equal to $R_{DS(ON)}$.

The magnitude of $R_{DS(ON)}$ determines the forward voltage drop and the internal power dissipation in the device, in its ON state, for a drain current I_D. These are given by

$$v_f = I_D R_{DS(ON)}$$

$$p = I_D^2 R_{DS(ON)}$$

1.6.5 Safe Operating Area (SOA)

A typical safe operating area (SOA) is shown in Fig. 1.32(b). The boundaries are set by (1) the maximum permissible drain current, (2) the maximum power dissipation and (3) the maximum drain-to-source voltage. These are shown in the figure. The limits for current and power will be somewhat higher for pulsed operation, as compared to continuous working. These will also depend on the duration of the pulse, as indicated in the figure. There is no second breakdown phenomenon in power MOSFETs. In estimating the maximum power, the power dissipation resulting from reverse current flow should not be over-looked, if such reverse current flow occurs in circuit operation.

1.6.6 Gate Electrode Capacitance

Since the gate electrode layer is insulated from the pellet by a silicon dioxide layer, the gate input current in the static ON state may be considered zero for all practical purposes. But the conducting surface of the gate layer has appreciable capacitance with the drain electrode metal layer and also with the source electrode metal layer. These capacitances cause charging and discharging currents to flow in the gate terminal during switching, and therefore affect the design of the gate control circuit.

Figure 1.33 shows the three interelectrode capacitances C_{gd}, C_{gs} and C_{ds}. Device data sheets usually specify the capacitances in the following form:

$$C_{iss} = C_{gs} + C_{gd}$$

$$C_{oss} = C_{ds} + C_{gd}$$

$$C_{rss} = C_{gd}$$

C_{iss} is the input capacitance of the gate terminal with the source and drain tied together. C_{oss} is the output capacitance measured with the gate tied to source. C_{rss} is called the reverse transfer capacitance. An aspect of circuit behavior that is very important in the design of the gate control circuit is the magnification of C_{gd} due to the "Miller effect." This is illustrated by the following example, in which the capacitance values are typical of a power MOSFET.

> **Illustrative example 1.12.** The power MOSFET in the circuit of Fig. 1.34 has interelectrode capacitance $C_{gs} = 4500\,\text{pF}$, $C_{gd} = 250\,\text{pF}$ and $C_{ds} = 800\,\text{pF}$. Assume that the device is otherwise ideal. A control gate pulse of amplitude 15 V with

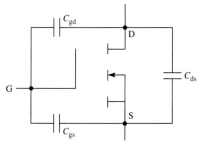

FIGURE 1.33
Interelectrode capacitances in a power MOSFET.

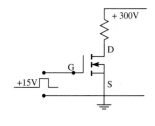

FIGURE 1.34

respect to source, is applied to implement turn ON. This voltage pulse rises from 0 to 15 V in 200 ns. The switching takes place with a linear fall of drain-to-source voltage during this interval from 300 to 0 V. Determine the peak gate current amplitude during the turn ON transition.

Solution. Let us focus attention on the gate-to-drain capacitance C_{gd} before and after the switching. Before turn ON, this capacitance is charged to 300 V with the drain side positive. After turn ON, the same capacitor is charged to 15 V, now with the drain side negative. Therefore the total change of voltage across this capacitor is given by

$$V_{gd} = 15 - (-300) = 315 \text{ V}$$

The total flow of charge due to this change of voltage is $Q = C_{gd}V_{gd} = 250 \times 10^{-12} \times 315 = 7.875 \times 10^{-8}$ C. This flow of charge is assumed to take place linearly in 200 ns. Therefore the amplitude of the current pulse during this interval will be

$$I_1 = \frac{7.875 \times 10^{-8}}{200 \times 10^{-9}} = 0.394 \text{ A}$$

During the same interval, the gate-to-source capacitance C_{gs} gets charged to 15 V. The charge required for this will be $15C_{gs} = 15 \times 4500 \times 10^{-12} = 6.75 \times 10^{-8}$ C. Therefore the amplitude of the gate current pulse needed to charge the gate-to-source capacitance will be

$$I_2 = \frac{6.75 \times 10^{-8}}{200 \times 10^{-9}} = 0.338 \text{ A}$$

The total amplitude of the gate current pulse in $I_1 + I_2 = 0.732$ A.

Comments. This current pulse of large amplitude lasts only for the duration of the switching transition. After the ON switching is completed, the current falls to zero. A similar reverse current pulse occurs during turn OFF switching transition. A significant part of the current pulse is due to the effective multiplication of the gate-to-drain capacitance (Miller effect) by a factor of

$$1 + \frac{V_{ds}}{V_{gs}} = 1 + \frac{300}{15} = 21$$

1.6.7   Power MOSFET Switching Times

Four time intervals associated with switching transitions are usually stated in the data sheet of a power MOSFET. They are to be understood in relation to the waveforms of the switching control voltage v_{GS} and the voltage across the

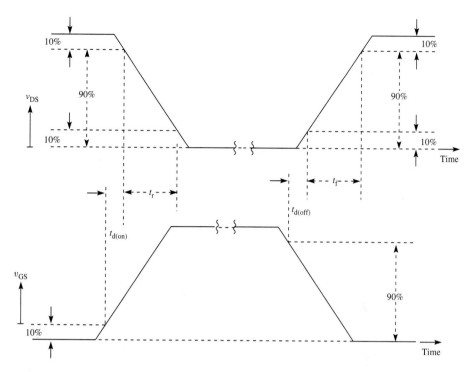

FIGURE 1.35
Switching time definitions.

switch v_{DS}. Typical waveforms of both are given in Fig. 1.35, approximated to linear shapes. The specified time delays are defined as follows:

$t_{d(on)}$ turn ON delay: this is the interval measured from the instant when the gate voltage V_{GS} has risen to 10% of its final value to the instant the drain voltage has fallen by 10% of its initial OFF state value

t_r rise time: this is the time interval during which V_{DS} falls from 90% to 10% of its initial OFF state value; during this time, the drain current rises between the corresponding limits

$t_{d(off)}$ turn OFF delay: this is the time interval during turn OFF switching from the instant V_{GS} has fallen to 90% of its ON state value to the instant V_{DS} has risen to 10% of its final OFF value

t_f fall time: this is the time during which V_{DS} rises from 10% to 90% of its final OFF state value; during this time, the drain current falls between the corresponding limits

The turn ON time t_{ON} is defined as:

$$t_{ON} = t_{d(on)} + t_r$$

The turn OFF time t_{OFF} is defined as

$$t_{OFF} = t_{d(off)} + t_f$$

Power MOSFETs, in general have shorter switching times than other power semiconductor switches. Actual values of t_{ON} and t_{OFF} depend on device ratings. Because of the faster switching capability, they are preferred when higher switching frequencies are required.

1.6.8 Switching Aid Circuits (Snubber Circuits)

In general, it is desirable to use a snubber circuit with a power MOSFET to protect it from excessive stresses during switching transitions. Such a circuit will be similar to those we have described for junction power transistors. Special considerations of relevance to power MOSFETs are discussed below.

Our earlier discussion of the effects of interelectrode capacitances shows that during turn OFF switching, the charging of the drain to gate capacitor results in a current flow in the direction from the drain terminal to the gate. The external gate circuit conditions can be such that this current raises the gate potential and leads to a spurious turn ON. Such an occurrence can result in device damage. Therefore the turn OFF transition is an interval during which the device may be overstressed because of the excessive dv/dt. Present-day devices, generally, have high dv/dt withstand capability. The snubber circuit components can be suitably chosen to limit the dv/dt if required.

The power MOSFET has an integral reverse diode whose recovery characteristics are similar to those we have described earlier for fast recovery diodes. Therefore transient over-voltages can occur, leading to excessive values of forward voltage across the device, during the recovery of the integral diode, in exactly the same way as we described for single diodes. This aspect also needs to be considered in the choice of the snubber components.

1.6.9 Gate Drive Circuits for Power MOSFETs

To exploit the fast switching capability of the power MOSFET, to realise high repetitive switching frequencies, it is necessary to provide steeply rising gate pulses, to achieve fast turn ON. We have seen that this results in a gate current pulse of large amplitude, because of the gate-to-drain Miller capacitance and the gate-to-source capacitance. Integrated circuit modules specifically designed as "drivers" for power MOSFETs are presently available. These have the capability to output fast rising voltage pulses of the required transient high current capability. They could be directly interfaced with TTL or MOS logic gates and also provide for inverted or noninverted operation. Inverted operation means that when the input to the driver goes to the low logic level, the output goes high, and vice versa.

Electrical isolation between the switching control circuit and the power circuit is generally needed in static power converters. This is achieved by the use of a pulse transformer, which should have the required rise time and pulse width capabilities.

A gate drive circuit with these features is shown in Fig. 1.36(*a*).

Two aspects of the circuit of Fig. 1.36(*a*) need careful attention before it can be made to work satisfactorily. First it should be noted that the pulses applied to the primary of the pulse transformer are unidirectional and will have a DC component. This can cause excessive saturation of the core unless the resulting DC component of the primary current is limited, say by means of a

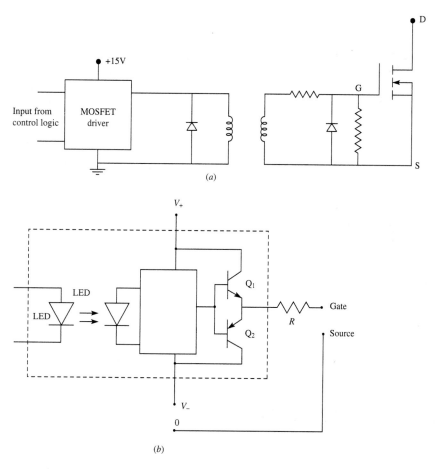

FIGURE 1.36
Gate drive circuits for power MOSFETs.

suitably placed resistance. Alternatively, a scheme for "resetting" the magnetic core of the transformer, by sending a reverse magnetizing current pulse through a third winding on the transformer, may overcome the difficulty. The second aspect is to ensure that the pulse transformer is designed to reproduce the primary pulse with fast enough rise time and low droop.

A popular method used for driving power MOSFETs (and IGBTs, which we shall be considering later, and whose drive requirements are similar) is to use an opto-coupler to provide the required isolation. Such a scheme needs a power supply for the gate circuit that is isolated from the control circuit power supply. Integrated circuit chips intended for the specific purpose of the isolated driving of power MOSFETs and IGBTs can be used. Figure 1.36(b) shows such a scheme. The circuit within the IC package is within the broken lines. To drive the power MOSFET, a current pulse of the required pulse width is sent through the LED shown. Light from this falls on a photo diode and causes the circuit block shown inside to turn ON the transistor labeled Q_1. This causes the positive power supply voltage labeled V_+, which may typically be about 12 V, to be applied to the gate, to turn ON the power device. The circuit shown also has the means to apply a small negative voltage to the gate, while turning OFF and during OFF periods. This may help to improve the switching performance, especially if the device being switched is an IGBT. This is achieved by keeping the transistor Q_2 ON when the LED is not energized. The external resistance R limits the initial charging gate current to a value safe for the transistors Q_1 and Q_2. The isolated power supply provides the positive voltage V_+ and the negative voltage (if needed) V_-. The common terminal of the isolated power supply is connected to the source terminal of the device.

1.7 INSULATED GATE BIPOLAR TRANSISTORS (IGBTs)

1.7.1 IGBT Compared with Power MOSFETs and Power BJTs

The IGBT has appeared on the scene relatively recently as a successful static power switch that combines several advantages of the power MOSFET and the bipolar junction power transistor. Like the power MOSFET, it is a voltage controlled switch, and its switching control requirements are practically the same as for a power MOSFET. Also like the power MOSFET, it is a continuously controllable device. But in static power converters it is always used as a switch. Its ON state voltage drop is typically lower than that of a power MOSFET. It is closer in this respect to a junction power transistor. Unlike the power MOSFET, the IGBT has no integral reverse diode. When such a diode is needed for free wheeling in any circuit application, a separate antiparallel power diode has to be used along with the IGBT. But single modules, in which IGBTs are packaged along with antiparallel fast recovery diodes with soft recovery characteristics, are available. The IGBT has no significant reverse voltage blocking capability. The maximum reverse voltage it can withstand is typically well below 10 V. IGBTs are manufactured in voltage and current ratings extending well beyond what are normally available in power MOSFETs. For example, at the high power end, devices with a voltage rating of 1200 V and current rating of 600 A are available.

The switching speeds of IGBTs are higher than those of bipolar power transistors. The turn ON times are about the same as in power MOSFETs. But turn OFF times are longer. Therefore, the maximum converter switching frequencies possible with IGBTs are intermediate between bipolar power transistors and power MOSFETs.

1.7.2   Junction Structure, Principle of Working

1.7.2.1   JUNCTION STRUCTURE. Figure 1.37(a) shows the junction structure of a typical IGBT cell. This should be compared with the structure of an n-channel power MOSFET given earlier in Fig. 1.31. It will then be noticed that there is only one difference. In the IGBT, there is an additional p^+ layer over the n^+ drain layer of the power MOSFET structure. This p^+ region constitutes the "collector" of the IGBT. As in the power MOSFET, the adjacent n region consists of an n^+ and an n^- region. The n^+ region serves to achieve better performance, but does not materially change the operating principle of the device. We shall therefore treat these two layers as together constituting a single n layer when explaining the switching operation of the IGBT.

The collector is a power terminal of the IGBT switch. The other power terminal is the terminal labeled "emitter" in Fig. 1.37. Comparison with the power MOSFET structure shows that the emitter's place in the structure is identical to that of the source in the power MOSFET. This statement applies also to the "gate," which is the control terminal in both devices. The switching

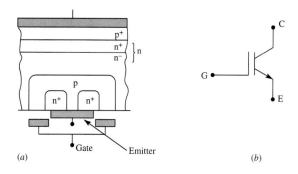

(a)

(b)

FIGURE 1.37
IGBT junction structure (a) and circuit symbol (b).

control voltage for the IGBT is applied across the gate and the emitter, and this controls the switching in exactly the same way as the control voltage applied across the gate and the source in the power MOSFET.

The circuit symbol generally used for the IGBT is shown in Fig. 1.37(*b*). It is similar to that of an npn bipolar junction power transistor, but with an insulated gate terminal in place of the base.

1.7.2.2 PRINCIPLE OF WORKING. The operation of the IGBT is very similar to that of the power MOSFET. The basic difference is that the resistance offered by the n region [on the top in Fig. 1.37(*a*)] when current flows through the device in its ON state is very much smaller in the IGBT. This decrease in resistance occurs because of the injection of holes from the top p^+ zone into this n zone. This effect is called conductivity modulation of the n region. We shall presently explain how the injection of holes into the n region is achieved. But we may state here that the resulting conductivity modulation significantly reduces the ON state voltage drop. Because of this, the current ratings go up from five to ten times in a chip of the same area, compared with a power MOSFET.

In practical use, the collector terminal voltage polarity will be positive with respect to the emitter. Figure 1.38(*a*) shows the current flow paths in an IGBT cell when a positive gate-to-emitter control voltage above the threshold level is applied. A positive gate voltage greater than the threshold value creates an n channel in exactly the same way as was depicted in Fig. 1.31(*b*) for a power MOSFET. The n channel so created in the IGBT is shown in Fig. 1.38(*a*). This channel connects the n^+ emitter zone of the IGBT to the middle n region. The top p^+ zone, the middle n region and the lower p island constitute a pnp transistor. The top p^+ region, which is the collector of the IGBT, functions as the emitter of this pnp transistor under the normal circuit voltage polarities. A circuit model on this basis is drawn in Fig. 1.38(*b*). The p,

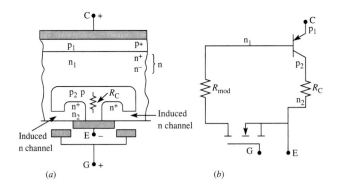

FIGURE 1.38
Current flow paths (*a*) and circuit model (*b*) in the ON state.

n and p regions of the transistor in this model are labeled with appropriate subscripts to identify them with the corresponding regions in the structure in Fig. 1.38(a). We notice that the middle n region constitutes the base of the pnp transistor. When a channel is formed by the application of a gate voltage, a current flow path is created from the collector terminal of the IGBT across the forward-biased top pn junction, across the middle n zone, and through the channel to the emitter terminal. The current flowing through the channel serves as the base current for our pnp transistor. Therefore this causes emitter current to flow in this transistor, resulting in the large scale injection of holes across the top pn junction. These injected holes are responsible for the conductivity modulation of the middle n zone. There are now two current flow paths to the emitter terminal. One is through the middle n zone and through the channel. The other is across the collector junction of our pnp transistor and through the lower p zone. The resistances in these two paths are shown separately in the circuit model of Fig. 1.38(b). Of these two resistances, it is the one labeled R_{mod} that is modulated by carrier injection from the top $\mathrm{p^+}$ zone.

In the above description of current flow, in the interests of clarity, we did not look at another parasitic transistor that is also present in the structure of the device in Fig. 1.38(a). This is an npn transistor whose collector, base and emitter are respectively the middle n region, the lower p region and the lower $\mathrm{n^+}$ region. From an examination of the junction polarities, we notice that the lower $\mathrm{n^+}$ region is the emitter of this transistor. Therefore, to complete our circuit model, we include this npn transistor also, and the revised model is shown in Fig. 1.39.

The two transistors of opposite types, interconnected in the manner shown in Fig. 1.39, constitute what may be described as a "regenerative" circuit, which creates difficulties in the design and manufacture of the IGBT. We shall later see how such an arrangement is put to advantage in another power semiconductor switch—the thyristor. In the circuit of Fig. 1.39 we notice that the collector of the npn transistor is connected to the base of the pnp, and the collector of the pnp transistor is connected to the base of the npn. Therefore an increment in the base current of the npn can cause an amplified increment in the base current of the pnp. This in turn can result in a further

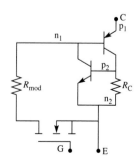

FIGURE 1.39
Modified circuit model of IGBT structure including both the pnp and npn transistors.

amplified increase in the base current of the npn. In this manner, the two transistors can mutually drive each other into the saturated ON state. Such an occurrence is called latch up. If latch up occurs, the device will stay in the ON state without the presence of a gate input. The gate has lost control and is unable to implement turn OFF switching. It is therefore a faulty condition. Earlier designs of IGBT suffered from latch up at relatively low values of ON state currents. But at the present time, the latch up problem has been overcome by suitable design of the device. This is achieved primarily by designing the device in such a way, that the pnp transistor has effectively a very low current gain.

1.7.3 Terminal Capacitances, Gate Drive Requirements, Switching Times

The terminal capacitances of the IGBT are specified in the same manner as was indicated for the power MOSFET. When looking at the capacitance specifications in the data sheet, it should be remembered that the terminal labels are different. The collector of the IGBT corresponds to the drain of the power MOSFET, and the emitter of the IGBT corresponds to the source of the power MOSFET. These differences will be reflected in the letter subscripts used in the labeling of the capacitances.

The gate-to-collector capacitance for the IGBT tends to be significantly smaller in comparison with the corresponding gate-to-drain capacitance of the power MOSFET. Since it is this capacitance that gets magnified during switching due to the Miller effect (see Example 1.12 earlier), this is an advantage resulting in a reduction of the effective input capacitance seen by the gate drive circuit. The gate drive circuits for IGBTs are similar to those for power MOSFETs. The circuit shown in Fig. 1.36(b) can be used for IGBTs also.

The switching times for the IGBT are specified in a manner similar to those of the power MOSFET. The switching performance at turn ON is very similar to that of the power MOSFET and the time specifications are about the same. However, there is a difference as regards turn OFF switching behavior. During turn OFF, the initial fall in current is steep, similar to that of the power MOSFET. But this is followed by a longer "tail" during which the decay takes place relatively slowly. Typically, the tail starts around 25% of the ON state current. The tail in the current decay waveform is because of the time needed for the excess minority carriers (injected holes) in the base region of the pnp transistor to disappear by recombination. Since there is no external terminal in contact with this n zone through which these excess carriers can be "sucked out," as can be done in a bipolar junction power transistor, the carrier lifetime by and large determines this tail duration. The upper n^+ zone of the n region [Fig. 1.37(a)] does help to speed up the recombination and shorten the tail. The overall turn OFF time is longer than in the power MOSFET. During the duration of the tail, the device has the ability to support the full forward voltage, while the tail current is flowing. But this causes increased switching power loss, and therefore limits the upper usable frequency for repetitive switching.

1.8 THE THYRISTOR

The thyristor, also known as the silicon controlled rectifier (SCR), was the first solid state power semiconductor device to be developed to function as a controlled static switch, with large current and voltage capability. It was the advent of the thyristor that gave start to a new era of major developments in static power conversion and control, which has made rapid strides in recent years, making "Power Electronics" a recognized technology in its own right.

1.8.1 Junction Structure, Packaging, Circuit Symbol

The junction structure is shown in Fig. 1.40(*a*). This is a four-layer structure with three internal jucntions shown labeled as J_1, J_2, and J_3. The device has three terminals. The "anode" (A) and "cathode" (K) are the power terminals of the switch. Control input is between the "gate" (G) and K. The anode metallic contact is on the outer p layer shown on top. The cathode contact is on the outer n layer. The inner p layer is the gate layer. During the fabrication of the device, the regions where the gate electrical contact is to be made are masked, before the cathode n layer is formed. Since the gate p layer lies sandwiched between two n layers, it is impossible to make external electrical contact with the entire gate layer area. Actually, the gate electrical contact is limited to a relatively small fraction of the total wafer area, the rest of it being

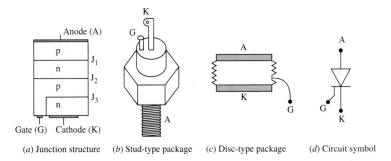

(*a*) Junction structure   (*b*) Stud-type package   (*c*) Disc-type package   (*d*) Circuit symbol

FIGURE 1.40
The thyristor structure, types of packaging and circuit symbol.

allocated for the cathode contact. The geometrical pattern of the gate contact area on the surface of the pellet significantly affects the switching characteristics of the device. We shall have occasion to highlight this later on. We shall see that the absence of direct electrical contact between the gate terminal and a major part of the gate p region is responsible for limiting the role of the gate to the turn ON function only. When a forward voltage (positive voltage polarity at the anode terminal) exists across the main terminals of the thyristor, a short current pulse from gate to cathode will "fire" the thyristor, that is, trigger it into the ON state. Once the thyristor is fired, the gate has no further control over the current flow through the device. During the subsequent conduction, it behaves like a diode. It cannot be turned OFF by a reverse current pulse on the gate.

The two commonly available types of casings in which thyristor pellets are packaged are shown in Figs. 1.40(*b*) and (*c*). The stud-type package is similar to the stud-type package for power diodes, except that there is an additional terminal—the gate. In a thyristor, the stud is always the anode. The cathode and gate leads, whose internal contacts with the pellet are on the side opposite to the anode, are brought out with insulation from the casing, as shown on the top in (*b*). The stud is used for mounting the device on a metallic heat sink. The disc-type package, shown in (*c*), is similar to the corresponding diode package, except that a gate lead is present. This comes out on the side through the ceramic housing that separates the flat anode and cathode terminals. In some designs, an additional cathode lead (the gate return) is also provided as a thin flexible cable, along with the gate lead, so that these twin leads can be conveniently connected to the external gate control circuit. Power connections are made through pressure contacts with the flat surfaces of the anode and cathode, using special mounting hardware. The flat surfaces provide large contact area, both for electric currents and for heat flow. For better cooling, it is advisable to use two separate heat sinks, insulated from each other, one on each side of the device.

Figure 1.40(*d*) shows the circuit symbol for the thyristor. This is derived from that for the diode, with the addition of the gate terminal. The gate terminal location near the cathode is in conformity with the internal geometry and the fact that the firing control input is always between the gate and the cathode.

1.8.2 Operating States of the Thyristor

The thyristor can exist in one of three alternative states in circuit operation. Two of these are OFF states—the reverse blocking OFF state and the forward blocking OFF state. The third is the forward conducting ON state. It can stay in each of these states without an electrical input being present on the gate terminal. The gate serves only to implement the transition from the forward blocking OFF state to the forward conducting ON state. Therefore in our

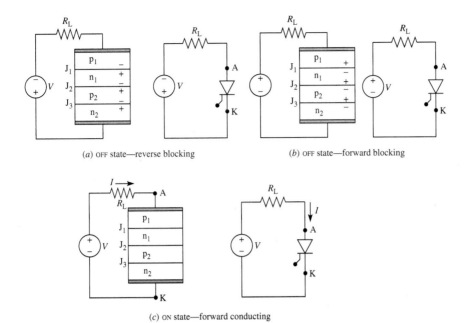

(a) OFF state—reverse blocking   (b) OFF state—forward blocking

(c) ON state—forward conducting

FIGURE 1.41
Operating states of the thyristor switch.

present discussion using Fig. 1.41, we are not showing any electrical connection to the gate terminal.

Figures 1.41(a), (b) and (c) show the three operating states of the thyristor. In each condition, figures are drawn—one showing the junction structure and the second showing the circuit with the circuit symbol. The thyristor is shown connected in series with a resistance to a voltage source with polarity appropriate to each condition.

In (a), the source polarity is such as to reverse-bias the thyristor. With such a polarity, the thyristor can only exist in the OFF state. The reverse blocking voltage between A and K is distributed serially across the three junctions. These junction voltage polarities are marked in (a). It will be seen that in this condition, J_1 and J_3 are reverse-biased and J_2 is forward-biased. Because of the reverse-biased junctions in series, the thyristor cannot conduct, except for the inevitable small leakage current, which we shall neglect. It would appear that the reverse voltage capability of the thyristor will be decided by the total reverse voltage capability of J_1 and J_3. In practical thyristors, the reverse voltage breakdown limit of J_3 is very small, usually well below 10 V. Therefore the reverse voltage rating of a thyristor is almost entirely determined by the breakdown limit of the junction J_1.

In (b), the source voltage polarity is such as to forward-bias the thyristor. Therefore the junction voltage polarities are opposite to those in (a). Now J_2

is reverse-biased, and the thyristor still cannot conduct because of this. We now have the forward blocking OFF state. The forward blocking voltage capability of a thyristor is therefore determined by the breakdown limit of the junction J_2. In practical thyristors, the forward blocking voltage rating (decided by J_2) is about the same as the reverse blocking voltage rating (decided by J_1). The thyristor is a "symmetrical voltage blocking" device. The symmetrical voltage blocking ability is a result of the structure for the following reason.

During the fabrication of the device, the junctions J_1 and J_2 are created simultaneously by a single diffusion operation, during which "p"-type impurity atoms are diffused into both sides of a "n"-type silicon wafer of high purity. The cathode n layer is formed subsequently by a second diffusion, this time only on one side of the wafer. During this second diffusion, some areas on the cathode side are also masked, to provide space for the metal deposition for the gate electrode. From this, it will be seen that J_1 and J_2 have similar characteristics and about the same breakdown limit. These breakdown limits are large, because of the high purity of the starting n type wafer. Since J_3 has higher impurity levels on both sides, its breakdown limit is much smaller.

1.8.3 Turn ON Switching, Two-Transistor Analogy

The turn ON switching of a thyristor is best explained using the "two-transistor" analogy. Figure 1.42(a) shows the thyristor junction structure as a composite of a pnp transistor T_1 and an npn transistor T_2, by visualizing an imaginary plane through the pellet as shown by the broken line.

For greater clarity, the two transistors are shown physically separated in (b), but with the common layers connected together. The thyristor is shown

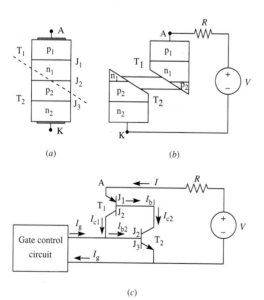

(a)　　　　　(b)

(c)

FIGURE 1.42
Two-transistor equivalent of the thyristor structure to explain turn ON switching.

forward-biased by an external source, which makes the anode positive with respect to the cathode. In the case of the pnp transistor, the anode layer p_1 functions as the emitter. In the case of the npn transistor, the cathode layer n_2 functions as the emitter. On this basis, the circuit is redrawn in (c) using the appropriate transistor symbols. Included in (c) are the voltage source V and a load resistor R. Also shown is a gate control circuit, whose output is connected between the gate and the cathode. This serves to send a gate current I_g in the direction shown, when desired.

Initially I_g is zero. Both transistors are OFF. If we now send a small gate current, this serves as the base current I_{b2} for the transistor T_2. Therefore a collector current I_{c2} results. Inspection of the circuit shows that I_{c2} serves as the base current I_{b1} for transistor T_1. Because of I_{b1}, a collector current I_{c1} is initiated. I_{c1} serves as additional base current for T_2, causing further increase in I_{c2}. This in turn causes further increase of the base current of T_2 and therefore of I_{c1}. In this way, a regenerative current build up process takes place, and both T_1 and T_2 drive each other into the saturated ON state. This happens in a matter of a few microseconds. Once turned ON, the two transistors mutually supply each other's base current, and there is no need for an external gate current to maintain the ON state. The thyristor stays in the ON state with a small forward voltage drop, which is usually in the neighborhood of 2 V for a high power device.

In fact, the gate current is needed only until the current builds up to a certain level, after which the regenerative process will continue even without the external gate current, and the turn ON switching will be completed. The minimum value of the current in the external circuit after which the turn ON switching will be completed without further gate supply being necessary is called the latching current. We notice that the gate serves only to implement the turn ON switching transition. After turn ON, the gate loses control and it is not possible to implement turn OFF switching by means of a reverse gate current.

In practical thyristors there is a minimum current necessary to maintain the device in the ON state. If we decrease V or increase R, the thyristor will turn OFF when the current tends to fall below this minimum level. The minimum current necessary to keep the thyristor in the ON state is called the "holding current." The holding current is lower than the latching current in practical thyristors. In an ideal thyristor, we shall assume the holding current to be zero. Such a thyristor behaves as a diode after it has been turned ON. The gate regains control only after the thyristor has turned OFF, typically by a reversal of the polarity of the anode-to-cathode voltage.

Illustrative example 1.13. The thyristor in Fig. 1.43 has a latching current of 300 mA. Neglect forward voltage drop across the thyristor from the instant of commencement of the gate pulse. Determine the minimum duration of the gate pulse necessary to ensure turn ON.

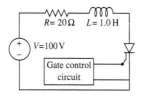

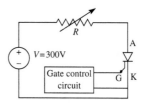

FIGURE 1.43
Circuit of Example 1.13.

FIGURE 1.44
Circuit of Example 1.14.

Solution. The loop equation for the power circuit from the instant of commencement of the gate pulse may be written as

$$L\frac{di}{dt} + Ri = V$$

The solution of this equation with $i = 0$ at $t = 0$ will be

$$i = \frac{V}{R}(1 - e^{-t/\tau}), \quad \text{where } \tau = \frac{L}{R}$$

Substituting given numerical values, we get

$$i = 5(1 - e^{-20t}) \quad \text{A}$$

The gate pulse should be present at least until the current i rises to the latching level, given as 0.3 A. Therefore

$$0.3 = 5(1 - e^{-20t})$$

This gives the minimum duration of the gate pulse to ensure turn ON as $t = 3.094$ ms.

Illustrative example 1.14. The thyristor in Fig. 1.44 has a holding current of 150 mA. When it was turned ON, R was at a low value. Now if R is progressively increased, at what value of R will the thyristor turn OFF? Neglect ON state forward drop.

Solution. The specified holding current implies that the thyristor will turn OFF if the current tends to fall below this value of 0.15 A. Therefore the highest value of R possible with the thyristor ON will be $R = 300/0.15 = 2000\ \Omega$.

1.8.4 Why Turn OFF is Impossible by Reverse Gate Pulse

We shall take a closer look at the reason why a thyristor cannot be turned OFF by a reverse gate current pulse. We do this for two reasons. First, examination of this question gives us a better understanding of an important parameter of a thyristor—its di/dt limit. Second, this prepares us for the understanding of another important static power switching device, the gate turn off thyristor

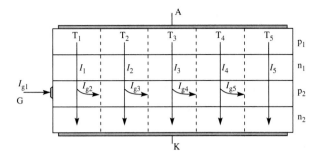

FIGURE 1.45
A single thyristor viewed as several individual cells in parallel.

(GTO) which we will consider later, and which has a similar junction structure, but can in fact be turned OFF by a reverse gate current pulse.

In Fig. 1.45, the single thyristor T is viewed as consisting of many in parallel, such as T_1, T_2, and so on. Of these, only T_1 is shown to have an external gate terminal, it being assumed that T_1 is the one nearest to the gate electrode. Therefore, when a gate current pulse I_{g1} arrives to implement turn ON switching, the first of these thyristors to turn ON is T_1. Thereby, a current I_1 is created in this thyristor. Part of this current flows through the gate layer of T_2 and thus provides the gate current I_{g2} to turn ON T_2. In a similar way, T_2 turns ON T_3 by providing its gate current I_{g3}. In this way, all the thyristors that constitute the device turn ON, one after the other, beginning with the one nearest to the gate electrode layer. Once this has taken place, the gate is extremely ineffective in influencing the current in the elementary thyristors which are in remote locations from the gate electrode. The main reason why the gate is unable to turn OFF the device is the fact that the gate electrode layer is not in close proximity to a large part of the cathode area.

The above consideration also shows that there is a finite time needed for the switching ON operation to spread across the entire area of the silicon wafer. This means that the thyristor switch behaves as though it consisted of several switches connected in parallel, all of which do not turn ON simultaneously. One turns ON, this causes the next to turn ON, and in this manner the switching progresses until every one comes ON. If we now look at the external power circuit, the current that flows does not depend on how many of the switches have turned ON at a given instant of time. Therefore, if the current in the external circuit rises at too fast a rate in comparison with the rate at which switching progresses across the area of the thyristor pellet, excessive local current density can occur in regions close to the gate electrode and cause permanent damage to the device. In other words, the thyristor has a di/dt limit that should not be exceeded to avoid damage to it. This is one of the parameters specified by the manufacturer in the data sheet of the thyristor.

Interdigitated gate geometry. In the practical design of thyristors, one technique often used to increase the di/dt rating of the device is to use a gate electrode pattern that results in the gate electrode boundaries being within

FIGURE 1.46
An interdigitated gate–cathode geometry to achieve a high di/dt capability. (Gate areas are shown shaded; the remainder is the cathode area.)

short distance of any part of the cathode area. This is achieved by distributing the gate electrode area over the pellet and interleaving the gate and cathode regions on it. Such a gate electrode geometry is referred to as an interdigitated gate structure. Figure 1.46 shows an example of an interdigitated gate pattern.

Illustrative example 1.15. The specified di/dt for the thyristor in Fig. 1.47 is 50 A/μs. The inductance L is included in the circuit for the purpose of protecting the thyristor from damage due to excessive di/dt. Find approximately the minimum value of L required.

Solution. Neglecting forward drop across the thyristor, the loop equation during turn ON switching, for the power circuit loop, will be

$$L\frac{di}{dt} + Ri = V$$

The initial di/dt with $i = 0$ at $t = 0$ will be

$$\frac{di}{dt} = \frac{V}{L} = \frac{500}{L} < 50/10^{-6}\,\text{A}$$

This gives the minimum value of L as 10 μH.

Comment. In the solution, we have assumed that di/dt at turn ON should not exceed the specified limit. Practical thyristors can in fact withstand a small initial step increase in current, after which di/dt should be limited to the specified value. This initial permissible step depends on how much of the cathode area is in immediate proximity to the gate electrode. It therefore depends on the gate

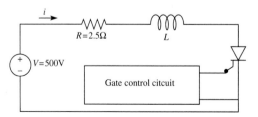

FIGURE 1.47
Circuit of Example 1.15.

geometry. It is this proximate area that turns ON first, after which the turn ON spreads to distant locations from the gate boundary.

1.8.5 Rate of Rise of Forward Voltage, dv/dt Rating

For reliable operation, a static thyristor switch should turn ON only when a gate pulse is applied to it, and for no other reason. One physical reason that can cause a thyristor to turn ON without a gate pulse is an excessive rate of rise of forward voltage (dv/dt). How this can happen is explained with reference to Fig. 1.48. The thyristor shown there is in the forward-biased OFF state. No gate connection is shown. Junction J_2 is reverse-biased and is blocking the source voltage. A reverse-biased pn junction, because of the charge distribution on either side of the junction boundary plane, behaves like a charged capacitance. If this capacitance is denoted by C, the associated charge will be

$$Q = CV$$

If the voltage V increases at a rate dV/dt, there will be an associated charging current of this capacitor, which is dependent on dV/dt. The higher the value of dV/dt, the larger will be the magnitude of this charging current. This charging current flows from the n_1 zone to the p_2 zone across the junction J_2, and is denoted by i_c in Fig. 1.48. The current i_c can cause the turn ON switching of the thyristor in exactly the same manner as a gate current pulse injected into the p_2 zone through the gate terminal, if it exceeds the threshold value necessary for turn ON. This means that the thyristor has a maximum dV/dt limit beyond which it will turn ON without the presence of a gate pulse. The dv/dt rating is one of the parameters of the thyristor specified in its data sheet.

When devices with high dv/dt are required, the technique used to achieve this is called the shorted emitter construction. This may be explained with reference to Fig. 1.49. This shows the cathode (n_2 region) as made up of

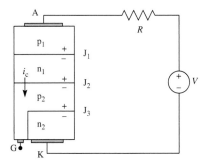

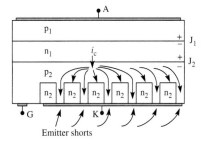

FIGURE 1.48
dv/dt limit of a thyristor due to junction capacitance charging current.

FIGURE 1.49
Shorted emitter construction of a thyristor to achieve a high dv/dt rating.

several n_2 islands instead of one continuous layer. The cathode electrode metal deposition, however, covers not only the n_2 islands, but also the p_2 areas between these n_2 islands. Therefore the cathode electrode metal layer short-circuits the p_2 and n_2 areas at the boundaries between these areas. In the two-transistor equivalent of a thyristor given earlier in Fig. 1.42, the p_2 layer functions as the base of the npn transistor, and the n_2 layer functions as the emitter. Therefore in this arrangement we are creating local short circuits between the emitter and some areas of the base. These areas are labeled as "emitter shorts" in Fig. 1.49. The capacitance current that is caused when there is a dv/dt is also shown in the figure labeled i_c.

We notice that these emitter shorts provide an alternative path through which i_c can flow to the cathode terminal, without crossing into the cathode n_2 zone. Therefore part of i_c is bypassed directly to the cathode terminal through the emitter shorts and does not contribute as a gate current to cause turn ON switching. In this way, the shorted emitter construction serves to raise the dv/dt threshold for turn ON switching. It should be noted that when we send a gate current pulse to fire the thyristor, some of this current is also bypassed through the emitter shorts. Therefore a thyristor with the shorted emitter construction has a comparatively higher gate current threshold for turn ON switching.

Illustrative example 1.16. The thryistor in Fig. 1.50 has a dv/dt rating of $500\,\text{V}/\mu\text{s}$. Determine the minimum value of C needed to avoid erratic turn ON when the power circuit is energized by closing the switch S.

Solution. The thyristor is initially OFF. We neglect its capacitance in comparison with C. We also neglect the forward leakage current through it. Therefore the loop equation immediately after S is turned ON is given by

$$v + RC\frac{dv}{dt} = V$$

where v is the voltage across the thyristor. Since $v = 0$ initially, the initial value of dv/dt will be

$$\frac{dv}{dt} = \frac{V}{RC} = \frac{500}{10C}$$

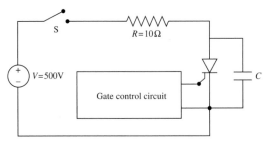

FIGURE 1.50
Circuit of Example 1.16.

Equating this to the maximum permissible value of $500\,\text{V}/\mu\text{s}$, we get the minimum value of C as $C = 0.1\,\mu\text{F}$.

1.8.6   Switching Characteristics—Turn ON Time

The turn ON time t_{ON} of a thyristor may be defined by reference to the circuit and waveforms of Fig. 1.51. The figure shows typical waveforms of the gate voltage and the thyristor current during a turn ON switching transition. The timing instants shown have the following significance.

t_1 Instant at which the gate voltage rises to 10% of the final value

t_2 Instant at which the thyristor current rises to 10% of the final ON state value

t_3 Instant at which the thyristor current rises to 90% of the final ON state value

Using the above as reference instants, we define the following:

$$t_d = t_2 - t_1 = \text{delay time}$$

$$t_r = t_3 - t_2 = \text{rise time}$$

The turn ON time is defined as

$$t_{\text{ON}} = t_d + t_r$$

The turn ON time is usually in the range of a few microseconds. The geometry of the gate structure has a significant effect on t_{ON}. An interdigitated gate reduces the time needed for the turn ON switching to spread over the entire cathode area. A shorter turn ON time is therefore usually associated with a higher di/dt rating.

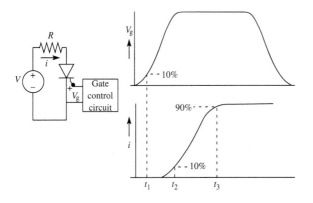

FIGURE 1.51
Definition of turn ON time for a thyristor.

1.8.7   Switching Characteristics—Turn OFF Time

The turn OFF switching of a thyristor occurs when a reverse voltage is made to appear across the main terminals of the switch, in the same manner as a diode gets turned OFF. The reader may refer to the waveforms of current and voltage during turn OFF for a power diode given earlier in Fig. 1.10. The corresponding waveforms for a thyristor are shown in Fig. 1.52(a). Comparison will show that both are similar until the instant labeled t_3 in both the figures. This is the instant at which each device is able to block the full reverse voltage.

In the case of a thyristor, the turn OFF switching is considered complete only when the device has also regained its ability to block forward voltages. So the question is whether the device is capable of blocking a forward voltage from the instant t_3.

To find the answer to this, we can experiment by applying a forward voltage across the device immediately after t_3, without any input to the gate terminal. It will then be found that the device, in fact, turns ON without a gate pulse, as if it were a diode. The thyristor actually needs a finite time after t_3 to regain its ability to block forward voltages. To ascertain the minimum time required for this, we can repeat the experiment several times, each time increasing the time delay in very small steps after t_3, before applying the

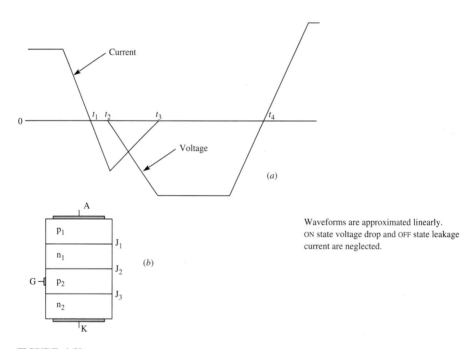

(a)

Waveforms are approximated linearly.
ON state voltage drop and OFF state leakage current are neglected.

(b)

FIGURE 1.52
Turn OFF time definition for a thyristor.

forward voltage. We do this until we arrive at the condition when the device just stops turning ON when forward-biased. Alternatively, we can begin with a long delay initially and repeat the test, progressively decreasing the time delay in small steps each time, until the device starts turning ON when forward-biased. Whichever way we do the test, it is this minimum delay condition that is shown in Fig. 1.52(a). The current and voltage waveforms of Fig. 1.52(a) are used for defining the turn OFF time of the thyristor.

By definition, t_{OFF} is the time interval from the instant t_1 to t_4 in Fig. 1.52(a). t_1 is the instant at which the forward current has fallen to zero during the turn OFF switching transition. t_4 is the instant at which the voltage across the thyristor crosses zero and the device begins to get forward-biased. This is the earliest instant at which the device is able to block forward voltages. It is at this instant that the gate terminal has regained control over subsequent turn ON switching.

The reason for the finite time delay necessary for the device to attain forward blocking ability, after recovering reverse blocking capability, may be explained by reference to the junction structure of the thyrstor shown in 1.52(b). The junction J_1 is primarily responsible for blocking reverse voltages. Therefore, when excess minority carriers on either side of this junction have disappeared, the thyristor has recovered the reverse voltage blocking capabiltiy. This happens when the reverse recovery current transient has ended at t_3. But the junction that provides the forward voltage blocking ability is J_2. It takes a finite time for the excess minority carriers on either side of this junction to disappear after the reverse recovery current transient. Actually, a reverse current for J_1 is a forward current for J_2. The excess minority carriers on the n_1 side of J_2 will be holes injected from the p_2 zone. Those on the p_2 side will be electrons from the n_1 side, but these will be relatively few because, in practical devices, the n_1 layer has a very high purity level. Therefore the primary reason for the delay to regain the forward blocking ability is the time needed for the excess holes in the n_1 zone to disappear by recombination. By speeding up the recombination process, it is possible to achieve a reduction in the turn OFF time. A technique used for this purpose in the manufacture of fast thyristors is to diffuse gold as an impurity. Gold atoms have been found to create "recombination centers," thereby shortening the lifetime of minority carriers. However, the use of gold diffusion also increases the ON state voltage drop, so that a trade off in this respect is necessary in the manufacture of fast thyristors.

If a forward voltage is applied before the excess holes have completely disappeared then these will be pushed across J_2. This will constitute a gate current in the same way as the dv/dt current described earlier. The thyristor will therefore turn ON in a manner similar to the dv/dt turn ON. From this, it is apparent that the dv/dt at the instant t_4 in Fig. 1.52(a) should affect the turn OFF time of the thyristor. In other words, if our dv/dt during the testing to determine the instant t_4 is lower, t_4 will occur earlier and vice versa. The dv/dt in this context is called the reapplied dv/dt. The reapplied dv/dt is one of the test conditions to be stated if the turn OFF time of a thyristor is to be specified in an exact manner, because t_{OFF} will be shorter if reapplied dv/dt is lower, and vice versa. The usual dv/dt specification of a thyristor is accurate only if the thyristor has been OFF for a long enough time for the junction J_2 to completely recover.

1.8.8 Thyristor Classification According to Switching Times and Thyristor Selection According to Converter Types

Based on switching times, thyristors are often listed under two categories. Fast switching thyristors (with short t_{OFF}) belong to one class, while slower thyristors belong to the other. The user has to choose the category, according to the type of converter where they are to be used. In this context, we may state here that thyristor converters are classified under two types: (1) line commutated converters and (2) force commutated converters. Line commutated converters generally work at relatively low power line frequencies such as 50 or 60 Hz. In such converters, the relatively longer turn OFF times of the slower category thyristors have negligible adverse effect on the converter performance. It is therefore always preferable to choose the slower category of thyristors, both from the point of view of cost and that of performance, because the faster switching times of the other category involve a trade off in other benefits, such as forward voltage drop. For this reason, this class of thyristors are often separately listed as "line commutated" types.

In general, force commutated converters operate at higher repetitive switching frequencies, and therefore require fast switching thyristors. Force commutated converters employ special "force commutation circuits" for the specific purpose of implementing the turn OFF switching of the thyristors by forcing a reverse voltage across the thyristor to be turned OFF, at the appropriate instant in the converter switching cycle. The sizes of the circuit elements needed in the force commutation circuits depend on how long the thyristor to be turned OFF has to be maintained in reverse bias, that is, on the turn OFF time of the thyristor. Fast thyristors are invariably chosen for force commutated converters, even if the repetitive switching frequency of the converter is low, because in this way the size and cost of the force commutation circuits will be less.

1.8.9 Gate Circuit Requirements for Thyristor

In this section, we discuss two aspects of gate control: (1) gate pulse level and (2) gate pulse duration.

1.8.9.1 GATE PULSE LEVEL. The threshold gate current pulse amplitude needed to fire (that is, turn ON) a thyristor is stated in the data sheet of the device. It depends on the thyristor voltage and current ratings, and is typically in the range of about fifty to several hundred milliamperes for medium and large size devices. Since the gate cathode junction is being forward-biased by the gate pulse, the gate-to-cathode voltage level is small, well below 5 V, typically around 1.2 V. In practice, the gate current pulse amplitude actually used for firing a thyristor is considerably higher than the threshold value, and its rise time is as short as possible, to ensure fast turn ON switching.

The reverse voltage capability of the gate cathode junction is very low. It is therefore important to carefully examine the gate control circuit and ensure that there is no possibility of a large reverse voltage occurring, even under any abnormal circuit conditions. An expensive high power thyristor can be permanently damaged by the application of a reverse gate voltage, which may be low, but still above the breakdown limit.

1.8.9.2 GATE PULSE DURATION. Gate firing of a thyristor will be successful only if the gate pulse lasts at least till the thyristor current rises to the latching level. Looking at this requirement solely from the point of view of the thyristor, the minimum time needed is small, typically in the neighborhood of 10 μs for a large power device. However, in practical converters, there are external power circuit conditions that will necessitate the use of much wider pulses. This could be one or both of the following.

1. External circuit conditions can cause a delay in the current rise. A situation of this kind was considered in Example 1.13, where the presence of a large inductance in the power circuit delayed the rise of the current to the latching level, necessitating a wide gate pulse.
2. There can be a delay in the forward-biasing of the thyristor.

A gate firing pulse can be effective only if the thyristor is forward-biased. In some converter circuits, the exact instant in the converter switching cycle when a thyristor becomes forward-biased varies according to load conditions, and it becomes very difficult to design the gate firing circuit to provide a properly timed narrow pulse. It is more convenient to provide a wide pulse that commences at the earliest instant at which the thyristor may turn ON. In this way, even if the thyristor is not forward-biased when the gate pulse commences, the gate pulse will still be present when it does become forward-biased and is ready to turn ON. We shall encounter such situations in the operation of some converters described in Chapters 2 and 5.

1.8.10   Timing Control and Firing of Thyristors

The gate control circuit of a thyristor converter is designed to provide the gate firing pulses to the thyristors at the appropriate instants of time. The switching control unit of a power electronic converter generally has a timing circuit in it.

This timing circuit generates the timing pulses at the correct instants of time at which each power device has to be switched. The timing circuits for different types of converters will be described when we consider the converters themselves in subsequent chapters. The timing circuit for a switching element normally provides a pulse whose duration is the ON time of the device. In general, such a pulse from the timing circuit is unsuitable for being directly applied to the control terminal of the switching device for the following reasons.

1. A latching device like a thyristor only needs a pulse of short duration, and not for the entire duration of the ON time. Therefore unnecessary gate power dissipation can be avoided by using a very short pulse that starts at the rising edge of the timing pulse.

2. It may need further amplification to be able to provide the required current and power to successfully turn ON the power switching device.

3. It is invariably necessary to provide electrical isolation between the switching control circuit and the power circuit of the converter. The power switching elements will generally be working at high and variable potentials, from which the control circuits will have to be isolated.

In the case of a thyristor, the first of the above three requirements can be met by using a monostable mulitivibrator chip that is triggered by the rising edge of the timing pulse. The multivibrator chip can be programmed to output a pulse of the required width for the thyristor.

To raise the power capability of a voltage pulse, we can use a "driver." Drivers are available as integrated circuits. They reproduce the input pulse at the output terminals with greatly increased current capability. Alternatively, we can use discrete elements as shown in Fig. 1.53 for pulse amplification.

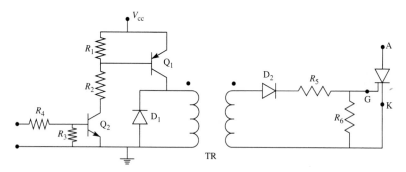

FIGURE 1.53
Gate firing circuit for a thyristor.

A pulse transformer can be used to provide electrical isolation. But careful design of the pulse transformer is usually needed to faithfully reproduce the input pulse at its output terminals. Opto-isolators are also used for electrical isolation. An opto-isolator chip consists of a light emitting diode (LED) that throws light on to a photodiode or phototransistor, causing it to conduct. If an opto-isolator is used for electrical isolation, it is normally required to have a floating (isolated) power source on the gate side, for pulse amplification, because it is impossible to transmit any appreciable power through the opto-isolator. A gate firing circuit for a thyristor using discrete transistors for pulse amplification and a pulse transformer for isolation is shown in Fig. 1.53. This circuit assumes that the input pulse available is of the required width. Therefore the monostable multivibrator chip is not included.

In this arrangement, the turn ON pulse to be amplified is applied to the base of the transistor Q_2, causing it to turn ON. R_1 and R_3 are high resistances to improve performance by providing ohmic paths between the respective base and emitter. These may be ignored for the purpose of our present description. When Q_2 turns ON, it provides an amplified base drive current for the pnp transistor Q_1 and causes it to turn ON. The value of the base drive current for Q_1 is decided by the value chosen for R_2 and the power supply voltage, labeled V_{cc} in the figure, provided that the input to Q_2 is sufficient to turn it fully ON. When Q_1 turns ON, it applies a pulse of amplitude approximately equal to V_{cc} to the primary of the pulse transformer. This pulse has the same duration as the input pulse. But it has a large current capability, which is determined by the base current provided to Q_1. In this circuit, the transistors are functioning as switches.

The secondary coil of the pulse transformer feeds the pulse to the gate terminal of the thyristor. The resistances R_5 and R_6 serve to limit the gate voltage and current. The diode D_2 prevents any reverse voltage on the gate from the transformer. R_6 also serves to provide an ohmic path from the gate to the cathode of the thyristor. This is a desirable feature, and makes the gate less sensitive to stray voltages. The diode D_1 is for the purpose of providing a freewheeling path through which the current in the primary of the transformer will freewheel when Q_1 turns OFF. This will avoid an excessive voltage occurring across this transistor due to the transformer inductance. Careful design of the pulse transformer is essential for the satisfactory working of this circuit. It has to have the required pulse width capability. Also, the unidirectional pulses in the primary will have a DC component that can lead to magnetic saturation of the core.

1.8.11 Thyristor Ratings and Protection

Unlike the power diode, the thyristor has two categories of voltage ratings— the forward blocking and the reverse blocking. For reasons explained in Section 1.8.2, these ratings are generally equal.

The current ratings are specified in a manner similar to what we described for power diodes. As in the case of all power semiconductor devices, current ratings are generally valid only if cooling is provided by the use of suitable heat sinks, to keep the device temperature within the specified limit. As in the case of power diodes, average, r.m.s., repetitive peak as well as surge current ratings are usually specified separately, especially for devices with large ratings. Power semiconductor category fuses may be used with each individual thyristor in a converter, to provide short-circuit protection to it. The main parameter on the basis of which the fuse is selected is the I^2t rating. To afford protection, the I^2t rating of the fuse must be less than the I^2t rating of the thyristor. Its continuous current rating should match that of the thyristor. The voltage rating of the selected fuse will be based on the maximum circuit voltage that can occur across it, in the event of a fuse blow.

Other major specifications of the thyristor, which are directly related to its safety, from the power circuit side, are the dv/dt and the di/dt ratings. These are expressed in terms of V/μs and A/μs respectively. A dv/dt failure causes an erratic turn ON of the device, and, depending on the circuit conditions, this may result in a short circuit in the system, with the possibility of damage to the device and other components. Exceeding the di/dt limit can directly damage the device because of excessive local current concentration in an area of the thyristor pellet.

A snubber circuit is invariably used to protect a thyristor from excessive stresses. Figure 1.54(a) shows the commonly used R–C snubber. This snubber circuit functions in a similar way as for a power diode, by limiting the over-voltage resulting from the reverse recovery current transient. Besides, by suitable choice of values for R and C, it can also serve to mitigate the dv/dt stress. But the snubber can be made more effective in limiting dv/dt by the addition of a diode as shown in Fig. 1.54(b).

To limit di/dt, when circuit conditions are such that there is a danger of exceeding the specification, an inductance may be added as shown in Fig. 1.54(c). This is usually a coil of a few turns, capable of carrying the full thyristor current. In some designs, this coil is wound on a small ferrite ring. This reduces the size of the coil for the same inductance value. Besides, the core gets saturated when the current exceeds its saturating level, and so the effective inductance falls to a low value from then.

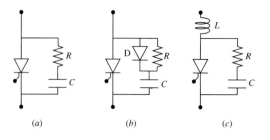

(a) (b) (c)

FIGURE 1.54
Snubber circuits for thyristors.

1.9 THE ASYMMETRICAL THYRISTOR

The asymmetrical thyristor, also known as the asymmetrical silicon controlled rectifier (ASCR) is a modified version of the thyristor. Its turn OFF time is much shorter. Therefore it can be used for switching at a higher repetitive frequency than the ordinary thyristor. The shorter turn OFF time is made possible at the cost of the ability to block reverse voltages. In choppers and inverters (studied in Chapters 4 and 5), the controlled switching element is used with an antiparallel power diode. Therefore, for such applications, there is no need for the device to block reverse voltages except the small forward voltage drop of the antiparallel diode.

The reason why reverse blocking ability has to be sacrificed to achieve faster turn OFF time, can be seen from an examination of the modified junction structure shown in Fig. 1.55.

Inspection of Fig. 1.55 shows that the junction structure is exactly the same pnpn four-layer structure of the thyristor, with one difference. The middle n layer now consists of a low resistivity (high impurity) region labeled in the figure as n^+, and the usual high resistivity (low impurity) region labeled as n^-. The reason for the long turn OFF time in the conventional thyristor is that, during the reverse recovery transient, the flow of reverse current causes holes to be injected across the junction J_2 from the p_2 to the n_1 layer. These holes have to disappear, mainly by recombination, before the junction J_2, which is the junction responsible for blocking forward voltages, recovers its blocking ability. In normal thyristors, this recombination process takes a longer time because of the high purity level of the n_1 layer. In the asymmetrical thyristor, the presence of the higher impurity n^+ layer speeds up the recombination process and so shortens the turn OFF time.

However, it is the junction J_1 that gives the conventional thyristor its ability to block large reverse voltages, since the reverse blocking ability of J_3 is very little. This large voltage blocking ability of J_1 is made possible by the high purity level of the n side of this junction. The ability to block large voltages is lost in the modified structure of Fig. 1.55, because of the n^+ region on the n side of J_1. Therefore, the shorter turn OFF time is achieved at the cost of a significant loss of reverse voltage blocking ability.

The presence of a small negative voltage at the gate terminal during the turn OFF switching transition will further help to reduce the total turn OFF time.

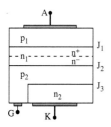

FIGURE 1.55
Asymmetrical thyristor junction structure.

When the gate firing circuit is designed to provide this small negative voltage during turn OFF, we call it gate assisted turn OFF. If gate assisted turn OFF is employed in an asymmetrical thyristor to shorten the turn OFF time, it is important to remember that the reverse voltage capability of the gate cathode junction (J_3 in Fig. 1.55) is very small and should not be exceeded, to avoid damage to the device.

Asymmetrical thyristors are often made with the shorted emitter construction, in the same way as many conventional thyristors. This is done for the same purpose, namely to increase the dv/dt capability.

1.10 THE REVERSE CONDUCTING THYRISTOR

In the power circuit configuration of DC/DC converters (generally called "choppers") and DC/AC inverters, which we will be studying in Chapters 4 and 5, each controlled semiconductor switching element is used with a power diode in "antiparallel," to enable the unrestricted flow of reverse current. This flow of reverse current is called, in the terminology of Power Electronics, "freewheeling." The reverse conducting thyristor has been developed to eliminate the need for a separate antiparallel diode for free wheeling. This is an integrated power device in which the reverse diode structure is integrated along with the main thyristor structure, in the same silicon wafer. The junction structure of the reverse conducting thyristor is symbolically shown in Fig. 1.56.

The junction structure shows that a part of the area on the silicon wafer functions as a single pn junction. This is shown on the right-hand side of the n_1p_2 junction in Fig. 1.56. The rest of the wafer area has the four-layer pnpn structure typical of the thyristor.

In effect, the reverse conducting thyristor has a "shorted emitter" structure on the cathode side and another such structure on the anode side. The anode side shorting takes away most of the reverse voltage blocking ability. This, together with the cathode side emitter shorting, totally removes the reverse blocking ability and enables unrestricted flow of reverse current through the device.

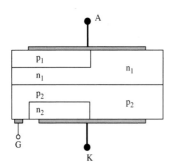

FIGURE 1.56
Reverse conducting thyristor junction structure.

Reverse conducting thyristors are generally manufactured for large current ratings, for use in high power applications. The diode region is designed for fast recovery, and the thyristor region is designed for short turn OFF time and high dv/dt rating.

1.11 LIGHT-FIRED THYRISTORS

A conventional thyristor is turned ON by providing a gate current pulse from its gate to its cathode. But a light-fired thyristor is turned ON by incident light on the gate region. Other than that, light-fired thyristors work like conventional thyristors. The reader may refer to the two-transistor equivalent circuit of the conventional thyristor in Fig. 1.42. In the light-fired thyristor, the equivalent npn transistor is a phototransistor. In the phototransistor, in place of a base current input, it is incident light on the base region of the device that causes current flow in the collector circuit. Therefore a light-fired thyristor can be turned ON by causing a narrow beam of light to fall on the base region of the thyristor pellet. This is done by using a fiber optic cable to transmit the light from a light-emitting device. The main reason for the development of light-fired thyristors may be explained as follows.

Thyristors are now being used in converters for high voltage direct current (HVDC) power transmission systems. In a HVDC system, power that is generated as AC, typically at a remotely located generating station, is first converted into DC at a very high voltage and transmitted over a long distance overhead transmission line or cable to the receiving location, where it is again converted to AC.

The converters at both ends of a HVDC link normally have to handle hundreds of amperes at several hundred kilovolts. The maximum possible voltage rating of thyristors at the present time is limited to a few kilo volts. Therefore the type of switching element used in HVDC converters consists of a string of series-connected thyristors to handle the required voltage. Such a string of a large number of thyristors is called, in HVDC terminology, a thyristor valve. The thyristor valve functions as a single switch. To fire a thyristor valve, each thyristor in it has to be given a gate pulse applied across its gate and cathode terminals. In the blocking state, the voltage across the valve may be something like 100 kV, which gets distributed from one end to the other of the thyristor chain. Therefore the gate firing pulses have to be applied at vastly different voltages levels. Also, the pulses have to be simultaneous, with negligible delay between the individual pulses of the thyristors. For example, if the pulse to one thyristor is delayed, it will not have turned ON, while all the others have already turned ON, and it will be called upon to do the impossible task of blocking the entire voltage. It is extremely difficult to overcome difficulties due to stray inductances and capacitances and the associated delays, and provide such accurately timed simultaneous pulses from a single firing circuit to different locations, which are mutually insulated for vast and different voltages. The development of light-fired thyristors has successfully overcome this problem. The gate firing pulses are beams of light, transmitted from a single source, to the gate of every thyristor, through fiber optic cables which are electrical insulators that can withstand the high voltages involved.

1.12 THE GATE TURN OFF THYRISTOR (GTO)

In the conventional thyristor, the gate serves only to implement the turn ON switching. It has no role to play in the turn OFF switching operation. But the GTO can be turned ON like a conventional thyristor, and can also be turned OFF by means of a reverse gate current pulse. The conventional thyristor has a symmetrical voltage blocking ability, which means that it has the ability to block forward as well as reverse voltages of approximately equal magnitude. Viewed from this aspect, there are two types of GTOs—the reverse blocking type, which has symmetrical voltage blocking capability, and the "anode short" type, which can block only large forward voltages. The reverse voltage blocking ability of the anode short type of GTO is very small, typically below 15 V. In recent years, the GTO has become a popular switching device for high power applications. At the higher end of the power range, a single GTO of the anode short type may have a voltage rating of 4500 V and a current rating of 300 A. The corresponding values for a symmetrical GTO will be 4500 V and 2500 A.

1.12.1 Junction Structures of Symmetrical and Anode Short Types of GTOs

The junction structures of both types of GTOs have evolved from the conventional thyristor structure, which is redrawn in Fig. 1.57 for ready reference.

In the conventional thyristor, the major part of the gate p layer, labelled p_2 in the figure, can be seen to be sandwiched between the cathode n layer

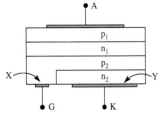

FIGURE 1.57
The conventional thyristor junction structure.

(n_2) and the middle n layer (n_1). The gate metallization, which is the electrical terminal connection to the gate terminal through which gate current flows into the gate layer, is limited to a small area of the gate layer, which is labeled as X in the figure. It is physically distant from areas such as Y of the cathode layer. Therefore, once the thyristor has been latched into conduction, a reverse gate current pulse of reasonable amplitude cannot influence the cathode current in remote regions of the cathode such as Y.

In an interdigitated gate geometry, such as shown in Fig. 1.46, the gate metallization areas and the cathode areas are interspersed in such a way that the maximum distances to the remotest regions of the cathode from the boundaries of the gate contact are very much reduced. Interdigitization is used in conventional thyristors for shortening the time for the turn ON switching process to spread across the cathode areas and so increase the dI/dt rating of the thyristor. But the extent of interdigitization is still not sufficient to give the gate the ability to significantly reduce the cathode current through the gate layer (p_2) to the level at which turn OFF switching can also be achieved by a reverse gate current pulse.

In the GTO the basic four-layer pnpn structure of the conventional thyristor is retained. But the concept of interdigitization of the gate and cathode areas is extended and implemented to such an extent that every location on the cathode is in close proximity to the boundaries of the gate metallization. Therefore, with a reverse gate current pulse, it is possible to reduce the current through the gate layer to the level at which the sequence of switching that was initiated by a positive gate current and resulted in the turn ON switching of the device can be made to take place in reverse, resulting in the turn OFF switching.

The turn ON switching was explained by using the two-transistor equivalent structure of the thyristor in which the two transistors regeneratively increase each other's base current, when once the gate current of the npn transistor is initially brought to a high enough level at which this mutual regenerative current build up can commence. The turn ON switching in a GTO takes place in a similar manner. But, in a GTO, the turn OFF switching is also implemented by lowering the current through the gate layer by a reverse gate current pulse to a level below which the two transistors begin to mutually reduce each other's base current and drive both to the OFF state.

There is, however, a difference between the minimum amplitude of the gate current pulse required to turn ON and to turn OFF in a GTO. The minimum gate current pulse amplitude to turn ON the device is independent of the actual ON state current through the device that flows after turn ON. But a similar statement is not true for the turn OFF switching. The minimum amplitude of the reverse gate current pulse to successfully turn OFF the device is dependent on the current I to be turned OFF. It is given by the following relationship:

$$I_g = I/\beta_{off}$$

where β_{off} is defined as the "turn OFF current gain." The turn OFF current gain

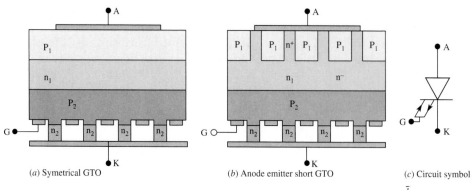

(a) Symetrical GTO   (b) Anode emitter short GTO   (c) Circuit symbol

FIGURE 1.58
GTO junction structures and circuit symbol.

of a GTO is low, and can be typically in the range 4–5. This means that, to turn OFF a current of 100 A, the minimum reverse current peak has to be in the range of 20–25 A. The turn OFF current gain depends also on circuit conditions such as rate of rise of reverse current. A higher $-di_g/dt$ will generally be associated with a higher turn OFF gain, but may lead to greater power dissipation. Detailed data sheets from the manufacturer will give the recommended range of values for this.

A consequence of the higher level of interdigitization in a GTO is that, although a GTO is turned ON by a positive gate current pulse of short duration, like a conventional thyristor, it also normally needs a continuous current of small magnitude lasting for the entire duration of the ON state, to maintain it stably in the ON state. The gate firing circuit should therefore be designed also to provide this continuous positive current, following the turn ON pulse of higher amplitude.

The junction structure of a symmetrical GTO is shown in Fig. 1.58(a). The GTO retains the basic pnpn four-layer structure of the conventional thyristor. The difference is in the higher level of interdigitization between the gate metallization and cathode areas. To achieve this high level of interdigitization, the cathode regions (n_2) are made to stick out as several individual islands, this being achieved by etching off the silicon surrounding these islands. A common cathode metal plate makes contact with all the cathode islands and serves as the cathode terminal. Surrounding these cathode islands, we have the gate metal deposition over the entire exposed areas of the gate p_2 layer. This gate metallization is a single metal area through which the cathode islands stick out, although they appear separated in Fig. 1.58(a) because this figure is a sectional view. This gate metallization area is connected to the gate terminal of the GTO. The anode metallization is exactly like in a conventional thyristor, on the outside of the top p_1 layer.

GTOs are also made with the shorted emitter structure on the anode side. This is shown in Fig. 1.58(b). This helps to shorten the turn OFF switching time, and thereby gives the GTO a higher switching frequency capability. But the emitter short on the anode side takes away the reverse blocking capability of the pn junction on the anode side. The burden of reverse voltage blocking now falls entirely on the gate cathode junction. This junction can block only a few volts, typically 15 V. Since reverse voltage blocking is not a requirement for DC/DC converters and DC/AC inverters, GTOs with anode side emitter short are more suitable for these.

The circuit symbol of GTO is shown in Fig. 1.58(c). This is a modification of the thyristor symbol, with the additional indication of the reverse gate current direction.

1.12.2 Gate Control Circuit for a GTO

The design of the gate control circuit will depend on the specifications of the GTO, in particular the positive gate current requirements for turn ON and the reverse gate current requirements for turn OFF. The positive gate current usually consists of a pulse of short duration at the commencement of turn ON, followed by a lower continuous current for the entire duration of the ON state. The amplitude of the reverse gate current will depend on the maximum anode current to be turned OFF and the turn OFF current gain of the GTO. The typical features of a gate control circuit are illustrated by the scheme shown in Fig. 1.59.

This circuit employs two isolated power supplies, labeled PS$_1$ and PS$_2$ in the figure. The common point of both the power supplies is connected to the

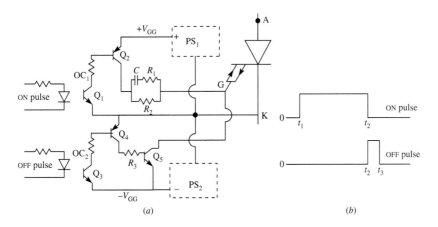

FIGURE 1.59
A gate control circuit for a GTO.

cathode of the GTO. PS_1 provides an isolated positive voltage labeled $+V_{GG}$, and PS_2 provides an isolated negative voltage labeled $-V_{GG}$, both with respect to the cathode of the GTO. The duration of the ON state of the GTO is shown in Fig. 1.59(b) as from t_1 to t_2. In this scheme, the switching control circuit of the converter, in which the GTO is used as a switching element, should provide a positive pulse lasting for the entire duration of the ON period, as shown by the top waveform in Fig. 1.59(b). The instant of initiation of the turn OFF switching is the instant labeled t_2. The switching control circuit should provide a positive pulse of short duration, as shown from t_2 to t_3 by the lower waveform in Fig. 1.59(b), to implement turn OFF. This scheme employs optocouplers to isolate the turn ON and turn OFF pulses. The turn ON pulse sends current through the light-emitting diode of the optocoupler module labeled OC_1. This causes the phototransistor of this module, labeled Q_1, to turn ON. The turn ON of Q_1 causes Q_2 to turn ON by providing base current to it. The collector current of Q_2 serves as the positive gate current input to the GTO. Initially, this current has a larger amplitude, because initially the capacitor C is uncharged and functions as a short circuit. Therefore the initial current magnitude is determined by the value of R_1, which is low. As the capacitor C gets charged, the current is diverted through R_2, which has larger value and therefore limits the gate current to the low value needed for the rest of the duration of the ON state.

The turn OFF switching is achieved through the optocoupler OC_2. The turn OFF switching signal current through the LED of the optocoupler causes turn ON of its phototransistor, labeled Q_3. This turns on Q_4 and thereby Q_5. The collector current of Q_5 will be determined by its base current input, which in turn will be determined by the value of the base resistance, labeled R_3. The collector current of Q_5 is the reverse gate current that turns OFF the GTO. In a high power converter, in which the ON state current through the GTO may be of the order of 1000 A, the magnitude of the reverse gate current may be several hundred amperes and it may be necessary to have several individual transistors in parallel to serve as Q_5. In such cases, the negative power supply PS_2 may be of higher current rating than the positive gate supply PS_1.

1.12.3   Protection

A snubber circuit is invariably used with a GTO. GTOs normally do not have the shorted emitter construction on the cathode side. Therefore their dv/dt capabilities are lower than thyristors of comparable rating. It is important to limit dv/dt during the turn OFF switching of a GTO below a safe limit. Otherwise a high dv/dt during the turn OFF switching transition can retrigger a GTO into the ON state and defeat the turn OFF attempt. Figure 1.60 shows the turn OFF snubber for a GTO. The capacitor C serves to limit dv/dt, and it has to have high enough value for this purpose. The snubber diode must have a high surge current capability, because, during turn OFF, the large GTO current is diverted through the diode to the capacitor. The snubber resistor must have

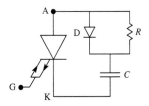

FIGURE 1.60
Turn OFF snubber for a GTO.

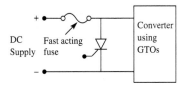

FIGURE 1.61
The crowbar protection scheme.

a small enough value to ensure that the snubber capacitor gets fully discharged during a shortest ON period of the GTO that can occur in operation. Each time the GTO comes ON, the stored energy of the snubber capacitor gets dissipated in the snubber resistor. For this reason, the power rating of the snubber resistor is usually high.

Every GTO has a "maximum controllable current" above which it is impossible to turn it OFF by a reverse gate current pulse. If, during the operation of a converter using GTOs as the switching elements, a faulty condition should occur (such as an accidental short circuit), causing the current to increase, the protective system should be designed to sense this and turn OFF the GTO for its own protection. If the fault current in the GTO is below its maximum controllable current, this can be achieved by a reverse gate current pulse of adequate amplitude. This scheme will fail if the fault current exceeds the maximum controllable current. In such a case the typical protection scheme used is the "crowbar protection." The underlying principle of this is to create a short circuit across the supply by firing a thyristor connected across it. The resulting fast rising current causes an extremely fast acting fuse to blow and so disconnect the supply. This is illustrated by Fig. 1.61.

1.13 THE TRIAC

The thyristor is a unidirectional device that permits current flow only in the anode to cathode direction through it. For the controlled switching of currents in an AC circuit, in which the current flow is bidirectional, we need to use two thyristors in "antiparallel" as shown in Fig. 1.62(a). In this circuit, a gate pulse applied across G_1 and K_1 of thyristor T_1 will serve to turn ON current flow in one direction, which we will call the positive direction, at the desired instant of time during the positive half-period of the AC. For turning ON during the negative half-cycle, for reverse current flow, we shall have to turn ON the thyristor T_2. For this, we shall have to provide the firing pulse across two separate terminals, namely G_2 and K_2, of the thyristor T_2. Such a scheme for control of bidirectional current flow is practical and is in use. The resulting current waveform in a resistive AC load is shown in Fig. 1.62(b), when the switching in each AC half-period is delayed by an angle labeled α.

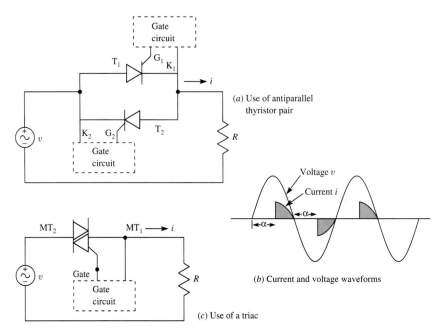

(a) Use of antiparallel thyristor pair

(b) Current and voltage waveforms

(c) Use of a triac

FIGURE 1.62
Controlled bidirectional switching in an AC circuit.

This type of switching scheme is widely used for control of AC heating and lighting loads. The "triac" is a three-terminal power semiconductor switching device with which such a switching scheme can be implemented more simply, using a single device. The triac is a bidirectional device that is functionally equivalent to two thyristors in antiparallel. However, it has only one gate terminal, and this serves to switch current in either direction. Reference to Fig. 1.62(a) shows that if we use two antiparallel thyristors, we have to provide gate pulses to two separate pairs of terminals, which do not have a terminal in common. The identical switching scheme using a triac, in place of two antiparallel thyristors, is shown in Fig. 1.62(c).

The bidirectional current flow feature of the triac is indicated by the arrows in both directions in the circuit symbol for the device, which is shown in Fig. 1.62(c). For this reason, the power terminals of the device are not called anode and cathode as in a thyristor. They are given the names "main terminal 1" (MT1) and "main terminal 2" (MT2). The control terminal is called "gate." The switching control terminals are the gate and MT1, irrespective of the direction of the current to be switched through the main terminals.

The triac is a latching device like a thyristor, and needs only a short pulse to latch into the ON state. The gate loses control once the device has been latched into conduction, and it continues to conduct like a diode as long as the current flow is in the same direction. It turns OFF when the current tends to reverse, due to reversal of the voltage polarity across the main terminals. The gate now regains control, and the triac can come ON again only when the next gate signal arrives. A triac can be turned ON either by a positive or a negative gate current input from the gate to MT1 for either direction of main terminal polarity. The thresholds of the gate current, however, will vary, depending on the gate input polarity and the main terminal polarity. In the normal use of a triac, the practice is to use a positive gate current to fire the triac during the half-cycle when MT2 is positive with respect to MT1, and a negative gate current to fire it during the negative half-cycle, when the main terminal polarity is reversed.

Although it is more convenient to use a triac instead of two antiparallel thyristors, the high frequency switching capabilities of triacs are inferior to those of thyristors. Triacs are seldom used in AC systems of over 400 Hz. Their most common use is at the power system frequencies of 50 and 60 Hz for the control of low and medium power loads, such as lighting and heating.

1.14 STATUS OF DEVELOPMENT OF POWER SWITCHING DEVICES

The power devices that we have described so far are well-established ones that have proved themselves in practical equipment and are generally available from manufacturers in production quantities. At the same time, there is development activity going on, to improve the switching characteristics of these devices and also to extend their voltage and current ratings. It is therefore always advisable to look at the latest updated catalogs from manufacturers before selecting power semiconductor switches for power electronic converters.

In addition to these well-established devices, there are some others that have appeared in recent years but that have not yet found wide use, or are not available in production quantities. This is either because they are still in the development stage or because of characteristics that narrowly limit their use to a specific application. But information regarding such devices is available in the technical literature. In this section, we shall give a brief review, highlighting the main features of the more important ones. The devices that we shall describe are (1) the static induction transistor (SIT), (2) the static induction thyristor (SITh) (a device of this type has also been called the field controlled thyristor) and (3) the MOS controlled thyristor (MCT) (a similar device goes under the name MOS–GTO).

1.14.1 The Static Induction Transistor

The SIT is actually a high power version of the junction field effect transistor (JFET). The SIT, however, is a vertical device—by which we mean that the current flow path is across the silicon pellet. The junction structure of a SIT is

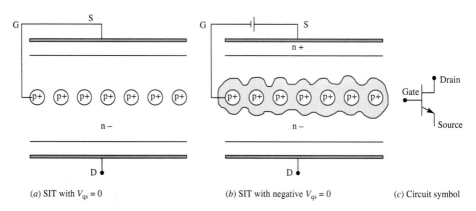

(a) SIT with $V_{qs} = 0$ (b) SIT with negative $V_{qs} = 0$ (c) Circuit symbol

FIGURE 1.63
The static induction transistor (SIT).

symbolically shown in Fig. 1.63(a). The bulk material is high resistivity (high purity) n type, labeled as n⁻ in the figure. There are low resistivity n regions adjacent to the drain and source electrodes, which are on opposite sides of the silicon pellet. The gate consists of low resistivity p zones, which are the p^+ regions shown in the figure as circles, all of which are connected together and constitute the gate terminal. If the gate-to-source voltage is zero, as is the case in Fig. 1.63(a), current flow between the drain and the source is possible, through the areas between the p^+ circles shown in the figure. If the gate-to-source voltage is made negative, thereby reverse-biasing the gate-to-source pn junction, then a depletion layer begins to surround the p^+ circles. With a sufficiently large reverse bias, the depletion zones outside the individual gate regions overlap, as shown in Fig. 1.63(b), thereby completely blocking the current flow path between the drain and source terminals. This is the OFF state of the SIT. Therefore the SIT is a "normally ON" switching device. With zero voltage input across the gate-to-source terminals, it is in the ON state. To turn it OFF, we need to apply a reverse gate-to-source voltage of sufficient magnitude. A circuit symbol used for the device is shown in Fig. 1.63(c).

The "normally ON" feature of the SIT is a disadvantage in power converter applications. If a converter employs normally OFF type of switching devices, initially during the "power up" period, all devices are OFF and there is no danger of a short circuit. This is not so if the devices used are of the normally ON type. Special steps must be taken to ensure that the devices are OFF initially before the operation of the converter commences. SITs with normally OFF characteristics are also reported to be under development. SITs at present have a large ON state voltage drop across them. This results in increased power loss in the device and reduces efficiency. The major advantage of the SIT is its higher switching frequency capability. Its turn ON and OFF switching transition times are very short even compared with power

MOSFETs, making it possible to use it for applications that demand higher switching frequencies.

1.14.2 The Static Induction Thyristor

Structurally, the SITh has close similarities with the SIT. The junction structure of the SITh is symbolically shown in Fig. 1.64(*a*) and a circuit symbol used for it is shown in Fig. 1.64(*b*). The main terminals are labeled "anode" and "cathode". The main difference from the SIT is that it has additionally a low resistivity p layer (labeled p^+ in the figure) adjacent to the anode. During forward conduction, holes are injected into the n region from this anode p^+ zone. This greatly increases the conductivity in this region and contributes to a significant reduction in the ON state voltage drop across the device. The operation of the SITh is in certain ways similar to that of the SIT.

Like the SIT, it is a normally ON device, although normally OFF devices are also reported to be under development. With zero voltage between gate and cathode, it is in the ON state. However, a small positive gate voltage is usually used to improve ON state performance. When the SITh is turned OFF from the conducting state, by the application of a reverse gate voltage, the resulting removal of charge through the gate is associated with a large reverse gate current pulse. This reverse gate current pulse, which is of short duration, is similar to the reverse gate current during the turn OFF switching of a conventional GTO. If we use the term "turn OFF current gain" to express the ratio of the anode current turned OFF to the amplitude of the reverse gate current pulse, in the same manner as we did for the GTO, the turn OFF current gain of a SITh is typically lower than 3, which is much lower than that of a GTO, which is typically in the range 4–5. This means that the reverse gate current pulse has a much higher amplitude in a SITh. Once the SITh has been turned OFF, the gate has to be maintained in reverse bias until it is turned ON

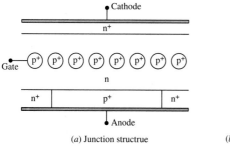

(*a*) Junction structrue   (*b*) Circuit symbol

FIGURE 1.64
The static induction thyristor (SITh).

again. The SITh is not a latching device, although the word "thyristor" in its name suggests that it is.

1.14.3 The MOS Controlled Thyristor (MCT)

At the time of writing, devices of this type have become available commercially, although data on experimental devices were published earlier. The MCT has several very good features as a power switching element.

The circuit symbol of the device is shown in Fig. 1.65(c). Its three terminals are labeled "anode," "cathode" and "gate." Unlike the conventional thyristor, the gate input for the MCT is applied between the gate and anode terminals. It is a latching device for both turn ON and turn OFF switching. It is latched into conduction by a short negative pulse on the gate with respect to the anode. It is turned OFF like a GTO, but with a short positive pulse on the gate with respect to the anode. It has an insulated gate similar to an IGBT or a

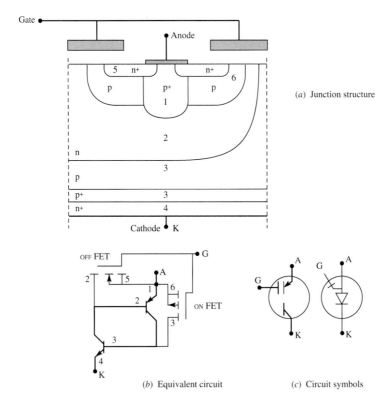

(a) Junction structure

(b) Equivalent circuit

(c) Circuit symbols

FIGURE 1.65
The MOS controlled thyristor (MCT).

power MOSFET. Therefore the gate currents are limited to the charging currents of the terminal capacitances associated with the gate. For this reason, the turn OFF current gain of a MCT is very much greater than that of a GTO. Its switching speed is high, comparable to that of an IGBT. At the same time, the forward voltage drop in the ON state is relatively lower than in an IGBT. It is not a symmetrical device as regards voltage blocking ability. Its reverse voltage blocking ability is very low.

The junction structure of a MCT is illustrated in Fig. 1.65(*a*). What is shown in the figure is a section through a single MCT cell. A single device actually consists of thousands of individual parallel connected cells, each of which has the same MCT structure. From this junction structure, we can derive the equivalent circuit of the device, which is shown in Fig. 1.65(*b*). To facilitate this, we have labeled the different regions in the junction structure by numbers, and the same numbers are shown for the corresponding locations in the equivalent circuit.

The regions labeled 1, 2, 3 and 4 give us the classical pnpn structure of the thyristor, which is replaced by the equivalent pnp and npn transistors, in the same manner as we did for the thyristor. Additionally, the junction structure has two MOSFETs, whose terminals correspond to the regions indicated by the numbers in the equivalent circuit. One of these is a p-channel MOSFET and the other an n-channel MOSFET. Both the MOSFETs have the same gate terminal, which is the gate of the MCT. Both have the same source terminal, which is the anode of the MCT. The p-channel MOSFET is responsible for the turn ON switching. It is described as the ON FET of the device. When a negative voltage appears on its gate with respect to the source, it conducts and provides a base current input to the npn transistor. This initiates the regenerative switching process of the npn and pnp transistors, in the same manner as in a conventional thyristor, and the two transistors drive each other into the saturated ON state. The n-channel MOSFET is the OFF FET of the MCT and responsible for the turn OFF switching. A positive voltage on its gate with respect to its source causes this transistor to conduct and thereby short-circuit the base of the pnp transistor to its emitter. As a result, the two transistors cause each other to go into the OFF state, in a manner similar to what happens during turn OFF in a GTO.

The MCT is a promising switching device. The first commercially available device has recently been released by the manufacturer Harris Semiconductor. It is rated at 600 V and 75 A. It is described as a P-MCT, and its junction structure is similar to that shown in Fig. 1.65. The "P" designation is because the blocking voltage capability of the MCT depends on the characteristics of the p zone in its structure. It is also the polarity of the ON FET, which is a p-channel FET. Although N-MCTs have not yet become commercially available, the manufacturer claims that such devices have been fabricated and tested in several applications. The P-MCTs have higher turn OFF capability. The peak controllable current specified in the data sheet of the 600 V, 75 A device mentioned above is 120 A.

An important advantage claimed for the MCT is its low forward voltage drop in the ON state, which is claimed to be as low as between one-third to one-half of a comparable IGBT. The switching power loss and switching frequency limits cannot match the best of today's IGBTs. However, the device has a high di/dt capability, which is specified as $2000\,A/\mu s$ for the device mentioned above.

The reference terminal for the gate control input for turn ON as well as turn OFF is the anode. From this point of view, the device is basically different from conventional thyristors and GTOs, for which the reference terminal is the cathode. The gate drive circuit needs to be designed to provide a negative pulse to implement turn ON and a positive pulse for implementing turn OFF, both with respect to the anode.

PROBLEMS

1.1. The ON state forward voltage drop of the controlled static switch in Fig. P1.1 is 2 V. Its forward leakage current in the OFF state is 2 mA. It is operated with a switching frequency of 1 kHz and a duty cycle of 30%. Neglect the switching transition times and determine

(a) the peak and average power dissipations in the switch;

(b) the proportion in which this power dissipation is shared between the ON state dissipation and the OFF state dissipation.

1.2. The ON state voltage drop of each controlled switch in Fig. P1.2(a) is 2 V. The forward leakage current of each is 5 mA. They are operated at a repetitive switching frequency of 500 Hz according to the timing diagram shown in Fig. P1.2(b). Neglect switching transition times.

(a) Sketch the waveforms of the load voltage V_{AN} and the switch voltage V_{s1}. Determine the peak blocking voltage capability needed for the switch.

(b) Determine the average and the peak power dissipations in a switch.

(c) Determine the overall efficiency of power conversion.

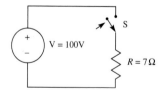

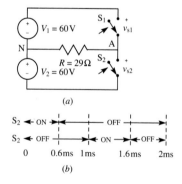

FIGURE P1.1   **FIGURE P1.2**

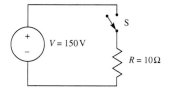

FIGURE P1.3   **FIGURE P1.4**

1.3. The turn ON and the turn OFF switching transition times for the controlled switch
S in Fig. P1.3 are respectively 2 and 4 μs. Assume linear variation of current and
voltage during the transitions. Neglect ON state voltage drop.
   (a) Derive the expression for the average switching power loss in terms of
       voltage, current, the switching transition times and the switching frequency.
   (b) Determine the average switching power loss at switching frequencies of 1, 10
       and 100 kHz.

1.4. The voltage and current variations may be assumed to be linear during switching
transitions for the static switch shown in Fig. P1.4. The switching transition times
are $t_{on} = 3$ μs and $t_{off} = 5$ μs. The ON state voltage drop is 2 V. The duty cycle of
switching is 40%.
   (a) Derive an expression for the average switching power loss.
   (b) Determine the total (both switching and static) power loss averaged over a
       complete switching cycle at switching frequencies of 1, 2 and 50 kHz.
   (c) Determine the frequency at which the switching power loss begins to exceed
       the static power loss.

1.5. In Problem 1.4, neglect the forward voltage drop of the switch in the expression
for the voltage during switching transitions. Determine how much percentage
error will result in the switching power loss calculated because of this
approximation.

1.6. In a nonideal static switching device, the current and voltage values during the
switching transitions follow parabolic laws as stated by the following equations:

$$i = I\left(\frac{t}{t_{on}}\right)^2, \qquad v = V\left(\frac{t_{on} - t}{t_{on}}\right)^2 \qquad \text{for the ON transition}$$

$$i = I\left(\frac{t_{off} - t}{t_{off}}\right)^2, \qquad v = V\left(\frac{t}{t_{off}}\right)^2 \qquad \text{for the OFF transition}$$

Assume that the static performance of the switch is ideal. Derive an expression
for the switching power loss at a frequency of f (in Hz).

1.7. A static switch has to switch ON and switch OFF 25 A in a 100 V circuit. The
switching transition times are $t_{on} = 2$ μs and $t_{off} = 10$ μs. The switching frequency
is $f = 5$ kHz. Determine the switching power loss under the following conditions:
   (a) both the current and voltage variations are linear during the switching
       transitions;
   (b) current and voltage both follow the parabolic laws as stated in problem 1.6.
   (c) If the device is mounted on a heat sink, so that the overall junction-to-
       ambient thermal resistance is 1.2°C/W, what reduction can be achieved in the

internal junction temperature if the voltage and current waveforms can be shaped to conform to case (*b*) instead of case (*a*).

1.8. The four static switches in the circuit of Fig. P1.8(*a*) are operated according to the timing diagram shown in (*b*). The forward ON state voltage drop of each switch is 2.5 V. Each switch has a thermal resistance from junction to casing equal to 1.2°C/W. Each is mounted on a heat sink, and the thermal resistance of device casing to heat sink surface is 0.8°C/W while that of heat sink surface to ambient is 0.5°C/W. The switching frequency is small, and therefore the switching power loss can be neglected. The forward leakage current in the device can also be neglected.
(*a*) Determine the peak and average power dissipations in each switch.
(*b*) Determine the junction temperature in the device and also that on the surface of the heat sink for an ambient temperature of 45°C.

1.9. The circuit shown in Fig. P1.9(*a*) is employed to obtain an AC output voltage of frequency 20 kHz from the dual DC voltage source shown. For this purpose, the controlled static switches S_1 and S_2 are operated according to the timing diagram shown in (*b*). The turn ON and turn OFF switching transition times of each switch are respectively 2 and 3 μs. The ON state voltage drop of each is 2.5 V. Neglect power loss due to forward leakage currents.
(*a*) Determine (1) the average switching power loss, (2) the average static power loss and (3) the peak instantaneous power loss in each switch.
(*b*) The thermal resistance from junction to casing of each switch is 0.4°C/W. The maximum ambient temperature is 50°C. The device junction temperature is

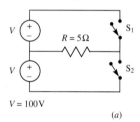

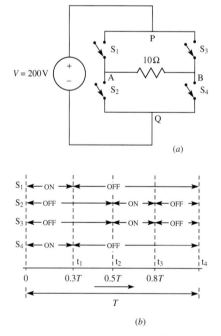

(*a*)

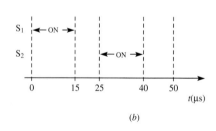

(*b*)

FIGURE P1.8   **FIGURE P1.9**

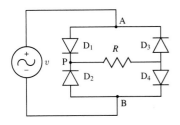

FIGURE P1.10

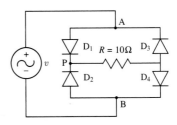

FIGURE P1.11

to be limited to 120°C. Determine the thermal resistance of the heat sink to be chosen for each device, from the device casing to the ambient.

1.10. For the rectifier circuit of Fig. P1.10, sketch the voltage waveform across the diode D_1 if the AC input is a sinusoidal 60 Hz source of r.m.s. value 120 V. Determine the peak reverse voltage and the voltage rating of the diode to be chosen for this rectifier circuit on the basis of a safety factor of 2.3.

1.11. In the rectifier circuit of Fig. P1.11, the sinusoidal AC input has a r.m.s. value of 440 V and a frequency of 60 Hz. Sketch the theoretical waveform of the current through a diode. From this, determine the following three categories of the minimum current rating needed for the diode: (*a*) Repetitive peak, (*b*) average and (*c*) r.m.s. currents.

1.12. In the circuit of Fig. P1.12(*a*) assume that L is very large and that the controlled switch CS has been ON for a long time initially, so that the current I can be treated

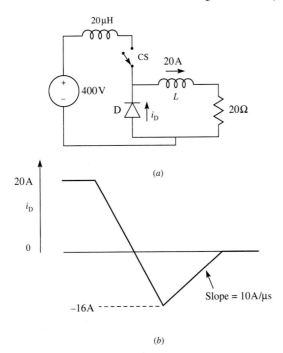

FIGURE P1.12

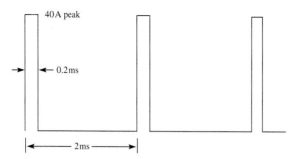

FIGURE P1.14

as constant at 20 A. The switch CS is now turned OFF for a short time, so that the current freewheels through the diode D. During the short OFF time of the switch, assume that the current in the inductance is unchanged. CS is turned ON again. The waveform of the current through the diode from the instant CS is turned ON is sketched in (b). For the values of the circuit parameters indicated in the figure and on the basis of the waveform given, determine the following:

(a) the initial rate of decay of the forward current through the diode when CS is turned ON;

(b) the turn OFF time of the diode;

(c) the approximate amplitude of the reverse voltage spike that is to be expected across the diode during its recovery;

(d) the maximum instantaneous magnitude of the current spike to be expected in the switch CS during its turn ON switching.

1.13. A diode is to be used for freewheeling function in the converter of Problem 1.12 at repetitive switching frequency f (in Hz). What should be the highest possible value of f for the following values of the recovery time t_{rr} of the diode: (a) 500 ns; (b) 1 μs; (c) 2 μs.

1.14. In a static converter working at a switching frequency of 500 Hz, a freewheeling diode has the current waveform shown in Fig. P1.14. Determine the required minimum current rating of the diode under each of the following categories:

(a) repetitive peak;

(b) average;

(c) r.m.s.

1.15. In the circuit of Fig. P1.15, the power BJT is being used as a static switch to switch a resistance load of 10 Ω in a 200 V DC circuit. Its parameters are $h_{FE} = 15$, $V_{CE(sat)} = 2$ V and $V_{BE} = 0.7$ V.

(a) Determine the minimum value of the driver output voltage labeled V_b in the figure necessary to drive the switch to obtain the maximum current in the load. Determine also the power dissipation in the switch under this condition.

(b) If V_b falls to 80% of the value determined in (a), what will be the change in the load current and the change in the power dissipation in the transistor?

1.16. In the circuit of Fig. P1.16, the transistor switch has been ON for a long time. It is turned OFF by removing the gate drive. Assume that the inductance L is very large. What will be the maximum voltage that the collector will have to withstand during the turn OFF switching?

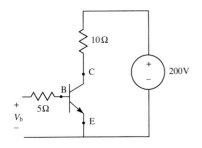

FIGURE P1.15

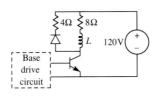

FIGURE P1.16

1.17. For the transistor switch in Fig. P1.17, the storage time and the fall time during turn OFF switching are 1.8 and 1.5 μs respectively. The ON state current is 20 A. Determine suitable values for the snubber capacitance C and the snubber resistance R that will ensure the following:

(a) the voltage rise across the transistor will not exceed 100 V at the instant when current falls to zero;

(b) the peak discharge current of the capacitor at a turn ON switching will not exceed 8 A.

(c) Determine also the snubber power loss at a switching frequency of 10 kHz for the chosen snubber circuit parameters.

1.18. Discrete devices are used to implement a power Darlington static switch. The h_{FE} of the main transistor is 12. What should be the h_{FE} of the driver transistor to be chosen to give an overall h_{FE} for the switch equal to 180?

1.19. The power MOSFET functions as a static switch in the circuit of Fig. P1.19. Its interelectrode capacitances are $C_{gs} = 5000$ pF, $C_{gd} = 200$ pF. The gate is being driven from a 12 V pulse source. The leading edge of the gate pulse rises from zero to 12 V in 180 ns. Assume that v_{ds} falls linearly in the same interval during the ON switching. Determine the peak value of the charging current to be supplied by the gate source at each turn ON switching.

1.20. In Problem 1.19, what is the increase in the effective gate-to-source capacitance due to the Miller effect.

1.21. Linearly approximated current and voltage waveforms during the turn OFF of an IGBT are given in Fig. P1.21. The ON state current before the commencement of turn OFF switching is 40 A. The current waveform has a "tail" that commences

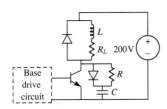

FIGURE P1.17

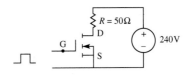

FIGURE P1.19

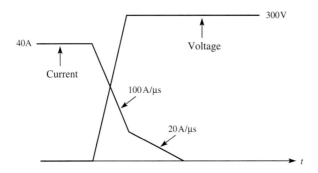

FIGURE P1.21

when it has fallen to 25% of the ON state magnitude. While the initial decay was at 100 A/μs, during the tail it is 20 A/μs. The voltage recovers to the full value of 300 V at the same instant as the commencement of the tail current. Assume a switching frequency of 20 kHz.

(a) Determine the average power dissipation in the device due to turn OFF switching only.

(b) How much of the power determined in (a) occurs during the initial fall of current and how much during the tail current flow?

(c) What would be the turn OFF switching power loss if the tail did not exist and the entire fall of current to zero occurred at the initial rate of decrease. Assume that the voltage reaches the final value at the instant the current falls to zero.

1.22. It has been stated in the text that an IGBT has a small reverse voltage blocking capability—typically below 10 V. Sketch the junction structure of an IGBT and identify the junction that provides this small reverse voltage capability.

1.23. The thyristor is used as static switch to turn ON current in the highly inductive circuit of Fig. P1.23. The latching current of the thyristor is 250 mA. What is the minimum duration of the gate drive pulse required to turn ON the thyristor switch?

1.24. The thyristor in the switching circuit of Fig. P1.24 is in the ON state. Its holding current rating is 150 mA. An additional resistance R is introduced in the circuit by opening the switch labeled S across this resistance. What is the lowest value of R that will cause the thyristor to turn OFF when S is opened?

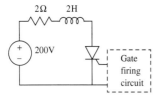

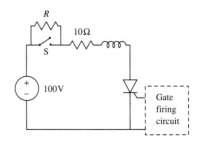

FIGURE P1.23   **FIGURE P1.24**

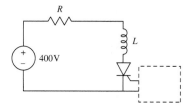

FIGURE P1.25

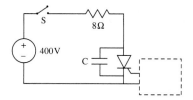

FIGURE P1.26

1.25. The thyristor in the switching circuit of Fig. P1.25 has a *di/dt* rating of 50 A/μs. Determine the minimum value of L needed to protect the thyristor from *di/dt* failure.

1.26. The *dv/dt* rating of the thyristor in the circuit of Fig. P1.26 is 100 V/μs. Determine the minimum value of the capacitance C necessary so that no erratic turn ON due to *dv/dt* will occur when the power is turned ON by closing the swtich S.

CHAPTER
2

LINE COMMUTATED CONVERTERS

2.1 INTRODUCTION

A static power converter makes use of a configuration of power semiconductor devices that function as switches. These are made to turn ON and turn OFF repetitively in such a way as to implement the required conversion function. Whenever a switch is turned OFF, the path of the current flow changes, that is, there is a "commutation" of current away from that switch. In the terminology of Power Electronics, it has become common to describe the turn OFF switching of the device itself as "commutation." A conducting diode is automatically commutated, that is, turned OFF, when a reverse bias voltage appears across it. A conducting thyristor is also commutated in a similar manner because the gate is ineffective to achieve commutation. The reverse voltage that serves to turn OFF a thyristor or diode is called the commutating voltage. We can turn OFF a thyristor by injecting a reverse bias across it and "force" it to turn OFF. For this, we can use a specially designed circuit specifically for injecting this reverse bias across the main terminals of the thyristor, namely the anode and the cathode, and maintaining this reverse bias for a time that is at least equal to the specified turn OFF time of the thyristor, which will be sufficient for the gate to regain control over the switching. In this way, the device will not turn ON again after the reverse bias disappears. When a thyristor is turned OFF in this manner by using a circuit specifically designed to force it into the OFF state

106

by a reverse bias of short duration, we describe this as "artificial commutation" or "forced commutation." In this book, we shall use the term "force commutation." To implement force commutation, we need a force commutation circuit to create the reverse bias pulses at the required instants of time in a switching cycle, when the device is to be turned OFF. We shall present some examples of force commutation circuits later in Chapter 4. Static converters that employ force commutation circuits for the turn OFF switching of thyristors are categorized "force commutated converters." The chopper converters studied in Chapter 4, which convert a DC input at one voltage level to a DC output at another voltage level, employ force commutation when thyristors are used as the switching elements. The same statement is true of the inverters studied in Chapter 5, which convert a DC at its input terminals to an AC output of the desired frequency.

2.1.1 Line Commutation

The alternative to force commutation is "line commutation." This is possible if one or more AC voltages are available in the converter, which can be made use of as the commutating voltage to implement the turn OFF switching of its thyristors. For example, if the converter is a rectifier, an AC supply is already available at the input terminals for conversion to DC. We can commutate a conducting thyristor by using this voltage, during intervals of time when it has the proper polarity. We shall illustrate this description of line commutation by reference to Fig. 2.1. This circuit is basically a three-phase rectifier circuit that converts a three-phase AC input into DC. In this figure, v_{an}, v_{bn} and v_{cn} are three identical AC voltages that differ in phase mutually by 120° and therefore constitute a balanced three-phase supply. In practice, these will be the secondary windings of a three-phase transformer, which are shown "Y" connected with the terminal labeled N in the figure as the neutral terminal. The switching elements are the three thyristors labeled T_a, T_b and T_c. The gate circuits of these thyristors are not shown in the figure for the sake of clarity.

The "line voltage" $v_{ba} = v_{bn} - v_{an}$ is an alternating voltage. Let us

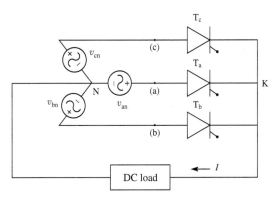

FIGURE 2.1
Circuit using line commutation.

assume that thyristor T_a has been turned ON during an interval in the AC cycle when it is forward-biased and that it is conducting current into the load from the (a) phase source. Let us assume that the other two thyristors are OFF. At any instant in the AC cycle when the line voltage v_{ba} is positive, that is, when the line (b) is at a positive potential with respect to line (a), the thyristor T_b is forward-biased and a gate pulse can turn it ON. If we do turn T_b ON then the line voltage will appear as a reverse voltage across the conducting thyristor T_a and commutate it. In this way, the line voltage serves as the commutating voltage to turn OFF an outgoing thyristor. From this, we see that by sequentially firing an incoming thyristor when the line voltage has the correct polarity, we can successively commutate each thyristor. Such a scheme does not require any special force commutation circuit to "artificially" generate and apply a reverse bias pulse to commutate an outgoing thyristor. The incoming thyristor automatically applies the line voltage in reverse to the outgoing thyristor, which is commutated "naturally." The term "natural commutation" has also been used to describe this type of commutation. We shall, however, use the term "line commutation."

Therefore we see two aspects of line commutation. First, line commutation is possible only in converters that are connected to an AC voltage bus, because the presence of an alternating voltage is necessary to serve as the commutating voltage. Second, to successfully achieve turn OFF switching by means of line commutation, it is essential that the associated line voltage, which serves as the commutating voltage, must have the polarity that will reverse-bias the outgoing thyristor. This in effect means that line commutation of a conducting thyristor is not possible at any arbitrary instant in an AC cycle. For each thyristor in a rectifier circuit, there is an associated AC line voltage that serves as the commutating voltage. Commutation of that particular thyristor is only possible during the 180° interval during which the waveform of its commutating line voltage has the correct polarity for reverse-biasing it. We shall later see that this 180° available for successful line commutation is the ideal limit. Practical limitations—of the circuit as well as the switching devices—actually make the available interval somewhat less.

For the above description, we chose the example of a thyristor rectifier. But in a diode rectifier also, the turning OFF of an outgoing diode occurs by line commutation. The incoming diode automatically turns ON when it starts to get forward-biased, and thereby impresses the commutating voltage across the outgoing diode to turn it OFF. Besides the rectifier, there are other converters that employ line commutation. In this introductory section, we shall present a brief overview of the different types of converters that belong to the line commutated category. The remaining sections of this chapter will be devoted to their detailed treatment.

> **Illustrative example 2.1.** In the converter circuit of Fig. 2.1 each phase voltage is sinusoidal with an amplitude of 100 V and a mutual phase displacement of 120°. Phase (b) is the lagging phase with respect to phase (a). Sketch the waveforms of

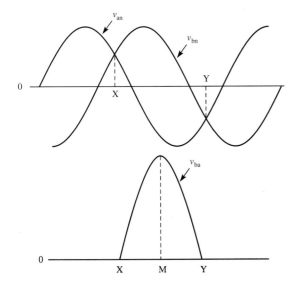

FIGURE 2.2
Waveforms in Example 2.1.

the phase voltages and also the line voltage v_{ba}. Identify the interval of time in a cycle when successful line commutation of the thyristor T_a is possible by the firing of thyristor T_b. What is the maximum possible magnitude of the commutating voltage? Locate on the waveforms the instant at which this maximum value occurs.

Solution. The waveforms of the phase voltages v_{an} and v_{bn} are sketched in Fig. 2.2. These are sine waves, each with amplitude 100 V. The line voltage v_{ba} is given by

$$v_{ba} = v_{bn} - v_{an}$$

The positive half-period of v_{ba} is also plotted separately in Fig. 2.2. This is also a sine wave of amplitude $\sqrt{3} \times 100 = 173.2$ V.

Line commutation of T_a by the turning ON of T_b is only possible in the interval when v_{ba} is positive. The limits of this interval are indicated by labels X and Y in the figure.

The maximum possible commutating voltage is equal to the maximum value of the line voltage v_{ba}. This is equal to 173.2 V and occurs at the instant labeled M in the figure.

2.1.2 Diode Rectifiers—Basic Configurations of the Switching Elements

There are two basic rectifier configurations which are most commonly used for large power applications. These are

1. the midpoint configuration and
2. the bridge configuration.

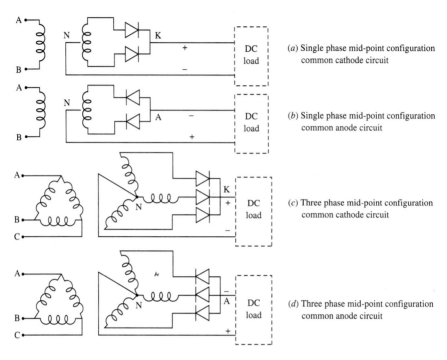

FIGURE 2.3
Midpoint configurations of rectifiers.

2.1.2.1 MIDPOINT CONFIGURATION. Figure 2.3 shows single-phase and three-phase midpoint rectifier circuit configurations. The midpoint configuration requires a transformer, in addition to the switching elements. The switching elements are on the secondary side of the transformer. The transformer has to have a "midpoint," and this midpoint is one terminal of the DC output. In single-phase rectifiers, the transformer has a center-tap on the secondary side, labeled N in (*a*) and (*b*), and this serves as the midpoint. In three-phase rectifiers, the secondary connection should be such as to provide a neutral terminal. This secondary neutral terminal, labeled N in (*c*) and (*d*), serves as the midpoint. The transformer has to handle the full power of the converter. Because of this, the midpoint configuration makes the converter larger and heavier. But the transformer provides two benefits. First, it provides electrical isolation between the input AC side and the output DC side. Second, the voltage ratio of the transformer can be suitably chosen to match the required DC side voltage, to the available AC bus voltage.

The switching elements, which are diodes in Fig. 2.3, have a common terminal connection, and this common point is the other DC output terminal. In the single-phase circuit (*a*) and the three-phase circuit (*c*) the cathode terminals are connected together. We have labeled the common cathode terminal by the letter K. In the single-phase circuit (*b*) and the three phase circuit (*d*), the anodes constitute the common terminal, and this we have labeled by the letter A. In a common cathode midpoint rectifier the common cathode point K is the positive terminal of the DC side and the midpoint is the negative. In the common-anode rectifier circuit, the common anode is the negative terminal and the midpoint is the positive one.

The above statements regarding the DC side terminal voltage polarities are strictly valid only when the switching elements are diodes. In "controlled rectifiers," we use thyristors instead of diodes as the switching elements. We usually also have a series-connected inductor on the DC side to minimize ripples in the DC current. In such a rectifier, as will be seen later, there could be short intervals in the AC cycle during which the terminal voltage polarities on the DC side are actually reversed. Converters that employ thyristors as switching elements can also be used to transfer power from the DC side to the AC side. This is called the "inversion mode" of operation, during which power is actually made to flow from a DC source to the AC bus. To operate the circuit as an inverter, the DC source has to be connected with reversed polarity. In other words, for inverter operation, the common cathode terminal K will have the negative voltage polarity. For the common anode circuit, similarly, the common anode terminal A will have the positive voltage polarity in the inversion mode.

However, whatever the operating condition, the current direction is unique for the common-cathode and common-anode types of connection. This is a consequence of the fact that the switching elements connected to the common terminal, be they diodes or thyristors, can conduct current only in one direction. Therefore we should look at the common-terminal polarity not always on the basis of the voltage, but on the basis of the direction of DC current flowing from it. If the positive current direction is defined as away from the terminal to the DC side, the common cathode K is always the positive current terminal, but not always the positive voltage terminal. Similarly the common anode A is always the negative current terminal, but not always the negative voltage terminal.

2.1.2.2 THE BRIDGE CONFIGURATION. Figure 2.4 shows the bridge configuration for single-phase and three-phase AC inputs. The bridge requires double the number of switching elements compared with the midpoint configuration. The bridge basically consists of a common cathode circuit and a common anode circuit. Half the total number of switching elements have their cathodes connected together at the common cathode terminal labeled K. This terminal is the positive current terminal, according to the sign assignment for current direction that we have just made for the midpoint circuit. The remaining half of the switching elements have their anodes connected together at the common anode terminal labeled A. This is the negative current terminal. When we analyze the bridge circuit later in this chapter, we shall treat the

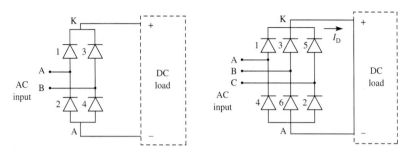

FIGURE 2.4
The bridge configuration.

bridge as a combination of a common cathode section and a common anode section.

In the bridge configuration, the DC current has to flow simultaneously through two switching elements—one belonging to the common cathode section and the other belonging to the common anode section. This is a disadvantage in comparison with the midpoint circuit. But the main advantage of the bridge circuit over the midpoint circuit is that there is no necessity to use a transformer. Since the transformer will have to have the full power rating of the converter, the elimination of the converter transformer will mean a large reduction in cost, weight and size. But a transformer may still be necessary if electrical isolation between the AC and DC sides is a requirement. It may also have to be used for stepping up or stepping down the AC voltage, as a means of overcoming a disparity in the available AC bus voltage and the required DC voltage.

2.1.3   Uncontrolled, Controlled and Semicontrolled Rectifiers

2.1.3.1   UNCONTROLLED RECTIFIERS. When all the switching elements of a rectifier are diodes, we call it an uncontrolled rectifier. We do not have a means of controlling the switching instants of the devices. Each diode automatically turns ON at the instant when it becomes forward-biased and turns OFF when it becomes reverse-biased. Uncontrolled rectifiers have a fixed voltage ratio between the DC output voltage and the r.m.s. AC input voltage. When load current flows on the DC side, there will be some internal voltage drop in the rectifier, depending on the value of the load current. Later in this chapter, we shall derive expressions for the no-load voltage ratio and for the output voltage drop with load. In Figs. 2.3 and 2.4, the switching elements shown in every rectifier circuit are diodes. Therefore they are all uncontrolled rectifiers. The circuits shown in these figures are what we can call the "power circuit block." Static converters generally also have a "control circuit block." The latter generates the switching pulses for the thyristors for their controlled switching with correct timing. Since all the switching elements of an uncontrolled rectifier are diodes, there is no need for a control circuit block for this category of rectifiers.

2.1.3.2 CONTROLLED RECTIFIERS. When all the switching elements of a rectifier are thyristors, we call it a controlled rectifier. We can change any of the rectifier circuits in Figs. 2.3 and 2.4 from the uncontrolled category to the controlled category by replacing every diode by a thyristor. In contrast to a diode, a thyristor does not automatically turn ON at the instant in the AC cycle at which it becomes forward-biased. After it has become forward-biased, it waits till a gate pulse is impressed on its gate terminal. The controlled rectifier has a control circuit block to generate and supply the "gate trigger pulse," also called the "gate firing pulse," to each thyristor at the appropriate instant in every switching cycle. Control of the DC output is achieved by adjusting the delay time of the gate firing pulse to each thyristor from the instant it would have turned ON had it been a diode. In other words, we are adjusting the "phase" of the gate firing pulse with respect to a reference instant, which for each thyristor is the instant at which it starts to get forward-biased. For this reason, this type of control is generally described as "phase control." In a controlled rectifier, since all the switching elements are thyristors, phase control can be exercised over every switching element.

2.1.3.3 SEMICONTROLLED RECTIFIERS. In addition to uncontrolled and controlled rectifiers, there is a third category, known as "semicontrolled" or "half-controlled" rectifiers. In a semicontrolled rectifier, some of the switching elements will be thyristors and the rest diodes. A typical semicontrolled rectifier uses the bridge configuration. As we have illustrated in Fig. 2.4, the bridge configuration has a common cathode section and a common anode section. We can obtain a semicontrolled bridge configuration by having all thyristors in one section and all diodes in the other. Such an arrangement is shown in Figs. 2.5(a) and (c). An alternative arrangement that can be used in single-phase bridges is shown in Fig. 2.5(b). Here we have two thyristors in one "leg" of the bridge rather than in one "section" of it. The two diodes constitute the other leg. In Fig. 2.5(a) for the case of the single-phase semicontrolled bridge and in Fig. 2.5(c) for the case of the semi-controlled three-phase, we have chosen the common cathode section for the thyristors. This may be convenient in some situations for the following reason. The gate firing pulse to a thyristor is always to be applied between the cathode and gate terminals. The cathode terminal is, in fact, the reference terminal for the gate pulse. So, if we have all the cathodes connected to a common terminal, we have a common reference point for the gate pulses of all the thyristors. This may help to simplify the gate firing circuit for the semicontrolled bridge.

The semicontrolled bridge is used primarily for reasons of economy, in situations that do not demand a fully controlled bridge. There will be a saving in cost, because diodes are less costly than thyristors. Since phase control of the bridge is limited to the thyristors only, the control circuit block for the

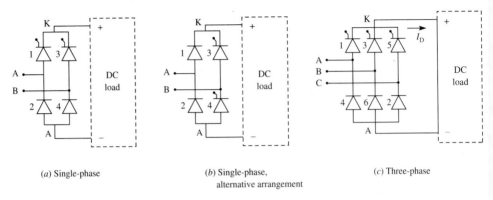

(a) Single-phase

(b) Single-phase,
alternative arrangement

(c) Three-phase

FIGURE 2.5
Configuration of the switches in semi-controlled bridge rectifiers.

semicontrolled bridge will also be economical, because it need be designed to provide gate pulses only to the thyristors in the bridge.

2.1.4 "Firing Angle"—the Means of Phase Control

We have stated in the previous section that phase control is exercised by delaying the ON switching of a thyristor by an adjustable delay time, after it has become forward-biased, that is, after the instant it would have turned ON had it been a diode. This time delay is usually expressed in angular measure, by treating one full period of the AC input as equal to 360°. This phase angle delay is commonly known as the "firing angle" of the switching element. In this book, we shall use α for the firing delay angle. Therefore the basic difference between the switching of a diode and that of a thyristor, in a line commutated converter, is that for a diode the firing angle α is always zero, whereas for a thyristor α is adjustable by phase control.

The analytical results that we shall derive in this chapter for the controlled rectifier will be applicable for the uncontrolled rectifier also by putting $\alpha = 0$. Therefore we shall not make a separate analysis of the uncontrolled rectifier. We use phase control primarily to vary the DC output voltage. The voltage conversion ratio, between the DC output and the AC input, will be shown to be a function of the firing delay angle α. For the uncontrolled rectifier, this ratio will be constant because α has a fixed zero value. There will, however, be a relatively small load-dependent drop in the DC output voltage when the converter supplies a load current.

2.1.5 Functional Circuit Blocks of a Line Commutated Converter

Figure 2.6 shows the main functional circuit blocks of a phase controlled line commutated converter. The input we chose for this is a three-phase AC, shown

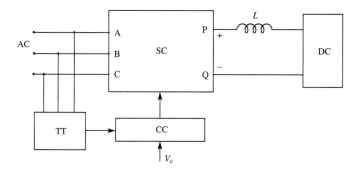

FIGURE 2.6
Functional circuit blocks of a typical line commutated converter.

connected at the AC terminals labeled A, B and C. The circuit block labeled SC is the power circuit of the static converter. This will contain the switching elements if the configuration is a bridge, or switching elements plus the transformer if the configuration is midpoint. The control circuit block, labeled CC provides the firing pulses to the thyristors inside SC. The control circuit block is also frequently referred to as the "firing circuit." The firing circuit is designed to make the firing angle adjustable. Typically, this adjustment is made by means of a control voltage, labeled V_c, which is shown as an input to the control circuit block in the figure. The firing delay angle α is adjusted by varying V_c.

The firing circuit also needs timing inputs from the AC supply, because it has to have the reference instants in every AC cycle with respect to which the firing angle delays are to be implemented. The timing reference instants are obtained from the timing transformer block, labeled TT in the figure. There will be only one small transformer in this block if the input is single-phase, but more if the input has more than one phase. This block TT provides a low voltage replica of the input AC waveform to the firing circuit.

Like all electronic signal processing circuits, the firing circuit may need low voltage regulated DC power. This is not shown in the figure. If the converter is an uncontrolled rectifier, in which all the switching elements are diodes, the timing transformer block and the control circuit block will be absent.

The voltage at the DC terminals, which are labeled P and Q in the figure, will not be a perfect DC, but will have a significant amount of AC "ripple." A "filter" is normally used to bring this ripple component down within acceptable limits. The filter element used for higher power applications is invariably an iron-cored inductor, labeled L in the figure. This inductor is part of the power circuit and must be capable of handling the full rated DC current of the converter without magnetic saturation.

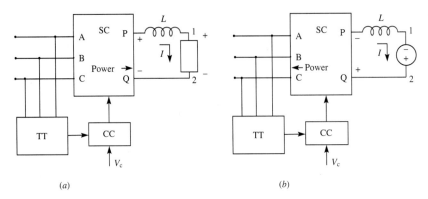

FIGURE 2.7
Phase controlled converter operation with (a) forward and (b) reverse power flow.

2.1.6 Direction of Power Flow—Inverted Operation

2.1.6.1 THE RECTIFICATION MODE OF OPERATION. Figure 2.7(a) shows a phase controlled line commutated converter, which fits the general description of Fig. 2.6, functioning as a rectifier and feeding power into a DC load. The terminals of the load are labeled 1 (positive) and 2 (negative). If the smoothing inductance is assumed to be ideal with zero resistance, the DC component of the voltage at the output terminals P and Q of the converter will be the same as the DC component of the voltage across the terminals 1 and 2 of the load. If the inductance L is very large then the voltage across 1 and 2 will be a near-perfect DC, with negligible ripple. What the inductance really does is to withstand the instantaneous voltage difference between the AC phase voltage that is driving the current (labeled I in the figure) and the voltage across the load terminals 1 and 2. The voltage that appears across the terminals P and Q of the converter is really that of the AC phase, appearing through the switches that are ON at the instant. The difference between this and the voltage across the terminals 1 and 2 is what the inductor has to withstand. If the inductance is sufficiently large, the ripple current resulting from the instantaneous value of this ripple voltage will be negligibly small, and we can assume that the DC side current has a constant magnitude.

If now we look at the polarity of the voltage across the load terminals 1 and 2, and the direction of the current I in Fig. 2.7(a), we see that power, which is the product of the current and the voltage, is flowing into the load. The direction of power flow is from the AC side to the DC side. This direction of power flow is indicated by an arrow in the static converter block SC in (a). This statement about the direction in which power flows can be confirmed if we look at the voltage polarity and current direction from the AC side. The current I that is flowing on the DC side is actually supplied by an AC phase through the static switches that are in the ON state during the conducting interval of this phase. From the waveforms to be derived later, it can also be seen that the voltage polarity of an AC phase during its interval of conduction is in conformity with power flow from this AC phase into the load.

2.1.6.2 THE INVERSION MODE OF OPERATION. If all its switching elements are thyristors, we can operate the same converter in the "inversion mode" also. For this mode, the flow of power is from the DC side to the AC side. The inverted operation is illustrated by Fig. 2.7(*b*). To supply power from the DC side, we need to have a DC source on this side, as shown in (*b*). Since the switching elements of the converter are unidirectional, the direction of current flow through them cannot change. This in effect means, that the direction of current at the terminals P and Q cannot change and must be the same as shown in (*a*) for the rectification mode. Therefore, for reversed power flow, the DC source must be connected with reversed polarity, as shown in (*b*). Since the instants at which thyristors are turned ON are at our control, we can turn them on in such a way that the conduction for each AC phase occurs when the segment of this AC phase voltage waveform has the reversed polarity. We are able to make use of the reversed polarity of the DC source voltage to forward-bias and make the thyristors conduct in opposition to the instantaneous AC phase voltage. In this way, we make power flow in the reverse direction on the AC side. This statement about power flow can be confirmed if we look at the waveforms of voltage and current on the AC side. The basic fact is that we are able to operate the converter in the inversion mode by reversing the DC side voltage polarity and at the same time choosing a different range of firing angles from the range of firing angles used for the rectification mode. The inversion mode is possible only if all the switching elements are thyristors. The semicontrolled bridge cannot operate in the inversion mode.

The need for the inductor L is greater in the inversion mode of operation than in the rectification mode. The function performed by the inductor is to absorb the ripple, which is the voltage difference between the constant DC side voltage and the instantaneous value of the conducting AC phase voltage that is presented at the terminals P–Q through the thyristor switches that are ON. In the rectification mode, if we do not have the inductor L, or if we have one but do not have a large enough value for it, this means that we shall have to put up with a large ripple in the DC side voltage across the terminals 1 and 2 in Fig. 2.7(*a*). In the inversion mode, what we have across the terminals 1 and 2 in (*b*) is a DC source that has no ripple voltage. This ripple-free DC voltage cannot balance the DC plus ripple voltage across the terminals P–Q. Without this inductor, there will be no circuit element present to prevent the ripple current from becoming arbitrarily large.

In the interests of accuracy, we may also make a clarification regarding the power flow directions shown for the rectifier mode in Fig. 2.7(*a*) and for the inversion mode in (*b*). To be exact, these directions should be considered as the directions of flow of average or resultant power. Because of the fluctuating nature of the conducting AC phase voltage, there could be instants during which voltage polarity reversals occur causing reversed power flow. But the average power flow for each mode of operation will be in the directions indicated in Fig. 2.7. The fluctuating power flow on the AC side and a constant or nearly constant power flow on the DC side are possible, because the inductor L is a reservoir of energy. It acts like a cushion, storing and releasing energy on the AC side while maintaining a nearly constant power flow on the DC side.

2.1.7   Phase Controlled Inverters as Distinct From Other Types of Inverters

When a phase controlled thyristor converter is made to function in the inversion mode, it may be called a "phase controlled inverter." In Power Electronics, by "inverter" we usually mean a static converter that generates an independent AC output from a DC input source. Such inverters are presented in Chapter 5. They can be designed to generate AC of the desired frequency. Phase controlled inverters do not belong to such a category of inverters.

The phase controlled inverter must already have an AC available at its AC terminals from an external bus. The frequency or waveform of this AC is not adjustable by manipulation of the control circuit of the inverter. The control circuit only changes the firing angle in such a way that power now flows from a source connected to the DC side, into the AC voltage bus on the AC side of the converter. Phase controlled rectification and phase controlled inversion are two different operating modes possible with the same converter, connected to the same AC bus. This AC bus voltage serves to implement line commutation in both the modes of operation.

2.1.8   Single-Quadrant and Two-Quadrant Operation of a Phase Controlled Converter

In Fig. 2.8 we show rectangular coordinates which can be used to locate an operating point on the DC side of a phase controlled converter. We have used

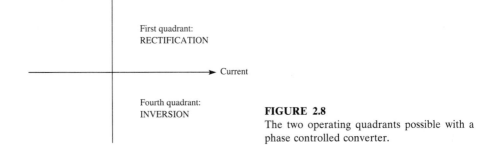

FIGURE 2.8
The two operating quadrants possible with a phase controlled converter.

the horizontal axis for the DC side current and the vertical axis for the DC side voltage. We shall make positive sign assignments of voltage polarity and current direction to correspond to the rectification mode of the converter, as shown in Fig. 2.7(a). An operation with a particular value of DC voltage and a particular value of DC current can be located on the graph using the respective axes for these two quantities. On the basis of our reference polarity for voltage and reference direction for current as stated above, all the operating points in the rectification mode will be located in the first quadrant—the quadrant for positive voltage and positive current. For the inversion mode shown in Fig. 2.7(b), all the operating points will be in the fourth quadrant—the quadrant for positive current, but negative voltage. The control circuit of a phase controlled converter can be designed to provide for operation in both modes, and to facilitate moving from one mode to the other. Such a converter is described as a two-quadrant converter.

An example of two-quadrant working—DC motor drive with regenerative braking. An important industrial application of phase controlled converters is for adjustable speed DC motor drives. In a large majority of such drive systems, the converter is used not only for driving the motor at an adjustable speed, but also for braking it electrically when moving to a lower speed or stopping. Figure 2.9(a) and (b) show the two modes of operation. In both figures, we have omitted the field circuit of the motor. The field circuit of the motor is typically supplied from a smaller converter, which is designed for the rectification duty only. This converter supplies the DC current, known as the excitation current, to magnetize the magnetic poles of the motor. This field converter is unnecessary for a motor that uses permanent magnets. Since the field circuit is not relevant to our present description, we have omitted it in Fig. 2.9.

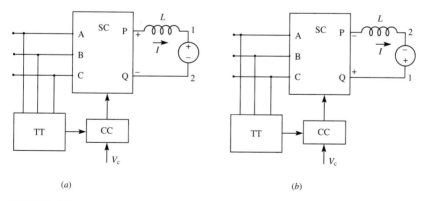

(a)   (b)

FIGURE 2.9
Two-quadrant phase controlled converter for a DC motor drive with regenerative braking.

In both (*a*) and (*b*), the terminals labeled 1 and 2 are the terminals of the armature circuit of the motor. The electrical power that gets converted to mechanical power to drive the load coupled to the motor is provided through its armature terminals. The armature terminals 1 and 2 are shown connected to the DC terminals P and Q of the converter, including also a smoothing inductance *L* in series.

Figure 2.9(*a*) is for the driving mode. The converter is operated as a rectifier to deliver DC power into the motor armature terminals. The speed of the motor is adjustable by adjusting the DC output voltage of the rectifier by phase control. When the motor spins, it generates an induced DC voltage in the armature circuit, which is known as the back e.m.f. The back e.m.f. polarity will be as shown in (*a*), making terminal 1 positive. The DC current has to be driven into the armature circuit against the back e.m.f. That is how the electrical power is pumped into the armature for conversion to the mechanical power needed for driving the load.

The motor will continue to have the back e.m.f. with the polarity shown, as long as it continues to spin in the same direction. To brake the motor that is spinning at a certain speed, we can momentarily disconnect the motor armature terminals from the converter and reconnect it again with the terminal connections reversed, as shown in Fig. 2.9(*b*). This can be done using a fast acting contactor used as a reversing switch. At the same time, we must also change the firing angles of the converter so that it now operates in the inversion mode. The machine will now be functioning as a generator and not as a motor. The generator receives its driving power from the stored kinetic energy of the mass in motion coupled to its shaft and its own rotational inertia. This energy flows back into the AC bus through the converter, which is now made to function in the inversion mode. This type of braking is called regenerative braking. As the motor slows down, the induced e.m.f. in its armature decreases, but we can suitably keep adjusting phase control to continue the regenerative braking down to low speeds. Regenerative braking overcomes the following disadvantages of friction braking. The latter involves generation of heat, creating heat dissipation problems, brake shoe wear and wastage of energy. There are many large industrial DC motor drives in which the motor has to go through repeated cycles of acceleration and braking. Rolling mill drives in the metallurgical industry are typical examples. Friction braking will mean expensive maintenance requirements and large wastage of energy for such drives.

2.1.9 Two-Quadrant Phase Controlled Converter—Second Example: Converter for HVDC Power Link

A second example of the two-quadrant working of phase controlled converters is in the area of bulk power transmission using high voltage direct current (HVDC). Most of the transmission of electric power, over long distances, is done universally using three-phase AC at very high voltages. There are, however, situations where it is more economical to use DC at an equally high or higher voltage to interconnect two AC power systems and transmit power from one system to the other. A notable example of this is the underwater high voltage DC cable across the English Channel, which interconnects the AC electrical power networks of Great Britain and France. This cross-channel DC cable is being used to supply electric power from the British power network to the French network, during the hours of the day when peak demands of power occur on the latter. Due to differences in industrial habits, the peak demands in power on the British and French networks occur at different times of the day. Therefore, when the peak power demand occurs on the British side, this demand is met by power from the French network. In this way, both networks can keep using the generators that are most economical to run, without having to bring in the less cost-effective ones to meet peak power demands, thereby achieving overall economy. The cross-channel HVDC link has been working successfully over the last several years and is continuing to do so. There are also some more examples of underwater HVDC cable links in operation in other countries. For power transmission by overhead lines also, HVDC has been shown to be more economical than AC, when the distances and power levels are higher than break-even limits. The "Pacific Intertie" in the United States, which is 850 miles long, and the "Nelson River Bipole" in Canada, which is 560 miles long, are working examples of overhead HVDC transmission links in North America. Figure 2.10 shows the block schematic of the main features of an HVDC link between two AC power networks, labeled #1 and #2. In the example of the cross-channel link between Great Britain and France, one of these may be the British power grid and the other the French power grid. The three terminals labeled A_1, B_1 and C_1 constitute the three-phase bus of the AC network #1. Similarly, A_2, B_2 and C_2 constitute the three-phase AC network #2. The block labeled "Converter #1" consists of transformers and thyristor bridges.

There is usually more than one bridge connected in series, so that their DC voltages add to give a large overall DC transmission voltage. Each bridge will be supplied from a separate transformer secondary. The smoothing inductance L is shown outside the converter block. The circuit at the other end of the line, namely at station #2, will be similar.

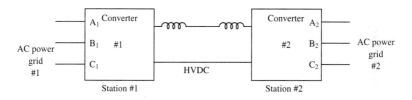

FIGURE 2.10
Two-quadrant operation of phase controlled converters for power transmission by HVDC.

When the power flow is to be from network #1 to network #2, the converter #1 will be working in the rectification mode (that is, in quadrant 1) and the converter #2 will be working in the inversion mode (that is, in quadrant 4). The operating modes of the two converter stations will be reversed when power flow from network #2 to network #1 is required. The coordinated phase control of both converters is achieved by using signals transmitted through communication channels between the two stations.

2.1.10 Static Four-Quadrant Operation

The two application examples outlined above, namely DC motor drives with regenerative braking and converters for HVDC links, make use of the two-quadrant operating capability of a phase controlled thyristor converter. We shall now look at the example of the DC motor drive and consider how operation in all the four quadrants can be achieved.

Figure 2.11(a) shows the connections of the motor armature terminals 1 and 2 to the converter DC terminals P and Q when the motor is being driven in the forward direction. This circuit is the same as in Fig. 2.9(a) when details other than the converter terminals and the motor terminals are omitted. The smoothing inductor L is shown in all the circuits in Fig. 2.9, since it is necessary to absorb the ripple voltage between the motor back e.m.f. which is essentially a ripple-free DC, and the converter output voltage, which is a DC plus ripple voltage. The inductance L may be treated as part of the motor circuit or the converter circuit. Since the following description is primarily concerned with the DC component of the voltages, we shall ignore the presence of the inductor, since its purpose is to absorb the AC component of the inverter terminal voltage.

To locate a *motor* operating point on the four-quadrant graph, we shall assign the positive motor voltage reference polarity as shown in Fig. 2.11(a), that is, when terminal 1 of the motor is positive with respect to terminal 2. For the motor current, we shall assign the positive reference direction as shown in the same figure, that is, when the current is flowing in the motor from terminal 1 to terminal 2. Therefore, for the motoring operation (driving) in the forward direction, corresponding to Fig. 2.11(a), the motor operating quadrant is quadrant I (positive current and positive voltage).

If now we want to slow down the motor and bring it to rest, while still spinning forward, we reverse the connections from the motor terminals to the converter terminals by means of a fast acting reversing switch, to give the new connection as shown in Fig. 2.11(b). We now operate the converter in the inversion mode by changing the firing angles of the thyristor switches in the converter. By doing so, we have reversed the terminal voltage polarity at the converter terminals. This happens because we are now using our DC voltage source, which is the motor now working as a generator, to force the current through the AC phases when their polarities are actually opposite to

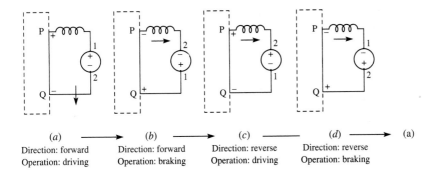

(a)	(b)	(c)	(d)	(a)
Direction: forward	Direction: forward	Direction: reverse	Direction: reverse	
Operation: driving	Operation: braking	Operation: driving	Operation: braking	

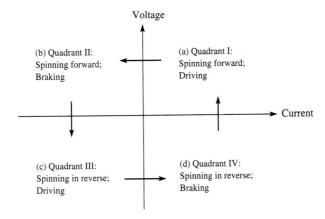

(e) Motor operating quadrants

FIGURE 2.11
Sequential changes in the operating quadrants in a reversible DC motor drive with regenerative braking.

what they were in the rectification mode. In this way, we are sending power into the AC bus and also causing a reversal of the DC voltage across the converter terminals P and Q. Therefore, from the converter point of view, we have moved into quadrant IV, that is, negative voltage and positive current. But, from the point of view of the motor, we have not reversed its voltage polarity, because terminal 1 continues to be positive with respect to terminal 2. But we have reversed the current through the motor, which is now negative because it is in the direction from terminal 2 to terminal 1 as shown in (b). Therefore, on the graph for the motor, the operating point has moved to quadrant II, that is, positive voltage and negative current as shown in Fig. 2.11(e).

If, after the motor stops, we now want to spin the motor in the reverse direction, we can again operate the converter in the rectification mode by changing the firing angles accordingly. The motor will now have DC voltage impressed across its terminal with reverse polarity (terminal 2 positive with respect to terminal 1), which is opposite of Fig. 2.11(a). Therefore the motor will spin in the reverse direction. By doing so, it will develop a back e.m.f. that makes terminal 2 positive with respect to terminal 1. Therefore both motor current and motor voltage will be negative. The operating quadrant for the motor will now be III. This is the motor operating quadrant when driving in the reverse direction. As far as the converter is concerned, the operating quadrant will now be I, corresponding to the rectification mode.

For regenerative braking while spinning in reverse, we shall again have to operate the reversing switch, to reverse the connections of the motor terminals 1 and 2, to the converter terminals P and Q, at the same time changing the firing angles of the converter switches to go into the inversion mode. The motor operating quadrant will now be IV, corresponding to positive current, but negative voltage, as indicated in Fig. 2.11(e). The converter operating point will also be in the fourth quadrant, corresponding to the inversion mode.

In this manner, operation in all the four quadrants can be achieved, for a reversible DC motor drive with regenerative braking. There are a large number of industrial drives requiring four-quadrant operation. A typical example is a reversible rolling mill drive in the metallurgical industry for rolling steel or aluminum sheets to the required thickness. In such a mill, the raw sheet is squeezed between rollers, and the DC mill motor spins the rollers, causing the sheet to move in one direction between them. Towards the end of this forward "pass," the motor is regeneratively braked and brought to rest. The motor is then rotated in the reverse direction for rolling the sheet in the reverse direction. At the same time, the pressure between the rollers is increased to further reduce the thickness of the sheet. Towards the end of this reverse pass, the motor is again regeneratively braked. One forward pass and one reverse pass complete one cycle of operation. These cycles are again continuously repeated until the sheet attains the required thickness. In one cycle of rolling, consisting of a forward pass and a reverse pass, the motor goes through all the four operating quadrants, in the sequence shown in Fig. 2.11(e).

From the above description, we notice that although the converter itself has only two quadrant operating capability, we were able to achieve four-quadrant operation of the motor, with the help of a reversing switch, which we used to reverse the connections between the motor terminals 1 and 2 and the converter terminals P and Q, every time we wanted to move from a driving quadrant to a braking quadrant. For industrial drives that have to continuously go through four quadrant cycles, as in the example of the continuously operating reversible rolling mill described above, it is undesirable to keep changing the connections mechanically using a reversing switch. Fully static converters, which provide four-quadrant operation without the use of mechanical contactors, are manufactured for such applications.

A static four-quadrant converter is also known as a dual converter, because it is essentially a combination of two two-quadrant converters. The

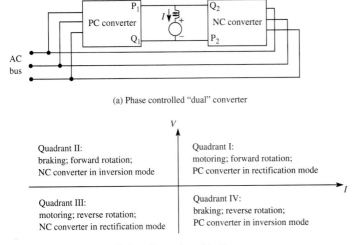

(a) Phase controlled "dual" converter

Quadrant II:
braking; forward rotation;
NC converter in inversion mode

Quadrant I:
motoring; forward rotation;
PC converter in rectification mode

Quadrant III:
motoring; reverse rotation;
NC converter in rectification mode

Quadrant IV:
braking; reverse rotation;
PC converter in inversion mode

(b) Operating quadrants "dual" converter

FIGURE 2.12
Phase controlled dual converter for fully static four-quadrant operation.

principle is shown in Fig. 2.12(a). The reason why we are unable to get four-quadrant operation with a phase controlled converter is that thyristors of the converter can conduct current only in one direction, making it impossible to reverse the direction of current through the converter.

In a dual converter, we use two converters—one to handle positive currents and the other to handle negative currents. The converter labled PC is the positive current converter. The converter labeled NC is identical in all respects to PC. The two converters are connected in "antiparallel," by which we mean that their terminal polarities are reversed with respect to each other. The P terminal of converter PC, which is labeled P_1 in the figure, is connected to the Q terminal of converter NC, which is labeled Q_2. In the same way, the Q terminal of converter PC is connected to the P terminal of converter NC. The load shown in Fig. 2.12 is a DC motor. A smoothing inductor is also shown in series with the load. The reference positive direction for load current and reference polarity for load voltage will be assigned as marked in Fig. 2.12. The two converters together constitute a "dual converter," and we shall assign the same reference voltage polarity and reference current direction for the output current of this dual unit.

For driving in the forward direction, we block all the gate pulses to the thyristors of the negative current converter and operate the positive current converter in the rectification mode. To brake the motor while spinning in the forward direction, we block the gate pulses to the positive current converter and operate the negative current converter in the inversion mode. In this way, we can reverse the current through the motor statically without the use of a

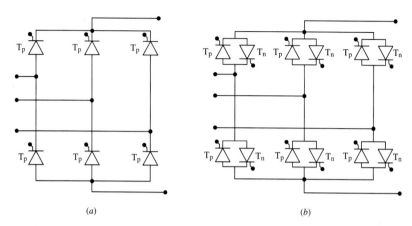

FIGURE 2.13
Arrangement of thyristors in single and dual-three-phase bridges.

reversing switch. For driving in the reverse direction, we continue to operate the negative current converter, but changing over to the rectification mode. To brake while spinning in the reverse direction, we stop the gate pulses to the negative current converter and operate the positive current converter in the inversion mode. These operating sequences are also indicated in the appropriate quadrants in Fig. 2.12(b).

We may point out here that although the positive current converter and the negative current converter, which together constitute a dual converter, are shown as separate circuit blocks in Fig. 2.12(a), the physical arrangement of the thyristors is usually much simpler than what this figure might suggest. Figure 2.13(a) shows a three-phase thyristor bridge for two-quadrant operation. The thyristors in this bridge are labeled with the subscript 'p.' This can be converted into a dual bridge for four-quadrant operation by adding an extra thyristor in "antiparallel" with each thyristor of Fig. 2.13(a). This arrangement is shown in Fig. 2.13(b). The additional thyristors constitute the negative current bridge. These are labeled with the subscript 'n.' With this arrangement, some protective features for the thyristors, such as fuses and snubbers, need not be duplicated for the two bridges. The dual converter unit will not be very much larger in dimensions than a single converter.

2.1.11   The Phase Controlled Cycloconverter— A Dual Converter Operating Cyclically

We have seen that the phase controlled dual converter can handle a load with all four combinations of voltage polarity and current direction. Therefore, with a suitable design of the control circuit, we can operate the converter in such a way that the current supplied to the load is alternating. The positive current converter is operated to supply positive current to the load. Then the negative current converter is operated for negative load current. These operations can be cyclically repeated at a low frequency, so that the current supplied to the load is alternating in nature. Since every combination of voltage polarity and current direction is possible, it is not necessary that the load current be in phase with the load voltage. When a dual converter is operated in this manner, to provide a low frequency AC output, we call it a phase controlled cycloconverter. Although the frequency of the AC output of the cycloconverter is adjustable, this frequency has to be considerably less than the frequency of the AC input. It should be remembered that the phase controlled cycloconverter is a line commutated converter, in which the AC input serves to provide line commutation. For satisfactory operation of the converter, the realistic upper limit of the output frequency is about a third of the input frequency. Therefore, when working from a 60 Hz AC bus, reasonably good operation will not be possible for outputs above about 20 Hz. Cycloconverters are used in practice for very large industrial AC motor drives, which have to operate at very slow speeds and therefore need be supplied with low frequency AC only. With a single dual converter, there could be only one output phase, irrespective of the number of input phases. More dual converters will be needed if more than one output phase are required.

2.1.12 Summary of Section 2.1

In this introductory section we have first explained that the term "commutation" is often used in Power Electronics terminology to describe the turn OFF switching of a power semiconductor device that is functioning as a switching element in a static power converter. To achieve the commutation of a power diode or a thyristor, it is necessary to apply reverse bias across the power terminals, which are the anode and the cathode of the device. If the converter is connected to an AC voltage bus, as in the case of a converter for rectification, an appropriate segment of this AC line voltage can be used as the commutating voltage, to apply reverse bias across the device to be turned OFF. In a converter that has several thyristors or diodes switching ON and OFF in sequence, the turning ON of an element automatically applies the commutating voltage across the device to be turned OFF. A converter in which the switching elements are turned OFF in this fashion is called a line commutated converter.

We have shown the two commonly used configurations in line commutated converters, which are the midpoint configuration and the bridge configuration the bridge being effectively a combination of two midpoint configurations—the common cathode and the common anode. The converters can be classified as uncontrolled, controlled or semicontrolled, depending on whether the switching elements are diodes, thyristors, or thyristors plus diodes. We showed the main functional circuit blocks of a line commutated converter, which in general includes a control circuit block in addition to the power circuit block. The control circuit block is needed to provide gate pulses to the thyristors, and gives us control over their switching. The means of control is actually the adjustable phase delay, commonly known as the firing angle, in the gate pulse provided to turn ON a thyristor. This type of control is called phase control.

By means of phase control, we can operate a thyristor converter both for rectification and inversion. In the inversion mode, we need a DC source connected in reverse on the DC side. We use this DC source to force current in opposition to an AC phase voltage, and at the same time use phase control to make the thyristors conduct during negative segments of the AC phase voltage waveforms. We cannot reverse the direction of current flow through the converter, because the thyristor switches are unilateral conducting devices. But we can reverse the voltage polarity by turning ON, by phase control, the thyristors in such a way that the segments of the AC phase voltage waveforms that sequentially appear at the DC terminals of the converter provide us with a DC voltage component of reverse polarity. These segments of the AC voltage waveform that appear at the DC terminals of the converter also result in an AC component, termed the ripple. In practical power converters, we use a sufficiently large inductor as a buffer between the output terminals of the converter, which give us a DC plus ripple voltage and a nearly ripple-free DC at the terminals of the external DC circuit. For the inversion mode, there is a greater requirement for this inductor, because the DC source generally has no ripple.

A phase controlled thyristor converter can therefore operate with positive or negative DC voltage but with only positive DC current. On a four-quadrant graph in which we use the X axis for current and the Y axis for voltage, these two modes give us operating points in the first and fourth quadrants. We have given two practical examples of such a two-quadrant converter, namely DC motor drives with regenerative braking and converters for HVDC links.

For a reversible DC motor drive with regenerative braking, which has to spin in both directions, we need four-quadrant operation for the motor. We have shown how this can be achieved with a two-quadrant converter with the help of a fast acting switch to reverse motor connections mechanically while moving from driving to braking. We have then shown that fully static four-quadrant operation can be achieved by using a dual converter consisting of a pair of two-quadrant converters connected in antiparallel.

We have then shown that we can utilize the four-quadrant operating capabilty of such a dual converter actually to provide a low frequency AC output by cyclically changing the operating quadrants at the desired AC frequency. We thus arrive at the principle of the phase controlled cycloconverter, which is practically put to use in low speed high power AC motor drives.

We have used this introductory section to provide an overview of the basic features and types of line commutated converters. We shall now proceed to analyse their operation in greater detail and derive waveforms and quantitative relationships that are necessary to implement them practically.

2.2 THE SINGLE-PHASE HALF-WAVE RECTIFIER

The single-phase half-wave rectifier uses only one switching element, which may be a diode or a thyristor. It is seldom used for high power applications. The more popular configurations for power rectification are the midpoint circuit and the bridge circuit. But the single-phase half-wave circuit brings out clearly certain aspects of rectifier operation. For this reason, we present a study of it in this section.

2.2.1 The Single-Phase Half-Wave Rectifier with Resistive Load

Figure 2.14(*a*) shows the circuit when the switching element is a diode, while (*b*) shows the same circuit with a thyristor as the switching element. In both circuits, the AC input is across A and N. This AC input voltage is assumed to be sinusoidal, and its waveform is shown by the broken lines in Figs. 2.14(*c*) and (*d*). The DC output terminals are labelled K and N. In Fig. 2.14, the load on the DC side is a resistance. We shall assume that the switching element, whether thyristor or diode, is ideal, and therefore the forward voltage drop across it during conduction is zero. In the diode circuit, the diode will turn ON at the instant when it becomes forward-biased, that is, the instant labeled 0 in Fig. 2.14(*c*). It will commutate (turn OFF) at the instant at which it starts to get reverse-biased, that is, the instant labeled π in Fig. 2.12(*c*). From π, it will be in the OFF state and remain reverse-biased till 2π. The cycle of switching will then again repeat. When the diode is ON, the AC voltage at the input terminals A and N appears at the output terminals K and N. Therefore the output voltage

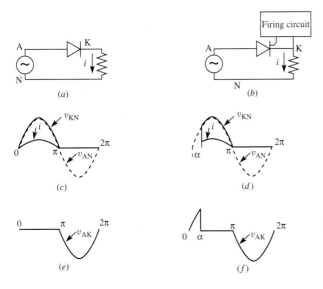

FIGURE 2.14
Single-phase half-wave rectifier with resistive load.

waveform v_{KN} will be the same as the input waveform v_{AN} from 0 to π, as shown by the thick line in Fig. 2.14(c). The current i through the load, which for this configuration is also the same as the input current on the AC side, will be equal to v_{KN}/R. The current waveform is also plotted in Fig. 2.14(c) on this basis.

When the switching element is a thyristor, as shown in Fig. 2.14(b), the turn ON switching is delayed until the firing pulse is received at the gate terminal. The phase angle delay, that is, the "firing angle," is labeled α in Fig. 2.14(d). By definition, the firing angle is the phase angle delay from the commencement of forward bias, that is, from the instant the device would have turned ON if it were a diode, to the actual instant of turn ON. The waveforms of the output voltage v_{KN} and the current i for firing angle α are plotted in Fig. 2.14(d).

Figures 2.14(e) and (f) show the waveform of the voltage across the switching element. In the case of the diode, it will be zero from $\theta = 0$ to $\theta = \pi$, and negative from $\theta = \pi$ to $\theta = 2\pi$. In the case of the thyristor [Fig. 2.14(f)], it will be noticed that it has to block a *forward* voltage from $\theta = 0$ to $\theta = \alpha$. Then it has also to block a *reverse* voltage from $\theta = \pi$ to $\theta = 2\pi$. The maximum blocking voltages can be found from the waveforms of Fig. 2.14(e) and (f). For the diode, this will be the peak negative voltage. Under the worst conditions, for the thyristor this will be equal to the peak positive and negative values of the AC input voltage. These will be the theoretical maximum. The actual voltage ratings of the device to be used will have to be higher, to take care of any transient over voltage that can occur in a practical circuit. The case we have considered above is for a resistive load R. In practice, inductive loads are more common. Therefore we shall now consider an R–L load. The results obtained will be applicable to resistive loads also by equating $L = 0$.

2.2.2   Half-Wave Rectifier Feeding an R–L Load

We shall consider only the circuit using a thyristor as the switching element, because it is more general and the results will be applicable to the diode circuit also, by putting the firing angle $\alpha = 0$. Figure 2.15(a) shows the circuit with the load resistance R in series with the inductance L. The AC voltage input v_{AN} is assumed to be sinusoidal, and its waveform is sketched as the broken lines in Fig. 2.15(b). The thyristor is turned ON at the instant labelled 0, which we shall take as the reference instant of time t and θ, where $\theta = \omega t$ is the angular measure of time phase. The firing angle α is always measured from the instant at which the thyristor starts to get forward biased, which in this circuit is the positive-going zero-crossing of the voltage waveform in Figs. 2.15(b), (c) and (d). With the thyristor ON, the loop equation may be written as

$$L\frac{di}{dt} + Ri = V_m \sin(\omega t + \alpha) \tag{2.1}$$

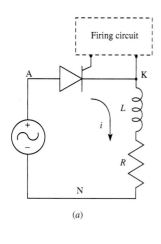

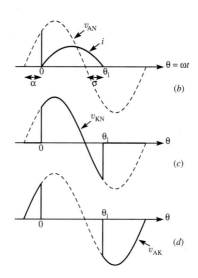

FIGURE 2.15
Half-wave rectifier with $R-L$ load.

The initial condition may be stated as

$$i = 0 \quad \text{at} \quad t = 0$$

Differential equations of this type, with a time-varying forcing function on the right-hand side, will also occur in our subsequent treatment of converters. We shall use the following straightforward method for writing the solutions directly. The complete solution will be the sum of two expressions— (1) the particular integral and (2) the complementary function. The particular integral is the steady-state current that would result if there were no subsequent switching. In our case, basic AC circuit theory tells us that the steady-state current for our $R-L$ circuit has a magnitude equal to voltage/impedance and a lagging phase angle equal to that of the complex impedance. Therefore we can immediately write the particular integral as

$$i_1 = \frac{V_m}{Z} \sin (\omega t + \alpha - \phi) \tag{2.2}$$

with impedance

$$Z = \sqrt{R^2 + X^2}$$

where

$$X = \omega L$$

$$\phi = \tan^{-1} \left(\frac{X}{R} \right)$$

The complementary function is the transient part of the current, which may have an initial amplitude, but will decay towards zero at a rate determined by the time constant of the circuit. This transient component is the solution of the equation obtained by equating the left-hand side of (2.1) to zero:

$$L\frac{di}{dt} + Ri = 0 \tag{2.3}$$

The solution of this is the expression for the decay of current in an R–L circuit. This gives the complementary function as:

$$i_2 = I_t e^{-t/\tau} \tag{2.4}$$

where $\tau = L/R$ is the time constant.

The complete solution of the loop equation will be $i = i_1 + i_2$, that is,

$$i = \frac{V_m}{Z}\sin(\omega t + \alpha - \phi) + I_t e^{-t/\tau} \tag{2.5}$$

We still have to determine the value of I_t, which is the initial value of the transient component. The transient term is determined to satisfy the initial condition for the complete solution. If the particular integral itself satisfies the initial condition, there will be no transient term. In our case, since the initial value of the current $i = 0$,

$$I_t = -\frac{V_m}{Z}\sin(\alpha - \phi) \tag{2.6}$$

Notice that there will be no transient term if $\alpha = \phi$.

We may therefore write the complete expression for the current as

$$i = \frac{V_m}{Z}[\sin(\omega t + \alpha - \phi) - e^{-t/\tau}\sin(\alpha - \phi)] \tag{2.7}$$

It is often more convenient to express time in phase angular measure ($\theta = \omega t$), and so we may write the expression for the current as

$$i = \frac{V_m}{Z}[\sin(\theta + \alpha - \phi) - e^{-(R/X)\theta}\sin(\alpha - \phi)] \tag{2.8}$$

where $X = \omega L$. Equation (2.8) will be valid for the current as long as the thyristor switch is ON, so that the circuit configuration does not change.

The question we now pose is that of when the thyristor will commutate. Will it be at the instant of reversal of the AC input voltage, or will the conduction extend into the negative half-period of the AC input waveform.

The answer is that the thyristor will continue to conduct as long as the circuit conditions can cause a forward current through it, even after the AC input voltage has reversed in sign. It must be remembered that, besides the

AC input voltage, there is additionally the induced voltage caused by the decay of current in the inductance L. This $L\,di/dt$ voltage opposes the decay of the current, and is of the polarity to forward-bias the thyristor. As long as the forward $L\,di/dt$ voltage is greater in magnitude than the negative AC voltage, the thyristor will continue to be forward-biased, and the current flow will continue. The instant of commutation of the thyristor will be obtained by equating the current to zero in (2.8).

The instant of commutation of the thyristor is labelled as θ_1 in Fig. 2.15(b). The conduction angle in the negative half-period is labeled σ. This is the interval from the negative-going zero-crossing of the voltage waveform to the actual instant of commutation. The determination of the conduction angle in the negative half-period is illustrated by Examples 2.2 and 2.3 below—one for the case of a thyristor and the other for the case of a diode.

Output voltage and voltage across the thyristor. For this configuration, as long as the thyristor is ON, the output voltage v_{KN} is the same as the input voltage v_{AN} appearing through the thyristor switch. When the thyristor is in the OFF state, both i and di/dt are zero, and therefore the output voltage v_{KN} is zero. On this basis, the waveform of v_{KN} is as sketched in Fig. 2.15(c). The DC component of the output voltage will be given by

$$V_{DC} = \frac{1}{2\pi}\int_{\alpha}^{\pi+\sigma} V_m \sin\theta\, d\theta$$

$$= \frac{V_m}{2\pi}(\cos\alpha + \cos\sigma) \tag{2.9}$$

Case of a diode. If the switching element is a diode, the firing angle $\alpha = 0$. Therefore the DC component of the output voltage will be given by

$$V_{DC} = \frac{V_m}{2\pi}(1 + \cos\sigma) \tag{2.10}$$

Resistive load. For a resistive load, $\sigma = 0$. Therefore, when the switching element is a thyristor and the delay angle is α

$$V_{DC} = \frac{V_m}{2\pi}(1 + \cos\alpha) \tag{2.11}$$

Thyristor voltage. The waveform of the voltage v_{AK} across the thyristor is sketched in Fig. 2.15(d). This will be zero during the ON state of the device, and, for this configuration, equal to the AC input voltage when it is in the OFF state, calling on both the forward blocking and reverse blocking capabilities of the device.

Illustrative example 2.2. In the circuit of Fig. 2.15(a), the AC input is 120 V r.m.s., $R = 30\,\Omega$ and $X = 40\,\Omega$. The firing angle is 30°. Determine
(a) the conduction angle in the negative half-period;
(b) the DC output voltage.

Solution. We have $Z = \sqrt{R^2 + X^2} = 50\,\Omega$, $\phi = \tan^{-1}(40/30) = 53.1°$ and $V_m = \sqrt{2} \times 120 = 169.7$ V.
(a) Substitution of these and the given numerical data with angles converted into radians in Eqn. (2.8) and equating it to zero, the instant at which the thyristor commutates will be given by the solution of the following equation:

$$0 = 3.39[\sin(\theta_1 - 0.4032) + 0.39e^{-0.75\theta_1}] \qquad (2.12)$$

The solution can be obtained by an iterative procedure using a programable calculator. This will give

$$\theta_1 = 3.5717 \text{ rad} = 204.6°$$

The conduction angle in the negative half-period, which is labelled σ in Fig. 2.15(b) will be given by

$$\sigma = \theta_1 + 30° - 180° = 54.6°$$

(b) The DC output voltage will be given by numerical substitution in (2.9) as $V_{DC} = 39$ V.

Illustrative example 2.3. The load on the DC side of an uncontrolled single-phase half-wave rectifier consists of a resistance of $20\,\Omega$ in series with an inductance of 100 mH. The input is from a 240 V (sinusoidal 60 Hz) AC. Determine
(a) the DC component of the output voltage;
(b) the reverse voltage rating of the power diode to be used, based on a safety factor of 2.3.

Solution
(a) We have

$$X = \omega L = 2\pi \times 60 \times 0.1 = 37.7\,\Omega$$

$$Z = \sqrt{20^2 + 37.7^2} = 42.7\,\Omega$$

$$\phi = \tan^{-1}\frac{X}{R} = \tan^{-1}\frac{37.7}{20} = 62.05°$$

$$V_m = \sqrt{2} \times 240 = 339.4 \text{ V}$$

Being an uncontrolled rectifier, the switching element is a diode, and so the firing angle is zero. To obtain the instant θ_1 at which the diode commutates, we therefore put $i = 0$ to equate the expression on the right-hand side of (2.8) to zero. This procedure gives

$$0 = \frac{V_m}{Z}[\sin(\theta_1 - \phi) + e^{-(R/X)\theta_1} \sin \phi]$$

Numerical substitution for the quantities with angles in radians in the above equation leads to

$$0 = 7.95[\sin(\theta_1 - 1.083) + 0.8834e^{-0.5305\theta_1}]$$

Solving the above using a programmable calculator gives

$$\theta_1 = 4.314 \text{ rad} = 247.2°$$

The conduction angle in the negative half-cycle for this instant of commutation will be

$$\sigma = \theta_1 - 180° = 67.2$$

The DC component of the output voltage for this conduction angle in the negative half-cycle will be given by (2.10). Numerical substitution in this equation gives

$$V_{DC} = 75 \text{ V}$$

(b) Referring to Fig. 2.15(d) and treating the firing angle as zero, we get the theoretical peak reverse voltage on the diode as is $\sqrt{2} \times V = 339.4$ V. Using a safety factor of 2.3, the diode voltage rating should not be less than $2.3 \times 339.9 = 780.6$ V. This exact voltage may not be listed in a manufacturer's catalog, and the practical diode chosen may have to have a rating of 800 V to satisfy the stated safety factor.

2.2.3 Use of a "Freewheeling Diode"

For the case of the inductive load, which we studied in the previous section, we found that the current flow continues into the negative half-cycle of the AC voltage, because of the stored energy in the inductance on the DC side. As a consequence, the output voltage waveform on the DC side has negative excursions in each switching period. These lower the output DC voltage for the same firing angle from what it would be for a resistive load. Also, they increase the peak-to-peak value of the voltage "ripple" on the DC side, resulting in a "less smooth" DC voltage output. During these reverse voltage conduction intervals, the current on the AC side is still positive, which means that power is actually flowing from the output side to the input side. Some of the stored energy in the inductor on the DC side is actually being fed back to the AC bus. We can overcome these limitations by using an additional power diode, connected in reverse, across the output terminals K and N, as shown in Fig.

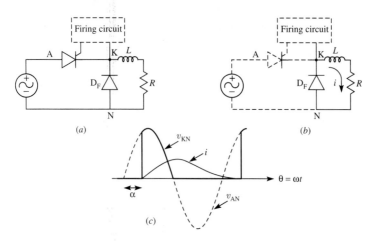

FIGURE 2.16
Half-wave rectifier with freewheeling diode.

2.16(a). At the instant when v_{KN} tends to reverse, this diode becomes forward-biased and turns ON. As a result, the thyristor becomes reverse-biased and turns OFF. The commutation results in the transfer of current from the thyristor to the diode. The circuit configuration after the commutation is shown in Fig. 2.14(b), where the conducting loop is shown by continuous lines and the nonconducting part of the circuit by broken lines. We notice that the current continues to flow in the load after the thyristor has turned OFF, solely because of the stored energy in the inductor, in much the same way as a bicycle "freewheels" due to stored kinetic energy after the rider has stopped pedalling. The diode, labeled D_F, that opens up the freewheeling path in Fig. 2.16 is called the freewheeling diode.

The waveforms of the input voltage v_{AN} and the output terminal voltage v_{KN} are shown respectively by broken and full lines in Fig. 2.16(c) for a firing angle α. The output voltage waveform will be the same as for the case of a resistive load. Its DC component will therefore be given by the same expression as (2.11).

The load current waveform is also shown in Fig. 2.16(c). The load current is the same as the input AC side current while the thyristor is ON. It is the same as the freewheeling diode current during freewheeling, when the current decays exponentially with a time constant $\tau = L/R$, starting with the initial value at the instant of commutation.

The freewheeling diode can be added across the DC output terminals in any of the rectifier configurations that we shall study later in this chapter, and will prevent negative excursions of the output terminal voltage of the rectifier. The half-wave configuration shows its effects in the most conspicuous manner. Example 2.4 below illustrates certain aspects of the working of a freewheeling diode in a phase controlled rectifier. It also shows how the currents during the initial switching cycles of a static converter can be different from what they will settle down to be after several cycles of switching, when "repetitive" conditions can be assumed to prevail.

Illustrative example 2.4. In the half-wave rectifier circuit with freewheeling diode of Fig. 2.16, the input AC voltage is 120 V at 60 Hz. The firing angle is 30°. $R = 5\,\Omega$ and $L = 0.1$ H. Determine
(a) the DC output voltage;
(b) the peak value of current in the freewheeling diode in the first cycle of switching;
(c) repetitive peak freewheeling diode current.

Solution. We have

$$V_m = \sqrt{2} \times 120 = 169.7 \text{ V}$$

$$X = 2\pi \times 60 \times 0.1 = 37.70 \,\Omega$$

$$Z = \sqrt{5^2 + 37.7^2} = 38.03 \,\Omega$$

$$\phi = \tan^{-1} \frac{37.1}{5} = 82.4°$$

$$\alpha = 30°$$

(a) The DC output voltage is given by (2.10). For a firing angle $\alpha = 30°$, this gives

$$V_{DC} = \frac{169.7}{2\pi} (1 + \cos 30°) = 50.4 \text{ V}$$

(b) In the first cycle of switching, the initial value of the current is zero. Therefore the initial condition used for deriving (2.8), which is recalled below, is satisfied and the current will be given by it:

$$i = \frac{V_m}{Z} [\sin(\theta + \alpha - \phi) - e^{-(R/X)\theta} \sin(\alpha - \phi)]$$

Numerical substitution in the above equation gives the following expression for the thyristor current

$$i = 4.46 \sin(\theta - 52.4°) + 3.53e^{-0.1326\theta}$$

The first commutation occurs at $\theta = 150° = 2.618$ rad. Substitution into the above equation gives

$$i = 6.92 \text{ A}$$

Commutation of the thyristor results in the transfer of this current to the freewheeling diode. Therefore the peak current in the freewheeling diode in the first switching cycle is

$$I_F = 6.92 \text{ A}$$

(c) During freewheeling, this current decays exponentially according to the following expression:

$$i = 6.92e^{-0.1326\theta}$$

θ being measured from the instant of commutation. Freewheeling continues until the thyristor is gated again, that is, until

$$\theta = 180° + 30° = 210° = 3.665 \text{ rad}$$

Therefore, when the thyristor is gated in the second switching cycle, the initial current will have a nonzero value given by

$$i = 6.92e^{-0.1326 \times 3.665}$$

$$= 4.26 \text{ A}$$

Therefore the initial current in the second cycle will be higher, and the second cycle will not be a repetition of the first. Similarly, there will be a further increase in the initial current in the third switching cycle, and so on. However, the increases between successive cycles will become less and less as the switching progresses. After several cycles, the increases between successive cycles will become negligible, and we may assume that the circuit has reached the repetitive state. To determine this repetitive initial current, we shall first represent it by the symbol I_0. Therefore the initial condition for $t = 0$ in (2.5) for determining I_t has to be modified as

$$I_0 = \frac{V_m}{Z} \sin (\alpha - \phi) + I_t$$

This will give I_t as:

$$I_t = I_0 + 3.53 \text{ A}$$

The new expression for i according to (2.5) for the current through the thyristor will now be

$$i = \frac{V_m}{Z} \sin (\theta_i + \alpha - \phi) + (I_0 + 3.53)e^{-(R/X)\theta}$$

Numerical substitution in this will give the following expression for the current through the thyristor:

$$i = 4.46 \sin (\theta - 52.4°) + (I_0 + 3.53)e^{-0.1326\theta}$$

Commutation of the thyristor will take place at:

$$\theta = 180° - 30° = 150° = 2.618 \text{ rad}$$

For this value of θ, we get the current at the instant of commutation as

$$I_F = 6.92 + 0.7067I_0$$

This current gets transferred to the freewheeling diode, and is the peak value of the freewheeling current. Its value is dependent on I_0, which remains to be determined.

During freewheeling, the current will be given by

$$i = I_F e^{-0.1326\theta} = (6.92 + 0.7067I_0)e^{-0.1326\theta}$$

In this equation, θ is measured from the instant of commutation. Freewheeling ends when the thyristor is fired again at

$$\theta = 180° + 30° = 210° = 3.665 \text{ rad}$$

Using this value of θ in the above expression, we get the current at the end of freewheeling as

$$I = 0.4347I_0 + 4.26 \text{ A}$$

At the end of freewheeling, this current gets transferred to the thyristor and becomes the initial value of the thyristor current. Since we have assumed that the prevailing conditions are repetitive, this should be the same as I_0. Therefore, because of the assumption of repetitive conditions,

$$I_0 = 0.4347I_0 + 4.26 \text{ A}$$

Solution of this equation gives

$$I_0 = 7.54 \text{ A}$$

The current I_F at the commencement of freewheeling will be obtained by substituting this value of I_0 into the previous equation for I_F. This gives the peak value of the freewheeling current under repetitive conditions as

$$I_F = 6.92 + 0.7067 \times 7.54 = 12.25 \text{ A}$$

2.3 THE MIDPOINT CONFIGURATION

The midpoint configuration has several AC phase voltage sources with respect to a common terminal, which is referred to as the midpoint. All the voltage sources will be identical in magnitude, but will differ in phase successively by the same angle. Figure 2.17 shows the phasor representation of the voltages in a general m-phase midpoint circuit. The voltage phasors are labeled $v_{1N}, \ldots, v_{mN}$. The magnitudes are equal and the phase difference between successive voltages is $2\pi/m$ rad.

In a practical midpoint converter, the m phase voltages are obtained from the secondary windings of a transformer. Figure 2.18(a), (b) and (c) show

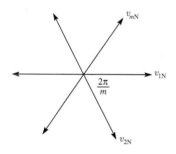

FIGURE 2.17
Voltage phasors in an m-phase midpoint rectifier.

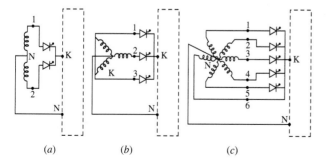

(a)   (b)   (c)

FIGURE 2.18
Midpoint rectifier circuits for $m = 2$, 3 and 6.

examples of midpoint rectifier circuits for $m = 2$, 3 and 6 respectively. For Fig. 2.18(a), $m = 2$, and we shall call this a biphase circuit. Here the midpoint is the center-tap of the secondary winding of a single-phase transformer. Figure 2.18(b) is a three-phase midpoint rectifier. The three phase voltages are provided by secondary windings of a three-phase transformer. The midpoint is the neutral terminal labeled N, obtained by Y-connecting these secondaries. In Fig. 2.18(c), $m = 6$. The six secondary phases are obtained from a three-phase transformer by having center-taps for each secondary. The three center-taps are tied together at N to give us the midpoint.

Each phase voltage terminal is connected to a static switching element, which may be a diode or thyristor. Thyristors are used in controlled rectifiers, and are shown in all the circuits of Fig. 2.18. Figure 2.18 shows the common cathode circuit. In this circuit scheme, each phase voltage connection to the thyristor is to its anode. The cathodes of all the devices are connected together to give us the common cathode terminal labeled K in the figure. The DC output terminals of the rectifier are the common cathode terminal K and the midpoint N. K has the positive voltage polarity and N the negative when the circuit works as a rectifier. In a common-anode configuration, the phase leads are connected to the cathodes, and the anodes constitute the common terminal. The common-anode midpoint circuit will have reversed voltage polarities and current directions on the DC side as compared with the common-cathode connection. Since the common-anode converter and the common-cathode converter function in the same manner, except for the reversed voltage polarities and current directions, we shall do the analysis only for one configuration, and for this we shall arbitrarily choose the common-cathode converter.

2.3.1 The Conduction Sequence

We shall consider the general case of an m-phase converter, which is illustrated for $m = 3$ in Fig. 2.19(a). Let us first assume that the firing delay angle is zero

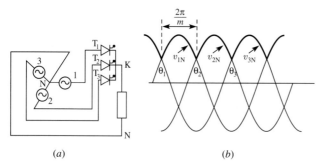

(a)   (b)

FIGURE 2.19
m-phase midpoint rectifier circuit and waveforms.

for every thyristor. This is equivalent to stating that the circuit operates as if all the switching elements are diodes. The waveforms of the phase voltages v_{1N}, v_{2N} and v_{3N} are sketched in Fig. 2.19(b). The instants at which these waveforms intersect on the positive side are labeled θ_1, θ_2 and θ_3 in the figure. In the interval θ_1 to θ_2, the phase-1 voltage v_{1N} has a higher positive magnitude than all the other phase voltages. Therefore, if thyristor T_1 is made to conduct in this interval, all the other thyristors will remain reverse-biased. This will be evident if we look at the resultant voltage polarity in a loop consisting of v_{1N} and any other phase. This situation will continue as thyristor T_1 continues to conduct, until the instant θ_2. After the instant θ_2, the phase-2 voltage v_{2N} becomes more positive than v_{1N}. Therefore, if we again examine the loop consisting of phases 1 and 2, we shall notice that thyristor T_2 is forward-biased. The instant θ_2 at which the waveforms of v_{1N} and v_{2N} intersect and after which $v_{2N} - v_{1N}$ becomes positive is that instant at which thyristor T_2 begins to get forward-biased. If we had used a diode in place of the thyristor T_2, it would automatically have turned ON at this instant. This therefore is the instant from which the delay angle α of the turn ON of the thyristor T_2 is to be measured. Our assumption that α is zero for all thyristors implies that T_2 receives its firing pulse at this instant and turns ON. By so doing, it automatically turns OFF T_1 by line commutation. The same sequence is repeated at the instant θ_3, when T_3 is turned ON in a similar manner, thereby commutating T_2.

We may summarize our observation as follows. Each thyristor, or diode as the case may be, conducts for an interval equal to $2\pi/m$ and is automatically turned OFF by line commutation at the end of this interval, when an incoming device turns ON. When a thyristor or diode is ON, the voltage v_{KN} at the output terminals, is the same as the corresponding AC phase voltage, because the device only serves to connect the phase voltage terminal to the output terminal K. Therefore the waveform of the output voltage will consist of segments of the AC voltage waveforms. On the basis of our assumed condition of zero firing delay, the output voltage waveform v_{KN} will consist of the peaks of the phase voltages, which are shown by the thick lines in Fig. 2.19(b). For the three-phase case for which we have drawn Fig. 2.19, there will be a succession of three pulses, each of 120° duration, in one period of the AC input. This is therefore called a three-pulse rectifier circuit. The general m-phase midpoint rectifier will be an m-pulse converter. If the switching elements are thyristors, and a firing delay has the same value α for all of them, the segments of the phase voltage waveforms from which the output waveform is built up will be delayed by a phase angle α, but we shall still get an m-pulse waveform, which will be similar to what is illustrated in Fig. 2.20(b) for $m = 3$.

Illustrative example 2.5. Taking the instant of the positive-going zero-crossing of the phase voltage waveform as the reference, express, in angular measure, the following for (a) biphase, (b) three-phase, (c) six-phase and (d) the general m-phase midpoint converters:

(1) the instant of reference from which the firing delay commences for the thyristor;
(2) the instant of commencement of conduction for a firing delay $\alpha = 20°$;
(3) the conduction interval of one thyristor.

Solution
(a) In the biphase circuit, the phase voltages differ by 180°, and therefore their waveforms intersect at zero-crossing.

    (1) Therefore the instant of reference for the firing delay angle is 0° with respect to the positive-going zero-crossing of the phase voltage waveform.

    (2) For a firing delay of 20°, the conduction will start at $\theta = 20°$ after the positive-going zero-crossing.

    (3) The duration of conduction for each thyristor will be 180° on the assumption that there is sufficient inductance in the DC circuit to force the current through the thyristor even after the phase voltage has crossed past the negative going zero-crossing.

(b) In the three-phase circuit, there are three phase voltages, and the instant of intersection on the positive side of successive phase voltages occurs at $\theta = 30°$ from the positive-going zero-crossing.

    (1) Therefore the instant of reference for firing angle delay is 30° from the positive-going zero-crossing.

    (2) For $\alpha = 20°$, conduction will start at $\theta = 30 + 20 = 50°$ from the positive-going zero-crossing.

    (3) The conduction interval for a thyristor will be $360/3 = 120°$.

(c) In the six-phase circuit, there will be six phase voltage waveforms, with a successive phase difference of 60°. The successive waveforms will intersect on the positive side at $\theta = 60°$ after the positive-going zero-crossing.

    (1) Therefore the instant of reference for firing delay will be $\theta = 60°$ from the positive-going zero-crossing.

    (2) For $\alpha = 20°$, conduction will start at $\theta = 60 + 20 = 80°$ from the positive-going zero-crossing.

    (3) The conduction interval will be $360/6 = 60°$.

(d) In the general m-phase circuit, there will be m phase voltage waveforms, with a successive phase difference of $(360/m)°$. The instants of intersection of successive phase voltage waveforms on the positive side will be at $(90 - 180/m)°$ with respect to the positive-going zero-crossing.

    (1) Therefore the instant of reference for firing delay will be at $\theta = (90 - 180/m)°$.

    (2) For $\alpha = 20°$, conduction will start at $\theta = [(90 - 180/m) + 20]°$ from the instant of positive-going zero-crossing.

    (3) The conduction interval will be $(360/m)°$.

2.3.2 Inductive Smoothing of DC

Figure 2.20(a) shows a midpoint rectifier feeding power into a DC load. The waveform of the voltage at the output terminals K–N of the rectifier will be made up of segments of the AC phase voltage waveforms as shown in Fig.

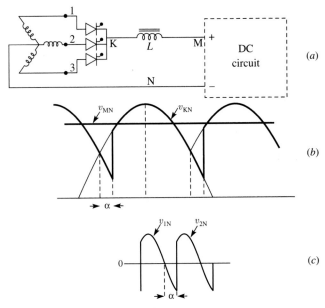

FIGURE 2.20
Inductive smoothing of DC.

2.20(b). This waveform has an AC "ripple voltage" component in addition to the DC component. At the terminals of the DC load circuit, which are labeled M and N in the figure, we like to have a nearly perfect DC in which the ripple component is negligible. The means employed for this purpose in high power applications is to use an inductor in series between the rectifier output terminals and the DC circuit, as shown by L in (a). Instead of using an inductor as a series smoothing filter element to "filter-off" the ripple, a capacitor could also be used in parallel, connected across the output terminals K–N. But the capacitor is not favored as the filter element for high power applications. The capacitor filtering takes place in a manner different from inductive filtering. The capacitor draws a charging current of large amplitude, gets charged to the peak value of the phase voltage segment that appears across the output terminals, and delivers this charge as a DC current into the load. It can be shown that, with capacitor filtering, the current on the AC side will consist of charging current pulses of large amplitude and short duration. Such currents are not permissible in a high power AC network because of the disturbances to the system bus voltages that they will create. But at low voltage and low power levels, capacitor filtering is more convenient. In the power supply unit of low power electronic equipment such as personal computers, capacitors are invariably used as the filter elements. We shall present such power supplies in Chapter 6. For large power applications, which are our concern in the present chapter, we shall always assume the use of a series inductor as the filtering element. This will be in conformity with actual industrial practice. Unless otherwise stated in a particular context, we shall assume that this inductor is large enough to make the ripple voltage at the terminals M–N of the DC load negligible—ideally zero. Mathematically, this implies an infinite value for the inductance L. In the actual equipment, the inductance will usually be a relatively large and heavy item of equipment. Its winding has to be designed to carry the full load current. It invariably has a magnetic core, which has to be designed such that it does not saturate magnetically at the large DC load current.

With infinite inductive smoothing, the voltage waveforms v_{KN} and v_{MN} for a three-phase rectifier with a firing angle α are as shown in Fig. 2.20(b). If α is large, v_{KN} can also have negative excursions. In other words, the conduction in a phase can continue even after the phase voltage has turned negative. This is made possible by the induced e.m.f. $L\,di/dt$, in the inductance, when the current tends to fall. This $L\,di/dt$ e.m.f. will continue to forward-bias the thyristor by overcoming the negative polarity of the phase voltage. In this way, conduction can continue into the negative region of the segment of the phase voltage waveform, until the next thyristor in the sequence is fired. In the case of a biphase rectifier, the excursion into the negative segment of the phase voltage occurs early, even with a small value of firing delay. The v_{KN} waveform for a biphase rectifier with firing delay is sketched in Fig. 2.20(c). Negative voltage excursions can also occur for higher phase numbers, but at larger firing delays.

2.3.3 DC/AC Voltage Relationship

We shall consider the general case of an m-phase or m-pulse rectifier. The DC component of the output voltage can be determined by taking the average over one repetitive period of the waveform of the output terminal voltage v_{KN}. We shall make the following ideal assumptions.

1. The phase voltage waveforms are ideal sine waves.
2. The switching elements are ideal, with zero ON state voltage drop and zero OFF state leakage.
3. Commutation is instantaneous.
4. There is infinite inductive smoothing on the DC side.
5. There is continuous conduction over the entire interval of $2\pi/m$ for each phase, whether or not there is excursion into the negative region of the phase voltage.

The waveform of the voltage v_{KN} at the output terminals of the midpoint rectifier of Fig. 2.20(a) is sketched in Fig. 2.21. In this figure, the reference zero

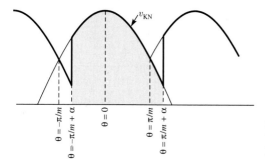

FIGURE 2.21
Voltage waveform for deriving the DC/AC voltage ratio.

used for the phase angles is the instant of positive peak of the phase voltage waveform. The phase angles are indicated in radians, assuming the phase number to be m. The DC component of the output voltage is obtained by finding the average of v_{KN} over one repetitive period of $2\pi/m$, corresponding to the conduction interval of one phase. We shall assume that the firing delay angle is α. To find the average, we shall integrate one phase voltage waveform over the repetitive interval $-\pi/m + \alpha$ to $\pi/m + \alpha$, which is the conduction interval for one phase. Taking the reference zero for phase angles as indicated in Fig. 2.21, that is the instant of positive peak, we may express the phase voltage as

$$v = V_m \cos \theta$$

The DC voltage component V_{DC} will be given by:

$$V_{DC} = \frac{1}{2\pi/m} \int_{-\pi/m+\alpha}^{\pi_m+\alpha} V_m \cos \theta \, d\theta = \frac{V_m \sin (\pi/m)}{\pi/m} \cos \alpha$$

$$= \left(\frac{\sqrt{2}\, m}{\pi} V_p \sin \frac{\pi}{m} \right) \cos \alpha$$

where V_p is the r.m.s. value of the phase voltage (voltage with respect to the midpoint). We may write this as

$$V_{DC} = V_{d0} \cos \alpha$$

where

(2.13)

$$V_{d0} = \frac{\sqrt{2}\, m}{\pi} V_p \sin \left(\frac{\pi}{m} \right)$$

Here V_{d0} is the DC voltage for zero firing delay—this will be the value of the DC voltage if the switching elements are diodes.

TABLE 2.1
DC/AC voltage ratio

Number of phases m	2	3	4	6	12	24	∞
V_{do}/V_P	0.9	1.17	1.27	1.35	1.40	1.41	1.41

The DC-to-AC voltage ratio V_{do}/V_p for different pulse numbers m, as given by (2.13), is shown in Table 2.1.

2.3.4 Ripple Voltage

The fundamental frequency f_r of the ripple component of the voltage is given by

$$f_r = mf$$

where f is the AC supply frequency. For an AC bus frequency of 60 Hz, the fundamental ripple frequency for a midpoint configuration will be 120 Hz for a biphase circuit, 180 Hz for a three-phase circuit and 360 Hz for a six-phase circuit. The size of the filter inductance is reduced as the ripple frequency increases. Therefore it is advantageous to have a high pulse number for the converter. Besides the fundamental component, the ripple voltage will have higher frequency harmonic components, which are integral multiples of the fundamental. Since the waveform of v_{KN} can be expressed analytically, it becomes a straightforward procedure to perform a Fourier analysis of the waveform, or other calculations, to evaluate the parameters relating to the ripple voltage. The following three numerical examples are intended to illustrate this.

Illustrative example 2.6. The secondary phase voltage of a three phase midpoint rectifier has an r.m.s. value of 100 V. The firing delay angle $\alpha = 20°$. Determine (1) the DC component of the output voltage and (2) the peak-to-peak voltage ripple.

Solution. Using the number for voltage ratio from Table 2.1 for $m = 3$, which is based on (2.13), we have

$$V_{DC} = 1.17 \times 100 \times \cos 20° = 109.9 \text{ V}$$

The peak magnitude of the phase voltage (at $\theta = 0$ in Fig. 2.21) $= \sqrt{2} \times 100 = 141.4$ V. Therefore the height of the positive peak above the DC level is $141.4 - 109.9 = 31.5$ V. For $\alpha = 20°$ the lowest value of v_{KN} occurs at $\theta = 60° + 20° = 80°$. This lowest value of v_{KN} is $\sqrt{2} \times 100 \times \cos 80° = 24.6$ V. Therefore the negative magnitude of the ripple voltage will be $109.9 - 24.6 = 85.3$ V. The positive amplitude of the ripple voltage (determined above) is 31.5 V. Therefore peak-to-peak ripple voltage is $31.5 - (-85.3) = 116.8$ V.

Illustrative example 2.7. For the converter considered in Example 2.5, determine the r.m.s. value of the total ripple voltage.

Solution. The r.m.s. values are related as follows:

$$V_{DC}^2 + V_{ripple}^2 = V_{total}^2$$

In the above relationship (from basic circuit theory), each voltage term is the r.m.s. value. V_{DC} is the r.m.s. value of the DC component, and it is the same as the DC value that we determined in the previous example as 109.9 V.

We shall now proceed to determine the r.m.s. value of the total voltage. For this, we first determine the "mean square" value of the total voltage v_{KN} in one repetitive period and then take its square root. The repetitive period that we choose for the integration is the interval $-\pi/m + \alpha$ to $\pi/m + \alpha$ in Fig. 2.21. In this interval,

$$v_{KN} = V_m \cos \theta$$

$$\text{mean square value of } v_{KN} = \frac{1}{2\pi/m} \int_{-\pi/m+\alpha}^{\pi/m+\alpha} V_m^2 \cos^2 \theta \, d\theta \qquad (2.14)$$

After integration and simplification, this can be written as

$$V_{total}^2 = \tfrac{1}{2} V_m^2 \left[1 + \frac{\sin(2\pi/m)}{2\pi/m} \cos 2\alpha \right]$$

By taking the square root, we get the r.m.s. value of the total voltage as

$$V_{total} = V_p \left[1 + \frac{\sin(2\pi/m)}{2\pi/m} \cos 2\alpha \right]^{1/2} \qquad (2.15)$$

where V_p is the r.m.s. value of the phase-to-neutral AC voltage. In our example, $V_p = 100$ V, $m = 3$ and $\alpha = 20°$. Substitution of these numbers into (2.15) gives

$$V_{total} = 114.7 \text{ V r.m.s.}$$

$$V_{ripple} = \sqrt{V_{total}^2 - V_{DC}^2} = \sqrt{114.7^2 - 109.9^2} = 32.8 \text{ V}$$

Therefore the r.m.s. value of the ripple voltage is 32.8 V.

Illustrative example 2.8. A three-phase midpoint rectifier employs diodes as the switching elements. The phase-to-neutral voltage on the AC side is 100 V at 60 Hz. Determine the frequency and amplitude of the fundamental frequency component of the output voltage ripple.

Solution. For this example, we shall refer again to Fig. 2.21. But this time we shall have to visualize the waveform to correspond to diode operation, that is, for $\alpha = 0$. Irrespective of whether or not $\alpha = 0$, for the three-pulse circuit, one repetitive period will be one-third of the period of the input AC. Therefore the fundamental frequency of the ripple voltage will be

$$f_r = 3 \times 60 = 180 \text{ Hz}$$

For $\alpha = 0$, the expression for v_{KN} in one repetitive period wil be

$$v_{KN} = V_m \cos\theta \quad \text{in the interval } -\tfrac{1}{3}\pi < \theta < \tfrac{1}{3}\pi$$

We shall write this in terms of β, which we shall use as the phase angle measure at the fundamental ripple frequency such that

$$\beta = 3\theta$$

On this basis, the above expression for v_{KN} becomes:

$$v_{KN} = V_m \cos\tfrac{1}{3}\beta \quad \text{for } -\pi < \beta < \pi$$

The amplitude of the fundamental frequency Fourier component will be given by (Fourier's theorem)

$$V_{r_1} = \frac{1}{\pi} \int_{-\pi}^{\pi} \cos\tfrac{1}{3}\beta \cos\beta \, d\beta$$

After integration and simplification, this will give us

$$V_{r_1} = 0.292 \, V_p$$

where $V_p = V_m/\sqrt{2}$ is the r.m.s. phase voltage on the AC side. Numerical substitution in this gives the amplitude of the fundamental component of the ripple voltage as

$$V_{r_1} = 29.2 \text{ V}$$

2.3.5   Current on the DC Side

The current on the DC output side will consist of a DC component and an AC ripple. We can use the principle of superposition to determine the components separately. For example, if the circuit on the DC side consists of a resistance plus the series smoothing inductance, the DC component will be given by the DC component of the voltage divided by the resistance. An individual frequency component of the ripple current will be the appropriate harmonic voltage magnitude of the ripple divided by the impedance of the $R–L$ circuit at that frequency. The following numerical example illustrates this.

Illustrative example 2.9. The rectifier of Example 2.8 feeds a load of resistance $6\,\Omega$. The smoothing inductance has a value of $20\,\text{mH}$. Determine
(a) the DC load current;
(b) the fundamental frequency component of the ripple current.

Solution
(a) The DC component of the output voltage is $1.17\,V_p$ (see Table 2.1) = $1.17 \times 100 = 117$ V. Therefore the DC current is $117/6 = 19.5$ A.
(b) The fundamental ripple frequency for the three-pulse midpoint circuit is $3f = 3 \times 60 = 180$ Hz. The amplitude of the ripple voltage component at this frequency, as determined in Example 2.7, is 29.2 V. The reactance X of L at this frequency is $2 \times \pi \times 180 \times 0.02 = 22.6\,\Omega$. The impedance of the circuit at 180 Hz is $\sqrt{(6^2 + 22.6^2)} = 23.4\,\Omega$. Therefore the amplitude of the 180 Hz

component of the current will be $29.2/23.4 = 1.25$ A. The r.m.s. value of this component will be $1.25/\sqrt{2} = 0.88$ A.

2.3.6 Waveforms with Large Inductive Smoothing

When the smoothing inductance has a large value—ideally infinite—the ripple component of the DC side current will be negligible—ideally zero. The ripple voltage still exists across the output terminals K and N of the converter. But at the terminals of the DC load, labeled M and N in Fig. 2.22, the ripple voltage will be zero. Ideally, the entire ripple voltage will be absorbed across the inductance L. In our subsequent treatment, unless otherwise stated in the context, we shall generally assume large inductive smoothing. The DC side current will be assumed to be ripple-free, and therefore have a constant magnitude, which we shall denote by I_{d}.

Let us now look at the flow of current in an AC phase. The current in an AC phase is the same as the current in the thyristor or diode, which is the switching element in that phase. In an m-phase midpoint rectifier, each phase conducts current for an interval of $1/m$ of a complete AC period. Therefore, on the AC side, the constant DC current will actually be made up of rectangular pulses, each of magnitude I_{d} and duration $2\pi/m$ in each individual AC phase. This situation is shown in Fig. 2.22 for $m = 3$ and $\alpha = 0$. Based on this waveform, the average, r.m.s. and peak currents in a secondary AC phase winding of the transformer, and also in the thyristor or diode used as the switching device, will be as follows:

$$I_{\text{average}} = I_{d}/m \tag{2.16}$$

$$I_{\text{rms}} = I_{d}/\sqrt{m} \tag{2.17}$$

$$I_{\text{peak}} = I_{d} \tag{2.18}$$

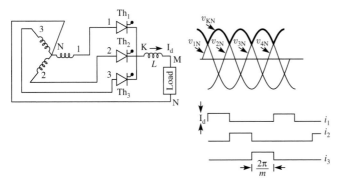

FIGURE 2.22
Waveforms of current in the thyristors and AC phases.

Illustrative example 2.10. A three-phase midpoint controlled rectifier has to be designed to provide a full load DC current of 50 A at 250 V DC. The firing delay angle under these conditions is to be 30°. Assume large inductive smoothing. Determine the following:
(a) the secondary r.m.s. phase voltage of the transformer;
(b) the secondary VA rating of the transformer;
(c) the r.m.s. average and peak currents of the thyristor.

Solution
(a) From (2.13),

$$V_{DC} = V_{d0} \cos \alpha$$

This gives $V_{d0} = 250/\cos 30° = 288.7$ V. From Table 2.1, $V_p = 288.7/1.17 = 246.8$ V. The r.m.s. voltage of the transformer secondary is therefore

$$V_p = 246.8 \text{ V}$$

(b) The r.m.s. phase current [see (2.17)] is $I_{dc}/\sqrt{3} = 50/\sqrt{3} = 28.9$ A. The volt–ampere rating of one secondary phase winding of the transformer is $246.8 \times 28.9 = 7132.5$ V A. There are three secondary phases. Therefore the total secondary VA rating of the transformer will be

$$3 \times 7132.5 = 21\,397.5 = 21.40 \text{ kVA} \quad \text{(rounded off to two decimal places)}$$

This is the total VA rating of the transformer. In conventional transformers, the primary-side VA rating and the secondary-side ratings practically have the same value, which is stated as the VA rating of the transformer. In rectifier transformers, there can be differences. We shall consider some primary side transformer connections later.
(c) The thyristor current ratings are

$$I_{average} = \tfrac{1}{3}I_d = 50/3 = 16.7 \text{ A}$$

$$I_{rms} = 28.9 \text{ A} \quad \text{(determined earlier)}$$

$$I_{peak} = I_d = 50 \text{ A}$$

2.3.7 Analytical Expression for the DC Side Current

Example 2.9 showed how the individual Fourier components of the DC side current can be determined. If, on the other hand, we wish to get an analytical expression for the current, we can obtain it using a different approach, in which we assume that repetitive conditions exist. This is shown in the following example.

Illustrative example 2.11. For a biphase midpoint rectifier, the r.m.s. phase voltage with respect to the midpoint is $200/\sqrt{2}$ V at 60 Hz. The circuit on the DC side is a resistance $R = 10\,\Omega$ in series with an inductance $L = 45.94$ mH. The firing delay angle is 45°. Obtain a mathematical expression for the DC side current for one repetitive period of the ripple.

Solution. We shall assume that the rectifier has been working for a sufficient number of switching cycles to justify the assumption that the current waveform has become repetitive. The waveforms of the output voltage and current are sketched in Fig. 2.23. Taking the reference zero of time as the instant of commencement of conduction in one phase, we can express the voltage at the output terminals for one repetitive period of the ripple as

$$v = 200 \sin \left(\omega t + \tfrac{1}{4}\pi \right)$$

The loop equation for the DC side will be

$$L \frac{di}{dt} + Ri = 200 \sin \left(\omega t + \tfrac{1}{4}\pi \right) \qquad 0 < \omega t < \pi \tag{2.19}$$

The solution will consist of a particular integral term i_1 and a complementary function term i_2. Here i_1 is the steady-state AC current given by AC circuit theory:

$$i_1 = \frac{200}{Z} \sin \left(\omega t + \tfrac{1}{4}\pi - \varphi \right)$$

where

$$Z = \sqrt{R^2 + (2\pi \times 60L)^2} = 20 \ \Omega$$

$$\varphi = \tan^{-1}(X/R) = 60°$$

Therefore

$$i_1 = 10 \sin (\theta - 15°), \quad \text{where} \quad \theta = \omega t$$

The complementary function term will be (see (2.8))

$$i_2 = I_t e^{-0.5774\theta}$$

where I_t is a transient amplitude term, to be determined using terminal conditions. The complete expression for the current becomes

$$i = i_1 + i_2$$

$$= 10 \sin (\theta - 15°) + I_t e^{-0.5774\theta} \tag{2.20}$$

I_0, the initial current at $\theta = 0$, becomes, from the above equation,

$$I_0 = I_t - 2.588 \ \text{A} \tag{2.21}$$

Since we have assumed that repetitive conditions prevail (one repetitive period

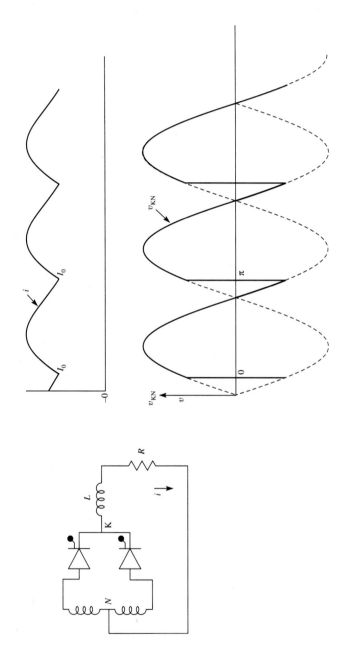

FIGURE 2.23
Circuit and waveforms in Example 2.11.

being $\theta = \pi$ in our biphase rectifier), at $\theta = \pi$ (180°), the current must again be I_0. Therefore

$$I_0 = 10 \sin 165° + (I_0 + 2.588)e^{-0.5774\pi}$$

From the above relationship, by solving for I_0, we get $I_0 = 3.596$ A. This gives, by (2.21), the value of the transient amplitude term I_t as 6.184 A. The required mathematical expression for the current will then be given by substituting this value of I_t into (2.20):

$$i = 10 \sin (\theta - 15°) + 6.184e^{-0.5774\theta} \quad \text{A} \tag{2.22}$$

2.3.8   The Discontinuous Current Flow Mode of Operation of Phase Controlled Converters

If the phase controlled rectifier is being used for a battery charging application or for the speed control of a DC motor, there will be a source of e.m.f. on the DC side. This is shown in Fig. 2.24, where the source on the DC side is labeled E. Under such a condition, the loop equation (2.19) in Example 2.10 has to be modified by adding an additional term $-E$. The new equation will be rewritten as

$$L\frac{di}{dt} + Ri = V_m \sin (\omega t + \alpha) - E \tag{2.23}$$

Based on the principle of superposition, we can state that the steady-state solution of this equation will add an additional term to the expression for the current that we obtained in (2.22). This will be a negative term equal to $-E/R$. If we take the case of a DC motor drive, the e.m.f. E, which is the induced e.m.f. in its armature when the motor rotates, will be initially zero at the start of rotation. Therefore (2.22) will be valid. The motor induced e.m.f. E is proportional to the speed of rotation, when the motor field flux is constant, as is the case for a "separately excited" motor or a motor with permanent magnet poles. Therefore, at low speeds (that is, at low value of E), the negative current term will be small. As the motor speeds up, E can become large

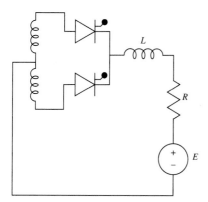

FIGURE 2.24
Midpoint converter with a source on the DC side.

enough to make the solution, including the $-E/R$ term, give us negative values at certain instants of time. Typically, this can happen when the motor runs at a high speed but does not draw a large current because the load torque is small. Now, inspection of the circuit (Fig. 2.24) shows that a negative instantaneous value is not possible for the DC side current, because the thyristors can conduct current only in one direction. What this means, in practice, is that the current flow will not be continuous over the entire allocated conduction period of a thyristor. The operation of the converter will change to the "discontinuous conduction" mode. In this mode, the current to the motor will flow in pulses that do not extend over the entire allocated conduction period of each thyristor. In the discontinuous conduction mode, the thyristor remains OFF when the current is zero. The thyristor switch has to be ON for the loop equation (2.23) to be valid. Therefore our earlier derivation of the expression (2.22) for the current will not be valid under these circumstances.

The discontinuous current mode of operation is not a desirable mode of operation for the motor, because the current flows in pulses, interrupted by zero values. It should also be noted that if the current flowing through a thyristor becomes zero, it turns OFF, and it will need a gate pulse to come ON again. Therefore if a possibility exists for the occurrence of the discontinuous current mode, it is advisable to design the gate drive circuit to maintain continuous gate drive over the entire allocated conduction period of the thyristor. The aim should be to ensure that the thyristor will be free to come ON and go OFF at the instants dictated by the circuit conditions, without being hampered by the absence of a gate drive to come ON at an instant when a positive current is possible.

2.3.9 Effect of Firing Angle on the AC Side Power Factor

We shall consider the case of an m-phase midpoint rectifier feeding a DC load with large inductive smoothing. Let us assume that the firing angle $\alpha = 0$. We have seen that the phase current waveform under these conditions ideally consists of rectangular pulses of amplitude I_d and duration $2\pi/m$. Because $\alpha = 0$, this rectangular pulse will be located symmetrically with respect to the positive peak of the phase voltage waveform, as shown in Fig. 2.22. We assume that the phase voltage waveform is sinusoidal. Fourier analysis of the rectangular current waveform will give us a component at the fundamental frequency, which is the same as the frequency of the phase voltage. Basic circuit theory tells us that, with a sinusoidal voltage, the net flow of power over a complete AC period requires a sinusoidal Fourier component of current in phase with the voltage. All other components only contribute to a pulsating flow of power that averages to zero over a complete period.

It is a straightforward procedure to determine the fundamental "in-phase" sinusoidal component of the current waveform by Fourier analysis. Under our stated assumption of zero firing delay, wherein the current pulse is symmetrically located with respect to the peak of the voltage waveform, the fundamental Fourier component of the current will be found to be in phase with the voltage. There will be no "quadrature" component. We shall leave the Fourier analysis as an exercise for the reader. We shall use a different approach to determine the fundamental "in-phase" component of the current.

Since we have assumed that the switching elements are ideal, there is no internal power loss in the power circuit of the converter. Therefore we can equate the average input power from the AC side to the DC output power. We shall denote the r.m.s. value of the fundamental component of the current, which is in phase with the voltage, as I_{p1}:

$$mV_{p1}I_{p1} = V_{d0}I_{d0}$$

This gives the following expression for I_{p1} when we use (2.13) for the voltage ratio V_{d0}/V_p:

$$I_{p1} = \frac{\sqrt{2}}{\pi} \sin\left(\frac{\pi}{m}\right) I_d \tag{2.24}$$

By definition, the volt–amperes (VA) in an AC circuit is the product of the r.m.s. value of the voltage and the r.m.s. value of the current, irrespective of the shapes of their waveforms. In our case, therefore, the phase VA is

$$\text{VA per phase} = V_p I_d / \sqrt{m}$$

By definition, the power factor in an AC circuit is the ratio of the average power to the VA. This gives the following expression for the power factor for our converter:

$$PF = \frac{V_p I_{p1}}{V_p I_d / \sqrt{m}}$$

Substituting the expression for I_{p1} from (2.24), we get

$$PF = \frac{\sqrt{2} \cdot \sqrt{m}}{\pi} \sin\left(\frac{\pi}{m}\right) \tag{2.25}$$

Distortion factor and displacement factor. The power factor values given by (2.25) for different phase numbers m are listed in Table 2.2. In a conventional AC circuit where both voltage and current are sinusoidal, the power factor has the highest value of unity when both are in phase. In our case, the voltage and the fundamental component of current are in phase, as can be verified by Fourier analysis of the current waveform. But the power factor is less than

TABLE 2.2
Distortion power factors for different phase numbers

Phase number m	2	3	6	12
Distortion power factor	0.64	0.68	0.55	0.40

unity. The reason for this is that, besides the fundamental frequency component, the current has harmonic components (and also a DC component), all of which contribute to increasing the r.m.s. magnitude of the current without contributing to resultant average power flow. Therefore the reason for the power factor of less than unity is the distortion of the current waveform from the ideal sine wave due to the presence of other components of nonfundamental frequency. We shall therefore call the power factor listed in Table 2.2 the distortion power factor.

We shall now consider what happens when a firing delay angle α is introduced. This will decrease the DC voltage by a factor equal to $\cos \alpha$. Let us assume that when we change the DC voltage by adjusting α, we also readjust the DC load circuit so that the current has the same value I_d. With these assumptions, if we now look at the AC side current waveform, we shall notice that on introducing a firing delay angle α, we are causing a shift in the waveform towards the lagging direction by exactly the same angle α. The waveshape of the current pulse remains the same, because we are keeping I_d unchanged. Therefore the fundamental and all the harmonics shift in the lagging direction by the same angle α measured in terms of the fundamental angular measure. The fundamental is the only current component that gives us resultant power on the AC side. AC circuit theory tells us that for a phase displacement of α between a sinusoidal voltage and sinusoidal current, the power factor is equal to $\cos \alpha$. So we arrive at a basic limitation of phase controlled converters, which is that we have to pay a price for the easy adjustability of output voltage by phase control. This price is the low power factor on the AC side. The reduction voltage on the DC side by a factor $\cos \alpha$ means an additional reduction in the AC side power factor by the same factor. This reduction in power factor occurs because of the displacement of the current waveform, and we shall call this the displacement power factor. The total power factor will be given by

$$\text{power factor} = \text{displacement power factor} \times \text{distortion power factor} \quad (2.26)$$

The final expression for the total power factor will therefore be, based on (2.25),

$$\text{resultant power factor} = \frac{\sqrt{2} \cdot Vm}{\pi} \sin\left(\frac{\pi}{m}\right) \cdot \cos \alpha \quad (2.27)$$

Illustrative example 2.12. In the three-phase controlled rectifier of Fig. 2.20, the secondary AC phase voltage is 200 V. The DC voltage is adjusted to 200 V by phase control. Determine the required value of the firing angle. Assume infinite inductive smoothing. Also determine the total power factor on the AC side and its break up into the displacement and distortion factors when the converter is supplying a load current.

Solution. The value of V_{d0} corresponding to zero firing delay, according to (2.13) (see also Table 2.1), will be $V_{d0} = 1.17 \times 200 = 234$ V. The firing delay angle required to lower the voltage to 200 V will be given by the relationship

$$\cos \alpha = \tfrac{200}{234} = 0.8547, \quad \text{giving } \alpha = 31.3°$$

The displacement power factor is

$$\cos \alpha = 0.8547$$

The distortion power factor for three pulse midpoint is 0.68 (see Table 2.2). The resultant power factor is $0.8547 \times 0.68 = 0.58$.

2.3.10 Voltage Ratings of the Switching Elements

The theoretical current ratings of the switching elements, be they thyristors or diodes, have been derived in (2.16)–(2.18). Voltage ratings are also needed for device selection. We can determine the theoretical waveform of the voltage across the switching element by following the procedure described below for a typical midpoint rectifier shown in Fig. 2.25(a). Figure 2.25(b) shows the individual phase voltage waveforms in broken lines during nonconducting intervals and in continuous lines during the conduction periods. The segments of the phase voltage waveforms during their conduction periods constitute the output voltage waveform v_{KN}. The voltage v_{AK} across a thyristor in any chosen phase at any instant is the difference between that phase voltage and the conducting phase voltage. Therefore, to obtain the voltage across the thyristor we subtract the value of v_{KN} from the value of the phase voltage at that instant. Therefore we can draw the waveform of v_{AK} across the thyristor in a particular phase, if we have the waveforms of that phase voltage and of v_{KN}, by taking the difference between them. Figure 2.25(c) shows the waveform of the voltage across the thyristor in phase 1 on this basis. From the statements made

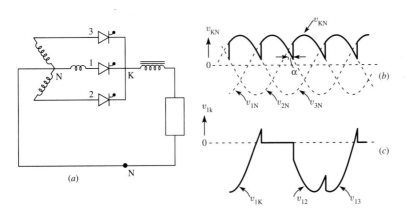

FIGURE 2.25
Voltage waveform across the switching element.

above and from the waveform of Fig. 2.25(c), we can conclude that the theoretical maximum reverse voltage that can occur across a thyristor is the maximum instantaneous magnitude of the voltage between two AC *lines*.

From Fig. 2.25(c), we see that the thyristor should also be capable of blocking forward voltages. The maximum forward voltage that has to be blocked depends on the firing delay angle α. In Fig. 2.25, this angle is small, and therefore the forward voltage blocking requirement is also small. But in the "inversion" mode of operation of the converter, α is large and greater than 90°. The thyristor will then be called upon to block large positive voltages. This is not a problem with conventional thyristors, because they have symmetrical voltage blocking capability. The asymmetrical thyristor and the asymmetrical gate turn off thyristor (GTO) have very little reverse blocking ability. They are unsuitable for rectifier duty.

The blocking voltage limit, which we determine from the waveforms as explained above, is the theoretical voltage limit based on ideal assumptions. In an actual piece of equipment, the device may be called upon to block larger than theoretical maximum values, because of the possible occurrence of transient overvoltages due to switching or disturbances in the power network. Therefore a safety factor is invariably employed in choosing the device voltage rating.

Illustrative example 2.13. An uncontrolled three-phase midpoint rectifier has to provide a DC output of 400 V. Determine the peak reverse voltage of the diodes to be used, based on a safety factor of 2.3.

Solution. The phase-to-neutral secondary phase voltage of the rectifier transformer may be determined using (2.13) or Table 2.1 for $m = 3$. This gives

$$V_p = 400/1.17 = 341.9 \text{ V}$$

The line voltage in the three-phase system for this value of the phase voltage is $\sqrt{3} \times 341.9 = 592.2$ V. These are r.m.s. values. The peak value of the line voltage is $\sqrt{2} \times 592.2 = 837.5$ V. This will be the theoretical peak reverse voltage that the diodes will be called upon to block under ideal conditions.

Based on a safety factor for voltage of 2.3, the selected diode must have a voltage rating of $2.3 \times 837.5 = 1926.3$ V. The diode to be chosen may have to have a voltage rating of 2000 V from the point of view of availability.

2.3.11 Commutation and Overlap

We had assumed till now that when an incoming thyristor is turned ON, the outgoing thyristor is instantly commutated and that the transfer of current from the outgoing to the incoming phase is instantaneous. Actually, this transfer of current needs a finite time, because each transformer phase has a

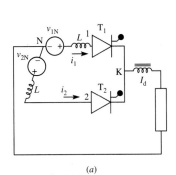

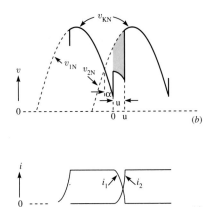

(a)

FIGURE 2.26
Overlap during line commutation.

finite leakage inductance. While this transfer of current from the outgoing to the incoming thyristor is occurring both the devices will be conducting. This interval is therefore called the overlap angle. Overlap period depends on the load current. It has a significant effect on how the terminal voltage of the converter falls as the output load current increases.

Figure 2.26(a) shows the two phases of an m-phase midpoint rectifier when they are undergoing the commutation process, which begins when thyristor T_2 in phase 2 is fired and ends when the current is completely transferred to phase 2 and thyristor T_1 turns OFF. All other thyristors are OFF during this time, and therefore the other phases are omitted in the figure for the sake of clarity.

Each phase is assumed to have a leakage inductance of L. This will be the total leakage inductance as seen from the transformer secondary. L is assumed to be the same for every phase. Figure 2.26(b) shows the relevant regions of the phase voltage waveforms v_{1N} and v_{2N}. They are assumed to be sinusoidal, with a phase displacement of $2\pi/m$ between successive phases. The resistance per phase is assumed to be negligible. A firing delay angle of α is assumed. For writing the mathematical expressions for the phase voltages, we shall take the reference zero at the instant labeled 0 in the figure, which is the instant at which T_2 is turned ON, thereby starting the overlap and commutation. We shall assume infinite inductive smoothing on the DC side, so that the DC current I_d can be considered constant.

$$v_{1N} = V_m \cos \left(\omega t + \frac{\pi}{m} + \alpha \right) \qquad (2.28)$$

$$v_{2N} = V_m \cos \left(\omega t - \frac{\pi}{m} + \alpha \right) \qquad (2.29)$$

When thyristor T_2 is turned ON at $t = 0$, the commutating loop voltage, which is $v_{2N} - v_{1N}$, will cause the current i_1 in the outgoing thyristor to decrease and the current i_2 in the incoming phase to increase. Since I_d is assumed constant,

$$i_1 + i_2 = I_d \tag{2.30}$$

and so

$$\frac{di_1}{dt} + \frac{di_2}{dt} = 0 \tag{2.31}$$

From (2.28) and (2.29), we get the commutating loop voltage as

$$v_{2N} - v_{1N} = 2V_m \sin\left(\frac{\pi}{m}\right) \sin(\omega t + \alpha) \tag{2.32}$$

The loop equation for the commutating loop when commutation is in progress will be

$$L\frac{di_2}{dt} - L\frac{di_1}{dt} = v_{2N} - v_{1N}$$

From (2.31) and (2.32), this may be written as

$$2L\frac{di_2}{dt} = 2V_m \sin\left(\frac{\pi}{m}\right) \sin(\omega t + \alpha) \tag{2.33}$$

By integration, we get the solution of (2.33) as

$$i_2 = \frac{2V_m \sin(\pi/m)}{2\omega L}[-\cos(\omega t + \alpha) + C] \tag{2.34}$$

where C is a constant of integration, which from the initial condition $i_2 = 0$ at $t = 0$ will be

$$C = \cos \alpha$$

In (2.34), the term $2V_m \sin(\pi/m)$ can be seen to be the peak value of the difference voltage that can appear in the loop if the commutation process fails and the two phases continue to remain short-circuited. Notice, that during overlap, the two phases are in fact short-circuited, because the two thyristor switches are ON. But this short-circuit is automatically cleared when the outgoing thyristor turns OFF at the end of the overlap. Also, the term $2\omega L$ is the total loop reactance that determines the short-circuit current in the loop. On this basis, the peak value of the short-circuit that will occur in the event of a commutation failure may be written as

$$I_{sp} = \frac{2V_m \sin(\pi/m)}{2\omega L} \tag{2.35}$$

We can use (2.35) to simplify the expression for the incoming phase current i_2 in (2.34) as

$$i_2 = I_{sp}[\cos \alpha - \cos(\omega t + \alpha)] \tag{2.36}$$

The outgoing phase current $i_1 = I_d - i_2$. This may be written as

$$i_1 = I_d - I_{sp}[\cos \alpha - \cos (\omega t + \alpha)] \tag{2.37}$$

Commutation is completed at the instant i_1 falls to zero. It is the same instant when i_2 rises to I_d and when T_1 can be treated as open. This instant is labeled u in Fig. 2.26(b). u is the overlap angle that is the interval from the commencement of commutation when the incoming thyristor is turned ON till the completion of it, when the current transfer to the incoming phase is completed. At $u = \omega t$, $i_1 = 0$ in (2.37) and $i_2 = I_d$ in (2.36). Both equations will yield the following relationship, which can be used for calculating the overlap angle:

$$\frac{I_d}{I_{sp}} = \cos \alpha - \cos (\alpha + u) \tag{2.38}$$

The waveforms showing the growth of the incoming current and the decay of the outgoing current according to (2.36) and (2.37) are sketched in Fig. 2.26(c).

During overlap, the expression for the output voltage v_{KN} can be written using either of the following two equations, as can be verified by reference to the circuit in Fig. 2.26(a):

$$v_{KN} = v_{1N} - L \frac{di_1}{dt} \tag{2.39}$$

$$v_{KN} = v_{2N} - L \frac{di_2}{dt} \tag{2.40}$$

By addition of these two equations and simplifying using (2.31), we get

$$v_{KN} = \tfrac{1}{2}(v_{1N} + v_{2N}) \tag{2.41}$$

We see from (2.41) that the output voltage during the overlap is the average of the conducting phases. The waveform of the output voltage based on (2.41) is also sketched in Fig. 2.26(b).

Illustrative example 2.14. For an uncontrolled three-phase midpoint rectifier, the theoretical short-circuit current that can be expected for a short circuit between the phase terminals is 1000 A r.m.s. Compute the overlap angle when the rectifier is supplying each of the following values of DC load current: (a) $I_d = 5$ A; (b) 10 A; (c) 50 A.

Solution. Since this is an uncontrolled rectifier, the firing angle $\alpha = 0$. The peak value I_{sp} of the short-circuit current to be used in (2.38) is

$$I_{sp} = \sqrt{2} \times 1000 = 1414 \text{ A}$$

The overlap angles are determined by substitution in (2.38) as follows:
(a) for $I_d = 5$ A,

$$u = \cos^{-1} \left(1 - \frac{5}{1414}\right) = 4.8°$$

(b) for $I_d = 10$ A,

$$u = \cos^{-1}\left(1 - \frac{10}{1414}\right) = 6.8°$$

(c) for $I_d = 50$ A,

$$u = \cos^{-1}\left(1 - \frac{50}{1414}\right) = 15.3°$$

2.3.12 Voltage Regulation due to Overlap

The phenomenon of overlap causes an internal voltage drop in the rectifier output DC voltage. This can be seen if we look at the waveform of the voltage v_{KN} at the DC output terminals in Fig. 2.26(a), which is plotted in Fig. 2.26(b). At no load, there is no overlap and commutation is instantaneous. Under such a condition, the waveform of v_{KN} from $t = 0$ should coincide with the phase voltage waveform v_{2N}. But, because of overlap, there is a decrease in voltage during the overlap interval. This decrease causes a reduction in the area under the v_{KN} waveform. The area lost is shown shaded in Fig. 2.26(b). Since the DC voltage is given by the average height of the v_{KN} waveform, there is a drop in DC voltage under loaded conditions. We can calculate the DC voltage regulation—by which we mean the fall in DC voltage on load—using the following procedure.

Equation (2.40) tells us that the instantaneous decrease in v_{KN} during the overlap inerval is equal to $L\, di_2/dt$. This may be expressed as $\omega L\, di_2/d\theta$, where $\theta = \omega t$. Therefore the loss of area [shown shaded in Fig. 2.26(b)] in one overlap interval will be

$$\int_{i_2=0}^{i_2=I_d} \omega L\, \frac{di_2}{d\theta}\, d\theta = \omega L I_d$$

There are m such areas lost in one AC period equal to 2π rad, because there are m commutations per cycle, m being the number of phases. Therefore the total loss of area in one AC period will be $m\omega L I_d$. The drop in DC voltage will be obtained by averaging the above loss of area over an interval of 2π. Therefore the drop in DC voltage will be given by ΔV, such that

$$\Delta V = \frac{m\omega L}{2\pi} I_d$$

$$= \frac{mX}{2\pi} I_d \tag{2.42}$$

where X is the leakage reactance per phase (in Ω). Equation (2.42) shows that the voltage drop due to overlap is proportional to the DC load current. The converter behaves as a DC voltage source with an internal resistance equal to $mX/2\pi$. The quantity $mX/2\pi$ should be treated as a fictitious resistance internally in series with the source. Overlap is actually caused by the inductive

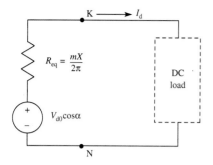

FIGURE 2.27
DC equivalent circuit showing fictitious internal resistance due to overlap.

reactance, and this fictitious resistance is an equivalent value that will cause the same voltage drop. We can construct a DC equivalent circuit using this fictitious resistance to make voltage drop calculations on load. Figure 2.27 shows this equivalent circuit.

In addition to the drop in DC voltage due to overlap represented by the fictitious resistance, there will be an additional voltage drop due to the real resistance of the transformer windings. A third cause of internal voltage drop will be the forward drop of the switching elements in the ON state.

Illustrative example 2.15. A three-phase diode rectifier has a no load DC voltage of 250 V. The leakage reactance per phase as seen from the secondary of the transformer is 0.4 Ω. Ignore all other causes of DC voltage drop.
(a) Determine the DC terminal voltage of the rectifier when it supplies 50 A.
(b) If the switching elements were thyristors and we had brought the no-load DC voltage to 100 V by phase control, what will be the DC terminal voltage when supplying the same DC load current of 50 A.

Solution
(a) The fictitious equivalent resistance in the DC equivalent circuit is $3 \times 0.4/2\pi = 0.19\,\Omega$. The resulting voltage drop for 50 A is $50 \times 0.19 = 9.5$ V. The converter terminal voltage on load is $250 - 9.5 = 240.5$ V.
(b) The equivalent resistance value is not dependent on the firing angle, and therefore will be the same as in (a). The voltage drop for the same DC current will also be the same. Therefore, for this load current, the terminal voltage will be $V_d = 100 - 9.5 = 90.5$ V.

2.3.13 Transformer Connections for Midpoint Converters

In a practical midpoint converter, a transformer is needed to provide the midpoint, which is one of the two DC output terminals, and the m individual phase voltages, which have the midpoint as the common terminal. The transformer has to be rated for the full power of the converter. It therefore becomes a major part of the equipment in terms of weight, dimensions and cost. There are some differences between the operation of a converter transformer for a midpoint converter and a conventional power transformer for transforming AC power.

In a midpoint converter, each phase current has a DC component, which has to flow through the windings that constitute the secondary. But a corresponding DC component of current does not flow in a primary winding, because the transformer does not transform DC. For the satisfactory operation of the transformer, steps are necessary to ensure that the resultant DC ampere turns of the secondary cancel by themselves. While it is possible to work with some level of DC saturation, it is not a good approach.

The AC power is normally available as either single-phase or three-phase. Single-phase power will be suitable only for the biphase midpoint circuit, for which the phase number $m = 2$. A higher phase number is always desirable, because that results in less peak-to-peak ripple and higher ripple frequency, making filtering easier to obtain a smooth DC output. Using a three-phase input for the primary, it is possible to get either three or a higher number of secondary phases.

In this section, we shall describe some typical examples of transformer connections for midpoint converters, which will bring into focus the above aspects and show how they can be resolved.

2.3.13.1 THE BIPHASE MIDPOINT CONVERTER WITH SINGLE-PHASE INPUT.

Figure 2.28(a) shows this circuit. The secondary and primary coils of the transformer are shown on the magnetic core. Careful attention should be paid to the direction of the magnetomotive force (ampere-turns) impressed on the magnetic core by the coil currents. The waveform of each secondary phase current as well as of the primary current are plotted in Fig. 2.28(b), assuming unit turns ratio between the primary and each section of the secondary. Large inductive smoothing is assumed on the DC side. Overlap is neglected. Therefore the secondary phase current waveforms are rectangular pulses of 180° duration. From the polarity of the windings, it will be noticed that the ampere turns created by the two secondary sections oppose each other and therefore the mmfs due to the DC component, as well as all harmonics which are in phase with each other, cancel. There is no resultant DC saturation due to the secondary currents.

From the waveform of the primary current we notice that the r.m.s.

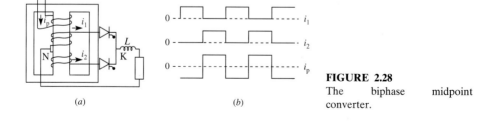

(a)   (b)

FIGURE 2.28
The biphase midpoint converter.

primary current for the assumed value of unity turns ratio between each secondary section and primary will be $I_{\text{p,rms}} = I_{\text{d}}$.

2.3.13.2 THREE-PHASE MIDPOINT CONVERTER. Figure 2.29(*a*) shows a conventional three-phase transformer being used to provide the phase voltages of a three-phase midpoint converter. This is not a satisfactory scheme. Ideally, each secondary phase current will consist of pulses of rectangular shape and amplitude I_{d} and duration 120°. These waveforms are shown by i_1, i_2 and i_3 in Fig. 2.29(*d*). Each of these currents such as i_1 has a DC component. In the case of the biphase circuit, we had another secondary coil on the same limb that canceled the DC m.m.f. of i_1. Such a situation is not possible in this transformer, because there is only one secondary winding on one limb. In a conventional transformer, the primary will be either in Y or Δ. The primary connections are not shown in Fig. 2.19(*a*). If we make it a Y connection, the following consequence should be borne in mind. If the neutral is isolated, as will be the case if the transformer is fed from a three-wire supply, it is not possible for current in an AC input line to flow in one phase only at any instant. So we have a situation that calls for current only in one phase coil on the secondary side, but more than one on the primary side, which is not conducive to satisfactory operation. This difficulty is overcome if the primary is delta connected. This may be a better choice, but we still have to live with the uncompensated DC ampere-turns on each limb of the core. For these reasons,

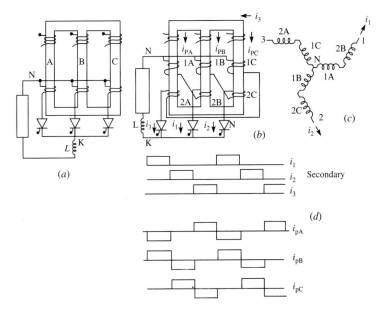

FIGURE 2.29
Transformer connection schemes for three-phase midpoint converters.

the three-phase transformer of Fig. 2.29(a) is not a recommended scheme for midpoint converters. We shall not deal with it further, but proceed to describe the alternative scheme that overcomes the above stated difficulties.

The alternative scheme is called the zig-zag connection. It is shown in Fig. 2.29(b). In this scheme, there are two identical secondary coils on each limb of the transformer. Each limb of the transformer corresponds to one phase. These phases are indicated by letters A, B and C in the figure. The two secondaries on the A-phase limb are labeled 1A and 2A. The other secondaries are labeled similarly. The coils 1A, 1B and 1C constitute one three-phase secondary group, which we shall call group 1. These are Y-connected at the neutral terminal labeled N. Each group 1 coil is connected in series to a group 2 coil with terminal polarities of the latter reversed. This can be understood better by reference to Fig. 2.29(c), where the secondary coils are all redrawn with each in alignment with the corresponding voltage phasor. For example the 1A coil is connected in series with the 2B coil in such a way as to give a resultant voltage

$$v_{1N} = v_{1A} - v_{2B}$$

By subtraction of the phasors, we find that the resultant phase voltage is

$$v_{1N} = \sqrt{3} \times \text{the voltage of one coil}$$

This arrangement gives balanced three-phase voltages on the secondary side and the required neutral terminal for the three-phase midpoint converter. The waveform of each phase current that flows through the switching element is plotted in Fig. 2.29(d). Under our assumptions of infinite inductive smoothing and negligible overlap, each is rectangular in shape and of duration 120°. We notice that each phase current has to flow through two secondary coils. For example, i_1 flows through phase winding 1A and also phase winding 2B in reverse. Careful examination of the resulting m.m.f. will show that it creates a downward m.m.f. on limb A and an upward m.m.f. on limb B. In this manner, with all the phase currents flowing, the resultant DC m.m.f. on every limb will be zero, because the number of turns are the same for each coil. This cancellation exists not only for the DC component of the phase currents, but for all of their harmonic components that have the same phase.

The three lower waveforms in Fig. 2.29(d) are those of the primary currents drawn on the basis of unit turns ratio between each secondary coil and the primary. For a turns ratio other than unity, the currents have to be scaled appropriately. These three primary phase current waveforms show that the current has no DC component, and that at every instant of time there is a return path for the current in a primary phase through another phase. There is a choice of Y or Δ available for the primary connection. The zig-zag connection overcomes both the difficulties that we pointed out for the simple Y-connected secondary.

2.3.13.3 SIX-PULSE MIDPOINT CONVERTER WITH FORK CONNECTED TRANSFORMER. The underlying concept of the zig-zag connection can be

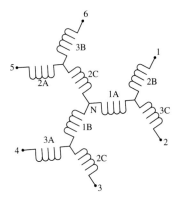

FIGURE 2.30
Six-pulse midpoint converter with fork connected transformer.

further extended to give a six-phase midpoint circuit by adding one more identical secondary coil on each limb of the transformer. There will now be nine secondary coils, and the manner of connection is shown in Fig. 2.30. Comparison of Figs. 2.29 and 2.30 will show that the latter is obtained by adding the three coils labeled 3A, 3B and 3C to the circuit of Fig. 2.29 without making any other changes.

This circuit operates in a similar manner to what we have described for the circuit of Fig. 2.29. Here the duration of each phase current will be 60°. But the coils of phase group 1 will carry currents for 120°, whereas the others will have currents only for 60°. The waveforms of the primary currents can be plotted by following the same procedure as we did for the zig-zag transformer connection. This is left as an exercise for the reader. This circuit has all the benefits that we described for the zig-zag connection and the additional advantage of a six-pulse output waveform instead of the three-pulse output of the zig-zag.

Illustrative example 2.16. A six-pulse, fork connected, midpoint thyristor rectifier has a no-load voltage of 250 V DC at a firing angle of 30°. The rated full load DC current is 60 A. Determine the total VA rating of the secondary windings of the transformer. Assume large inductive smoothing and neglect overlap.

Solution. For zero firing angle, $V_{do} = 250/\cos 30° = 288.7$ V. For $m = 6$, the phase-to-neutral AC voltage from Table 2.1, $288.7/1.35 = 213.9$ V. The voltage of each coil will be $1/\sqrt{3}$ of this, that is, $213.9/\sqrt{3} = 123.5$ V. The six coils of phase groups 2 and 3 carry currents for one-sixth of a period. Therefore the r.m.s. current in each is $60/\sqrt{6} = 24.5$ A. The three coils of phase group 1 carry current for one-third of a period. Therefore the r.m.s. current of each is $60/\sqrt{3} = 34.6$ A. Therefore the total VA rating of secondary is $6 \times 123.5 \times 24.5 + 3 \times 123.5 \times 34.6 = 30.974$ kVA.

2.3.14   Use of the Interphase Transformer

An interphase transformer, also called an interphase reactor, is used for the purpose of operating two converter groups in parallel when there is a phase shift between their waveforms. Its operating features are best illustrated by describing a circuit configuration that uses it. We shall choose the double three-phase converter with interphase transformer.

2.3.14.1 MIDPOINT DOUBLE THREE-PHASE CONVERTER CONFIGURATION WITH INTERPHASE TRANSFORMER.

In this, the transformer has two identical seocndary coils on each phase limb, as we described for the zig-zag connection, but they are connected differently. These six coils are shown on the core limbs in Fig. 2.31(a), and the coils are numbered $1, \ldots, 6$. Coils 1, 3 and 5 constitute one phase group, which we shall call group 1. These are Y-connected at the neutral terminal labeled N_1. Coils 2, 4 and 6 constitute the second three-phase group, which is also Y-connected, but at a different neutral terminal, labeled N_2. The "interphase transformer" is connected between the two phase groups, across their neutrals N_1 and N_2. It has a magnetic core similar to that of a single-phase transformer and has two identical coils connected in series to give a midpoint at the terminal labeled N. It may also be viewed as a reactor coil with a center-tap at N. The midpoint or center-tap of the IPT is the midpoint of the converter and one pole of the DC output. The voltage phasors of all the secondary coils are shown in Fig. 2.31(b), each with respect to the neutral of the group to which it belongs. Each coil has a switching element, which may be a diode or thyristor in the typical fashion of the midpoint configuration. In the figure, they are shown as diodes to make the waveforms simpler for the purpose of explaining the working principle. Each diode is labeled by the same number as the phase in which it switches.

Each phase group with its switching elements and neutral terminal constitutes a three-phase midpoint converter. Therefore the circuit configuration has two three-phase midpoint converters linked together by the interphase transformer. The basic purpose of the interphase transformer is to enable the two three-phase midpoint converters to work in parallel on the DC side, each

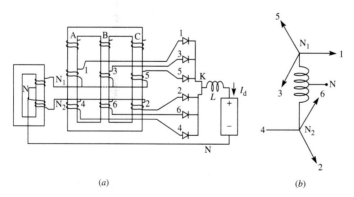

(a) (b)

FIGURE 2.31
Double three-phase midpoint configuration with interphase transformer.

sharing the DC load current equally. This happens in the manner described below.

If the interphase transformer is not present and the two neutrals are linked together directly at N then the circuit is a conventional six-phase midpoint configuration, and will function as such. Each phase will conduct for 60°. The commutations will be in the order 1 to 2, 2 to 3, and so on. Each commutation will involve a phase in phase group 1 and a phase in phase group 2. Even when the interphase transformer is present, the circuit will function in the same manner at no load or low values of DC current. In such an operation, we notice that the interphase transformer will come in series in the commutating loop at each commutation, because each commutation will be between a phase belonging to group 1 and a phase belonging to group 2. Because of the large inductance presented by the two coil sections of the interphase transformer, the overlap angle will be large for the commutations between two phases in different phase groups. When this overlap angle becomes large, there will be a change in the mode of operation, as explained below.

The waveforms of the individual phase voltages are sketched in Fig. 2.32. During the six-pulse midpoint operation at light loads, the output waveform will consist of the top positive 60° segments of these phase voltages. When overlap occurs between the phase groups, the voltage v_{KN} will be the average of the voltages of the overlapping phases, as sketched in the figure. The instant when diode D_2 comes ON in six-phase operation is labeled t_1 in the figure. Suppose the overlap of phases 2 and 1 is large and is not over at the instant labeled t_2. At t_2, v_3 begins to be more positive than v_1. Therefore, at this instant, D_3 will begin to be forward-biased and will turn ON, thereby commutating D_1. The previous overlap will continue as an overlap between 2 and 3. This overlap will again continue till the instant labeled t_3, when there will be a commutation from D_2 to D_4, and so on. The point to note is that now the commutations will be within a phase group and not between the phase groups. The interphase reactor does not come in the path of a commutation in the individual phase group, but only in intergroup commutations. The DC load current at which the commutations start to happen in the same phase group is called the transition load. Up to the transition load, the converter functions as a six-phase midpoint converter. It starts with a correspondingly large no-load

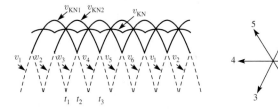

FIGURE 2.32
Waveforms with interphase transformer.

DC voltage, equal to $1.35V_p$, which, according to Table 2.1, is the no-load DC voltage for six-pulse operation. But, because of the large overlap in the intergroup commutations caused by the interphase transformer, the voltage steeply falls till the transition load. After the transition load has been crossed, each phase group works as a three-pulse midpoint converter. The DC voltage corresponding to three-phase operation will be below $1.17V_p$, as seen from Table 2.1. Subsequent overlap angles will be determined by the small leakage reactance of the phases in each group. Further voltage regulation will be much less steep as the DC load current further increases. The two phase groups will be effectively working in parallel, each supplying half the total DC load current.

2.3.14.2 VOLTAGE AND CURRENT RATINGS OF THE INTERPHASE TRANSFORMER. After the transition load has been exceeded, we shall notice that at every instant of time, there will be two conducting phases—one in each phase group. By examining the loop consisting of these conducting phases, we can see that the instantaneous voltage across the interphase transformer is the difference voltage between the conducting phases in the two phase groups. The waveform of the voltage across the interphase transformer can be plotted by plotting this difference, using the waveforms for the phase voltages given in Fig. 2.32, during the appropriate conducting intervals. This is left as an exercise for the reader. The peak value of the difference voltage during the conducting interval can be seen to be equal to

$$V_m - V_m \cos 60° = 0.5V_m$$

The wave shape will be found to be approximately triangular, with a fundamental frequency of three times the AC mains frequency. The DC current in each section of the interphase transformer will be half the total DC load current. Because of this AC voltage, which is applied to the outer terminals of the interphase transformer, there will be a magnetizing current that will flow to support this voltage. The path of this AC magnetizing current of the interphase transformer lies in the unidirectional current flow path of the phase groups. Therefore, for the unrestricted flow of the AC magnetizing current, it is essential that the DC current $\frac{1}{2}I_d$ in each section exceed the peak value of the AC magnetizing current. We can therefore look upon the DC "transition" load current as twice the peak magnetizing current of the interphase transformer. This shows that, by designing the interphase transformer with a small magnetizing current requirement, we can make the transition load smaller. The interphase transformer completely fulfills its purpose of making the two phase groups operate in parallel only after the transition load has been crossed. Therefore it is better to make the transition load small by using a larger interphase transformer. We notice that there is no net DC saturation in the magnetic core of the IPT because the DC ampere-turns of the two sections cancel.

There is no net DC saturation due to the secondary currents in the main transformer either, because there are two secondary coils on each limb, whose ampere-turns oppose each other. Primary side can be connected in Y or Δ. Although we obtain the same benefits with the zig-zag connection, there we get only three-pulse operation. With the interphase transformer, although each group operates as a three-pulse converter, the resultant output is a six-pulse waveform, which is a significant advantage from the point of view of smoothing. The interphase transformer technique is a general one, which can be used not only for the midpoint configuration but also for bridge configurations, to enable two converter groups to work in parallel, giving an output of a higher pulse number when there is a phase difference between their waveforms.

Illustrative example 2.17. A double three-phase rectifier with interphase transformer has diodes as the switching elements. The DC voltage at the transition load is 240 V. The full-load DC current is 80 A. Determine the maximum instantaneous voltage across the interphase transformer and its r.m.s. current rating.

Solution. The AC phase voltage (r.m.s.) corresponding to three-pulse midpoint operation is $240/1.17 = 205$ V. Peak value of phase voltage $= \sqrt{2} \times 205 = 289.9$ V. The peak value of the interphase transformer voltage is $\frac{1}{2} \times 289.9 = 145$ V. The r.m.s. current rating is $\frac{1}{2}I_d = 40$ A.

Illustrative example 2.18. A six-pulse diode rectifier uses the fork connection. Its no-load DC voltage is 400 V and its rated full load current 100 A. Determine the secondary VA rating of the transformer. Determine the saving, if any, in the total secondary VA rating of the main transformer if, instead of the fork connection, we use the double Y with interphase transformer.

Solution

Fork connection. The phase-to-neutral voltage on the secondary is $400/1.35 = 296.3$ V. The voltage of each secondary coil section is $296.3/\sqrt{3} = 171.1$ V. The r.m.s. current of the outer coils is $100/\sqrt{6} = 40.8$ A. The r.m.s. current of the inner coils is $100/\sqrt{3} = 57.7$ A. Therefore the total secondary VA rating is

$$6 \times 171.1 \times 40.8 + 3 \times 171.1 \times 57.7 = 71.5027 \text{ kVA}$$

Double Y with IPT. Assuming that the 400 V DC is after the transition to double three-pulse operation, the phase-to-neutral secondary voltage will be $V_p = 400/1.17 = 341.9$ V. The r.m.s. phase current is $50/\sqrt{3} = 28.9$ A. Therefore the total secondary VA rating is $6 \times 341.9 \times 28.9 = 59.2855$ kVA. There is a saving in main transformer VA of $71.5027 - 59.2855 = 12.2172$ kV A.

A note on 12-pulse operation. For achieving 12-pulse midpoint operation, we need 12 phase-to-neutral voltages with a mutual phase displacement of 30°. A technique used for this is to use two three-phase power transformers, one of which is Y connected and the other Δ-connected on the primary side. Both transformers are connected to the same three-phase bus on the primary side. With this connection, the primary-side phase voltages are shifted in phase by 30°, and so all the secondary phase voltages of one transformer will be shifted in phase by the same 30° with respect to the corresponding phases in the other. There will also be a difference in the magnitude of the primary phase voltages. But this can be taken care of by having a different turns ratio. In this way, the transformers can be designed to give identical secondary voltages.

2.4 THE BRIDGE CONFIGURATION

The bridge configuration is popular because, unlike the midpoint circuit, the bridge does not need a transformer. This results in a substantial saving in weight, size and cost of the converter equipment. A transformer if used with a bridge will be for one or both of the following purposes: (*a*) to provide electrical isolation between the AC bus and the load circuit; (*b*) to provide a voltage transformation because the available AC bus voltage is unsuitable to meet the voltage requirement of the load. The voltage level problem can in many instances be resolved by actually designing the load circuit to match the available AC bus voltage. For example, if the converter is to drive a DC motor, the motor voltage can be suitably chosen to match the DC voltage that can be provided by the bridge from the available AC bus voltage.

Figure 2.33 shows the bridge configuration for single-phase and three-phase converters.

The bridge has two circuit "sections." Half of the total number of switching elements constitute the common cathode section, which is on the top in each bridge in Fig. 2.33. The cathodes of all these elements are tied together at the common cathode terminal labeled K. The other, the common anode section, is shown at the bottom, where all the anodes are tied together at the common anode terminal labeled A. Each "leg" of the bridge, shown vertically, has one device belonging to the common anode section at the bottom and another device belonging to the common cathode section at the top. A bridge will have as many legs as there are input AC lines, each of which is connected to a leg. In the single-phase and three-phase examples shown, there are respectively two and three identical legs. Bridges with more number of phases can be built by adding additional legs.

The DC terminals of the bridge are K and A. The common cathode K is the positive voltage terminal in the rectification mode and the negative in the

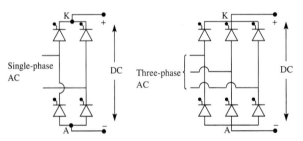

FIGURE 2.33
The bridge configuration.

inversion mode. It is always the positive current terminal when the positive current direction is defined as away from the terminal into the external circuit.

2.4.1 Basic Building Blocks—The Two Midpoint Circuit Sections

We generally used the common-cathode circuit in our description of the midpoint converter in the previous section. A common-anode midpoint circuit functions in the same manner except for reversed directions of currents and reversed polarities of voltages. If we split the bridge into its common cathode and common anode sections, we shall be able to use the same approach that we used in our study of the midpoint circuit and apply the same results. Figure 2.34 has been drawn to describe the manner in which the two midpoint configurations form parts of a three-phase bridge circuit.

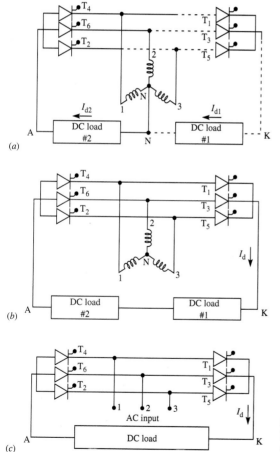

FIGURE 2.34

Synthesis of the bridge from the constituent midpoint sections.

The power supply to the bridge is shown as the Y-connected secondary windings of a three phase transformer. This is for the purpose of providing the neutral terminal for the midpoint circuits. The actual bridge does not require the transformer, and we shall eliminate the transformer and the neutral terminal when we complete our description. The bridge requires only the three line voltages of the three-phase system. If these are from a Δ-connected system, there will be no physical neutral terminal. But we can always replace such a system by an equivalent Y-connected source, in which case the neutral terminal becomes a hypothetical one to be used as reference terminal for expressing voltages.

In Fig. 2.34(a) the common cathode section is identified by broken lines. The neutral terminal N, whether real or hypothetical, serves as one DC pole for both the common cathode and the common anode sections, the negative terminal for the common cathode midpoint section, and the positive terminal for the common anode midpoint section. The two midpoint sections shown in (a) can work independently. The DC voltages of each section will depend on the firing angle α for that section. These need not be the same for both. Two separate loads are shown for the sections labeled #1 and #2. Each load is independently adjustable, and the load currents of the two sections can be different. If they are different, the difference current will flow through the link labeled NN in the figure.

In Fig. 2.34(b), this link has been removed. Thereby, the DC load current is forced to be the same in both sections. At some location in the load between the outer terminals K and A of the load, there will be a point at which the potential is the same as the transformer neutral. This will be the point that divides the load into its sections belonging to each midpoint section.

With the neutral connection removed, we can in fact treat both the loads as constituting a single load circuit. This has been done in the circuit of Fig. 2.34(c), where we have also eliminated the transformer because we no longer need the neutral terminal for any external connection. N can remain as a hypothetical point for use as a reference point for expressing voltages for the purpose of analysis. The bridge is in its final form in (c).

2.4.2 Voltage and Current Relationships

The total DC voltage at the DC terminals of the bridge can be seen from Fig. 2.34 to be the sum of the DC voltages of the two individual midpoint sections of the bridge. These will be determined by the firing angles α_k and α_a of the two midpoint sections. The resultant DC voltage will be given by

$$V_d = V_{d0} \cos \alpha_k + V_{d0} \cos \alpha_a$$

There are typically two ways in which the firing angles α_k and α_a are adjusted to implement phase control.

They may be both simulatnaeously adjusted to be always equal. This may be called simultaneous control or symmetrical control. Alternatively, only the firing angle of one section is adjusted at a time, to give one range of DC voltage, and then the other is adjusted to give a different range. This scheme of phase control may be described as asymmetrical control or sequential control. In our subsequent treatment, unless otherwise stated, we shall generally assume symmetrical control, implying that α is the same for both sections. In such a case, the total DC voltage will be twice the DC voltage of one section. For an m-phase bridge converter, the DC voltage is given by

$$V_d = 2V_{d0} \cos \alpha$$

where V_{d0}, the voltage from one section for zero firing delay is given by (2.13). Therefore, for the bridge,

$$V_{d0} = \frac{2\sqrt{2}m}{\pi} V_p \sin\left(\frac{\pi}{m}\right)$$

For the particular cases of the single-phase bridge, this will give the following relationships expressed in terms of "line" voltages V_L on the AC side: for single-phase ($m = 2$, $V_L = 2V_p$),

$$V_{d0} = 0.9V_L \tag{2.43}$$

while for three-phase ($m = 3$, $V_L = \sqrt{3}\,V_p$),

$$V_{d0} = 1.35V_L \tag{2.44}$$

In (2.43) and (2.44), V_{d0} is the DC voltage of the bridge for zero firing delay, which is twice that of one section given by (2.13).

Illustrative example 2.19
(a) A 3-phase bridge rectifier is required for a DC motor drive. The AC supply available is at a line voltage of 440 V. What voltage rating will you choose for the DC motor so that the bridge can be used without the need for a transformer.
(b) Repeat (a) if the available supply is single-phase AC at 240 V and a single-phase bridge is to be used.

Solution
(a) For three-phase input, $V_d = 1.35V_L = 1.35 \times 440 = 594$ V. No transformer will be needed if a motor rated 594 V or less is chosen. The bridge can provide 594 V maximum and lower voltages by phase control, if needed for speed control.
(b) For single phase operation, the corresponding voltage will be $0.9 \times 240 = 216$ V.

Bridge current relationship. The current flowing in an AC line will consist of the current supplied to the common cathode section and that to the common

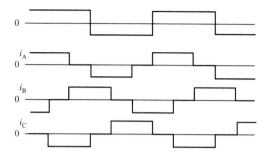

FIGURE 2.35
Line current waveforms for the single-phase and three-phase bridges.

anode section. One will be positive and the other negative. The waveforms of the line current for the single-phase and three-phase bridges are sketched in Fig. 2.35, assuming large inductive smoothing and neglecting overlap. Notice that the current has no DC components. The waveforms give the following values for the r.m.s. line current: for single-phase,

$$I_{rms} = I_d \tag{2.45}$$

while for three phase

$$I_{rms} = \sqrt{\tfrac{2}{3}}\, I_d = 0.82 I_d \tag{2.46}$$

2.4.3   Bridge Voltage Waveforms

The output voltage waveform of the single-phase bridge is similar to that of the biphase midpoint circuit with a pulse number of two. However, the three-phase bridge with symmetrical phase control gives an output with a pulse number six, which is twice that of the three-phase midpoint. This is a significant advantage of the three-phase bridge. In Fig. 2.36(b) and (c), we have plotted the voltage waveforms for the bridge circuit shown in (a). The steps involved in plotting the waveform are explained below.

The three-phase AC voltage bus may be replaced by an equivalent set of Y-connected sources with respect to a neutral terminal N, which may be real or hypothetical. These phase voltages with respect to this neutral are labeled v_{aN}, v_{bN} and v_{cN}, and are plotted in the figure as three sine waves with a mutual phase displacement of 120°. A finite value of the firing delay angle α approximately equal to 18° is assumed, this being the same for both sections. Overlap is neglected for the sake of simplicity. On this basis, the output voltage v_{KN} of the common cathode midpoint section has been plotted in the same way as earlier explained for the midpoint configuration. This is shown by the continuous lines on the top segments of the phase voltage waveforms. The output voltage v_{AN} of the common anode midpoint section is plotted by following the same procedure on the negative segments of the phase voltage waveforms. Having obtained the individual waveforms of the two consistuent

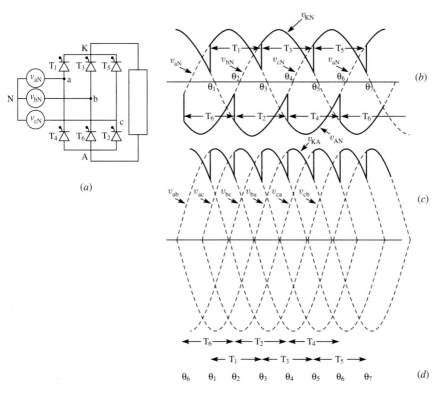

FIGURE 2.36
Three-phase bridge voltage waveforms.

midpoint sections, the bridge output waveform v_{KA} will be given by the following relationship:

$$v_{KA} = v_{KN} - v_{AN} \qquad (2.47)$$

The final waveform of the bridge output voltage plotted in Fig. 2.36(c) should conform to the above relationship. But it is really not necessary to measure the section voltages and do the subtraction shown by (2.47) every time, if we follow the procedure explained below.

The waveforms v_{KN} and v_{AN} of the two midpoint sections will give us the conduction periods of each individual thyristor in the bridge. The timings of these thyristors so obtained are shown in d) using the thyristor labels given in the bridge circuit (a) The subscript labels of the six thyristors indicate the sequence in which they conduct. From this, it will be noticed that at every instant of time, there are two conducting devices—one in the common cathode section and the other in the common anode section. What this pair does is to connect two AC lines and so cause the voltage across those lines to appear across the output terminals K and A of the bridge. There are six commutations at 60° intervals in one AC cycle—three in each section. At each commutation, a new thyristor replaces one in the conducting pair. In this manner, 60° segments of different line voltage waveforms are made to appear at the output terminals. These constitute the output voltage waveform of the bridge. When we consider the voltage of each line with respect to the other two, there are six line voltages, which may be listed as v_{ab}, v_{ac}, v_{bc}, v_{ba}, v_{ca} and v_{cb}. We can draw six sine waves, of amplitude equal to that of the line voltage, at 60° intervals and take 60° segments from these waveforms to give us the output voltage, if we can identify the conducting thyristors in the intervals. The final output waveform shown in (c) has been plotted on this basis. The appropriate line voltages are identified in this figure. The nonconducting intervals are shown by broken lines, and the 60° conduction v_{KA} segments by continuous lines.

The six-pulse output waveform with only three-phase input is a consequence of the fact that the peaks of the phase voltage waveforms do not occur at the same instants on the positive and negative sides of the phase voltage waveforms. They are displaced by exactly 60°. This is a benefit we get because the number of AC input phases is the odd number three.

In view of what was stated above and also for better insight into the working of the popular three phase bridge, it will be instructive if we take a closer look at the commutation sequence and the conduction intervals of the thyristors in the three-phase bridge of Fig. 2.36(a).

The conduction intervals in angular measure are displayed in Fig. 2.37. The reader may verify this by referring to Fig. 2.36. The instants of commutation are labeled $\theta_1, \ldots, \theta_6$ in Fig. 2.36(b) and also in Fig. 2.37. The sequence of commutation and the intervals of conduction for each device will be as listed below, which also the reader may verify by reference to Figs. 2.36 and 2.37:

θ_1 T$_1$ is turned ON, causing the commutation of T$_5$ in the common cathode section

$\theta_1 - \theta_2$ T$_1$ and T$_6$ are the conducting pair

θ_2 T$_2$ is turned ON, thereby commutating T$_6$ in the common anode section

$\theta_2 - \theta_3$ T$_1$ and T$_2$ are the conducting pair

θ_3 T$_3$ is turned ON, thereby causing the commutation of T$_1$ in the common cathode section

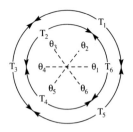

FIGURE 2.37
Commutation sequence and conduction intervals of thyristors.

$\theta_3-\theta_4$ $\quad$ T_3 and T_2 are the conducting pair

θ_4 $\quad$ T_4 is turned ON, thereby commutating T_2 in the common anode section

$\theta_4-\theta_5$ $\quad$ T_3 and T_4 are the conducting pair

θ_5 $\quad$ T_5 is turned ON, thereby commutating T_3 in the common cathode section

$\theta_5-\theta_6$ $\quad$ T_5 and T_4 are the conducting pair

θ_6 $\quad$ T_6 is turned ON, thereby commutating T_4 in the common anode section

$\theta_6-\theta_7$ $\quad$ T_5 and T_6 are the conducting pair

θ_7 $\quad$ T_1 is turned ON, thereby commutating T_5; one AC period (360°) is completed; the above sequence repeats from this instant, which is the same as θ_1 in the sequence.

At each commutation, there is a change in the thyristor switches that are in the ON state. Therefore the active part of the circuit configuration will be changing every 60°. The sequential changes in the active part of the circuit are shown in Fig. 2.38.

Device voltage and current ratings. The waveform of the voltage across a switching element in the bridge configuration can be plotted using a similar approach to that for the midpoint circuit. The theoretical peak reverse voltage possible will be found to be the peak value of the line voltage for both the single-phase and three-phase bridges. This should be multiplied by a safety factor when choosing the actual device, in order to take care of voltage

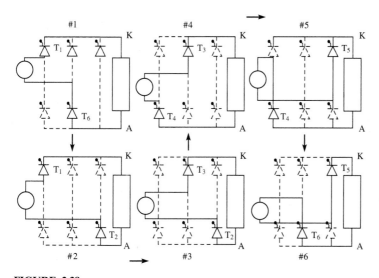

FIGURE 2.38
Changes in the circuit configuration as the devices switch sequentially in a three-phase bridge.

transients. For the single-phase bridge, the theoretical device current ratings will be as follows:

average current $I_d/2$
r.m.s. current $I_d/\sqrt{2}$
peak current I_d

For the three-phase bridge, the theoretical ratings will be as follows:

average current $I_d/3$
r.m.s. current $I_d/\sqrt{3}$
peak current I_d

The actual device ratings chosen will depend on overload limits as well as the thermal resistance of heat sinks used and whether or not there is going to be forced air cooling.

2.4.4 Gate Pulse Requirements for Starting of the Bridge

The sequence of switching described above for the three-phase bridge shows that the six thyristors are to be turned ON, one at a time, at 60° intervals. If narrow gate pulses are used for the purpose, the timing of the pulses as shown in Fig. 2.39 will meet this requirment. In this figure, the top row (a) shows the

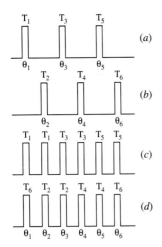

FIGURE 2.39
The double-pulse gating scheme for three-phase bridges.

pulses for the common cathode section according to the labels for the thyristors in Fig. 2.36. The next row (b) shows the pulses for the common anode section. Such a scheme of pulses, however, will not be sufficient for starting of the bridge. The starting difficulty arises because two thyristors—one in the common cathode and the other in the common anode section—must conduct for the DC current to flow. In the gating pulse scheme shown by (a) and (b), only one thyristor receives the gating pulse at a time. Therefore, when the first gate pulse arrives at a thyristor, it cannot turn ON, because its partner in the other section has not yet received its gate pulse. When later, after 60°, the partner does get the gate pulse, its own pulse has disappeared, assuming that the pulse width is less than 60°. There are two solutions to the starting problem. One is to use pulses of long duration, so that a pulse on a thyristor gate does not end before the next pulse arrives on the gate of the thyristor in the other section. The second method is to provide double pulses to each thyristor. The idea is to simultaneously provide a pulse to the conducting partner in the opposite section whenever a pulse is given to any thyristor. This purpose is satisfied by the scheme of pulse sequences shown in Fig. 2.39(c) and (d). Figure 2.39(c) is for common cathode thyristors and (d) for common anode ones. Every thyristor will be seen to get two consecutive pulses—the second one timed at 60° from the first. A continuous gating scheme in which each thyristor receives gate drive for the entire duration of its allocated conducting interval will also overcome the starting difficulty. A discontinuous mode of operation can also occur in a bridge, as we explained for the midpoint converter when the load circuit has a DC source large enough to cause it. Under such a condition, this type of continuous gate drive will be desirable, because it will also enable each thyristor to come ON as dictated by the circuit conditions after an interruption of current.

2.4.5 Voltage and Current Ripples on the DC Side

An advantage of the symmetrically controlled three-phase bridge is the reduction in the peak-to-peak ripple voltage and lower values of the fundamental ripple voltage resulting from the six-pulse output waveform, although the input is only three-phase. For making the ripple voltage and current calculations, we can use the same approach that we followed in Sections 2.3.4 and 2.3.5 for the case of the midpoint converter. The following numerical examples illustrate this.

Illustrative example 2.20. A three-phase diode bridge rectifier is fed from a 440 V three-phase bus. Determine the DC output voltage of the bridge and the peak-to-peak voltage ripple at the DC terminals.

$$\text{DC output voltage} = V_d = 1.35 V_L = 1.35 \times 440 = 594 \text{ V}$$

Solution. Reference to the voltage waveform in Fig. 2.36(c) and suitably modifying it for zero firing angle, we can see that the output waveform will consist of 60° segments of the AC line voltage, which are centered at the peak magnitude. The positive peak will have a magnitude equal to $\sqrt{2} \times 440 = 622.3$ V. The lowest magnitude of the voltage will occur at 30° from the peak. The magnitude of this will be $622.3 \cos 30° = 538.9$ V. The peak-to-peak voltage ripple is $622.3 - 538.9 = 83.4$ V.

Illustrative example 2.21. If in the previous example we had thyristors instead of diodes and the firing angle was 30°, how would the peak-to-peak ripple voltage change? If the AC bus frequency is 60 Hz, what will be the fundamental ripple frequency?

Solution. If we again look at the voltage waveform in Fig. 2.36 and suitably modify it for a firing angle of 30°, we shall notice that the new 60° segment will commence at the peak of the line voltage. Therefore the maximum value of the voltage will remain the same as before at 622.3 V. The minimum value of the voltage will now be the line voltage magnitude 60° after the peak, which will be $622.3 \cos 60° = 311.1$ V. The new peak-to-peak ripple voltage will therefore be $622.3 - 311.1 = 311.2$ V. The new DC output voltage will be, $V_d = 1.35 \times 440 \times \cos 30° = 514.4$ V. The fundamental ripple frequency will be six times the AC supply frequency, that is, 360 Hz, because of the six-pulse waveform.

Individual frequency components of the ripple voltage and current. The following two examples show how the individual frequency components in the ripple voltage and the ripple current can be evaluated. The method is similar to that used in the case of the midpoint converter and in Examples 2.8 and 2.9.

Illustrative example 2.22. A three-phase diode bridge rectifier is fed from a 440 V three-phase 60 Hz bus. Determine the frequency and amplitude of the fundamental component of the ripple voltage in the output.

Solution. The output voltage waveform will consist of 60° segments of the AC line voltage waveform, which are centered at the peak. Therefore we can express one repetitive period as follows:

$$v = V_m \cos \theta \quad \text{in the interval } \theta = -\tfrac{1}{6}\pi \text{ to } \theta = \tfrac{1}{6}\pi$$

We shall rewrite this in terms of β such that $\beta = 6\theta$, where β will be our phase angle measure in terms of the fundamental ripple frequency:

$$v = V_m \cos \tfrac{1}{6}\beta \quad \text{in the interval } \beta = -\pi \text{ to } \beta = \pi$$

The amplitude of the fundamental Fourier component will be given by

$$V_{r1} = \frac{1}{\pi} \int_{-\pi}^{\pi} V_m \cos \tfrac{1}{6}\beta \cos \beta \, d\beta$$

After integration and evaluation, this becomes

$$v_{r1} = 0.0546 V_m$$

where V_m, the peak magnitude of the AC line voltage, is $\sqrt{2} \times 440 = 622.3$ V. This gives the amplitude of the fundamental component of the ripple voltage as 34.0 V. Its frequency is $6 \times 60 = 360$ Hz.

Illustrative example 2.23. The bridge in the previous example supplies power to a DC motor whose resistance is $0.5\,\Omega$. What should be the minimum total inductance in the DC circuit to limit the fundamental ripple current to within 1 A r.m.s.

Solution. Let X denote the reactance of the inductor at the fundamental ripple frequency. The impedance to AC at this frequency presented by the load circuit will be

$$Z = \sqrt{R^2 + X^2}$$

The r.m.s. value of the fundamental frequency ripple voltage, from the previous example, is

$$V_{r1,rms} = 34/\sqrt{2} = 24 \text{ V}$$

The AC impedance needed to limit the current to 1 A r.m.s. will be

$$Z = 24/1 = 24\,\Omega$$

This gives $X = \sqrt{24^2 - 0.5^2} = 24\,\Omega$ at the ripple frequency of 360 Hz. The corresponding inductance value will be

$$L = \frac{24}{2\pi \times 360} = 10.6 \text{ mH}$$

2.4.6 Commutation and Overlap

When a bridge supplies DC load current, overlap occurs during commutations in the same manner as we explained for the midpoint circuit. In the case of the single-phase bridge, the commutations in the common cathode and common anode sections are simultaneous. These simultaneous commutation may be treated as one single commutation involving two series-connected pairs of thyristors with each pair switching together. In the case of the three-phase bridge, the commutations in the common cathode and common anode sections are not simultaneous, but are spaced at intervals of 60°. Each commutation can be treated as an independent one and can be analyzed in the same manner as we did for the midpoint circuit in Section 2.3.11. Equation (2.36) for the growth of current in the incoming line and (2.37) for the decay of current in the outgoing line will also be valid for the bridge. The peak value I_{sp} to be used in these equations will be defined in the same manner. We shall apply the relations to the bridge configuration in the following illustrative numerical example.

Illustrative example 2.24. A symmetrically controlled three-phase thyristor bridge is working from a 440 V, 60 Hz three-phase bus. The firing angle is 40°. The AC bus has an inductance of 2 mH per phase and negligible resistance. The DC load current is 100 A. Assume large inductive smoothing.
(a) Determine the overlap angle at each commutation.
(b) Obtain expressions for the growth and decay of current during commutation. Sketch the waveform of the AC line current and state the duration of a line current pulse.

Solution
(a) The line-to-line short-circuit reactance is $2\omega L = 2 \times 2 \times \pi \times 60 \times 0.002 = 1.5\,\Omega$. The peak value I_{sp} of the theoretical short-circuit current between two lines is $\sqrt{2} \times 440/1.5 = 414.8$ A. Then

$$\frac{I_d}{I_{sp}} = 100/414.8 \qquad \alpha = 40°$$

Substitution of these values into (2.38) gives the overlap angle as $u = 18.3°$.
(b) The expression for the growth of current in the incoming phase is given by (2.36), in which the reference zero of time t is the instant of commencement of the commutation. Numerical substitution in this equation gives

$$i = 414.8[\cos 0.6981 - \cos(\omega t + 0.6981)] \quad A$$

Notice that for consistency of units we have converted the firing angle α into radians.

The expression for the decay of current will be given by (2.37), in which the reference zero of time is the instant of commencement of the next commutation in the same section. Numerical substitution into this equation gives

$$i = 100 - 414.8\,[(\cos 0.6981 - \cos(\omega t + 0.6981)] \quad A$$

The waveform of the line current is sketched in Fig. 2.40. The growth portion of the current on the left side and decay portion on the right will be given respectively by the above two equations. In between, it will have a constant magnitude of 100 A. The total duration of the current pulse will be $120 + u = 138.3°$. In the figure the corresponding phase-to-neutral voltage is also plotted for reference.

DC voltage drop on load due to overlap. While a commutation is in progress in one section of the bridge, the potential at the DC terminal of the section will be the mean of the phase voltages involved, as we saw in the case of the midpoint converter. Therefore, in a similar manner, this will cause a decrease

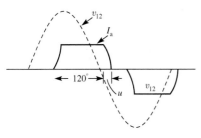

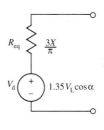

FIGURE 2.40
Line current with overlap in a three-phase bridge.

FIGURE 2.41
DC equivalent representation of a three-phase bridge.

in the area under the output voltage waveform resulting in a corresponding decrease in DC voltage. In a three-phase bridge there are six commutations in one AC cycle, and therefore, in conformity with (2.42), the net decrease in DC voltage will be

$$6\frac{X}{2\pi}I_d$$

On this basis, for three-phase bridge, the fictitious resistance value to be used for DC calculations corresponding to what was given in Fig. 2.27 will be $(3/\pi)X$. The equivalent DC representation of a three-phase bridge as a source with internal resistance will be as shown in Fig. 2.41.

> **Illustrative example 2.25.** A symmetrically controlled three-phase bridge is fed from a 208 V three-phase bus. The reactance per phase of the AC bus is $2\,\Omega$. The bridge is supplying power to a DC motor that has an internal resistance of $0.8\,\Omega$. The firing angle of the bridge is zero. The motor back e.m.f. is 250 V. Determine the motor current.
>
> **Solution.** The DC output voltage of the bridge is $1.35 \times 208 = 280.8$ V. The equivalent resistance for overlap is $(3/\pi)2 = 1.91\,\Omega$. Then
>
> $$I_d = \frac{280.8 - 250}{1.91 + 0.8} = 11.37 \text{ A}$$

2.4.7 The Semicontrolled Bridge Converter

This type of converter also goes by the names "half-controlled bridge" and "semiconverter." In a semicontrolled bridge, only half the total number of switching elements are thyristors. The other half are diodes. In a multiphase bridge, one of the two midpoint sections will be thyristors and the other, diodes. In a single-phase bridge, it is also possible to have an alternative arrangement with two diodes in one leg and two thyristors in the other. Such a circuit also behaves in the same manner as the arrangement in which diodes

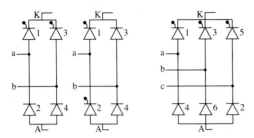

FIGURE 2.42
Semicontrolled bridge configurations.

and thyristors are in separate sections. Figure 2.42 shows the circuit configurations of single-phase and three-phase semicontrolled bridges. In the following treatment, we shall generally assume that the thyristors and diodes are in separate midpoint sections.

2.4.7.1 DC VOLTAGE RATIO. The semicontrolled bridge is a particular example of a bridge with asymmetrical phase control, in which the firing angles for the two midpoint sections are different. In a semicontrolled bridge, the firing angle of the diode section is always zero, and adjustment is possible only for the thyristor section. In a bridge with asymmetrical control, assuming continuous conduction, the DC output voltage will be the sum of the voltages of the individual sections, and therefore

$$V_d = V_{d0} \cos \alpha_k + V_{d0} \cos \alpha_a$$

where α_k and α_a are the firing angles of the respective sections and V_{d0} is the voltage of one section for $\alpha = 0$, being half the voltage of the bridge for zero firing angle. For the three-phase semicontrolled bridge, with $\alpha = 0$ for one section, the voltage relationship becomes

$$V_d = 0.675 V_L (1 + \cos \alpha) \tag{2.47}$$

2.4.7.2 ADJUSTMENT RANGE FOR THE DC VOLTAGE. Because control is possible for only one of the two midpoint sections, it might lead us to think that the range of adjustment of the DC voltage is limited to half of the maximum value. This is not correct, because we can increase the firing angle of the controlled section to more than 90°, which will make $\cos \alpha$ negative in (2.47). Ideally, commutation is possible over 180°. Therefore the ideal adjustment range is from zero to the maximum of the bridge voltage. The significance of firing angles greater than 90° will be explained in Section 2.5, where we describe the inversion mode. We shall see that in the inversion mode, the DC terminal voltage polarity reverses owing to α being greater than 90°. The mathematical expression for the voltage in the semicontrolled bridge is as if the thyristor midpoint section is operating in the inversion mode for $\alpha > 90°$. However, flow of power cannot reverse in the semicontrolled bridge, for the following reason. For reversal of power flow, the DC terminal polarity has to reverse. This is not possible, because if the DC side terminal polarity tends to reverse, a diode in the diode section will automatically get forward-biased and the DC current will begin to freewheel through this diode and the thyristor of that limb, conducting in series. This will be more evident if we examine the voltage and current waveforms presented below.

2.4.7.3 VOLTAGE AND CURRENT WAVEFORMS OF THE SEMI-CONTROLLED BRIDGE.

The waveforms for the three-phase semicontrolled bridge are shown in Fig. 2.43 for an arbitrarily chosen firing delay angle for the thyristor section. The thyristor section is assumed to be the common cathode section. Large smoothing is assumed, and overlap is neglected. The voltage waveforms are obtained using the same approach as we used for the fully controlled bridge. We first draw the waveforms of the individual midpoint sections, and thereby identify the conducting devices in each section at different intervals. We then draw the sine waves representing the line voltages (six line voltages) for the three-phase bridge. We then take the appropriate segments from these line voltages on the basis of the switches that are ON in an

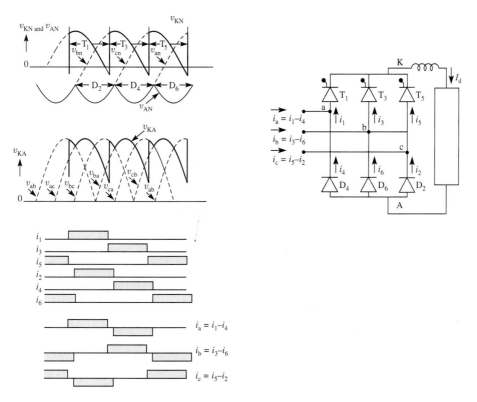

FIGURE 2.43
Voltage and current waveforms in the three-phase semicontrolled bridge.

interval. The current waveforms are also plotted in the same figure. The individual current waveforms may be identified by the labels given in the accompanying circuit diagram. The first six rows show the device currents labeled $i_1, \ldots, i_6$. These are rectangular pulses because of our assumption of large inductive smoothing and negligible overlap. The interval of each pulse will be the interval during which the corresponding switching element is conducting, as seen from the voltage waveform. The conducting device in each section is indicated on the voltage waveform. Each of the three AC line currents is determined by taking the difference between the device currents in the same leg. The line current will be seen to be an AC current without any DC component.

We notice that the repetitive frequency of the voltage ripple is three times the AC frequency and not six times the AC frequency. This is a disadvantage of the semicontrolled three-phase bridge compared with the fully controlled bridge, from the point of view of filtering.

Figure 2.44 shows the waveforms of the single-phase semicontrolled bridge—a firing angle of approximately 50° has been assumed for the common

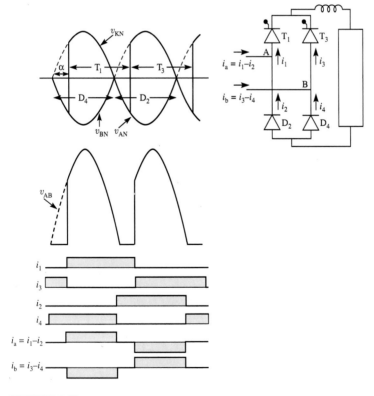

FIGURE 2.44
Waveforms in the single-phase semicontrolled bridge.

cathode section. The procedure followed in obtaining the waveforms is the same as that used for the three-phase bridge. The individual section waveforms are plotted using the midpoint as the reference potential N. Therefore each of the section voltage sine waves correspond to half the total single-phase line voltage. In the waveform for each section, we have identified the conducting device in each interval in the common cathode and the common anode sections. These switching elements are indicated in the figure. The total bridge output voltage has the amplitude of the total line voltage. The waveform of the total output voltage is obtained by drawing the sine wave sections and taking segments from them according to the switches that are ON at the instant. We notice that the total voltage is zero during some intervals. During these, the conducting devices both belong to the same leg of the bridge. The DC current is actually freewheeling through two switching elements on the same leg of the bridge. During these freewheeling intervals, the current in the AC line is zero. Therefore we notice that the duration of the line current is shorter than that of the current in each element.

The individual device currents are sketched below the voltage waveforms. We have assumed large inductive smoothing, and have neglected overlap for the sake of simplicity. Because of the freewheeling mentioned above, the line current pulse duration is shorter than the device current pulse by the angle α. However, the positive and negative line current pulses are equal, and therefore there is no DC current component in the lines.

2.4.7.4 AC SIDE POWER FACTOR CONSIDERATIONS. We have seen that phase control has the disadvantage of lowering the power factor on the AC side by introducing a displacement factor because of the phase displacement of the fundamental component of the AC current towards the lagging direction. A phase controlled converter therefore consumes lagging reactive power from the AC bus. A lagging reactive power is generally undesirable, because it creates larger voltage drops in the AC bus. In a semicontrolled bridge, there is no displacement created in the diode section of the bridge, whereas in a fully controlled bridge with symmetrical control, the displacement occurs in both sections. Therefore the semicontrolled bridge is advantageous from this point of view. It provides a DC voltage adjustment range from the maximum to pretty close to zero. The displacement is greatest at the middle of this range when the firing angle is 90°. Further increase in firing angle beyond 90° actually reduces the displacement and thereby increases the displacement power factor. In the case of a fully controlled bridge, as we go towards lower DC voltages, the firing angle approaches 90°, and therefore the power factor progressively decreases—the lowest displacement power factor occurring at the lowest end of the DC voltage range.

Better power factor and reduced cost are the reasons favoring the choice of the semicontrolled bridge. Cost saving is because of the reduced number of thyristors, which are more expensive than diodes, and also because the firing circuit need be designed for a smaller number of thyristors. The firing circuit design is also simplified by the fact that all thyristors will have a common cathode reference terminal for gate pulses if they are all in the common cathode section.

2.5 THE INVERSION MODE

The inversion mode is an operating mode of the phase controlled converter that is distinct from the rectification mode. In the inversion mode, power is made to flow from the DC side to the AC side of the converter. It is widely used in adjustable speed DC motor drives to achieve "regenerative braking" of the motor. During braking, the DC motor is made to function as a DC generator powered by the stored kinetic energy of the moving mass, and at the same time the converter is made to operate in the inversion mode, making it possible to return much of the kinetic energy to the AC bus without having to dissipate it as heat by means of friction braking. In the introductory section, we also gave another practical example of the inversion mode of operation at very high power levels in a high voltage DC power link. Here a phase controlled inverter is used in the inversion mode to transfer DC power at the receiving end of the link to the AC network at this end.

The inversion mode is possible only if all the switching elements in a converter are thyristors, so that their turn ON switching can be controlled. Uncontrolled and semicontrolled converters cannot be made to operate in the inversion mode. We shall first explain how reversal of power flow is possible, even though thyristors can conduct current only in one direction. We shall then proceed to describe the implementation aspects.

2.5.1 How Reversal of Power Flow is Possible

Figure 2.45(a) shows the conducting phase in a midpoint rectifier. During the conduction interval of this phase, the polarity of the AC phase voltage will be as marked. This polarity is such as to forward-bias the thyristor and so it is

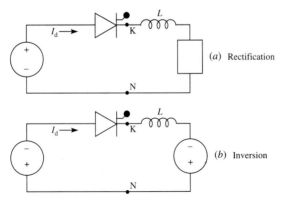

(a) Rectification

(b) Inversion

FIGURE 2.45
Reversal of power flow in inversion mode.

possible to make the thyristor conduct. If we look at the source voltage polarity and current direction, we notice that power is flowing out of the AC source. Successive commutations will cause the active phase voltage source to keep changing in Fig. 2.45. Therefore, at the terminal labeled K, we shall get a succession of segments from different phase voltage waveforms. Although the instantaneous value keeps changing, the average value will have the positive voltage polarity as shown, with respect to our reference terminal N.

Figure 2.45(b) shows a conducting phase in the inversion mode. To operate the converter in the inversion mode, we need to have a DC source on the DC side. This source must be connected with the voltage polarity as shown, which is the opposite of what we have in the rectification mode. In the inversion mode, we delay the firing instant so that the conduction of a phase takes place when its voltage polarity is the opposite of what we had in the rectification mode. We notice that the instantaneous AC voltage polarity is such as to reverse-bias the thyristor. But we have the DC source voltage, which is of a polarity such as to forward-bias the thyristor. Also, whenever the current tends to decrease, the induced $L \, di/dt$ voltage of the smoothing inductor will also be of a polarity that forward-biases the thyristor. Therefore it is possible to maintain the thyristor in conduction even during segments of the phase voltage waveform when the polarity is such as to reverse bias it. Now, for this polarity of voltage and the unchanged direction of current, we notice that the direction of power flow is from the DC source and into the AC source. As the thyristors switch in sequence, the active phase in (b) changes sequentially. The voltage waveform that appears across K–N will consist of negative segments from the individual phase voltage waveforms, resulting in an average that has the negative polarity shown in the figure—this polarity of v_{KN} being the opposite of what we had in the rectification mode. Therefore the average voltage presented at the terminals by the AC sources is one that opposes the DC source in the conducting loop. By making this opposing or "back-voltage" presented to the DC source by the inverter smaller than the DC source voltage, we can cause a net DC current to flow in the direction indicated.

2.5.2 Voltage Relationship

For a firing angle α, we had earlier derived the relationship $V_d = V_{d0} \cos \alpha$, where V_{d0} is the DC voltage for zero firing delay. This derivation will be found to be valid irrespective of the actual value of the firing angle, as long as the switches remain ON for their full conduction intervals. In the inversion mode, we make α larger than 90°. The commutation sequence and the conducting intervals remain the same. The DC source voltage and the smoothing inductance L must be adequate to ensure continuous conduction. For $\alpha > 90°$, we find that the voltage reverses in sign. This is in agreement with what we stated earlier. Therefore the change of the operating mode into inversion from the rectification mode involves two steps. A voltage source with reverse

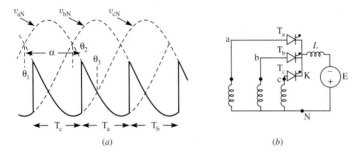

FIGURE 2.46
DC terminal voltage waveform in inversion mode.

polarity should be introduced on the DC side. Second, the firing delay angle should be increased to more than 90°. With $\alpha > 90°$, the converter presents an opposing voltage to the DC source in the loop consisting of the converter and the source. This opposing voltage has a DC magnitude equal to $V_{d0} \cos \alpha$.

The reversal of the polarity of the average voltage across the DC terminals for $\alpha > 90°$ can be seen by inspection of the waveform of the voltage v_{KN} across the DC terminal of the converter for such firing delays. Figure 2.46(a) shows this waveform, under ideal conditions of large inductive smoothing and negligible overlap for the converter shown in (b). The firing delay angle is approximately 140°.

The instant at which the a-phase " thyristor T_a begins to be forward-biased" is labeled θ_1 in the figure. Because of the large firing delay, T_a does not receive its gate pulse until the instant labeled θ_2, about 140° later. But when the gate pulse does arrive at θ_2, it is able to turn ON and commutate T_c, which is the one that is to be commutated in the sequence of switching. This is possible at any instant in the interval of 180° from θ_1 because the a-phase voltage v_a is more positive than v_c in this interval. Even if the actual value of v_a happens to be negative, T_a is still forward-biased if we look at the commuating loop, and we notice that the commutating voltage has the right polarity to turn OFF T_c. The conduction intervals of T_a, T_b and T_c are indicated against the waveform of v_{KN} in each interval. It will be noticed that the segments from each phase that constitute the v_{KN} waveform are mostly negative. Therefore the area under the v_{KN} waveform can be seen to have a resultant negative magnitude, the negative zones being much larger than the positive areas. Thus the waveform gives us a graphical verification of the reversal of the DC terminal voltage polarity for delay angles more than 90°. For a delay angle exactly equal to 90°, the positive and negative areas will be equal making the DC component of the voltage equal to zero. We conclude that, in the inversion mode, as a result of the large firing delay, the converter presents an opposing e.m.f. to the DC source in the conducting loop, equal in magnitude to $V_{d0} \cos \alpha$.

2.5.3 Limit of α For Successful Commutation

Reference to the waveforms of v_{aN} and v_{cN} in Fig. 2.46 shows that v_{aN} is more positive than v_{cN} in the 180° interval starting at θ_1. The end of this 180° limit is labeled θ_3 in the figure. During this interval, if we look at the voltage loop consisting of the a and the c phases, we notice that the thyristor T_a is forward-biased when T_c is ON. Therefore T_a can be turned ON at any instant in this interval, and will commutate T_c if the commutation is completed before the end of this interval at θ_3.

In the inversion mode, it is often more convenient to express the firing angle in terms of the "angle of advance" from the end limit θ_3 of the interval available for successful commutation than as the delay α from the beginning of the interval. If we denote the angle of advance by β then

$$\alpha + \beta = 180°$$

and so

$$\beta = 180 - \alpha, \quad \cos \alpha = -\cos \beta$$

To complete the commutation before the limit available when there is overlap, β should be greater than the overlap angle. The overlap angle u is variable and dependent on the DC load current. To take care of this, and also the turn OFF time of the thyristor, it is usual to provide a minimum angle of safety, which we shall denote by γ. This is also called the extinction angle. The minimum value of β must satisfy the relationship:

$$\beta \geqslant u + \gamma$$

It is important to avoid a commutation failure. Reference to the waveform in Fig. 2.46 shows that if the commutation from T_c to T_a fails then T_c will continue to conduct after θ_3. This will cause the current to build up in an uncontrolled manner because the c-phase voltage is moving towards positive values, which will add to the DC voltage in the conducting loop instead of opposing it. A typical waveform at the DC terminals of a three-phase, midpoint, phase controlled inverter with overlap is sketched in Fig. 2.47. The angle of advance, overlap angle and the safety margin are indicated in this figure.

Illustrative example 2.26. A three-phase bridge inverter is fed from a 208 V three-phase AC bus, whose reactance per phase is 1.5 Ω. Resistance may be

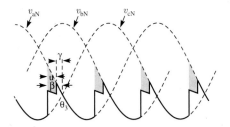

FIGURE 2.47
Terminal voltage with overlap in the inversion mode.

neglected. It is operating in the inversion mode with a DC source current of 20 A. The angle of advance $\beta = 40°$. Determine the safety margin available against commutation failure.

Solution. Although in our description we had used the midpoint configuration, the statements made can also be applied to the bridge. We shall first calculate the overlap angle using (2.38). The peak short circuit current $I_{sp} = 98.05$ A, $I_d = 20$ A and $\alpha = 180 - 40 = 140°$. Substitution in (2.38) gives $u = 25.94°$. The safety margin is

$$\gamma = \beta - 25.94 = 40 - 25.94 = 14.06°$$

2.5.4 Causes of Commutation Failure

An excessive value of the DC current can increase the overlap angle and eliminate the safety margin, and in this way cause a commutation failure. This could happen because of an increase in the DC voltage feeding the inverter or a decrease in the AC bus voltage. Voltage disturbances on the AC line resulting in a shift in the instant θ_3, which is the limit for successful commutation, can also cause commutation failure. One commutation failure can trigger more failures by increasing the current. For stable inverter operation when the possibility of commutation failure exists, it is desirable to employ a closed-loop gate firing scheme, which will automatically ensure a safe minimum extinction angle.

2.5.5 DC Equivalent Circuit in the Inversion Mode

We have seen that in the rectification mode the effect of overlap on the DC voltage is to decrease the DC voltage by the addition of a negative area under the DC terminal voltage waveform. This happens in the inversion mode also, but in the inversion mode, since the total area without overlap itself is negative, there is actually an increase in the magnitude of the negative terminal DC voltage. The area to be added is shown shaded in Fig. 2.47. The equivalent circuit of a 3 phase bridge for DC calculations is shown in Fig. 2.48. It is important to remember that this circuit assumes large inductive smoothing and continuous current flow. A discontinuous DC current flow mode can occur if the DC voltage is low or if the angle of advance β is decreased.

Illustrative example 2.27. In Example 2.26, determine the value of the DC source that feeds the inverter. Assume large inductive smoothing and continuous conduction of DC current.

Solution. The equivalent DC e.m.f. presented by the inverter is

$$V_{do} \cos \beta = 1.35 \times 208 \times \cos 40° = 215.1 \text{ V}$$

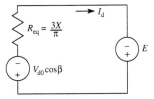

FIGURE 2.48
DC equivalent circuit for 3 phase bridge inverter.

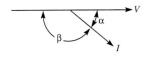

FIGURE 2.49
Fundamental current phasor.

This equivalent resistance introduced by overlap is $3 \times 1.5/\pi = 1.43\,\Omega$. The DC source voltage (see Fig. 2.48) is $215.1 + 20 \times 1.43 = 243.7$ V.

2.5.6 AC Side Power Factor Considerations in the Inversion Mode

We have seen in Section 2.3.9 that a firing delay angle α makes the fundamental component of the AC current lag by the same angle α, thereby introducing a displacement power factor equal to $\cos\alpha$. This lagging angle will be actually $>\alpha$ if we take into account the changes in the current waveform due to overlap, which we neglect here for the sake of simplicity. In Fig. 2.49, the current phasor is the fundamental sinusoidal component of the AC phase current. It is shown at a lagging angle α with respect to the phase voltage. The lagging reactive power consumed by the converter is proportional to $\sin\alpha$. If we progressively increase α, assuming that the DC current remains the same and so does the AC fundamental current magnitude, we notice that the reactive power consumption increases at first, reaching a maximum at $\alpha = 90°$. When we go into the inversion mode by further increase of α, there is a decrease in reactive power consumption. The reactive power consumed is seen to be proportional to the sine of the angle of advance β. It is desirable to minimize lagging reactive power consumption on an AC power network. From this point of view, it is desirable to operate the inverter with a small β whenever possible. In the converter that is functioning as the inverter in a HVDC power link's receiving end, this is particularly important because of the very large reactive power that the AC network at the receiving end has to supply. Such an inverter is therefore operated at the lowest possible value of the angle of advance by keeping the extinction angle (safety margin) γ at the fixed minimum, consistent with safety against commutation failure.

2.6 GATE CIRCUIT SCHEMES FOR PHASE CONTROL

To implement phase control in a converter, we need to have a control circuit that will generate the gate drive pulses to each thyristor. The design of this control circuit will give considerable scope for exercising ingenuity to the circuit designer, who may use analog or digital techniques or a combination of both. It is not our purpose here to describe these design details. We shall limit our treatment to highlighting the requirements to be met, by outlining some possible methods that may be used.

2.6.1 Gate Drive Methods—A "Gated Carrier Scheme"

The gate drive circuit design basically involves two steps. The first is the generation of the timing pulses for each thyristor. These pulses define the time intervals when the thyristor must have a pulse on its gate. The timing pulses need to be amplified to the required power level to drive the gates. Isolation between the gate and the control circuit is also a usual requirement. The pulse amplification and isolation scheme shown in Fig. 1.53 in Chapter 1 may be suitable for pulses of short duration up to tens of microseconds. Figure 2.50 shows a scheme, which we shall call the gated carrier scheme, that can be used for long duration pulses without creating problems of DC saturation in the pulse transformer. In this, the circuit block labeled "high frequency master square wave oscillator" generates a high frequency square wave and amplifies it to provide adequate power capability. The timing pulse of the thyristor is applied to the base of the gating transistor shown. When this transistor is ON, the carrier square wave is applied to the primary of a high frequency isolating transformer. On the secondary side, this is rectified and applied to the gate of the thyristor. The resistors shown in the gate circuit limit the voltage and current of the gate to the required levels. When the converter has several thyristors, the same high frequency carrier source can be used for all thyristors, with a separate gating transistor and transformer for each.

2.6.2 Generation of Timing Pulses for a Biphase Midpoint Converter

The circuit that generates the timing pulses must also provide a means by which these pulses can be shifted, to exercise phase control. For this purpose,

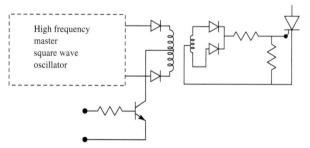

FIGURE 2.50
Gated ciarrier scheme.

the timing circuit has an input terminal for a DC control voltage, which we shall label V_c. Shifting of the pulses, as is needed to exercise phase control, is usually done by varying this control voltage. In a typical application such as the speed control of a DC motor, the variation of the firing angle may be effected by varying the control voltage either manually or by closed-loop control. V_c also serves for closed-loop control. In this, the control voltage V_c is obtained from an error amplifier, which compares the actual output speed or other parameter to be controlled against a reference and amplifies the error, to phase-control the converter in such a way as to eliminate the error.

Figure 2.51(a) shows a method for generating the timing pulses for a biphase midpoint converter shown in (c). The technique can be extended to three-phase midpoint and to bridge configurations. Basically, the circuit generates a linear ramp voltage, which is synchronized to correspond to the 180° available during which an incoming thyristor should be fired to commutate the outgoing thyristor. The figure shows the functional circuit blocks. The circuit needs a small timing transformer, two ramp generators which can

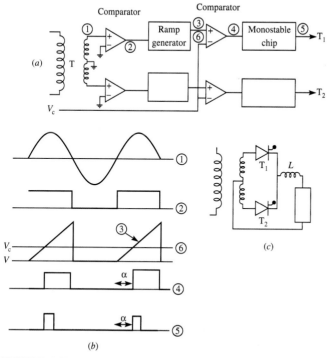

FIGURE 2.51
A scheme for generating gate timing pulses for phase control.

be assembled using operational amplifier chips together with a few external components), four voltage comparators (for which either comparator chips or operational amplifiers can be used) and two monostable multivibrator chips.

The timing transformer is a small one, with its primary connected to the same single-phase source as the main converter transformer. Its secondary has a center-tap, which is grounded at the ground of the control circuit power supply. Each outer terminal of the secondary gives a small voltage in the neighborhood of 4 or 5 V. We shall follow the circuit, beginning from one of the outer terminals of the secondary, towards the output, where we get the timing pulse for one of the thyristors. The circuit channel from the other section of the secondary to its output is identical, and ends up with the timing pulse for the second thyristor. At the location labeled "1" in the figure, the timing transformer provides the timing sine wave, which is sinusoidal, as shown by the waveform drawn for this location in Fig. 2.51(b). This is converted to a square wave by inputting it to one of the inputs of a comparator chip, whose other input is at ground (zero) potential. The square wave output at the output terminal of the comparator is also shown in the figure, labeled "2". The square wave output of the comparator is used to synchronize a linear ramp, which is generated by the circuit block so labeled. The ramp generation can be done using an operational amplifier, used as an integrator, to integrate a fixed voltage, with a transistor switch across the integrating capacitor. We may use the square wave at terminal 2 to open this shorting switch across the integrating capacitor when the square wave is positive, and thereby generate the ramp at the output location labeled "3" in the figure. The waveform at this location is indicated by the same label in the column of waveforms in (b). This ramp has a duration equal to the 180° interval of the alternate half-cycles, within which the thyristor T_1 is to be fired. The timing circuit can initiate a firing pulse at a desired instant in this interval. The adjustment of the firing instant is to be made using the control voltage labeled V_c in the figure. The ramp and the control voltage constitute the two inputs to a second voltage comparator chip. At the output terminal of the comparator, labeled "4," the wave form will be as shown by the same label in (b). This is a square wave, whose leading edge lags by the angle labeled α in the figure, with respect to the instant of commencement of the linear ramp. Since the ramp commences at the instant when thyristor T_1 starts to get forward-biased, this will be the desired firing angle delay if the firing pulse can be initiated at these leading edges of the comparator output. The waveform 4 itself can be used as the timing pulse for the gate of T_1. But the width of this pulse is variable when we vary the control voltage V_c. If a fixed pulse width is desirable then this purpose is achieved by using the monostable chip shown. In our circuit, the monostable chip will be programmed for positive edge triggering and to give the required pulse width. The output of the monostable chip, at the location labeled "5," is shown by the waveform with the same label. This output is the required timing waveform. It is a train of pulses of predetermined pulse width, whose leading edges are at the required firing angle α. This firing angle will vary linearly with the control voltage V_c. To fire the thyristor, this timing pulse can be used with the single-pulse amplification and isolation scheme shown on Fig. 1.53 in Chapter 1 or the gated carrier scheme shown on Fig. 2.50—the latter being more suitable for long pulse widths. The gate pulse for the second thyristor, labeled T_2, is produced by the second circuit channel commencing from the opposite output terminal of the timing transformer. In this channel, the waveforms are displaced by 180° with respect to the waveforms of the first channel shown, if the same control voltage V_c is used in the comparator circuit, as is shown in Fig. 2.51(*a*).

2.6.3 Timing Pulses for Single-Phase Bridges and Three-Phase Converters

The circuit described above can also be used for a single-phase fully controlled bridge. In such a bridge, the thyristors in the diagonally opposite locations on the two legs are to be fired at the same instant. Therefore each timing pulse provided by the circuit of Fig. 2.51 must be used for two thyristors.

To adapt the scheme to a three-phase midpoint circuit, we have to generate three timing ramps, each of which starts at the instant when the thyristor in that phase begins to get forward-biased. This instant will be the positive-going zero-crossing of the line voltage of the incoming line with respect to the outgoing line. We can get voltages that have the same phase as the line voltages if we have three timing transformers connected in Δ on the primary side and connected to the three lines. The timing waveforms to generate the three ramps can be obtained from each secondary phase voltage of these transformers. If the converter is a three-phase bridge, we can get the six timing sine waves by having a center-tap for the secondary of the same timing transformers. Alternatively, we can use an inverter logic gate to get another set of square waves of opposite polarity from each comparator output. These can be used to generate six ramps at the required intervals. If the phase control is symmetrical, with the same firing angle for both sections, simpler arrangements are possible. Asymmetrical control can be implemented, if required, by having a separate control voltage for each set of three ramps that are used for the two sections.

In the scheme that we have described, because we have used a linear ramp for comparison with the control voltage, we get a linear relationship between the control voltage and the firing delay angle α. Since the DC voltage of the converter is proportional to $\cos \alpha$, the relationship between the control voltage V_c and the DC voltage of the converter is nonlinear. It is possible to get a linear relationship between the DC voltage of the converter and the control voltage V_c by another scheme in which we use a cosine wave instead of a linear ramp for comparison. This is called the cosine wave crossing scheme. If the segment of the cosine wave and the range and zero level of the control voltage are suitably chosen, the phase angle variation can be made to have an inverse cosine relationship with respect to the control voltage V_c. This will give a linear relationship between V_c and the DC voltage of the converter. A feature of the cosine wave crossing scheme is that it has a self-regulating property for the DC output voltage of the converter when the AC line voltage fluctuates. If the AC bus voltage varies, the amplitude of the cosine reference waveform also varies, because the cosine wave is derived from it. If the control voltage remains unchanged under these conditions, the resulting change in the delay angle will be such as to maintain the DC voltage of the converter constant.

2.7 CHAPTER 2 SUMMARY

The term "commutation" is used to mean the turn OFF switching of a static power switch in a power electronic converter. In an AC/DC converter that has diodes or thyristors as the switching elements, the turning ON of an incoming device automatically causes an AC line voltage to be applied to reverse bias and turn OFF the outgoing device. Turn OFF switching that takes place in this manner is called line commutation. The two basic configurations of switching elements in line commutated converters are the midpoint and the bridge. Delaying the turning ON of an incoming thyristor by an adjustable phase angle and thereby delaying the turning OFF of an outgoing thyristor by the same interval is called phase control. The voltage at the DC terminals of a line commutated converter, whether midpoint or bridge, can be varied by phase control. We have derived the DC/AC voltage relationships for line commutated phase controlled converters as a function of the firing delay angle. We have also considered the waveform aspects on the DC side, and have seen how the ripple currents are minimized on the DC side by the use of a large series inductance for absorbing the voltage ripple. In the mathematical analysis of the converter circuit configurations, we have generally assumed the presence of an arbitrarily large smoothing inductance on the DC side. On this basis we have neglected the ripple content in the DC side current. The price to be paid for the easy adjustability of the DC voltage by phase control is a lower power factor, caused by the displacement of the fundamental component of the AC side current in the lagging direction. Therefore phase control makes it necessary for the AC bus to supply lagging reactive power to the converter. This is a major disadvantage of phase control.

When line commutation takes place, the current shifts from the conducting device to the incoming device. Therefore the current in the outgoing device has to fall to zero from its initial value, and the current in the incoming device has to rise from zero to the final value. These changes of current need finite time, because of the inductance in the AC sources. During this finite time needed for commutation to be completed, both the devices remain ON. Therefore there is an overlap of conduction for the two thyristors that are involved in a commutation. Overlap introduces some changes in the voltage waveform on the DC side, and changes the magnitude of the DC voltage. The duration of overlap (overlap angle) increases with the DC load current. In the rectifier mode of operation, there is a decrease in DC voltage from the no-load value. This decrease is proportional to the DC current. This can be modeled by treating the converter viewed from the DC side as a DC source with a fictitious internal resistance. We have derived this DC equivalent circuit and the mathematical expression for the fictitious internal resistance under the assumed conditions of continuous current flow and large smoothing inductance. This circuit is not valid under conditions of discontinuous current flow on the DC side. Operation with discontinuous current flow can occur in both the rectification and inversion modes of operation of a converter, depending on the value of a DC source present on the DC side.

Phase control makes it possible to operate a converter whose switching elements are all thyristors in the inversion mode. In this mode, which is distinct from the rectification mode, power is made to flow from the DC side to the AC bus. This reversal of power flow is achieved by reversal of voltage polarity and not by reversal of current direction. Two important applications of the inversion mode have been indicated. In an HVDC power transmission link between two AC networks, the converter at the receiving end of the link operates in the inversion mode to transfer the received power from the DC link to the AC bus at this end. Another common application is in reversible DC motor speed control applications. By operating the converter in the inversion mode, we are able to achieve the braking of the DC motor. During this time, the motor operates as a generator and returns the energy recovered through braking back to the AC bus through the converter.

A phase controlled converter that is designed to work in the rectification and the inversion modes at will is called a two-quadrant converter. A fully controlled thyristor converter, which may be of the midpoint or the bridge configuration, is a two-quadrant converter. A semicontrolled bridge has both thyristors and diodes as switching elements. It can only operate as a rectifier, and is therefore a single-quadrant converter. In a reversible DC motor drive with regenerative braking, the motor has to spin in both directions. Braking will be needed for each direction of spin. This calls for four-quadrant operating capability. Four-quadrant operation of the motor is possible, using the two-quadrant converter, if we use a reversing switch, when required, to reverse the connections between the motor and the converter. To achieve fully static changeover between operating quadrants, without using a reversing switch, we need a dual converter. The dual converter makes it possible to provide operation with all four combinations of voltage polarity and current direction on the DC side. A dual converter consists of two two quadrant converters in "antiparallel." We have explained the basic principle of the dual converter in the introductory section, but have not treated its circuit aspects in detail. Such a converter is most commonly used for reversible DC motor drives, and therefore its further treatment is reserved for Chapter 7, which deals with these drives.

The dual converter can drive currents in either direction through a load. Therefore, by cyclically operating a dual converter, it can be made to supply AC current into a load at low frequency. A dual converter operated in this manner to supply AC is called a phase controlled cycloconverter. Practical applications of cycloconverters are generally for large low speed AC motor drives. Further treatment of the cycloconverter is reserved for Chapter 8, which is devoted to AC motor drives.

PROBLEMS

2.1 The DC motor in Fig. 2.12 is being regeneratively braked with a braking current of 20 A. The motor voltage is 120 V, with the top terminal positive. Identify the converter through which the power is being returned to the AC bus. Assuming ideal conditions, what is the power being returned?

2.2 In the example in Problem 2.1, the active converter has a firing angle of 145°. What should be the firing angle of the other converter that will give the same DC voltage at the converter terminals? Both converters are identical and are connected to the same AC bus.

2.3 In the circuit shown in Fig. P2.3, the AC voltage is 120 V at 60 Hz. The firing angle of the thyristor is 45°. Determine the conduction angle during which current will flow in the negative half-cycle of the AC.

2.4 Figure P2.3 is modified by the addition of a freewheeling diode as shown in Fig. P2.4. All other data remain the same as in Probelm 2.3. Determine
(*a*) the peak magnitude of the current in the freewheeling diode in the first cycle of switching;
(*b*) the peak value of the current in the freewheeling diode under repetitive conditions.

2.5 Determine the peak reverse voltage ratings of the thyristor and the freewheeling diode to be selected in the circuit of Problem 2.4 if a safety factor of 2.3 is to be used. Determine also the maximum forward voltage that the thyristor has to block for the stated firing angle in the problem.

2.6 Determine the DC component of the output voltage of the single-phase half-wave rectifer in Problems 2.3 and 2.4:
(*a*) for the case without a freewheeling diode in Problem 2.3;
(*b*) for the case with freewheeling diode in Problem 2.4.

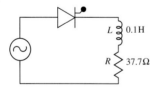

FIGURE P2.3

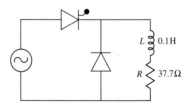

FIGURE P2.4

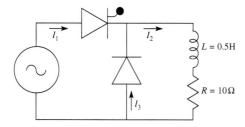

$L = 0.5H$

$R = 10\Omega$

FIGURE P2.7

2.7 In the circuit of Fig. P2.7, determine the lowest instantaneous value of the current I_2 for the given values of the circuit constants. The AC voltage is 120 V r.m.s. at 60 Hz and the firing angle is 30°. Assume repetitive conditions.

2.8 In Problem 2.7, if the value of the inductance is made arbitrarily large, sketch the waveforms to be expected for the currents labeled I_1, I_2 and I_3 in Fig. P2.7.

2.9 A biphase midpoint rectifier is to be designed to give 250 V DC at a firing angle of 25°. Determine the r.m.s. voltage per secondary section of the rectifier transformer. Assume large inductive smoothing on the DC side.

   If a single-phase bridge configuration is used, fed from one secondary section of the above transformer, what will be the DC voltage for the same firing delay angle?

2.10 A three-phase midpoint rectifier is suppled from a transformer that is zig-zag connected on the secondary side. The no-load DC voltage of the rectifier is 200 V at a firing angle of 30°. Determine the r.m.s. voltage of each secondary section of the transformer.

   What should the primary winding voltage be if it is to be Y-connected and fed from a 460 V three-phase bus?

2.11 An adjustable speed DC motor drive is proposed to be supplied from a three-phase bridge rectifier. The three-phase AC bus line voltage is 460 V. What will be the highest rating of the DC motor to be selected so that the bridge can be used without the need for a transformer?

2.12 In Problem 2.11, what will be the r.m.s. line current drawn from the three-phase AC bus when the motor is drawing 40 A? Assume ideal inductive smoothing and neglect overlap.

2.13 In Problem 2.11, determine the r.m.s. average, and peak current ratings of the thyristors if the full load motor current is to be 40 A. What should be the voltage rating of the thyristors, on the basis of a safety factor of 2.2?

2.14 In an m-phase converter, show that the voltage ripple frequency will be the same for both the bridge and the midpoint configurations when m is an even number. How will they be related if m is odd?

2.15 A three-phase diode bridge is supplied from a 460 V, 60 Hz three-phase bus. Determine the frequency and amplitude of the fundamental component of the voltage ripple.

2.16 In Problem 20.15, the DC load circuit consists of a resistance of 12 Ω in series with an inductance of 2 mH. Determine the amplitude and frequency of the fundamental component of the current ripple on the DC side.

2.17 A symmetrically controlled three-phase bridge converter is working from a 208 V three-phase AC bus. It is supplying 50 A DC at 200 V. Assume large inductive smoothing and neglect overlap. Determine the r.m.s. value of the AC line current. Also calculate the following;

(*a*) the firing angle of the bridge;

(*b*) the fundamental component of the AC line current;

(*c*) the input power factor; split this into displacement power factor and distortion power factor.

2.18 A three-phase bridge with symmetrical control is fed from a 208 V three-phase AC bus. The firing angle is 30°. Neglect overlap. Determine the r.m.s. value of the total ripple component of the voltage at the DC terminals.

2.19 An uncontrolled three-phase bridge rectifier has to supply 80 A at 600 V DC. What should be the AC input line voltage? What should be the voltage and the current ratings of the diodes?

2.20 An uncontrolled three-phase bridge rectifier is supplied from a 460 V three-phase bus. The total leakage reactance per phase is 0.5 Ω. Determine the overlap angle for a DC load current of 75 A.

2.21 A three-phase bridge is being supplied from a 460 V three-phase AC bus. The reactance per phase of the AC bus is 2 Ω. Determine the overlap angle if the bridge is working with a firing delay angle of 30° and supplying 10 A DC.

2.22 In a three-phase symmetrically controlled bridge, supplied from a 208 V three-phase AC bus, the no-load DC voltage is 200 V. The AC line reactance is 0.4 Ω per phase. Determine the voltage drop for a DC load current of 12 A due to overlap. Give an equivalent circuit for the bridge for DC voltage drop calculations.

2.23 A six-pulse fork-connected midpoint thyristor rectifier has a no-load voltage of 300 V DC at a firing angle of 25°. The rated full load current is 80 A. Determine the total VA rating of the secondary windings of the rectifier transformer.

2.24 A double three-phase configuration with interphase transformer is used to give the same voltage and current of Problem 2.23 at the same firing angle. Determine the total secondary VA rating of the main transformer.

2.25 In a six-pulse double three-phase diode rectifier with interphase transformer, the phase-to-neutral voltage of each secondary winding of the transformer is 250 V. Determine the DC voltage at no load and the transition load.

2.26 In Problem 2.25, what will be the total VA rating of the main transformer if the rated DC load current is 60 A. Also determine the total secondary VA rating that will be needed for the rectifier transformer if we had used the fork connection in place of the double three-phase with interphase transformer. Assume that the rated DC voltage and current are the same in both schemes.

2.27 A three-phase bridge inverter is fed from a 208 V three-phase AC bus. The reactance per phase of the AC lines is 1.2 Ω. $I_d = 50$ Amps. The angle of advance of the inverter is $\beta = 45°$. Determine the available extinction angle.

2.28 In Problem 2.27, determine the magnitude of the DC voltage source that feeds the inverter. Assume continuous current flow and large inductive smoothing.

CHAPTER 3

AC SWITCHING CONTROLLERS

3.1 INTRODUCTION—BIDIRECTIONAL SWITCHES

Most power semiconductor switching devices have capability for controlled switching for only one direction of current flow. However, employing a combination of devices, it is possible to use them as bidirectional switches in AC circuits. In this chapter, we shall present some practical switching control techniques that are used in AC circuits using such switches. We shall first describe how AC switches can be implemented using conventional power devices.

3.1.1 AC Switches Using Thyristors

The triac is a bidirectional thyristor that can be turned ON to conduct current in either direction. Its circuit symbol and terminal labels are shown in Fig. 3.1(a). The power terminals of the triac are labeled main terminal 1 (MT1) and main terminal 2 (MT2). The gate is the control terminal. The gate pulse to turn ON the triac is to be applied between the gate and MT1. A triac can be turned ON by either a positive or a negative gate current pulse for either direction of current. Typically, for positive current direction, which we shall consider as from MT2 to MT1 in the triac, a positive current pulse is used. A negative

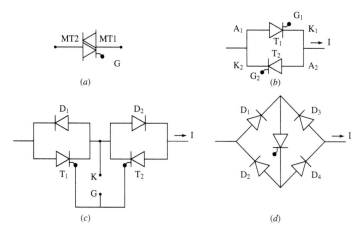

FIGURE 3.1
Bidirectional switches.

gate current pulse is typically used for negative current. The triac is a latching device like a thyristor. Just a short gate pulse is sufficient to turn it ON. It turns OFF like a diode or thyristor. As in the case of a thyristor, the gate has no capability to turn OFF the device. After the device has turned ON, the gate loses control over the switching. It regains control after the current has fallen to zero, and needs another gate pulse to turn ON again. In these respects, it is like a thyristor. However, its dynamic switching characteristics are inferior to those of a thyristor. Also, triacs are not available in such high current and voltage ratings as thyristors. Therefore, for many switching applications, AC switches using thyristors may have to be used instead of triacs.

Figures 3.1(b), (c) and (d) show three AC switch implementations using thyristors. The arrangement in (b) consists of two thyristors in "anitparallel." Thyristor T_1 is to be gated for currents in the direction shown in the figure, which we shall treat as positive. T_2 should be gated during the negative half-cycle, for negative currents. In this scheme, the gate drive input terminal pairs are separate for the two thyristors: K_1, G_1 for thyristor T_1; and K_2, G_2 for thyristor T_2. This may be a disadvantage from the point of view of the gate firing circuit.

The scheme shown in (c) uses two diodes in addition to the two thyristors. This can be implemented using two switching modules, each switching module consisting of a thyristor and an antiparallel diode across it. In this scheme, the two thyristor gates can be tied together, and so also the two cathodes, so that there is only one pair of terminals for the gate drive of both thyristors. This may be convenient from the point of view of control circuit design. As can be verified from the figure, the positive current flow path is through T_1 and D_2, whereas the negative path is through T_2 and D_1. The current has to flow through two devices—a thyristor and a diode—which will involve more power loss, because of the forward voltage drop of two devices, instead of one in the circuit of (*b*).

Figure 3.1(*d*) shows an arrangement that needs only one thyristor for both directions of switching. As can be seen, the positive current path is through the diodes D_1, D_4 and the thyristor, whereas negative currents flow through D_2, D_3 and the thyristor. Control is simple, because there is only one thyristor to be fired for both positive and negative currents. But current has to flow through three devices, causing added power loss.

3.1.2 AC Switches Using Other Devices

Switching devices other than thyristors may also be employed to implement bidirectional switches. Figure 3.2 shows some practical arrangements. Figure 3.2(*a*) uses two switching modules in reverse series, each module consisting of a bipolar junction power transistor and an antiparallel diode. For each direction of current, the path consists of one transistor and the antiparallel diode of the other transistor. The base terminals of the two transistors can be tied together, and so can the emitters, so that there is only one pair of terminals for driving the switch. The scheme shown in (*b*) is similar, but uses power MOSFETs, instead of BJTs. With the power MOSFETs, the antiparallel diode can be eliminated because the "body diode" of the power MOSFET will serve the same function. The bridge circuit arrangement shown in (*c*) is similar to the scheme in Fig. 3.1(*d*). In Fig. 3.2(*c*), a unidirectional switching device is shown without indicating any specific type of device. Other types of unidirectional switching devices like IGBTs or GTOs can also be used in the circuit arrangements shown in Figs 3.2(*a*) and (*c*). Unlike the switches shown in Fig. 3.1, those in Fig. 3.2 have the advantage that turn OFF switching can be

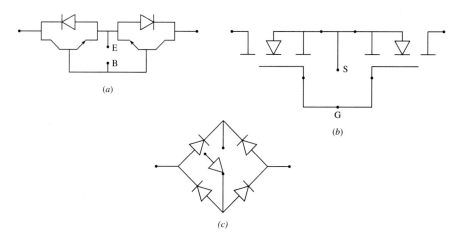

(*a*)

(*b*)

(*c*)

FIGURE 3.2
AC switch implementaton using other devices.

performed by means of the control terminal without waiting for the current to fall to zero naturally.

3.2 SERIES SWITCHING CONTROL OF AC LOADS

The current and therefore the power flowing into a load fed from an AC source can be varied by having an AC switch in series with it. This switch is repetitively operated with an adjustable ON time. There are two practical switching control schemes that may be used. These are "integral half-cycle control" and "phase control." Examples of these will be presented in this section.

3.2.1 Single-Phase Resistance Load—Integral Half-Cycle Control

Figure 3.3 shows a resistance load fed from an AC voltage source that is being controlled by putting an AC switch in series. The AC switch shown uses two thyristors in antiparallel. It could be a triac or any of the bidirectional switches shown in Fig. 3.1. Such a switch will have to be gated in each AC half-cycle, since it turns OFF at the end of each half-cycle of current. In the integral half-cycle control scheme, the switch is gated for an integral number of half-cycles and then kept OFF for another fixed number of half-cycles. This sequence of switching is repeated. If the total number of half-periods in a switching cycle is defined as the base period, we may define the duty cycle (which is also called the duty ratio) of the switch as

$$D = \frac{\text{number of ON half-cycles in a base period}}{\text{total base period in half-cycles}}$$

The load power adjustment is made by varying the duty cycle D. With a resistive load, the power is dependent on the r.m.s. value of the voltage applied to the load. It is therefore of interest to find how the r.m.s. load voltage is dependent on the duty cycle.

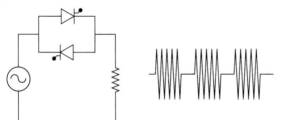

FIGURE 3.3
Integral half-cycle control.

The mean value of the square of the load voltage will be proportional to the duty cycle D:

$$\text{mean square value of the load voltage} = V^2 D$$

where V is the r.m.s. AC source voltage. Therefore

$$\text{r.m.s. load voltage} = V\sqrt{D}$$

Illustrative example 3.1. A resistance heating element fed from an AC bus is controlled by an AC switch with integral half-cycle control. The base period is 20 half-cycles of the AC supply. Determine the number of half-cycles for which the AC switch should be gated to bring down the heating power to 30% of the maximum.

Solution. The average power in the resistance element is proportional to the square of the r.m.s. voltage:

$$P = k(V\sqrt{D})^2 = kV^2 D$$

Therefore, for 30% power, $D = 0.30$. The number of ON half-cycles $0.30 \times 20 = 6$ half-cycles.

Integral half-cycle control involves fewer switching operations than the alternative scheme, which is phase control. In the latter the switches are operated in every half-cycle. There is less distortion of current in integral half-cycle control. It is suitable when the load time constant is large, such as in a heating load with a large thermal time constant. In such a load, the relatively longer time intervals between successive ON periods of the switch may not seriously affect the response.

3.2.2 Phase Control—Resistance Load

Figure 3.4 illustrates phase control of a resistance load supplied from a single-phase AC source. Here the AC switch is gated in every half-cycle. The instant at which it is turned ON is delayed by an angle labeled as α in the

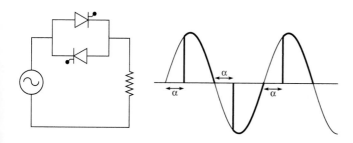

FIGURE 3.4
Phase control of a resistance load.

figure. The power control is achieved by variation of this firing angle α. Since the power in the load is proportional to the square of the r.m.s. value of the voltage, we shall obtain a relationship between the firing angle α and the r.m.s. load voltage as follows. For this, we first determine the mean square value:

$$V_{mean}^2 = \frac{1}{\pi} \int_\alpha^\pi V_m^2 \sin^2 \theta \, d\theta = \frac{V_m^2}{2\pi} \int_\alpha^\pi (1 - \cos 2\theta) \, d\theta$$

$$= \frac{V_m^2}{2\pi} (\pi - \alpha + \tfrac{1}{2} \sin 2\alpha)$$

$$V_{rms} = V \left(1 - \frac{\alpha}{\pi} + \frac{\sin 2\alpha}{2\pi} \right)^{1/2} \tag{3.2}$$

Illustrative example 3.2. A single-phase resistance load is supplied from a 120 V AC bus in series with a phase controlled static AC switch. The firing delay angle is 60°. Determine the r.m.s. value of the load voltage.

If we had used integral half-cycle control with a base period of 20 half-cycles, what would be the equivalent number of ON half-cycles for the same load power.

Solution. Putting $\alpha = \tfrac{1}{3}\pi$ in (3.2) gives

$$V_{rms} = 0.8969 \text{ V}$$

If we had used integral half-cycle control, the required duty cycle to give the same r.m.s. load voltage would be given by $\sqrt{D} = 0.8969$. Therefore

$$D = 0.8969^2 = 0.8045$$

The number of ON half-cycles would be $0.8045 \times 20 = 16$, to the nearest integer.

With phase control, the AC load current can be seen to be nonsinusoidal. The fundamental frequency of the distorted current wave is the same as the AC mains frequency. The fundamental and harmonic components of the current can be easily determined by Fourier analysis. Problems 3.4, 3.5 and 3.6 at the end of the chapter involve exercises on this. In integral half-cycle control, the repetitive switching frequency of the load current waveforms is below the AC frequency. Therefore this technique creates what may be called "subharmonic" frequency currents in the system.

3.2.3 The Triac Light Dimming Circuit

This circuit is very widely used for the dimming control of incandescent filament lamps. It is essentially a pracitcal example of the phase control of a single phase resistance load which was described above. The AC switch used is a triac. Another bidirectional switching device, which goes by the name "diac" is commonly used in the triggering circuit of triacs in light dimming controllers. The diac is a two terminal device that turns ON when the voltage across it

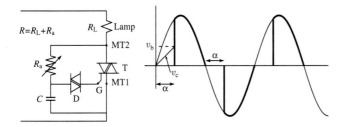

FIGURE 3.5
Triac light dimmer circuit.

exceeds its breakover voltage limit. The circuit arrangement is shown in Fig. 3.5. In each half-cycle of the AC, the capacitor C gets charged through the charging resistor R. The charging rate will be determined by the time constant CR, and will be slower the higher the value of R. When the capacitor voltage crosses the breakover voltage of the diac, the diac switches ON and causes the capacitor to discharge into the gate of the triac and so turn it ON. The firing angle at which the diac turns on is adjustable by varying the value of the charging resistor R, which is the intensity control for the lamp.

Using known values of the diac breakover voltage, the firing angle for given values of R and C can be determined using the following approach.

We shall assume that the switching devices are ideal with zero ON state voltage drop. Because of the forward-biased path from the gate terminal to the main terminal MT1 of the triac, the capacitor voltage will be treated as equal to the diac voltage. It will be zero at the commencement of a half-cycle, because the triac would have been conducting till the end of the previous half-cycle. The capacitor begins to get charged through R from the commencement of a new half-cycle, at which the triac is OFF.

In Fig. 3.5, R_L is the resistance of the lamp to be controlled. The adjustable resistance R_a in the charging circuit of the capacitor is usually in the range of several kilo-ohms. The hot resistance of a 120 V, 100 W filament lamp, when it is consuming 100 W of power, will be 144 Ω. At lower power dissipation it will be less, the temperature being less. Therefore the lamp resistance has negligible effect on the charging time constant. In any case, we shall denote the total charging resistance by R, such that

$$R = R_\mathrm{a} + R_\mathrm{L}$$

The AC supply voltage will be assumed to be sinusoidal, with a peak magnitude $V_\mathrm{m} = \sqrt{2}V$, where V is the r.m.s. value. The charging current of the capacitor will be $C\,dv_\mathrm{c}/dt$, where v_c is the capacitor voltage. On this basis, the charging current loop equation may be written as

$$RC\frac{dv_\mathrm{c}}{dt} + v_\mathrm{c} = V_\mathrm{m} \sin \omega t \tag{3.3}$$

The solution of this differential equation may be written in terms of the particular integral and the complementary function. The particular integral is written using steady state AC circuit theory as

$$v_c = \frac{V_m}{\sqrt{1 + \omega^2 C^2 R^2}} \sin (\omega t - \phi)$$

where

$$\phi = \tan^{-1} \omega CR$$

The complementary function is the solution of (3.3) with its right-hand side put equal to zero. This solution can be obtained by integration, and is the equation for the discharge of a capacitor through a resistor. It may be written as

$$v_c = V_1 e^{-t/CR}$$

where the integration constant V_1 has to be determined by using the initial condition in the complete solution. The complete solution is then

$$v_c = \frac{V_m}{\sqrt{1 + \omega^2 C^2 R^2}} \sin (\omega t - \phi) + V_1 e^{-t/CR}$$

To determine V_1, we use the initial condition $v_c = 0$ at $t = 0$. This initial condition is valid because the triac was in conduction till the end of the previous half-cycle. This procedure gives V_1 as

$$V_1 = \frac{V_m}{\sqrt{1 + \omega^2 C^2 R^2}} \sin \phi$$

The complete solution becomes

$$v_c = \frac{V_m}{\sqrt{1 + \omega^2 C^2 R^2}} [\sin (\omega t - \phi) + e^{-t/CR} \sin \phi]$$

This may be written as

$$v_c = \frac{V_m}{\sqrt{1 + \omega^2 C^2 R^2}} [\sin (\theta - \phi) + e^{-\theta/\omega CR} \sin \phi]$$

where

$$\theta = \omega t, \quad \phi = \tan^{-1} (\omega CR)$$

The firing of the triac occurs at the instant when the capacitor voltage rises to the breakover voltage of the diac. Therefore $\theta = \alpha$ when v_c is equal to the breakover voltage v_b, where α is the firing angle. Making these substitutions, we may write the equation for v_b as

$$\frac{v_b}{V_m} = \frac{1}{\sqrt{1 + \omega^2 C^2 R^2}} [\sin (\alpha - \phi) + e^{-\alpha/\omega CR} \sin \phi] \tag{3.4}$$

We can determine the firing angle for specific values of R and C and the breakover voltage by means of the above equation. The solution of this equation may need the use of a computer or a programmable calculator.

Illustrative example 3.3. The triac light dimmer circuit of Fig. 3.5 is used to adjust the intensity of a 120 V, 100 W incandescent filament lamp working from 120 V, 60 Hz mains. $C = 0.33\ \mu F$ and $R = 3.33\ k\Omega$ (a 5 kΩ potentiometer set at 3.33 kΩ). The breakover voltage of the diac is 40 V. Determine the firing delay angle α under these conditions.

Solution. We have

$$\omega CR = 2\pi \times 60CR = 0.4143$$

$$\phi = \tan^{-1} 0.4143 = 0.3928\ \text{rad}$$

Numerical substitution into (3.4) then gives the following:

$$0.2551 = \sin(\alpha - 0.3928) + 0.3828e^{-2.4137\alpha}$$

Computer solution of this equation gives

$$\alpha = 0.546\ \text{rad} = 31.3°$$

3.2.4 Single-Phase $R-L$ Loads

The presence of an inductance in the load circuit creates significant differences in the operation of the AC switching controller from the case of a resistance load, which need to be highlighted. Figure 3.6(a) shows the circuit. The AC source waveform, which we shall assume to be sinusoidal, is sketched in Fig. 3.6(b). We shall assume that the thyristor bilateral switch shown is gated at the instant labeled 0, the delay angle of this instant from the positive-going

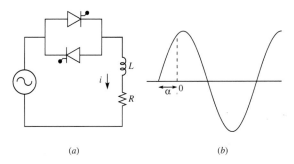

(a)   (b)

FIGURE 3.6
Switching control of an inductive AC load.

zero-crossing being the angular interval labeled α in the figure. Taking the reference zero of time at the instant labeled 0 in the (b), we can write the expression for the AC voltage and the loop equation as

$$v = V_m \sin(\omega t + \alpha)$$

$$L\frac{di}{dt} + Ri = V_m \sin(\omega t + \alpha)$$

We shall assume that this is the very first switching cycle, so that the initial condition can be stated as

$$i = 0 \quad \text{at } t = 0$$

As we have been doing before in similar cases, we shall write the solution of this differential equation in terms of the particular integral and the complementary function. The particular integral is given by the steady-state current that would exist if the switch stayed ON continuously. This, from AC circuit theory, is a sinusoidal current determined by the AC impedance of the circuit, lagging at an angle ϕ with respect to the voltage. This may be written as:

$$i = \frac{V_m}{Z} \sin(\omega t + \alpha - \phi)$$

where

$$\phi = \tan^{-1}\left(\frac{X}{R}\right), \quad Z = \sqrt{R^2 + X^2}, \quad X = 2\pi f L$$

The complementary function is the transient part, which is the solution of the differential equation with the right-hand side equal to zero. This is given by the exponential expression for the decay of current in an R–L circuit, and may be written as

$$i = I_1 e^{-t/\tau}$$

where the time constant $\tau = L/R$ and the integration constant I_1 has to be determined by using the initial condition in the complete solution. The complete solution may therefore be written as

$$i = \frac{V_m}{Z} \sin(\omega t + \alpha - \phi) + I_1 e^{-t/\tau}$$

To find I_1, we put $i = 0$ at $t = 0$ in this expression, and we get

$$I_1 = -\frac{V_m}{Z} \sin(\alpha - \phi)$$

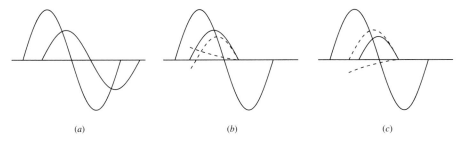

(a)   (b)   (c)

FIGURE 3.7
Controlled $R-L$ load. Limits on firing angle.

The complete solution may be written as

$$i = \frac{V_m}{Z}[\sin(\omega t + \alpha - \phi) - e^{-(R/L)t}\sin(\alpha - \phi)]$$

By putting $\theta = \omega t$ and $V_m/Z = I$, this may be written as

$$i = I[\sin(\theta + \alpha - \phi) - e^{-(R/X)\theta}\sin(\alpha - \phi)] \qquad (3.5)$$

In Figs 3.7(b) and (c), we have sketched the two parts of the solution, namely the particular integral which we shall call the steady state term, and the complementary function, which we shall call the transient term, separately by broken lines. Three different values of α or ranges of α are considered. Figure 3.7(a) is the case where $\alpha = \phi$; that is, the firing delay is the same as the phase angle of the load. For this case, the value of I_1 is zero; therefore the transient term is zero. The initial amplitude I_1 of the transient term will be the negative of the steady state term at $t = 0$, because only in this way will the condition of zero initial current be satisfied. When $\alpha = \phi$, the steady state term itself satisfies the initial condition, and so the need for the transient term does not exist.

Figure 3.7(b) shows the case when α is less than ϕ. In this case, the transient amplitude I_1 has to be positive, because the steady state term has a negative magnitude at $t = 0$. Notice that the positive half-period of the steady state term is one half-cycle of the AC, equal to π rad or 180°. Therefore we can see that when α is less than ϕ, the duration of positive current flow will be greater than π. This creates a problem in the gating. Normally, the firing delay for the positive half-cycle will be the same as the firing delay for the negative half-cycle. If this is the case, and if the gating pulses are of short duration, the gating pulse for the negative current will have arrived for the appropriate thyristor of the AC switch while the positive current has not yet come down to zero. Because the pulse is of short duration, it might have disappeared before the positive current falls to zero. If this happens, there will be no gate pulse for the negative current thyristor of the AC switch, after the forward current has fallen to zero and the thyristor AC switch is ready to conduct. In this manner, the negative current flow will be inhibited. This difficulty can be overcome by using gate drives of long duration for the switch. The gated carrier system described in Chapter 2 (Fig. 2.50) may be appropriate. Let us assume that this indeed is the case. If so, the reverse current flow will start as soon as the positive current flow ends. It will be as if the AC switch is continuously ON. The current will soon settle down to the steady state value after a few cycles of switching. We therefore conclude that for $\alpha \leqslant \phi$, no phase control is possible. The circuit will behave as if the AC switch is continuously ON.

Figure 3.7(c) shows the case where α is greater than ϕ. In this case, the transient has a negative amplitude. Therefore the conduction period will be less than π. Control will be available up to $\alpha = \pi$. The necessity for long gate drives for the switch also does not arise.

Illustrative example 3.4. An AC load circuit consists of a resistance of $30\,\Omega$ in series with an inductance of $0.2\,\mathrm{H}$. It is fed from a $120\,\mathrm{V}$, $60\,\mathrm{Hz}$ AC source. Determine the firing angle range in which the load voltage can be controlled.

Determine the conduction interval in a half-cycle for a firing delay of $100°$, and sketch the voltage waveform across the load for this firing angle.

Solution. We have $R = 30\,\Omega$, $X = 2\pi \times 60 \times 0.2 = 75.4\,\Omega$ and $X/R = 2.5133$, so that

$$\phi = \tan^{-1} 2.5133 = 68.3°$$

The control range for the firing angle is $68.3°$–$180°$.

To find the conduction interval when $\alpha = 100°$, we use (3.5) for $i = 0$. Numerical substitution after converting angles to radians will give

$$\sin(\beta + 0.5532) - 0.5255e^{-0.3979\beta}$$

where the conduction interval for one half-period is denoted by β. Computer solution of this equation gives $\beta = 2.3834\,\mathrm{rad} = 136.6°$.

The load voltage waveform and current waveform are sketched in Fig. 3.8. The load voltage will be zero during the zero-current intervals.

3.2.5   Thyristor Controlled Inductor (TCI)

The thyristor controlled inductor, also called the thyristor controlled reactor (TCR), is of considerable practical importance in an application area of Power Electronics in AC power systems that is commonly described as static VAR

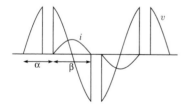

FIGURE 3.8
Load voltage and current waveforms.

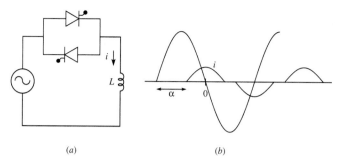

FIGURE 3.9
Thyristor controlled inductor.

compensation. VAR stands for "volt amperes reactive." In AC power networks, lagging reactive currents are undesirable because they cause excessive voltage drops and adversely affect stable operation. Therefore steps are taken to compensate for lagging reactive currents by introducing leading reactive currents. In such schemes, it is now common to use thyristor controlled inductors in a manner that we shall explain. Practical systems are three-phase. But it is best to describe the technique using the single-phase version.

The thyristor controlled inductor is shown in Fig. 3.9. It is a particular case, for $R = 0$, of the R–L load controlled by an AC switch that we considered above.

When $R = 0$, the phase angle of the load $\phi = \frac{1}{2}\pi$. Therefore, from the results of the case of R–L load, we find that phase control range of α is from $\frac{1}{2}\pi$ to π. The waveform of the source voltage is sketched in Fig. 3.9(b). Taking the reference zero of time as the instant labeled 0 in (b) and treating resistance as negligible, the voltage and the loop equation may be written as

$$v = -V_m \sin \omega t$$

$$\frac{di}{dt} = -\frac{V_m}{L} \sin \omega t$$

The expression for the current can be obtained by integration as

$$i = \frac{V_m}{\omega L} \cos \omega t + A$$

where the constant A can be obtained from the initial condition, which may be stated (for $\alpha > \frac{1}{2}\pi$) as

$$i = 0 \quad \text{at} \quad \omega t = -(\pi - \alpha)$$

This gives

$$A = \frac{V_m}{\omega L} \cos \alpha$$

With the constant A determined on this basis, the final expression for the current becomes

$$i = \frac{V_m}{\omega L}(\cos \alpha + \cos \omega t) \tag{3.6}$$

The TCI is used primarily as a means of obtaining an adjustable value of the fundamental lagging current. The waveform of the current is sketched in Fig. 3.9(b). From the symmetry considerations, we can see that Fourier analysis will give a cosine component that is a current lagging by $\frac{1}{2}\pi$ with respect to the voltage. The amplitude of this fundamental component may be found by Fourier analysis as follows:

$$I_1 = \frac{2}{\pi}\int_{-(\pi-\alpha)}^{\pi-\alpha} \frac{V_m}{\omega L}(\cos \alpha + \cos \theta) \cos \theta \, d\theta$$

$$= \frac{V_m}{\omega L}\frac{2}{\pi}(\tfrac{1}{2}\sin 2\alpha + \pi - \alpha) \tag{3.7}$$

From this, we see that the maximum amplitude of I_1 occurs for $\alpha = \frac{1}{2}\pi$ and is equal to $V_m/\omega L = \sqrt{2}V/\omega L$. This corresponds to an r.m.s. value of $V/\omega L$. The minimum value is zero at $\alpha = \pi$. Between these limits, the lagging current and so the lagging reactive VA are adjustable by variation of α.

Illustrative example 3.5. A thyristor controlled inductor used in a static VAR compensation circuit in a 11 kV, 60 Hz AC network has an inductance of 5 H. Determine the range of adjustable VAR using this inductor.

Solution. Using (3.7), we get the r.m.s. value of the largest possible fundamental lagging current as $11\,000/2\pi fL = 5.84$ A. The adjustable range is from zero at $\alpha = \pi$ to the maximum at $\alpha = \frac{1}{2}\pi$. Therefore the corresponding range for the adjustment of VAR in this single-phase circuit is from zero to $11\,000 \times 5.84 = 64.24$ KVAR.

3.2.6 Static VAR compensation Circuit

Figure 3.10 shows the typical method of static VAR compensation. The leading reactive current necessary for VAR compensation is actually supplied

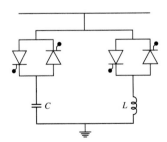

FIGURE 3.10
Static VAR compensation scheme using TSC and TCI.

by connecting capacitor banks across the AC lines. Before the development of thyristors, this was done using electromechanical circuit breakers. With the advent of thyristors, bidirectional AC switches using them were developed for connecting and disconnecting the capacitor banks. A capacitor bank connected in this way is called a thyristor switched capacitor (TSC). The problem to be faced when connecting the capacitor to the high voltage lines is the large inrush current that will flow at the moment of contact if there is a large instantaneous voltage differential between the line and the capacitor. It is relatively easy to overcome this difficulty, because the TSC uses static thyristor switches that can be gated at any desired instant in an AC cycle by means of a switching controller that will sense the voltage differential and gate the thyristor at the correct instant in the AC cycle when the voltage differential is within permissible limits.

If only TSCs are employed for VAR compensation, the leading VAR to be introduced can only be adjusted in steps, because the switching is to be done one capacitor bank at a time. For precise adjustments of VAR, as dictated by the system requirements, a continually variable feature is desirable. This is achieved in the scheme in Fig. 3.10 by having a thyristor controlled reactor in parallel with the capacitor bank as shown in Fig. 3.10. If the maximum lagging current drawn by the TCI is equal to the leading reactive VA of the capacitor, the two will cancel and the net reactive VA will be zero. From this point, the lagging VAR of the TCI can be progessively decreased by phase control, thereby increasing the net leading VAR. After reaching the maximum, further increase can be made by switching in another capacitor bank. In this manner, the TSCs provide VAR in steps, while the TCI will provide the continuous adjustment between steps. This scheme enables precise and fast automatic adjustment of the VAR by means of closed loop control. A practical system will be invariably a three-phase network, although we have used the single-phase equivalent to describe the technique. Also, in a practical high voltage system the TSCs and TCIs may be connected to the secondary side of a transformer. In this way, the maximum voltage requirements, of the thyristors and the capacitors, can be lowered.

3.3 STATIC CONTROL BY SEQUENTIAL TAP CHANGING

A major disadvantage of phase control using AC switches is the waveform distortion and the consequent generation of harmonics and lowering of the power factor. The technique described in this section minimizes the adverse effects, but requires a transformer for its implementation. It has been described as sequence control and synchronous tap changing control. Basically, the technique is tap changing of a transformer using static thyristor switches and continuous adjustment by phase control between taps.

Figure 3.11 shows the circuit scheme. The power is supplied through a transformer. This power transformer has several taps on the secondary, each tap being connected to the output through a bidirectional static switch using two thyristors. Ideally, the output voltage can be adjusted from zero to the total voltage of all the secondary coils. This adjustment can be made continuously without step changes over the entire range. The method of control can be understood if we confine our attention to the two lowest taps labeled 1 and 2. The same procedure is repeated for the subsequent taps. For description we shall assume a resistive AC load.

The lowest voltage is obtained from tap 1 through the AC switch labeled S_1. By phase control using this switch, the voltage is adjustable from zero to the full voltage v_1 of this tap. Further increase of voltage is achieved by increasing the voltage during parts of each half-cycle using the voltage from tap 2. The method of doing this can be understood by reference to the waveform of Fig. 3.12. This figure shows the voltage waveforms v_1 and v_2 respectively from taps 1 and 2. In the operating condition shown in the figure, the thyristor labeled 1P of the AC switch S_1 is ON from the instant labeled 1, which is the beginning of the positive half-cycle. All other thyristors are OFF. At the instant labeled 2 in the figure, the thyristor labeled 2P of the AC switch S_2 is gated. This thyristor will be forward-biased at this time because v_2 is more positive than v_1 at this

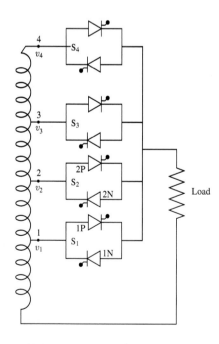

FIGURE 3.11
Multistage sequential tap changing.

FIGURE 3.12
Waveform with sequential tap changing control.

instant. Therefore the turning ON of 2P will automatically commutate 1P. The voltage at the output terminals from this instant will be v_2, as shown in the figure. It is important to ensure that the negative current thyristor labeled 1N of the switch S_1 remains OFF when 2P is ON, because otherwise there will be a short-circuit through 2P and 1N. The latter is gated from the instant labeled 3 in the figure, and we move to v_2 in the negative half-cycle at the instant labeled 4 in the same manner as we did during the positive half-cycle. For this, we have to fire 2N at the instant 4, making sure that 1P is not gated at this instant, to avoid a short circuit. The instants of changeover from v_1 to v_2, which are labeled 2 and 4 in the figure, can be progressively advanced by phase control until the output voltage is fully v_2. For further increase of voltage the change over to the next tap 3 is implemented on the same lines as the changeover from 1 to 2. In this manner, the voltage adjustment up to the total voltage of all the taps can be achieved continuously.

Use of bypass switches. A scheme that permits any sections to be combined in a transformer with several secondary sections is shown in Fig. 3.13. In this technique, there is a bypass AC switch across each secondary section in addition to a series switch. One of the sections can be earmarked for phase control. We may call this the secondary section for vernier adjustment. This will serve as an adjustable voltage, which could be added to the voltage of any combination of the other secondary sections. By suitable choice of voltages for the different secondaries, it will be possible to get continuous control over a wide range of voltages by adding sections in series, and using phase control on the vernier winding only between step changes. Figure 3.13(b) shows how the

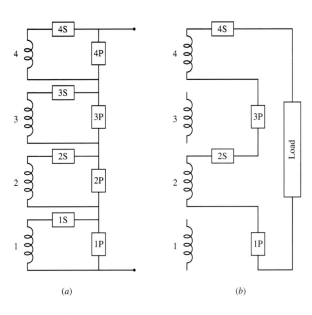

(a)

(b)

FIGURE 3.13
Sequential tap changing control using series and bypass switches.

sections can be combined independently by opening the respective bypass switches of the sections to be included and opening the series switches of those that are not to be included.

3.4 THREE-PHASE APPLICATIONS OF SWITCHING CONTROL

The AC switching control application in three phase circuits are basically adaptations of the single-phase techniques that we have described. We shall confine our treatment to a description of some of the three-phase adaptations by presenting the circuit configurations of the switches and the three-phase loads.

3.4.1 Series AC Line Switches With Y- or Δ-Connected Loads

In this scheme, there is a bidirectional AC switch in series with each line. The gating circuits of the thyristors should be designed to give the required range of control. A proper return path should be available at each instant for the current in each line at each instant. This return path may be through one or both of the other two lines. the scheme may be used with either the Y- or Δ-connected loads shown in Fig. 3.14. With Δ-connected loads, if the individual

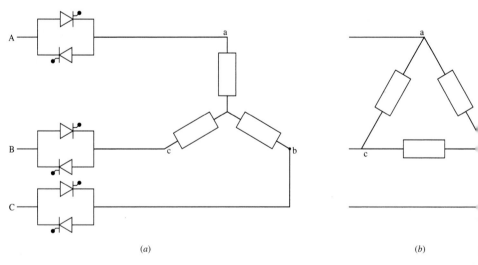

(a)   (b)

FIGURE 3.14
Three-phase switching controller with line switches in series.

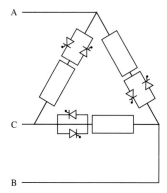

FIGURE 3.15
Series switching controller for a Δ-connected load.

phase circuits can be separated, then an alternative arrangement shown in Fig. 3.15 that can be used. In this circuit, the bidirectional switch is in series with each phase, inside the Δ configuration. Therefore the current in each switch is the phase current and not the line current. For this reason, the current ratings of the thyristors needed may turn out to be smaller—which is an advantage.

3.4.2 Neutral Point Controller

In the Y-connected load circuit of Fig. 3.14(a), we notice that each load phase is in series with an AC line switch. Therefore the operation of the circuit will be unaffected if the positions of the switch and the load phase are interchanged. Such an arrangement is shown in Fig. 3.16(a). Figure 3.16(b) shows an alternative circuit in which the switches are Δ-connected. In these two schemes, the switching control is located in the posiition of the neutral. A further simplification of the neutral point controller can be done by using only three thyristors instead of six, by the circuit shown in Fig. 3.16(c). If the gating circuit is suitably designed, a controlled bidirectional current path can be provided for each line current as by switches of the circuits (a) and (b).

3.4.3 Reversing Switch for Three-Phase Motors

The direction of rotation of three-phase induction and synchronous motors is reversed by reversing the phase sequence of the three phase AC connection to the motor terminals. A static switching scheme for doing this using AC thyristor switches is shown in Fig. 3.17. In this scheme, each of the switches labeled $S_1, \ldots, S_5$ is a bidirectional thyristor switch. S_1, S_2 and S_3 are to be ON

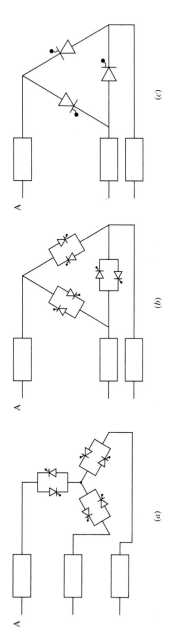

FIGURE 3.16
Neutral point switching control schemes.

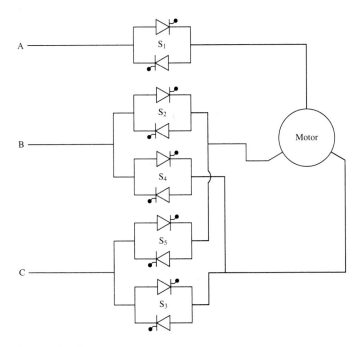

FIGURE 3.17
Static reversing switch for a three-phase motor.

for one direction of rotation. The switches that are to be ON for the reverse direction are S_1, S_4 and S_5. It is important to ensure that S_2 and S_5 are never simultaneously ON at any instant. The same statement applies to S_4 and S_3. Such an event is a short-circuit across the lines B and C.

3.5 CHAPTER 3 SUMMARY

With the exception of the triac, power semiconductor switching devices provide the controlled switching facility only for one direction of current flow. Triacs are not available in such high voltage and current ratings as thyristors. Thyristors are also available with superior dynamic switching characteristics, such as dv/dt, di/dt and turn OFF times. Therefore the use of triacs is generally limited to low power, low frequency applications such as lighting and heating control. We have described methods of implementing bidirectional AC switches using thyristor pairs or thyristors along with diodes. AC switching can be done using other devices such as power BJTs, power MOSFETs and IGBTs by using them in pairs along with power diodes. The circuit schemes for implementing such AC switches have been described. But the circuit techniques presented in this chapter have generally assumed the use of bidirectional thyristor switches.

The simple case of a resistance load supplied from an AC source was considered. Two control techniques have been described, namely integral half-cycle control and phase control. The circuit of the triac controller for incandescent filament lamps is a very common example of the phase control of a resistance load. A mathematical analysis of this circuit has been presented for the purpose of designing the gating circuits using a diac, for the brightness control of incandescent filament lamps and similar applications.

Many practical load circuits can be modeled as series R–L circuits, and therefore we have analyzed the phase control of such load circuits. It is found that control is possible only for firing angles larger than the load impedance angle.

The special case of an R–L load where $R = 0$ is of considerable practical interest in static VAR compensation circuits in AC power networks. This circuit device, known as a thyristor controlled inductor has been analyzed. The fundamental current component is a lagging reactive current, whose value can be adjusted from maximum to zero by phase control over the range $\frac{1}{2}\pi$. The method of using a TCI along with thyristor switched capacitor banks (TSC) for Static VAR compensation has been described.

The use of thyristor AC switches for adjustment of load voltage by sequential tap changing on a power transformer has been described. A second method using both series and parallel AC switches in each transformer secondary section has also been presented. In this scheme, the desired number of secondary sections of the transformer can be independently combined. When this is done with one section included for phase control, the arrangement enables step changes of voltage with vernier control between steps.

The adaptation of single-phase series phase control schemes to three-phase circuits has been considered. Of the two circuit schemes described, one uses AC switches in the AC lines, while the other uses switching at the neutral point.

Another three-phase application described is the reversing switch for a three-phase AC synchronous or induction motor. In this bidirectional thyristor pairs are used to interchange two line connections to the motor.

PROBLEMS

3.1. A 3 kW resistance heating element, working from a 120 V single-phase AC bus, is controlled by an AC switch with integral half-cycle control, with a base period of 12 half-cycles. Calculate the number of ON half-cycles in a base period if the power is to be adjusted to 2 kW.

3.2. If the power in the heating element in Problem 3.1 is being adjusted by phase control, determine the required firing angle.

3.3. A single-phase resistance load is supplied from a 120 V AC bus. A phase controlled AC switch is used in series to adjust the power in the load. The firing delay angle is 45°. Determine the power in the load in terms of the maximum.

3.4. In the circuit of Problem 3.3, determine the r.m.s. value of the total harmonic content in the voltage waveform across the load for the stated firing angle.

3.5. A resistance load of 12 Ω is supplied from a 120 V AC bus. It is phase controlled, with a firing angle of 60°. Determine by Fourier analysis the magnitude and phase of the fundamental sinusoidal component of the load current.

3.6. A resistance of 40 Ω is fed from a 200 V AC source through a phase controlled AC switch. The power in the resistance is adjusted to 400 W by phase control. Determine the firing angle.

For this value of the firing angle, determine the r.m.s. values of the fundamental and the next two higher-order harmonics in the load current. Determine also the phase angle of the fundamental current with espect to the AC source voltage.

3.7. A triac light dimming circuit is to be designed for a 200 W filament lamp working from a 120 V, 60 Hz source. The breakover voltage of the diac is 50 V. The capacitor used is 0.1 μF. Determine the resistance value of an adjustable resistor R, for intensity control that will give control up to a maximum firing angle of 140°.

3.8. A 60 W filament lamp working from a 120 V, 60 Hz supply is to use a triac dimmer circuit. The diac used has a breakover voltage of 40 V. The capacitor is 0.2 μF. Determine the value of the adjustable resistance that can give firing angles up to 150°.

3.9. A 2 kW heating element of a 120 V, 60 Hz electric stove is to be controlled by a triac circuit. The gate circuit of the triac uses a 0.1 μF capacitor and a diac with a breakover voltage of 45 V. Determine the adjustable resistance value to give control down to an r.m.s. voltage of 20 V.

3.10. A light dimmer circuit for a 100 W filament lamp, to be used in a 120 V, 60 Hz supply, is to be designed using a 0.05 μF capacitor, an adjustable resistor R and a diac with a breakover voltage of 40 V. Determine the lowest r.m.s. voltage down to which adjustment is possible if $R = 3$ kΩ.

3.11. A single-phase AC load consists of a resistance of 20 Ω in series with an inductance of 50 mH. It is fed from a 240 V, 60 Hz AC supply through a static AC switch consisting of two antiparallel thyristors. Determine the firing angle range within which it is possible to vary the voltage across the load. For a firing angle of 80°, determine the duration of conduction in each direction of current.

3.12. An AC load is supplied from a 240 V, 60 Hz supply. The load consists of a resistance of 10 Ω in series with an unknown inductance. When a phase controlled AC thyristor switch is used with it, no change in the AC current or voltage is noticed for firing delay angles up to 45°. What is the expected value of the inductance?

For a firing angle of 80°, determine the conduction angle in each half-cycle and sketch the waveform of the voltage across the load.

CHAPTER
4

CHOPPERS

4.1 INTRODUCTION

In AC applications, the transformer serves to convert electric power efficiently from one voltage level to another. Static DC/DC converters presented in this chapter achieve a similar function in DC. The operation of AC transformers is based on an alternating magnetic field. But in DC/DC converters, the voltage conversion is achieved by power semiconductors, which function as static switches, switching at a high repetitive frequency. Static DC/DC converters using the switching principle are also known as choppers.

The circuit configuration of a chopper converter can be designed either to step down from a higher input voltage to a lower output voltage, or to step up from a lower input voltage to a higher output voltage. When the voltage conversion ratio a (= output voltage/input voltage) is less than 1, we call it a voltage step down chopper. If $a > 1$, it is a voltage step up chopper. There are differences between the step down and step up choppers in the internal circuit configuration of their static switches. A chopper converter can also be designed in such a way that its circuit incorporates both the step down and step up switching schemes. In such a converter, it is possible to move between the step down and step up modes at will. Such a chopper is called a two-quadrant chopper. In a two-quadrant chopper, the power flow is from the high voltage side to the low voltage side in the step down mode, and in reverse in the step up mode. Therefore, the input terminal pair and the output terminal pair interchange their roles when moving from one mode of operation to the other. Single- and two-quadrant choppers constitute the subject matter of this chapter.

In any chopper converter, the voltage conversion ratio is determined by the switching times of the static switches that constitute the chopper. It is therefore easy to vary the voltage conversion ratio smoothly and continuously, by means of an adjustable voltage input into the chopper control circuit, to suitably modify the timing of the switching control pulses to the power switching elements. This can be done in both the step down and step up modes of operation. AC power transformers do not have such a facility for static control of voltage conversion ratio. It is the easy controllability of the voltage conversion ratio, statically (that is, without the operation of switches having moving contacts), by means of an adjustable control voltage, that gives the chopper its great usefulness as a power controller.

To highlight the usefulness of this feature, we shall take a brief initial look at the manner in which the chopper conversion principle is employed in some important practical applications.

Figure 4.1 shows how a chopper is used in a typical DC motor speed control application. Figure 4.1(a) shows the driving mode. Here the chopper functions in the voltage step down mode. The power source is a fixed DC voltage source V_1 connected to the high voltage terminals of the chopper, which are A_1 and A_2. In the driving mode, the motor receives its power from the output terminals B_1–B_2 of the chopper. The motor speed is continuously adjustable by varying the chopper voltage conversion ratio, and therefore the voltage applied to the motor. The direction of power flow through the chopper is from the high voltage to the low voltage side. With the positive reference polarity for motor voltage and positive reference direction for motor current as shown in Fig. 4.1(a), it is possible to locate any operating point of the motor on the graph shown in Fig. 4.1(b). It is evident that all operating points in the driving mode will be located in the first (positive I_2, positive V_2) quadrant of the graph.

By using a two-quadrant chopper, it is possible to incorporate the braking feature. For braking, the chopper operation is changed to the voltage step up mode. This is shown in Fig. 4.1(c). The purpose to be achieved by braking is to get rid of the stored kinetic energy of the moving mass, which includes the rotating motor and the load mass coupled to it. If we employ friction braking for this purpose, we dissipate the stored energy as heat generated at the brake shoes. This results in mechanical wear and temperature rise. In contrast, electrical regenerative braking enables recouping of the stored energy and avoids temperature rise due to dissipated heat. It also enables easier programming of the braking torque, to achieve a smooth braking characteristic without jerks. During braking, the motor functions as a DC generator [see Fig. 4.1(c)], driven because of the stored kinetic energy of the mass in motion. The motor delivers electrical power into the terminals B_1–B_2 of the chopper, which now become its input terminals. The chopper steps up the input voltage and returns power into the DC voltage source V_1. If V_1 is a battery, it will be recharging during the braking mode. As the motor slows down, its generated voltage keeps falling. This is taken care of by continuously adjusting the

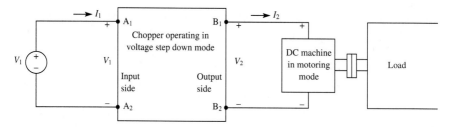

(a) Driving: chopper operation in voltage step down mode

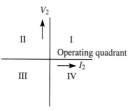

(b) Driving mode: operation in the first quadrant

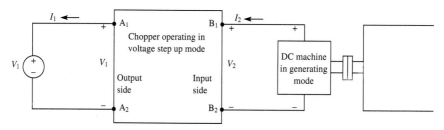

(c) Braking: chopper operation in voltage step up mode

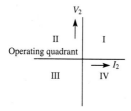

(d) Braking mode: operation in the second quadrant

FIGURE 4.1
Two-quadrant chopper application for DC motor speed control with regenerative braking.

voltage step up ratio of the chopper. We notice from Fig. 4.1(c) that the motor current gets reversed during the braking mode. The operating points during braking are therefore located in the second quadrant (positive V_2, negative I_2), as depicted in Fig. 4.1(d).

Another example where the chopper principle is used with advantage is in the design of "switch mode power supplies" (SMPS). These are widely used at the present time to provide constant stabilized DC voltage for a wide range of electronic equipment, including computers. In an SMPS, the chopper principle makes it possible to maintain constant output voltage by automatically changing the voltage conversion ratio as needed. This is done by closed-loop control, by comparing the output voltage against a fixed voltage reference. Any difference is amplified by an "error amplifier." The output voltage of the error amplifier is used to correct the error by suitably varying the voltage conversion ratio. Switch mode power supplies are dealt with in Chapter 6.

4.2 VOLTAGE STEP DOWN CHOPPER

4.2.1 Power Circuit Configuration and Working Principle

Figure 4.2 shows the circuit configuration of the voltage step down chopper. The chopper is delivering adjustable DC power into a resistive load from a fixed DC voltage source. The chopper power circuit is shown boxed inside broken lines. It consists of two power semiconductor devices, which function as static switches. Of these, the switch labeled S_1 is a controlled switching device. It could be any one of the several types of power semiconductor switching devices. The device can be turned ON or turned OFF at will, by the appropriate control signal on its control terminal. For example, if the device selected for S_1 is a gate turn off thyristor (GTO), it is turned ON by a short positive current pulse on its gate terminal and turned OFF by a short negative current pulse on the same terminal. If on the other hand, S_1 is an n-channel power MOSFET, the turn ON switching is achieved by a positive voltage pulse on its gate terminal lasting for the entire duration of its ON time. The OFF condition of the switch is implemented by making the gate voltage zero. Besides the power circuit shown in the figure, the chopper has a control circuit (not shown for the sake of clarity). It is this control circuit of the chopper that provides the switching signals to operate the power semiconductor static switch S_1 to turn it ON or OFF as required. We shall describe typical control circuits later. For the present, we shall assume the presence of the control circuit, without showing it in the circuit diagram, so that the switchings of S_1 take place in the manner to be described. The second switch shown in the figure, labeled S_2, is actually a power diode. The power diode is a static switch without a control terminal. It

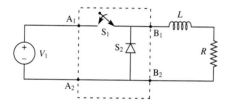

FIGURE 4.2
Voltage step down chopper.

automatically turns ON whenever "forward" current flow is possible, and turns OFF to prevent reverse currents. Both S_1 and S_2 will be considered as ideal. This implies that (1) their transitions between the ON and OFF states are instantaneous without time delays and (2) the voltage drop across the switch terminals in its ON state is zero.

The input terminals of the chopper are A_1 (positive) and A_2 (negative). The output terminals are B_1 (positive) and B_2 (negative). The input DC voltage has a fixed value V_1. The load resistance R is connected to the output through an inductance L. The purpose of this inductance is to smooth out fluctuations in the output current caused by the switching processs in the chopper. We shall see that if L is sufficiently large, the voltage across the load and the current through it will be substantially DC, with negligible AC ripple. It should be remembered that the inductance L, which performs the function of filtering off AC components, is a power element. It is usually iron-cored and should be capable of carrying the full DC load current without magnetic saturation.

To operate the chopper, we turn ON and turn OFF the static switch S_1 at a high repetitive frequency. To derive quantitative relationships, we shall make the following definitions.

f	repetitive switching frequency of the chopper (Hz)
T	cycle time of chopper $= 1/f$ (s)
T_{ON}	ON time of switch S_1 (s)
T_{OFF}	OFF time of switch S_1 (s)
D	duty cycle of switch S_1, defined as $D = T_{ON}/T$

When S_1 is turned ON, we notice two consequent effects. First, the voltage V_1 is applied in reverse across the power diode. Therefore the power diode, which we have labeled as the static switch S_2, must stay OFF as long as S_1 remains ON. The ON state of S_1 always implies the OFF state of S_2. Therefore, the circuit configuration when S_1 is ON will be as shown in Fig. 4.3(a). The second consequence of S_1 being ON is the build up of current in the load resistance. This load current is labeled i_2 in the figure. The growth of i_2 occurs exponentially because of the inductance L. The waveform of the voltage across the output terminals B_1–B_2 is shown in Fig. 4.3(c) for individual chopper switching cycles. This will be equal to V_1 when S_1 is ON. The waveform of the current i_2 in the load circuit is shown in Fig. 4.3(d). It will be noticed that in the first chopper cycle, the current builds up exponentially from an initial zero value.

The switch S_1 is kept ON for a time interval T_{ON}, and then turned OFF. At the instant when S_1 is turned OFF, i_2 has a finite value. We have labeled this current magnitude as I_{p1}. It is the peak value of the output current during the first chopper cycle. This peak current occurs at the instant of turn OFF of the switch S_1. This current cannot instantly fall to zero, because of the presence of the inductance L. The decay of i_2 causes an induced voltage $L\,di_2/dt$ to appear

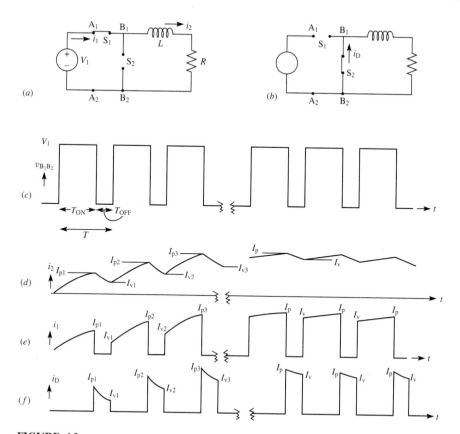

FIGURE 4.3

Voltage step down chopper: (a) circuit configuration during T_{ON} (S_1 ON, S_2 OFF): (b) circuit configuration during T_{OFF} (S_1 OFF, S_2 ON); (c) waveform of voltage across B_1-B_2; (d) waveform of current i_2; (e) waveform of current i_1; (f) waveform of freewheeling current i_D.

across the inductance. Because of this voltage, the diode gets forward-biased and causes the current flow to continue. The term "freewheeling" is commonly used to describe the flow of current in this manner without being caused by a voltage source, but solely due to the stored energy in the inductance. The purpose of the diode S_2 is to provide the freewheeling path when S_1 is turned OFF. Therefore the turning OFF of S_1 automatically causes the turning ON of S_2 when an inductance with stored energy is present. The new circuit configuration that results is shown in Fig. 4.3(b). The voltage across B_1-B_2 is zero, and the current i_2 decays exponentially. The decay of i_2 continues as long as S_1 remains OFF, that is, for a duration T_{OFF}. The lowest value to which the current falls at the end of the first chopper cycle is labeled in Fig. 4.3 as the valley magnitude I_{v1}. The output voltage and current waveforms during the first chopper cycle will therefore be as shown by Figs. 4.3(c) and (d) respectively, for this cycle. Examination of the circuit configurations during the ON and OFF periods of S_1 show that, during T_{ON}, the output current i_2 is the same as the input current i_1. The freewheeling diode current, labeled in the figure as i_D, is the same as i_2 during T_{OFF}. The waveforms of these two currents are shown separately in Figs. 4.3(e) and (f).

The second chopper switching cycle commences when S_1 is turned ON again at the end of the first T_{OFF}. The circuit configuration again changes to that of Fig. 4.3(a), and the current again starts to build up. There is already an initial current equal to I_{v1}, and therefore the second peak I_{p2} will be larger than I_{p1}. Consequently, the valley magnitude I_{v2} at the end of the second cycle will also be larger than I_{v1}. In this way, as the switching progresses, both the peak and valley magnitudes progressively increase. However, we shall see presently that the differences between successive cycles become less and less. After several cycles, the differences between successive cycles become negligibly small. We say that the circuit conditions have reached the repetitive state or steady state. This means that the peak current is effectively the same in successive cycles. A similar statement is true for the valley currents. The repetitive conditions are shown on the right-hand side for all the waveforms in Fig. 4.3. The repetitive peak and valley currents are labeled I_p and I_v respectively in Fig. 4.3. Notice that the waveform of the voltage across B_1–B_2 is repetitive from the first cycle itself.

4.2.2   Voltage Relationship

Figure 4.3(c) shows the waveform of the voltage $v_{B_1B_2}$ at the output terminals B_1–B_2 of the chopper. This is a train of rectangular pulses of duration T_{ON}. This voltage consists of a DC component and an AC component. The AC component is the "ripple voltage." The purpose of using the inductance L is to absorb the ripple voltage across it and present only the DC component to the load resistor R. Notice that the fundamental ripple frequency is the same as the chopper switching frequency. Later we shall show that, with a sufficiently large inductance and a high switching frequency, the ripple component of the output voltage can be made negligibly small. The magnitude of the output DC voltage at the load terminals will be given by the average height of the waveform of Fig. 4.3(c). This will be

$$V_2 = V_1 \frac{T_{ON}}{T} = V_1 D \qquad (4.1)$$

D being the switching duty cycle of the chopper, defined as the ratio of ON time to total cycle time. Therefore the voltage conversion ratio a of the chopper, defined as the ratio of output to input voltage, will be

$$a = V_2/V_1 = D \qquad (4.2)$$

The switching duty cycle can be varied ideally in the range 0–1 by variation of the ON time. It is therefore possible to operate the chopper with any desired voltage conversion ratio below unity and to vary the ratio according to requirements by adjusting the duty cycle.

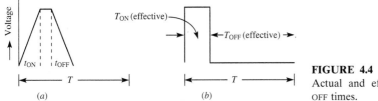

FIGURE 4.4
Actual and effective ON and
OFF times.

4.2.3 Duty Cycle Range of Practical Choppers

In practical choppers, it is impossible to achieve voltage ratio variation over
the full 0–1 range, because the chopper switch, which is a power semiconductor, is nonideal. A minimum effective ON time will be needed for the practical
operation of the power semiconductor switch, and this will depend on the
switching characteristics of the particular device employed as the switch. By
"effective ON time," we mean the ON time of an ideal switch, with zero
switching transition times, that will give the same voltage conversion ratio, as
explained with reference to Fig. 4.4.

Figure 4.4(a) shows the voltage waveform at the output terminals of the
chopper when the static switch has finite switching transition times, labeled as
t_{on} and t_{off} (we have used lower case letter t for the switching transition time, as
against the upper case T to represent the ON and OFF times of the chopper
switch.) Figure 4.4(b) shows the waveform with an ideal switch that has zero
switching transition times. The ON time T_{ON} and the OFF time T_{OFF} in Fig.
4.4(b) are such that the output DC voltage has the same value (same area
under the waveform) as in Fig. 4.4(a). T_{ON} and T_{OFF} of this ideal switch are
what we have defined as the "effective" ON and OFF times. We notice that in
the case of the nonideal switch, there will be a minimum finite effective ON time
even when the turn OFF switching is initiated with zero delay after the turn ON
switching is completed. Therefore the lowest possible voltage conversion ratio
will be finite. If a still lower output voltage is needed, it becomes necessary to
reduce the chopper switching frequency. This is illustrated by the following
example.

> **Illustrative example 4.1.** A chopper controlled electric train is powered from a
> 1500 V DC supply. The power semiconductor switching element has a minimum
> effective ON time of 40 μs. During starting and slow speed running, the output of
> the chopper has to go as low as 15 V. What is the highest chopper frequency
> possible to satisfy this requirement?
>
> **Solution.** The minimum duty cycle needed is $15/1500 = 0.01 = D$. The minimum
> possible T_{ON} is 40 μs. Since $DT = T_{ON}$, we have $T = T_{ON}/D = 40/0.01 = 4000$ μs.
> Therefore the maximum possible chopper frequency is $1/(4000 \times 10^{-6}) = 250$ Hz.

Note. A reduction in the chopper switching frequency increases the output
current ripple, thereby increasing losses and heating in the DC motor.

Therefore it is usually the practice in the chopper controlled trains to change over to a higher switching frequency at higher vehicle speeds, when the requirement of low output voltage no longer exists. The chopper in such a case may be designed to work at two (or more) discrete frequencies.

Illustrative example 4.2. If, in Example 4.1, the chopper switching frequency is increased to 2000 Hz, what is the minimum possible motor voltage?

Solution. For $f = 2000$ Hz, $T = 1/f = 500$ μs. The minimum T_{ON} is 40 μs. So the minimum possible duty cycle is $40/500 = 0.08$. The minimum output voltage that the chopper can provide to the motor is therefore $1500 \times 0.08 = 120$ V.

Maximum limit of voltage conversion ratio. If the chopper has to be turned ON and OFF in every switching cycle then the minimum effective OFF time also imposes a limit on the maximum duty cycle and therefore on the maximum voltage conversion ratio possible. The minimum effective OFF time is the OFF time of an ideal switch, with zero switching transition times, that will give the same voltage ratio as the real switch when the real switch commences its turn ON at the beginning of a chopper period and completes the turn OFF switching at the end of the same chopper period. This limitation of having to have a maximum voltage ratio of less than unity is usually overcome in chopper controlled vehicles by using an electromagnetic contactor (mechanical switch), which is turned ON to bypass the chopper at high vehicle speeds. The chopper switchings may be suspended when the chopper bypass contactor is turned ON. These statements are illustrated by the following example.

Illustrative example 4.3. The chopper in Example 4.1 has a minimum effective OFF time of 30 μs. The chopper frequency is 2000 Hz.
(*a*) What is the maximum duty cycle and maximum output voltage possible?
(*b*) If the full input voltage is to be made available to the load, show how this can be done by using a bypass contactor.

Solution.
(*a*) The chopper cycle time is $1/2000 = 500$ μs. The maximum effective ON time possible is $500 - 30 = 470$ μs. Therefore the maximum duty cycle possible is $470/500 = 0.94$ and the maximum output voltage possible is $0.94 \times 1500 = 1410$ V.
(*b*) To apply the full 1500 V to the motor, a chopper bypass contactor, connected as shown in Fig. 4.5, should be turned ON. When the contactor is ON, the switchings of the chopper can be suspended.

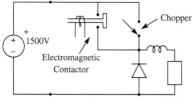

FIGURE 4.5
Use of chopper bypass contactor.

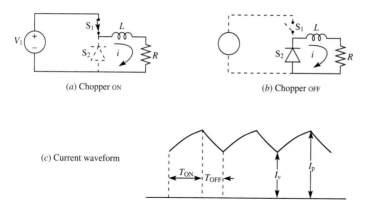

FIGURE 4.6
Current waveform under repetitive conditions.

4.2.4 Current Relationships

If the chopper has been working for a sufficient number of cycles at constant duty cycle and load conditions, repetitive conditions will prevail from cycle to cycle, as we shall presently verify. After repetitive conditions can be assumed, the load current waveform will be as shown in Fig. 4.6(c). The paths of the load current during T_{ON} and T_{OFF} are shown separately in Fig. 4.6(a) and (b).

During T_{ON}, the load current rises from the valley magnitude labeled I_v in Fig. 4.6 to the peak magnitude labeled I_p. During T_{OFF}, the current decays from I_p to I_v. This sequence is repeated during successive cycles. In the following analysis, we shall obtain expressions for I_v and I_p and also observe the progress of the chopper towards repetitive conditions, starting from zero initial current.

Figure 4.7 shows graphically the growth of load current during the progress of switching. The individual chopper cycles are labeled $1, \ldots, n$ in

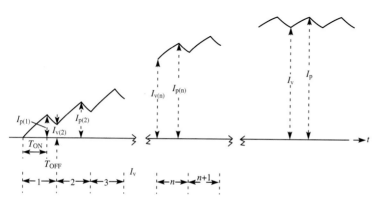

FIGURE 4.7
Growth of load current as switching progresses.

the figure. We shall assume that the current is zero initially at the commencement of the first cycle. We shall turn our attention to the nth cycle. The initial current at commencement of T_{ON} in the nth cycle is labeled as the valley current $I_{v(n)}$. At the end of T_{ON} in this cycle, the current reaches the peak magnitude labeled $I_{p(n)}$. For convenience, we shall first take our reference zero of time ($t = 0$) at the commencement of T_{ON}. The loop equation during T_{ON} [see Fig. 4.6(a)] will be

$$L\frac{di}{dt} + Ri = V_1$$

with $i = I_{v(n)}$ at $t = 0$. The solution of this gives, during T_{ON},

$$i = I_{v(n)}e^{-t/\tau} + \frac{V_1}{R}(1 - e^{-t/\tau}) \tag{4.3}$$

where $\tau = L/R$ is the time constant of the output circuit. At the end of the ON period ($t = T_{ON}$), $i = I_{p(n)}$, with

$$I_{p(n)} = I_{v(n)}e^{-T_{ON}/\tau} + \frac{V_1}{R}(1 - e^{-T_{ON}/\tau}) \tag{4.4}$$

We shall now consider the OFF period in the nth cycle. Again, for convenience, we shall take a new reference zero of time ($t = 0$) at the commencement of the OFF period. The loop equation during T_{OFF} [see Fig. 4.6(b)] will be

$$L\frac{di}{dt} + Ri = 0$$

with $i = I_{p(n)}$ at $t = 0$. The solution of this gives, during T_{OFF},

$$i = I_{p(n)}e^{-t/\tau} \tag{4.5}$$

At the end of the OFF period ($t = T_{OFF}$), $i = I_{v(n+1)}$, with

$$I_{v(n+1)} = I_{p(n)}e^{-T_{OFF}/\tau} \tag{4.6}$$

Substituting the value of $I_{p(n)}$ from (4.4)

$$I_{v(n+1)} = I_{v(n)}e^{-T/\tau} + \frac{V_1}{R}(e^{-T_{OFF}/\tau} - e^{-T/\tau}) \tag{4.7}$$

where $T = T_{ON} + T_{OFF}$ is the chopper period. Equation (4.7) is a recursive relationship between the valley currents in successive chopper cycles, which may be written as

$$I_{v(n+1)} = A + BI_{v(n)} \tag{4.8}$$

where

$$A = \frac{V_1}{R}(e^{-T_{OFF}/\tau} - e^{-T/\tau}) \tag{4.9}$$

$$B = e^{-T/\tau} \tag{4.10}$$

For the first cycle ($n = 1$), $I_{v(n)} = 0$. Therefore we can list the values of $I_{v(n)}$ as follows:

$$I_{v(2)} = A$$

$$I_{v(3)} = A(1 + B)$$

$$I_{v(4)} = A(1 + B + B^2)$$

$$\vdots \qquad \vdots$$

$$I_{v(n+1)} = A(I + B + B^2 + \cdots + B^{n-1})$$

Substituting the expression for the sum of the geometric series inside the parentheses, we get

$$I_{v(n+1)} = A\frac{1 - B^n}{1 - B} = \frac{V_1}{R}\frac{e^{-T_{OFF}/\tau} - e^{-T/\tau}}{1 - e^{-T/\tau}}(1 - e^{-nT/\tau}) \tag{4.11}$$

When n is large so that $e^{-nT/\tau} \ll 1$, we can assume repetitive conditions, and (4.11) gives the repetitive value of the valley current as

$$I_v = \frac{V_1}{R}\frac{e^{-T_{OFF}/\tau} - e^{-T/\tau}}{1 - e^{-T/\tau}} \tag{4.13}$$

Substitution of this value of I_v into (4.6) will give us the value of the peak current I_p after the assumption of repetitive conditions is justified. This will be

$$I_p = \frac{V_1}{R}\frac{1 - e^{-T_{ON}/\tau}}{1 - e^{-T/\tau}} \tag{4.14}$$

Notice that we could have obtained the repetitive values of I_v and I_p more quickly if we had performed the analysis starting from the assumption of repetitive conditions. But we chose the longer procedure to show how the currents grow towards the repetitive values.

Peak-to-peak ripple current. This is expressed as

$$I_{pp} = I_p - I_v \tag{4.15}$$

where I_p and I_v are given by (4.14) and (4.13).

DC component of the load current. The load current waveform after repetitive operation of the chopper prevails is shown by the right-hand section of Fig. 4.7. It consists of a DC component with a superposed AC ripple. Since the chopper is used basically to convert DC, it is the DC component of the output current that is of primary interest to us. This can be determined by finding the average height of the current waveform on the right-hand side in Fig. 4.7. For example, we can find the area under the waveform for one chopper period by integrating the expressions for the current given by (4.3) and (4.5) for the respective intervals. Division of the total area so obtained by the chopper period will

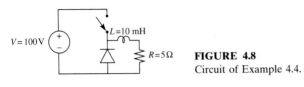

FIGURE 4.8
Circuit of Example 4.4.

yield the DC component I_d. But the same result can be more simply obtained by using the principle of superposition. For this, we divide the DC component of the output voltage by the load resistance R. Thus

$$I_d = \frac{V_2}{R} = \frac{DV_1}{R} \tag{4.16}$$

Using (4.15) and (4.16), we could express the AC ripple current as a percentage of the DC output current.

Illustrative example 4.4. The chopper shown in Fig. 4.8 is switching at a frequency $f = 1\,\text{kHz}$, with a duty cycle of 50%. Determine
(a) the DC load current I_d;
(b) the peak-to-peak ripple current (as an absolute value and also as a percentage of the DC value).

Solution.
(a) The DC component of the output voltage $V_2 = DV_1 = 50\,\text{V}$. So the DC component of the output current $I_d = V_2/R = 10\,\text{A}$.
(b) The chopper period $T = 1/f = 10^{-3}\,\text{s}$. $T_{ON} = 0.5T = 0.5 \times 10^{-3}\,\text{s}$. $T_{OFF} = (1 - 0.5)T = 0.5 \times 10^{-3}\,\text{s}$. The time constant of output circuit $\tau = L/R = 2 \times 10^{-3}\,\text{s}$. Then, from (4.14)

$$I_p = \frac{V_1}{R}\frac{1 - e^{-T_{ON}/\tau}}{1 - e^{-T/\tau}} = 11.24\,\text{A}$$

and, from (4.15)

$$I_v = \frac{V_1}{R}\frac{e^{-T_{OFF}/\tau} - e^{-T/\tau}}{1 - e^{-T/\tau}} = 8.75\,\text{A}$$

so that

$$I_{pp} = I_p - I_v = 2.49\,\text{A}$$

and the percentage peak-to-peak ripple current is then $100 \times I_{pp}/I_d = 100 \times 2.49/10 = 24.9\%$.

Illustrative example 4.5. In the previous example, other data remaining the same, the chopper frequency is increased four times (to 4 kHz).
(a) Determine how the percentage peak-to-peak ripple current is affected.
(b) Instead of changing the frequency, the smoothing inductance magnitude is increased four times (to 40 mH). What will be the percentage ripple current? Compare the results of (a) and (b), and comment.

Solution.

(a) The new numbers are: $T = 2.5 \times 10^{-4}$ s, $T_{ON} = 1.25 \times 10^{-4}$ s and $T_{OFF} = 1.25 \times 10^{-4}$ s. Substitution of these numbers in the same expressions as before gives $I_p = 10.31$ A, $I_v = 9.69$ A and so

$$I_{pp} = 0.62 \text{ A}$$

I_d is the same as before, 10 A and so the percentage peak-to-peak ripple current is

$$(0.62/10) \times 100 = 6.2\%$$

Notice that a large reduction of ripple current has been achieved by increasing the chopper frequency.

(b) Increasing L four times increases the time constant τ four times. But since the frequency is not increased, the ratios: T/τ, T_{ON}/τ and T_{OFF}/τ will be exactly the same as in (a). Since the duty cycle is unchanged, the DC current will also be the same as before. Therefore the values of I_p, I_v and I_{pp} will be the same as in (a).

The above results show that the same reduction in ripple current can be achieved either by increasing the chopper frequency or by increasing the filter inductance value. In practice, a higher switching frequency is preferred, since this enables the use of a smaller inductance, for the same ripple magnitude. The inductance is usually iron-cored, designed to carry the full DC current without the core getting saturated. Therefore it could be a heavy piece of equipment.

4.2.5 Dependence of Ripple Current on Duty Cycle

Variation of the duty cycle is the practical means employed to vary the current in a load. Therefore it is of interest to find how the ripple current depends on the duty cycle, the load circuit parameters (R and L) remaining the same.

The peak-to-peak ripple current given by (4.15) may be written as

$$I_{pp} = \frac{V_1}{R} \frac{1}{1 - e^{-T/\tau}} [(1 + e^{-T/\tau}) - (e^{-DT/\tau} + e^{-(1-D)T/\tau})] \qquad (4.17)$$

In this equation, the quantity that varies when the duty cycle alone is changed is that enclosed in the last pair of parentheses. Inspection shows that I_{pp} will be maximum when this term is minimum. Therefore the duty cycle at which maximum I_{pp} occurs can be found by differentiating the expression within these parentheses with respect to D and equating to zero. This procedure will give the value of D at which the minimum value of the quantity $e^{-DT/\tau} + e^{-(1-D)T/\tau}$ occurs in (3.17) as

$$D = 0.5 \qquad (4.18)$$

This gives a useful result that the maximum absolute value of the peak-to-peak ripple current occurs at 50% duty cycle. This being the worst-case condition, we can use this duty cycle value for determining the smoothing inductance L to limit the peak-to-peak current ripple below any specified limit.

Maximum value of the peak-to-peak current ripple. The maximum value of I_{pp} will be given by putting $D = 0.5$ in (4.17). On simplification, this may be written as

$$I_{pp} = \frac{V_1}{R} \frac{(1 - e^{-0.5T/\tau})^2}{1 - e^{-T/\tau}}$$

We may replace $1 - e^{-T/\tau}$ by the product of its factors $1 - e^{-0.5T/\tau}$ and $1 + e^{-0.5T/\tau}$ and simplify, giving

$$I_{pp} = \frac{V_1}{R} \frac{1 - e^{-0.5T/\tau}}{1 + e^{-0.5T/\tau}} \qquad (4.19)$$

Illustrative example 4.6. Use (4.19) to determine the peak-to-peak ripple current and check with the value calculated in Example 4.4.

Solution. Substitution of the appropriate numbers into (4.19) gives

$$I_{pp} = \frac{100}{5} \frac{1 - e^{-0.5 \times 10^{-3}/2 \times 10^{-3}}}{1 + e^{-0.5 \times 10^{-3}/2 \times 10^{-3}}} = 2.49 \text{ A}$$

which agrees with the previous result.

4.2.6   Choice of Filter Inductance Value and/or Chopper frequency to Limit the Ripple Current within a Specified Limit

We can use (4.19) to determine the inductance value and/or choose the chopper frequency so as to limit the maximum value of the peak-to-peak ripple current as required. Equation (4.19) gives

$$e^{-0.5T/\tau} = \frac{1 - a}{1 + a} \qquad (4.20)$$

where

$$a = \frac{I_{pp}R}{V_1} \qquad (4.21)$$

This may be written as

$$\frac{T}{\tau} = 2 \ln \frac{1 + a}{1 - a}$$

where $T = 1/f$, with f the chopper frequency, and the load circuit time constant $\tau = L/R$. Making these substitutions, we get

$$fL = \frac{R}{2 \ln [(1 + a)/(1 - a)]} \qquad (4.22)$$

An important practical implication of (4.22) is that the ripple current can be kept below a specified limit, either by increasing the value of L or by proportionately increasing the chopper frequency. This explains why chopper converters should be designed to work at high switching frequencies to minimize filter inductance requirements.

> **Illustrative example 4.7.** In Example 4.6, it is desired to limit the peak-to-peak ripple current over the entire range of duty cycle to 0.2 A.
> (a) What is the lowest chopper frequency at which this will be possible for the given values of the load circuit parameters.
> (b) If the highest switching frequency possible with the selected switching element is 5 kHz, what should be the value of the filter inductance, to satisfy this ripple current specification?

> *Solution.*
> (a) Substitution of the given value of I_{pp} into (4.21) and (4.22) gives $fL = 125$. For $L = 10^{-2}$ H, this gives the lowest permissible chopper frequency, namely $f = 12.5$ kHz.
> (b) If the highest frequency capability of the switching element is only 5 kHz, the value of the inductance should be increased proportionately. The required value will be $L = 125/5 = 25$ mH.

4.2.7 Chopper Operation when Load Presents a "Back e.m.f."

Speed control of DC motors is a very important application of choppers. In this and some other applications of chopper control, an opposing voltage source may be present in the load circuit of the chopper converter. When a chopper is used to vary the speed of a DC motor, the motor has a "back e.m.f.," which is the voltage induced in the motor circuit because of its rotational speed. The chopper has to drive the load current in opposition to this induced e.m.f. of the motor. In a battery charging application, the chopper has to drive the charging current in opposition to the battery e.m.f. Figure 4.9 shows a chopper feeding a load that has such an e.m.f. We can make a general analysis of the chopper operation with such a load and obtain relationships similar to what we obtained earlier for operation with R–L loads. These

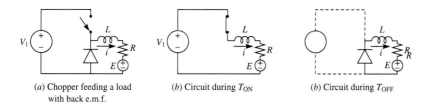

(a) Chopper feeding a load with back e.m.f.

(b) Circuit during T_{ON}

(b) Circuit during T_{OFF}

FIGURE 4.9
Load with back e.m.f.

relationships will be applicable to earlier case also by making the value of the e.m.f. zero. Since the manner in which the current builds up to the final repetitive state has already been illustrated by the earlier analysis, we shall limit our analysis here to steady state operation, that is, operation after repetitive conditions have been attained. Assumption of repetitive conditions implies that the load current waveform will be repeated in successive chopper cycles as shown typically in Fig. 4.6. The peak values, which are the same in successive cycles, will be labeled I_p and the valley magnitudes will be labeled I_v.

During the ON state of the chopper switch (T_{ON}), the circuit configuration will be as shown in Fig. 4.9(b). The loop equation will therefore be

$$V_1 = L\frac{di}{dt} + Ri + E \tag{4.23}$$

the initial condition being $i = I_v$ at $t = 0$, taking the reference zero of t as the commencement of T_{ON}.

During the OFF state of the chopper switch, that is, during the interval T_{OFF}, for which we shall take a new reference zero of t at the commencement of T_{OFF}, the circuit configuration will be as shown in Fig. 4.9(c). The corresponding loop equation will be

$$0 = L\frac{di}{dt} + Ri + E \tag{4.24}$$

the initial condition being $i = I_p$ at $t = 0$.

The solution of these two differential equations can be done in exactly the same manner as for the earlier case of $E = 0$, and is omitted here, being left as an exercise for the reader. This will yield the following results:

$$I_p = -\frac{E}{R} + \frac{V_1}{R}\frac{1 - e^{-T_{ON}/\tau}}{1 - e^{-T/\tau}} \tag{4.25}$$

$$I_v = -\frac{E}{R} + \frac{V_1}{R}\frac{e^{-T_{OFF}/\tau} - e^{-T/\tau}}{1 - e^{-T/\tau}} \tag{4.26}$$

where $\tau = L/R$ is the load circuit time constant. Notice that these two expressions can be obtained from those for steady state I_p and I_v derived earlier for chopper operation with an R–L load (without back e.m.f.) by adding the negative current term $-E/R$.

The peak-to-peak ripple in the output current will be given by:

$$I_{pp} = I_p - I_v$$

This will give the same expression for I_{pp} as we obtained earlier for the case of an R–L load.

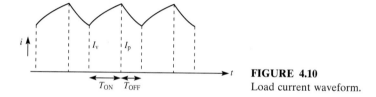

FIGURE 4.10
Load current waveform.

Figure 4.10 shows a typical waveform of the load current based on the above analysis.

Illustrative example 4.8. An electric car employs a DC motor fed through a chopper from a 200 V battery. The chopper operates at a fixed frequency of 2 kHz. The motor resistance is 0.04 Ω and the total inductance in the load circuit is 0.1. mH. At a speed of 40 miles/h, the motor devlopes an induced e.m.f. of 60 V. The chopper duty cycle D while cruising at this speed on level road is 33.2%. Determine
(a) the peak-to-peak ripple current in the motor;
(b) the DC component of the motor current.

Solution
(a) the load time constant is $L/R = 0.1 \times 10^{-3}/0.04 = 2.5 \times 10^{-3}$ s. The chopper period $T = 1/f = 5 \times 10^{-4}$ s. Therefore $T_{ON} = DT = 0.332 \times 0.5 \times 10^{-3} = 1.66 \times 10^{-4}$ s and $T_{OFF} = (1-D)T = 3.34 \times 10^{-4}$ s. Also, $V_1 = 200$ V and $E = 60$ V. Substituting the above numbers into the expressions for the peak and valley current gives

$$I_p = 272 \text{ A}, \quad I_v = 50.4 \text{ A}$$

the peak to peak ripple current is then $I_p - I_v = 221.6$ A.
(b) The DC component of the output voltage is $DV_1 = 0.332 \times 200 = 66.4$ V. Therefore the DC component of the output current is

$$(66.4 - E)/R = 160 \text{ A}$$

4.2.7.1. CONTINUOUS AND DISCONTINUOUS CURRENT MODES. Example 4.8 shows that, in this case, the peak-to-peak ripple current has a very large value compared with the DC current. To reduce the ripple, we can either choose a higher chopper frequency or increase the value of the inductance. This statement can easily be verified if we rework Example 4.8 with a higher value of either the inductance or the chopper frequency or both.

If, however, we reduce the inductance, the valley current will decrease further and finally become zero. A still further decrease in L will cause the current to fall to zero even before the completion of the OFF period (T_{OFF}) of the chopper. We shall illustrate this by reworking Example 4.8 with the inductance value changed to 0.06 mH, all other data remaining the same. Substitution of the new value of L in the expressions for the peak and valley current gives

$$I_p = 347.9 \text{ A}, \quad I_v = -21 \text{ A}$$

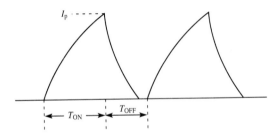

FIGURE 4.11
Discontinuous flow of load current.

We notice that the expression for I_v gives a negative value for the current. This is impossible, because of the diode in the freewheeling path [see Fig. 4.9(c)]. The diode turns OFF at the instant the current falls to zero and tends to reverse. Therefore the original loop equation that gave us the expression for I_v becomes invalid from the instant the diode switch turns OFF. Therefore, after the current has fallen to zero, it stays at zero for the rest of the OFF period of the chopper. The current again builds up from an initial zero value during T_{ON} in the next and every other chopper cycle. The current waveform is sketched in Fig. 4.11.

We notice that the current flows in pulses, and is not continuous. We say, that the chopper is operating in the discontinuous load current mode. Such an operation is generally not desirable, particularly in the speed control of DC motors. It is generally possible to avoid this mode of operation by suitable choice of chopper frequency or load circuit time constant, or both.

The limiting condition for the onset of the discontinuous current mode can be found by equating the expression for I_v to zero as shown below. Equating $I_v = 0$, we get

$$\frac{E}{V_1} = \frac{e^{-T_{OFF}/\tau} - e^{-T/\tau}}{1 - e^{-T/\tau}} \tag{4.27}$$

The above condition for the onset of discontinuous current mode tells us that the determining parameters are

1. the ratio of the back e.m.f. to the chopper voltage, E/V_1
2. the ratio of the chopper period to the inductive time constant $\tau = L/R$ of the load.

These parameters determine the duty cycle below which the chopper operation will be in the discontinuous load current mode.

The use of the limiting condition derived above is illustrated by the following numerical example.

Illustrative example 4.9. The speed of a DC motor is controlled by a chopper that works from a 200 V DC supply at a fixed frequency of 2 kHz. The motor circuit has a resistance of $0.2\,\Omega$ and inductance of $0.2\,\text{mH}$. When operating at a particular speed, the motor develops a back e.m.f. of 150 V. What is the minimum duty cycle of the chopper below which the operation will be in the discontinuous current mode?

Solution. The load circuit time constant $\tau = L/R = 0.2 \times 10^{-3}/0.2 = 1$ ms, the chopper period $T = 1/f = 0.5$ ms, and $E/V_1 = 150/200 = 0.75$. We substitute these numbers into the condition for the commencement of discontinuous current as given by (4.27). This gives $T_{OFF}/\tau = 0.103\,55$. Therefore $T_{OFF} = 0.103\,55$ ms and $T_{ON} = T - T_{OFF} = 0.3965$ ms. This gives the limiting duty cycle below which current will be discontinuous as

$$D = 0.3965/0.5 = 0.793$$
$$= 79.3\%$$

4.2.7.2 CHOPPER OPERATION WITH INFINITE INDUCTIVE SMOOTHING.

We notice from the above treatment that, from the point of view of avoiding a discontinuous load current, it is desirable to have a high value of inductance in the load circuit. We also see that the peak-to-peak ripple of the load current becomes less and less as the value of the inductance is increased. When the inductance has a very large value (ideally infinite), we get the best steady state operation of the chopper with zero ripple. The load current waveform will be a perfect DC. Figure 4.12 shows the current waveforms in the load, the freewheeling diode and the power semiconductor chopper switch for such an ideal operation with infinite inductive smoothing.

Ideal operation with a large (infinite) smoothing inductance in the load circuit is often assumed in a simplified study of the chopper converter. We have already shown how an exact analysis can be performed for a real chopper with finite load time constant, both for steady state operation and when progessively moving towards the steady state repetitive conditions. Therefore we shall generally limit our subsequent treatment to the ideal operation (unless otherwise stated in the particular context), so as to highlight the basic aspects of the chopper converter.

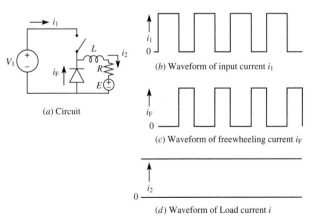

(a) Circuit

(b) Waveform of input current i_1

(c) Waveform of freewheeling current i_F

(d) Waveform of Load current i

FIGURE 4.12.
Current waveforms with infinite inductive smoothing.

4.2.8 Use of Filter on the Input Side

An inspection of the current waveforms in Fig. 4.12 shows that, although the load current waveform is a perfect DC for the case of ideal output filtering, the current drawn by the chopper on its input side is not so. We notice that the input current flowing from the DC source actually consists of a train of pulses. Such a current flow is far from satisfactory for a DC source, such as a battery or a DC power supply bus. The major disadvantages of such a pulsed current flow are as follows.

1. The source has to handle a larger peak current.
2. There is a higher power loss in resistive paths due to the higher value of the r.m.s. current.
3. There is electromagnetic interference due to the high frequency components in the current and the sharp rising and falling edges of the current pulses.

It is therefore desirable to provide a filter circuit on the input side also, with the objective of keeping the peak-to-peak ripple current in the DC source within acceptable limits. Figure 4.13 shows a filter circuit that will serve this objective.

For the chopper shown in Fig. 4.13, we shall assume infinite inductive smoothing on the load side. We have shown a back e.m.f. on the load side, labeled E. This will be zero for a load without back e.m.f. The filter on the input side is shown boxed inside the broken lines, and consists of an inductance labeled L_1 and a capacitance labeled C_1.

Let us assume the load current labeled i_2 in the figure has negligible ripple. When the chopper switch S_1 is ON, this current is the sum of the capacitance current i_c and the DC source current i_1, the latter being the same as the current flowing through the inductance L_1. The capacitance current will be in the direction shown, and therefore the capacitor will lose some charge during T_{ON} of the chopper, with a corresponding drop in the voltage applied to the chopper. This drop will be small if the capacitor has a sufficiently large

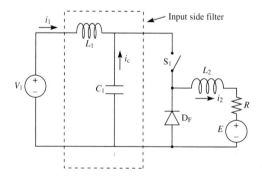

FIGURE 4.13
Chopper with input side filter.

value. During the OFF state of S_1 (T_{OFF}), current flow in the inductance will continue, but this current will now be the charging current of the capacitor. We therefore notice that the capacitor will be charging during T_{OFF} and discharging during T_{ON} of the chopper switch. Therefore the capacitor current will be an alternating current during the steady state working of the chopper. Therefore there will be a corresponding AC ripple superimposed on the capacitor voltage, which is the same as the input voltage to the chopper. In making the above statements, we are referring to the situation when the chopper is operating in the steady (repetitive) state. Prior to the commencement of the operation of the chopper switch, adequate time should be provided for the capacitor to charge to the full value of the DC source voltage V_1. If we neglect the resistance of L_1, the capacitor voltage will have the same DC value during the steady state operation of the chopper also. This is evident from the fact that there cannot exist a steady DC voltage across an ideal inductance. The ripple in the capacitor voltage can be kept low, within acceptable limits, by choosing a sufficiently large value for C_1. Fortunately, it is economical to use large values, because we can use the electrolytic type of capacitors, whose polarized nature is acceptable for DC circuits. Such capacitors are available in compact sizes for large capacitance values. With a sufficiently large value of C_1, it is easy to keep the ripple in the chopper input voltage within acceptable limits. This is illustrated by the following numerical example.

Illustrative example 4.10. The chopper of Fig. 4.13 is working at a frequency of 1 kHz and a duty cycle of 60%. The DC suppy voltage is 250 V. The output load current is 20 A, and may be assumed to be ripple-free because of perfect inductive smoothing. The filter inductance L_1 on the input side may be assumed to be large enough to justify treating the source current i_1 as having negligible ripple. Determine approximately the value of the input side filter capacitance necessary to keep the peak-to-peak ripple voltage on the chopper input side within 4% of the DC source voltage. Treat all circuit elements and switching devices as ideal.

Solution. For a duty cycle of 0.6, the output DC voltage is $0.6 \times 250 = 150$ V. The output power is $150 \times 20 = 3000$ W. Because all circuit elements and switches are ideal, there is no internal power loss in the chopper. Therefore

$$\text{input power} = \text{output power} = 3000 \text{ W}$$

This gives the value of the current drawn from the source V_1, which is assumed to be constant and ripple-free because of the large value of L_1, as

$$i_1 = 3000/250 = 12 \text{ A}$$

During the OFF period of the chopper, this 12 A is the charging current of the capacitor. The total charge flowing into the capacitor is therefore

$$Q = 12T_{OFF}$$

where

$$T_{OFF} = (1 - \text{duty cycle}) \times \text{chopper period} = 4 \times 10^{-4} \text{ s}$$

This gives $Q = 4.8 \times 10^{-3}$ C. Therefore the rise in the capacitor voltage will be

$$4.8 \times 10^{-3}/C_1 \quad 1 \text{ V}$$

During the ON period of the chopper, the capacitor current will be $20 - 12 = 8$ A. The capacitor will discharge during this interval. The quantity of charge that flows out of the capactor during the ON period will be seen to be exactly the same as the quantity flowing into it during the OFF period, which we calculated above. Therefore the voltage drop V that occurs during the ON period will be exactly the same in magnitude as the voltage rise during the OFF period. Any one of the two may be used to determine the peak-to-peak capacitor ripple voltage.

The specified maximum value of the peak-to-peak capacitor ripple voltage is $0.04 \times 250 = 10$ V. Therefore

$$4.8 \times 10^{-3}/C_1 = 10 \text{ V}$$

This gives $C_1 = 480 \ \mu\text{F}$.

The value of the filter capacitance calculated above is approximate, because of the assumption of ideal elements and ripple-free currents in the source and the load. But the procedure is simple and serves to quickly determine realistic values of C_1. A larger value of C_1 will make the ripple voltage less. Notice that the ripple voltage across the capacitor is also the ripple voltage across the filter inductance L_1, because the other end of the inductor is tied to a fixed ripple-free voltage source. Therefore the use of a larger value of C_1 will correspondingly reduce the input ripple current, which in the above approximate calculation we treated as negligible.

4.2.9 Ideal Chopper with Perfect Filters on Both Input and Output Sides

In the chopper circuit shown in Fig. 4.14, the load circuit time constant is assumed to be very large, so that the load current can be considered as ripple-free. The filter inductance L_1 and the filter capacitance C_1 on the input side are assumed to be very large, so that the input source ripple current and the input chopper ripple voltage are negligible.

With the above conditions, the chopper functions as an ideal voltage step

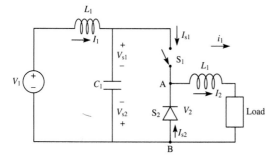

FIGURE 4.14
Chopper with ideal filters on both input and output sides.

down transformer for DC, similar to an ideal AC voltage stepdown transformer. The corresponding relationships between primary and secondary quantities are as follows:

$$\text{voltage ratio} \quad V_2/V_1 = D$$

$$\text{current ratio} \quad I_2/I_1 = 1/D$$

$$\text{Input power} = V_1 I_1 = \text{output power} = V_2 I_2$$

The voltage ratio of our DC transformer is adjustable in the range from zero to one through the simple means of adjusting the duty cycle D of the chopper. The different voltages and currents of interest to us are labeled by appropriate symbols in Fig. 4.14, and their waveforms are sketched in Fig. 4.15. These include the currents through the switching devices (the controlled chopper switch S_1 and the uncontrolled freewheeling diode switch S_2) and the voltages across them. The voltage and current waveforms of a switching device are

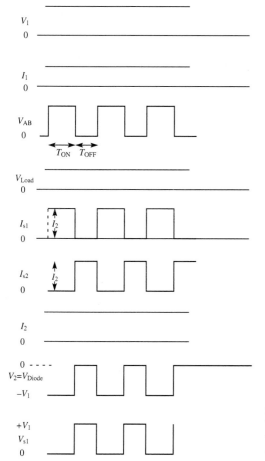

FIGURE 4.15
Waveforms of voltage step down chopper with ideal filters on both input and output.

needed to determine the actual ratings of the device to be chosen. It should however, be remembered, that the waveforms shown in the figure are based on ideal assumptions. It is always advisable to use a relatively large safety factor, both as regards the voltage rating and as regards the current rating, when choosing the switching elements. The safety factor in the voltage rating is necessary to take care of unavoidable overvoltage spikes, which may occur owing to switching transients. The safety factor in the current rating takes care of overcurrents due to abnormal operating conditions, overloads etc. Also, current ratings are dependent on the degree of cooling provided for the device, that is, the type of heat sink used with it.

Illustrative example 4.11 For the ideal chopper of Fig. 4.14, $V_1 = 100$ V, $V_2 = 60$ and $I_2 = 10$ A. Determine the following:
(a) the duty cycle;
(b) the input current I_1;
(c) the average and r.m.s. values of the current I_{s1} in the static switching element S_1;
(d) the average and r.m.s. values of the current I_{s2} in the freewheeling diode;
(e) the repetitive peak forward and reverse voltages of the static switching elements S_1;
(f) the repetitive peak forward and reverse voltages of the freewheeling diode (switching element S_2).

Solution.
(a) The duty cycle is equal to the voltage ratio. Therefore $D = 60/100 = 0.6$.
(b) The input current $I_1 = DI_2 = 0.6 \times 10 = 6$ A.
(c) The average value I_{s1} is

$$I_2 T_{ON}/T = 10 \times 0.6 = 6 \text{ A}$$

The average value of I_{s1}^2 is

$$I_2^2 T_{ON}/T = DI_2^2$$

so the r.m.s. value of I_{s1} is

$$\sqrt{DI_2^2} = \sqrt{D}I_2 = \sqrt{0.6} \times 10 = 7.7 \text{ A}$$

(d) The average value of I_{s2} is

$$I_2 T_{OFF}/T = I_2(1 - D)$$
$$= 10 \times 0.4 = 4 \text{ A}$$

The average value of I_{s2}^2 is

$$I_2^2 T_{OFF}/T = I_2^2(1 - D)$$

so the r.m.s. value of I_{s2} is

$$\sqrt{1 - D}I_2 = \sqrt{0.4} \times 10 = 6.3 \text{ A}$$

(e) The repetitive peak forward voltage of S_1 is $V_1 = 100$ V and the repetitive peak reverse voltage is 0 V.

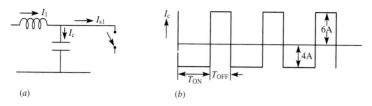

(a) (b)

FIGURE 4.16
Capacitor current waveform in Example 4.12.

(*f*) The repetitive peak forward voltage of S_2 is $0\,V$ and the repetitive peak reverse voltage is $V_1 = 100\,V$.

Illustrative example 4.12 In the previous example determine the waveform of the current flowing into the filter capacitor C_1.

Solution. The capacitor current I_c is the difference between the DC source current I_1 and the chopper switch current I_{s1} [see Fig. 4.16(*a*)]. $I_1 = 6\,A$ (see the solution Example 4.11).

During T_{ON}, $I_{s1} = I_2 = 10\,A$. Therefore, during T_{ON}, $I_c = 6 - 10 = -4\,A$.
During T_{OFF}, $I_{s1} = 0$. Therefore $I_c = 6 - 0 = 6\,A$.
The waveform of I_c is sketeched in Fig. 4.16(*b*) on the basis of the above.
Note that this example illustrates the fact that the capacitor current is an AC, without any DC component, under repetitive operating conditions.

4.3 VOLTAGE STEP UP CHOPPER

4.3.1 Power Circuit Configuration

The circuit configuration of the chopper converter used for stepping up a DC voltage is shown in Fig. 4.17. The difference in the switching circuit configuration is that in the voltage step up scheme, the positions of the power diode and the controlled switching element are interchanged as compared with the circuit of the voltage step down chopper described earlier. To operate the chopper, the controlled switch, which is labeled S_2 in Fig. 4.17, is repetitively turned ON and OFF at the chosen chopper frequency, in the same manner as in

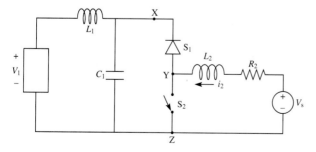

FIGURE 4.17
Voltage step up chopper with filter.

the case of the voltage step down chopper. The voltage step up ratio is determined by the switching duty cycle.

In Fig. 4.17, the power input is from the DC voltage source V_s, which is shown on the right, and it feeds power into the load, which is at a higher voltage V_1 and is shown on the left. The chopper serves to raise the voltage from the source level to the output level. Also shown in Fig. 4.17 is a filter circuit, identical to the one we described previously for the high voltage side for the voltage step down chopper. This filter, consisting of the elements C_1 and L_1, is used on the high voltage side here also. But for the step up mode, the high voltage side is the load side. With this filter, the current flowing into the load will have reduced ripple. We will assume that these filter elements have large enough values to make the load ripple current negligible.

Figure 4.17 also shows an inductance labeled L_2 on the low voltage side, similar to the smoothing inductance we used on the low voltage side for the voltage step down chopper, to smooth out the low voltage side ripple current. It is an essential requirement for the voltage step up mode of operation. In the step up mode, it functions as an interim reservoir of energy, drawing energy from the DC source during the ON time of the chopper and feeding the same energy into the source at a higher voltage during the OFF time of the chopper. It also serves to smooth out ripple current on the low voltage side, which in this case is the input side. With a large enough value of L_2, the input source current will have negligible ripple. The resistance, labeled R_2 in Fig. 4.17, is the unavoidable resistance of the inductive coil L_2, and includes the source resistance and wiring resistance.

4.3.2 Analysis of the Voltage Step Up Chopper

We shall simplify our analysis by assuming ideal circuit elements and switches and restricting it to the following operating conditions:

1. use of ideal filters on the high voltage side (large values for L_1 and C_1);
2. chopper operating at a constant duty cycle D;
3. chopper has operated for sufficient number of cycles to justify assumption of repetitive conditions.

The large value of L_1 implies that the ripple in the load current is negligible. Therefore, if the load is resistive, the voltage across it, which is labeled V_1 in Fig. 4.17, will have a constant value. Alternatively, V_1 could also be a voltage source, such as in a battery charging application. The voltage across the filter capacitor will have the same DC magnitude as V_1, because, under steady state conditions, a DC voltage cannot exist across the ideal inductance L_1. If C_1 is very large, the voltage ripple across it will also be negligible. Therefore, the voltage appearing across the chopper on its high voltage terminals, which are labeled X–Z in Fig. 4.17, can be represented by a

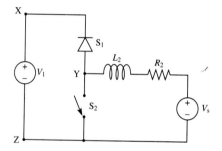

FIGURE 4.18.
Voltage step up chopper operating with fixed DC voltage on the high voltage side.

constant voltage V_1. Figure 4.18 shows the same circuit as Fig. 4.17. But to make our treatment simple, the filter elements L_1 and C_1 are not shown explicitly in Fig. 4.18.

4.3.2.1 CIRCUIT CONFIGURATIONS DURING THE ON TIME AND OFF TIME (T_{ON} AND T_{OFF}). Figure 4.19(a) shows the circuit configuration during the ON period of the chopper. The voltage across the switch S_2 will be zero, and therefore the diode switch S_1 will be reverse-biased by a voltage of magnitude V_1 and will remain OFF. During this interval, current will build up in the inductance to a peak magnitude, which we shall label I_p, as a consequence of which some energy will be transferred from the voltage source V_s into the inductor L_2.

The circuit configuration during the OFF period is shown in Fig. 4.19(b).

When the chopper is turned OFF, the decay of current in the inductor L_2 causes a voltage $L \, di_2/dt$ to occur across it. This adds to the source voltage V_s to forward-bias the diode switch S_1 and turn it ON. Current flows into the high voltage side, thereby transferring power to the latter from the low voltage side. During this time, the current decays to a valley magnitude, which we shall denote by I_v. Depending on the circuit parameters and the duty cycle, the

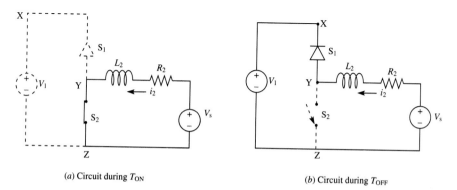

(a) Circuit during T_{ON} (b) Circuit during T_{OFF}

FIGURE 4.19
Chopper circuit configurations during T_{ON} and T_{OFF}.

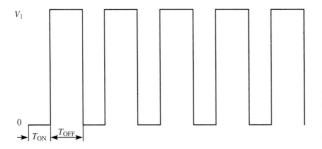

FIGURE 4.20
Voltage waveform at the low voltage terminals Y–Z of the chopper converter.

current may decay to zero even before the end of the OFF period of the chopper switch (T_{OFF}). In such a case, the current flowing in the voltage source V_s will be discontinuous. For the present, we shall assume that the chopper is operating in the continuous current mode.

4.3.2.2 VOLTAGE RATIO. Assuming continuous current flow mode, the waveform of the voltage occurring across the low voltage terminals Y–Z of the chopper is shown in Fig. 4.20, on the basis of the above description. This shows that during the ON period of the chopper switch, the voltage has zero magnitude. During the OFF period, because of the continuous current flow through the diode, the voltage will be equal to V_1, which we have assumed to be constant.

We notice from Fig. 4.20 that the voltage appearing across the low voltage terminals of the chopper (Y–Z) is a train of pulses. This waveform has a DC component and a ripple. The ripple will be substantially absorbed across the inductance L_2, and it is the DC component that is our main concern. This DC component will be labeled V_2, and is given by the average height of the waveform in Fig. 4.20. Therefore

$$V_2 = \frac{V_1 T_{OFF}}{T_{ON} + T_{OFF}} = V_1(1 - D) \tag{4.28}$$

Therefore the voltage conversion ratio a of the chopper, defined as the voltage at the output terminals (X–Z) divided by that at the input terminals (Y–Z), will be given by

$$a = \frac{V_1}{V_2} = \frac{1}{1 - D} \tag{4.29}$$

Equation (4.29) tells us that, ideally, the step up voltage ratio of the chopper is continuously adjustable in the range from 1 ($D = 0$) to infinity ($D = 1$) by choice of the appropriate duty cycle.

4.3.2.3 CURRENT RELATIONSHIPS. The voltage relationships derived above is valid from the commencement of the chopper operation, even if sufficient number of cycles of operation have not been completed to justify the assumption of repetitive conditions, provided the operation is in the continuous current flow mode. But the current relationships derived below, assume both repetitive conditions and continuous current flow mode.

The loop equation during T_{ON} for the circuit configuration shown by 4.19(a) is

$$L_2 \frac{di_2}{dt} + R_2 i_2 = V_s \tag{4.30}$$

the initial condition being $i_2 = I_v$ at $t = 0$, and the reference zero of t being the instant of commencement of T_{ON}.

The loop equation during T_{OFF} for the circuit configuration shown in Fig. 4.19(b) is

$$L_2 \frac{di_2}{dt} + R_2 i_2 = V_s - V_1 \tag{4.31}$$

the initial condition for this equation being $i_2 = I_p$ at $t = 0$, and the reference zero for t being the start of the OFF period.

I_p and I_v are the peak and valley magnitudes of i_2, which are assumed to be repeated in successive chopper cycles.

The solution of (4.30) with the stated initial condition gives, at $t = T_{ON}$,

$$I_p = \frac{V_s}{R_2}(1 - e^{-T_{ON}/\tau}) + I_v e^{-T_{ON}/\tau} \tag{4.32}$$

where $\tau = L_2/R_2$. The solution of (4.31) gives, at $t = T_{OFF}$,

$$I_v = \frac{V_s - V_1}{R_2}(1 - e^{-T_{OFF}/\tau}) + I_p e^{-T_{OFF}/\tau} \tag{4.33}$$

equations (4.32) and (4.33) are two simultaneous equations for I_p and I_v. Their solution gives

$$I_p = \frac{V_s}{R_2} - \frac{V_1}{R_2}\frac{e^{-T_{ON}/\tau} - e^{-T/\tau}}{1 - e^{-T/\tau}} \tag{4.34}$$

$$I_v = \frac{V_s}{R_2} - \frac{V_1}{R_2}\frac{1 - e^{-T_{OFF}/\tau}}{1 - e^{-T/\tau}} \tag{4.35}$$

The waveform of i_2 is sketched in Fig. 4.21.

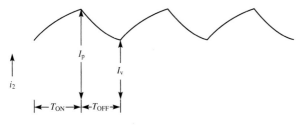

FIGURE 4.21
Waveform of i_2.

The peak-to-peak ripple in the input current i_2 will be given by

$$I_{pp} = I_p - I_v$$

Therefore, using the expressions (4.34) and (4.35) for I_p and I_v,

$$I_{pp} = \frac{V_1}{R_2} \frac{1}{1 - e^{-T/\tau}} [(1 + e^{-T/\tau}) - (e^{-DT/\tau} + e^{-(1-D)T/\tau})] \qquad (4.36)$$

Equation (4.36) for the peak-to-peak ripple on the low voltage side of the voltage step up chopper is exactly the same as the expression we derived earlier for the peak-to-peak ripple current on the low voltage side of the voltage step down chopper [see (4.17)].

Since the expressions for I_{pp} are the same for the low voltage side ripple current for both step up and step down choppers, two other relationships, which we derived earlier, based on the expression for I_{pp}, are also valid for the step up chopper. These are as follows.

1. The duty cycle D at which the peak-to-peak ripple current on the low voltage side has the maximum value is 50% ($D = 0.5$) [see (4.18)].
2. The maximum value of this peak-to-peak ripple current is given by the same expression, namely (4.19).

4.3.3 DC Component of Input (Low Voltage Side) Current

The input current supplied by the voltage source V_s fluctuates between the peak magnitude given by (4.34) and the valley magnitude given by (4.35). A typical waveform is sketched in Fig. 4.21. This has a DC component and a ripple component. The ripple current exists because we have assumed only a finite, not infinite, value for the inductance on the low voltage side. The derivations of (4.34) and (4.35) are on the basis of a finite value for L_2. We are primarily interested in the DC component. This can be determined from the average height of the current waveform in Fig. 4.21 by the integration of the appropriate expressions. A simple method of determining the DC component will be by using the principle of superposition, and we shall follow this method.

The DC component of the voltage appearing across the terminals Y–Z has been shown to be

$$V_2 = V_1(1 - D) \qquad (4.28)$$

Therefore the DC current I_2 will be given by

$$I_2 = \frac{V_s - V_2}{R} = \frac{1}{R}[V_s - (1 - D)V_1] \qquad (4.37)$$

Illustrative example 4.13 In a battery powered car, the battery voltage is 120 V. It is driven by a DC motor and employs chopper control. The resistance of the motor circuit is 0.2 Ω. During braking, the chopper configuration is changed to the voltage step up mode. While going down hill at a certain speed, the induced e.m.f. of the machine is 110 V and the braking current is 10 A. Determine the chopper duty cycle. Assume continuous current operation.

Solution. In (4.28), the DC component of the voltage across Y–Z labeled V_2, is

$$V_2 = V_s - I_d R = 110 - 10 \times 0.2 = 108 \text{ V}$$

From (4.28)

$$V_2 = (1 - D)V_1$$

so

$$108 = (1 - D)120$$

This gives $D = 0.1$ or 10%.

4.3.4   Operation with Discontinuous Current Flow

With low values of inductance on the low voltage side, the input current can become zero even before the end of the OFF period of the chopper. The limiting condition for the onset of discontinuous current flow can be obtained by equating $I_v = 0$ in the expression (4.35) for the valley current I_v. This gives

$$\frac{V_s}{V_1} = \frac{1 - e^{-T_{\text{OFF}}/\tau}}{1 - e^{-T/\tau}} \tag{4.38}$$

Illustrative example 4.14. In Example 4.13, take the total inductance on the low voltage side (consisting of the motor inductance and any additional inductance inserted for filtering) to be 300 μH. The chopper frequency is 1 kHz.
(a) For the induced voltage of 110 V (as in Example 4.13), during regenerative braking, determine the duty cycle below which discontinuous current flow will occur.
(b) For the above condition, what is the minimum DC braking current possible with continuous current flow?
(c) If the vehicle is being driven (not braked), with the same value of the motor-induced EMF, what is the limting duty cycle below which the motor current would be discontinuous? What is the DC motor current at this duty cycle?

Solution.
(a) We have

$$T = 1/f = 10^{-3} \text{ s}$$
$$\tau = L_2/R_2 = 0.3 \times 10^{-3}/0.2 = 1.5 \times 10^{-3} \text{ s}$$
$$V_s = 110 \text{ V}, \quad V_1 = 120 \text{ V}$$

Substituting these numbers into (4.38), we get $T_{OFF} = 8.86 \times 10^{-4}$ s, which gives the limiting duty cycle as

$$D = \frac{T - T_{OFF}}{T} = 0.114, \quad \text{or } 11.4\%$$

(b) We have

$$V_2 = V_1(1 - D) = 120 \times (1 - 0.114) = 106.3 \text{ V}$$

so

$$I_2 = \frac{V_s - V_2}{R_2} = \frac{110 - 106.3}{0.2}$$
$$= 18.5 \text{ A}$$

Note that comparison of this result with Example 4.13 shows that the chopper frequency (which is not specified in Example 4.13) has to be greater than 1 kHz to justify the assumption of continuous conduction in Example 4.13.

(c) While driving, the chopper functions in the voltage step down mode. For this mode, the limiting condition for the commencement of discontinuous conduction is given by (4.27). This condition is rewritten, using V_s in place of E, as

$$\frac{V_s}{V_1} = \frac{e^{-T_{OFF}/\tau} - e^{-T/\tau}}{1 - e^{-T/\tau}}$$

Therefore

$$\frac{110}{120} = \frac{e^{-T_{OFF}/\tau} - e^{-T/\tau}}{1 - e^{-T/\tau}}$$

Substitution of numbers in the above expression gives $T_{OFF} = 6.2 \times 10^{-5}$ and $T_{ON} = 10^{-3} - 6.2 \times 10^{-5} = 9.38 \times 10^{-4}$ s. Therefore the limiting duty cycle for commencement of discontinuous conduction will be

$$D = 0.938/1 = 0.938, \quad \text{or } 93.8\%$$

The DC component of the motor current at this duty cycle will be

$$I_d = \frac{D \times 120 - 110}{0.2} = 12.8 \text{ A}$$

4.3.5 Effect of Chopper Frequency on the Onset of Discontinuous Current in the Motor

Example 4.15 illustrates the effect of increasing the chopper frequency on the limiting duty cycle for the onset of discontinuous conduction. We notice that we could go down to lower duty cycles and lower braking currents without encountering discontinuous currents in the machine if we increase the chopper frequency, other conditions remaining the same. The same is true of the driving mode. We could go down to lower duty cycles in this mode also, and lower driving currents, without going into the discontinuous current flow in the motor. Since the relevant terms appearing in the expression for the limiting condition appear as ratios of the chopper period to the inductive time constant of the motor circuit, increasing this time constant has exactly the same effect as increasing the chopper frequency.

Illustrative example 4.15. In Example 4.14, the chopper frequency is increased to 6 kHz, all other data remaining the same. What is the new value of the limiting duty cycle? What will be the corresponding value of the DC braking current? Repeat also part (c) of Example 4.14 (relating to the driving mode), using the new value of the chopper frequency.

Solution. The new value of the chopper period is $1/6000 = 0.1667$ ms. The other data are as in Example 4.14. Substitution of numbers into (4.38) gives $T_{OFF} = 0.1520$ ms. Therefore $T_{ON} = 0.1667 - 0.1520 = 0.0147$ ms. This gives the new limiting duty cycle as

$$D = 0.0882 = 8.82\%$$

We have $V_2 = V_1(1 - D) = 120 \times (1 - 0.0882) = 109.42$ V, so the corresponding value of I_d will be

$$I_d = \frac{V_s - V_2}{0.2} = \frac{110 - 109.42}{0.2} = 2.9 \text{ A}$$

For part (c), during driving, the chopper is functioning in the voltage step down mode. The limiting condition for the onset of discontinuous current flow in the motor is given by (4.27):

$$\frac{V_s}{V_1} = \frac{e^{-T_{OFF}/\tau} - e^{-T/\tau}}{1 - e^{-T/\tau}}$$

Substitution of numbers into this expression gives

$$T_{OFF} = 1.3203 \times 10^{-5} \text{ s}$$
$$T_{ON} = T - T_{OFF} = 0.1667 \times 10^{-3} - 1.3203 \times 10^{-5}$$
$$= 0.1535 \text{ ms}$$

so

$$D = \frac{T_{ON}}{T} = \frac{0.1535}{0.1667} = 0.9208 = 92.08\%$$

The DC component of the motor current at this duty cycle will be

$$I_d = \frac{120 \times 0.9208 - 110}{0.2} = 2.48 \text{ A}$$

4.3.6 Waveforms of the Voltage Step Up Chopper with Perfect Filters on Both Low and High Voltage Sides

In our treatment of the voltage step up chopper so far, we have assumed an ideal L–C filter on the high voltage side but a finite value for the smoothing inductance on the low voltage side. We did this primarily to highlight the onset of discontinuous conduction that occurs with inadequate filtering on the low voltage side, both in the voltage step up and the voltage step down modes of operation of the chopper converter. We also wanted to highlight the existence of the AC ripple in the low voltage side current and its dependence on the value of the smoothing inductance and/or the chopper frequency. In a chopper controlled vehicle drive, the driving motor is on the low voltage side and the battery on the high voltage side. We alternate between the voltage step down and the voltage step up modes of operation of the chopper when we move between driving and braking of the vehicle. For better operation of the motor in both modes, it is highly desirable to have a large value for the smoothing inductance L_2 on the low voltage side. For good operation of the battery, it is also desirable to have a filter with large L_1 and C_1 values on the high voltage side. Having drawn attention to these factors, we shall simplify our subsequent treatment by assuming that our chopper functions with such ideal filters, unless otherwise stated. We shall also assume that the chopper employs ideal switching elements and therefore has no internal power loss.

For such an ideal chopper, when used for a DC motor control application such as a vehicle drive, both the DC source current and the motor current will be perfect DC, without ripple.

The power on the high voltage side is $V_1 I_1$, while that on the low voltage side is $V_2 I_2$. Equating the two, we get the current relationship for the voltage step up chopper as $I_1/I_2 = 1 - D$.

The waveforms for such an ideal chopper are shown in Fig. 4.22. For the labels of currents and voltages in this figure see Fig. 4.24(b).

Illustrative example 4.16. A battery powered, chopper controlled car is being regeneratively braked. The driving motor, which is now functioning as a DC generator, is supplying 25 A. The chopper duty cycle is 20%. Assume that the chopper is ideal and the filtering is perfect. What is the charging current flowing into the battery?

Solution. The current ratio is given by battery current/motor current $= 1 - D = 0.8$. Therefore the battery current is $25 \times 0.8 = 20$ A.

Illustrative example 4.17. The chopper frequency in the previous example is 4 kHz. Sketch the waveform of the capacitor (C_1) current. Does this current have a DC component?

Solution. Refer to Figs. 4.22 and 4.23(a) for the labeling of currents. During T_{OFF}, the capacitor current

$$i_c = I_{s1} - I_1 = 25 - 20 = 5 \text{ A}$$

During T_{ON},

$$i_c = -I_1 = -20 \text{ A}$$

The chopper period $T = 250 \ \mu s$, $T_{ON} = 50 \ \mu s$ and $T_{OFF} = 200 \ \mu s$. The capacitor current waveform is sketched in Fig. 4.23(b) on the above basis. The area under

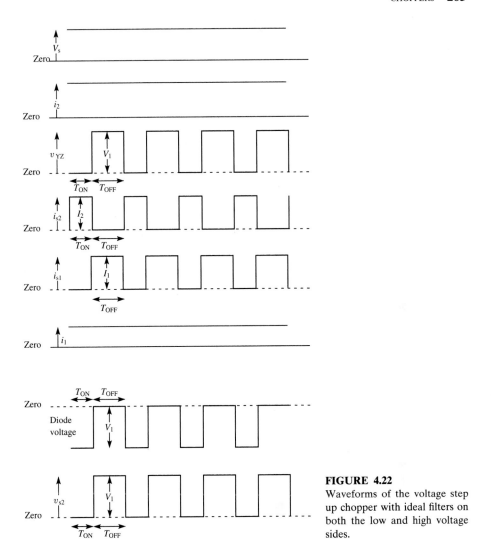

FIGURE 4.22
Waveforms of the voltage step up chopper with ideal filters on both the low and high voltage sides.

the positive current interval is equal to the area under the negative current interval for this waveform. Therefore the DC component is zero.

4.4 TWO-QUADRANT CHOPPER

We have described the two modes of chopper operation: the voltage step down mode and the voltage step up mode. We have studied the power circuit configuration for each mode of operation. These two configurations of the static switching elements are shown again in Figs 4.24(*a*) and (*b*). Each mode of operation needs two power semiconductor switching elements: a controlled

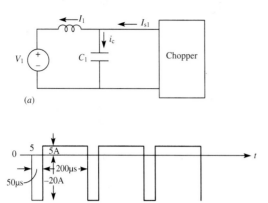

(a)

(b)

FIGURE 4.23
Waveform of the filter capacitor current.

power semiconductor switch and an uncontrolled power diode switch. The circuit difference between the two modes is that the positions of the diode and the controlled switch are interchanged.

Figure 4.24 is typical of a DC motor speed control application. The motor is on the low voltage side of the chopper and the DC supply is on the high voltage side. The polarity of the induced voltage in the motor depends on its direction of rotation. In both Figs. 4.24(a) and (b), the motor is spinning in the same direction, and therefore has the same polarity of terminal voltage. However, in Fig. 4.24(a), the motor is driving a load, and is therefore drawing power through the chopper. It is therefore drawing current in the direction shown, which we shall consider as the positive direction. In Fig. 4.24(b), the motor is being regeneratively braked, and therefore the current is negative. If we plot on a graph the motor voltage along the vertical axis and the motor current along the horizontal axis, all the operating points on this graph for the driving mode [Fig. 4.24(a)] will lie in the first quadrant (positive I and positive V). For the braking mode, corresponding to Fig. 4.24(b), all the operating points will be in the second quadrant (negative I and positive V). This is

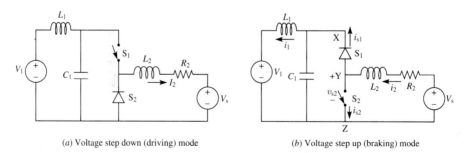

(a) Voltage step down (driving) mode   (b) Voltage step up (braking) mode

FIGURE 4.24
The two operating modes of the two-quadrant chopper.

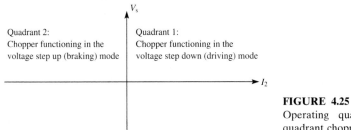

FIGURE 4.25
Operating quadrants of a two-quadrant chopper.

illustrated by Fig. 4.25. A chopper designed for changeover from one mode to the other at will is called a two-quadrant chopper.

In a two-quadrant chopper, the changeover between driving and braking can be implemented in one of two possible ways.

(i) Mechanical change-over scheme. We can physically change the connections to the controlled switch and the power diode. In this case, we need to have only one controlled switching element and one power diode for both modes of operation of the chopper. For example, we can implement the circuit change by using either a manually operated or an electromagnetically operated switch. Such a scheme is illustrated by Fig. 4.26.

In Fig. 4.26(a) the three terminals of the circuit that are connected to the static switching elements of the chopper are labeled X, Y and Z. The terminals of the static switching elements of the chopper are shown in Fig. 4.26(b), with the terminals of the controlled switch labeled 1 and 2 and those of the power diode labeled 3 and 4. The changeover between driving (voltage step down mode) and braking (voltage step up mode) is implemented using a four-pole double-throw switch, which may be designed for either manual or relay operation. In one position of the switch, the terminals X, Y and Z are connected to 1, 2 3 and 4 as shown in Fig. 4.26(c). This will give us the configuration for driving, that is, the voltage step down mode of the chopper. The interconnections between X, Y and Z and 1, 2, 3 and 4 as shown in the second position of the switch in Fig. 4.26(d) will give us the braking or voltage step up mode.

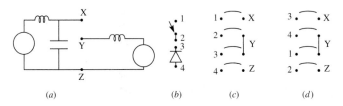

FIGURE 4.26
Mechanical changeover scheme between voltage step down (driving) and voltage step up (braking) modes for a two-quadrant chopper.

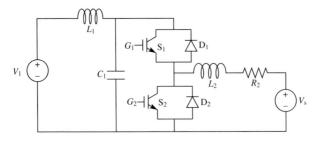

FIGURE 4.27
Two-quadrant IGBT chopper with static changeover between voltage step up and voltage step down modes.

(ii) Static change-over scheme. In this scheme, we use two controlled switches and two power diodes. Each controlled switch has a power diode in "antiparallel" with it as shown in Fig. 4.27. A controlled switch together with its antiparallel diode constitutes a "switching block." For example switching block 1 in Fig. 4.27 consists of the controlled switch S_1 and its antiparallel diode D_1. Similarly, switching block 2 consists of S_2 and D_2. The controlled switches shown in Fig. 4.27 are IGBTs, but they could be any type of controlled power switching device such as power BJTs, power Darlingtons or GTOs. If the switch is a power MOSFET, there is generally no need to have a separate antiparallel diode, because a power MOSFET functions as a diode for reverse current flow.

To operate the two-quadrant chopper of Fig. 4.27 in the voltage step down (driving) mode, we block the switching control pulses to the gate of S_2 and channel them to the gate of S_1. S_2 and D_1 will be inoperative during this mode of operation. To change over to voltage step up (braking) mode, we block the gate pulses to S_1 and channel them to S_2. In this mode, S_1 and D_2 will be inoperative. The inhibition of pulses to one switching block and the channeling of them to the other can be done in the gating control circuit of the chopper. There will be no need for mechanical switching in the power circuit.

If a battery powered car is powered by a DC motor with two-quadrant chopper control, it is relatively easy to implement the changeover with the driving foot pedal itself. When the latter is pressed down, the gate pulses to S_2 are inhibited and they are channeled to S_1. The duty cycle will be controlled by the downward position of the pedal. Upward movement of the pedal, beyond a limiting position, will block the switching pulses to S_1 and channel them to S_2, thereby moving into regenerative braking. The duty cycle, and therefore the braking torque, will be dependent on the upward displacement of the pedal. This makes it possible to do normal driving and braking using a single pedal only, and use friction braking only for stopping and emergencies.

4.5 MULTIPHASE CHOPPERS
4.5.1 Circuit Configurations and Switching Sequences

There are some advantages if, instead of a single chopper, we employ several in parallel to supply current to a load. All the choppers will be identical and

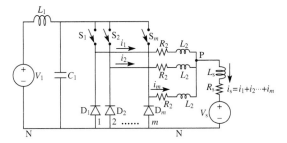

(a) m-phase voltage step down chopper

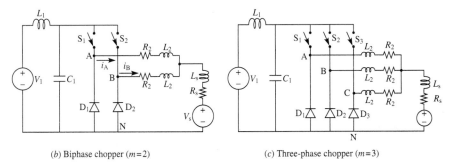

(b) Biphase chopper (m=2)   (c) Three-phase chopper (m=3)

FIGURE 4.28
Multiphase chopper configurations.

operating with the same switching frequency and the duty cycle. However, their switching periods will be phase-displaced with respect to each other. The primary advantage is the higher ripple frequency in the ripple output, making it easier to filter off the ripple currents. Such a scheme, consisting of two or more choppers operating in parallel, with mutual phase displacement in their switching periods, is called a multiphase chopper. Figure 4.28(a) shows the general m-phase voltage step down chopper. Two specific examples of multiphase choppers—the biphase chopper $(m = 2)$ and the three-phase chopper $(m = 3)$—are shown respectively in Figs. 4.28(b) and (c). In Fig. 4.28, the input filter consisting of L_1 and C_1 is shown in each circuit. Each chopper has a separate but identical output smoothing inductance labeled L_2. The resistance labeled R_2 is the unavoidable resistance of this inductive coil. An additional common inductance labeled L_s is also shown in the load circuit. To make the circuit more general, we have also included a "back e.m.f." labeled V_s, such as will be present in a DC motor speed control application, in addition to the load circuit resistance R_s.

Each chopper in the general m-phase chopper will have the same switching frequency and duty cycle. However, the start of the switching period of chopper 2 will be delayed with respect to the start of the switching period of chopper 1 by an interval equal to T/m seconds, which will correspond to a phase delay of $360°/m$ in angular measure. Similarly chopper 3 will have a

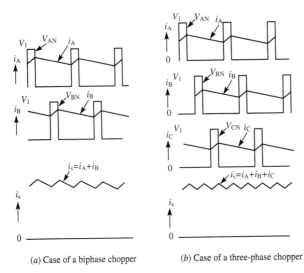

FIGURE 4.29
Waveforms of individual phase currents and resultant load current in multiphase choppers.

(a) Case of a biphase chopper

(b) Case of a three-phase chopper

phase shift of $360°/m$ with respect to chopper 2 and so on. In the biphase chopper of Fig. 4.28(b), for which $m = 2$, the phase displacement wil be 180°, while for the three-phase chopper of Fig. 4.28(c), it will be 120°, that is, one-third of a chopper period.

The effects of this phase shift on the current waveforms of the individual choppers and on the resultant load current are illustrated in Figs. 4.29(a) and (b) for the biphase and three-phase choppers. The first set of waveforms of Fig. 4.29(a) show (1) the voltage at the output terminal of chopper 1, which is the terminal labeled A in Fig. 4.28(b), with respect to the common negative terminal labeled N, and (2) the waveform of the current i_A supplied by this chopper. Similarly, the second set of waveforms show the output voltage and current of chopper 2. The waveform shown at the bottom in this figure is that of the output load current i_s, which is the sum of the individual chopper currents. We notice from these that the ripple frequency of the resultant current is two times the chopper frequency. The same procedure is repeated for the three-phase chopper in Fig. 4.29(b), and here we notice that the frequency of the output current ripple is three times the chopper frequency.

4.5.2 Analysis of the m-Phase Voltage Step Down Chopper

We shall limit our analysis to the case of an $R-L$ load without back e.m.f. This means that V_s is zero in Fig. 4.28. Except for this, the circuit that we use for our analysis will be the same as Fig. 4.28(a). We shall also assume that several cycles of switching have already taken place, and therefore repetitive conditions prevail. The number of chopper switches that will be simultaneously ON at any given instant of time will depend on the duty cycle D.

For $0<D<1/m$, there will be an interval when one chopper switch will be ON. All other phases will be freewheeling. The current in the phase that is ON will be rising and all other phase currents will be decreasing. This is the interval when the resultant current will be rising towards a peak magnitude, which we shall label I_p. This interval will be followed by another interval when all the chopper switches will be OFF and all phase currents will be freewheeling. During this interval, the resultant current will be decaying towards a valley magnitude, which we shall label I_v. The chopper circuit configuration will alternate between these two modes of conduction, that is, one chopper switch ON and the load current rising and, next, $1-1=0$ switches ON and the load current falling.

For $1/m<D<2/m$, similar considerations as above show that, in this case also, there will be two modes of conduction, between which the circuit configuration will alternate:

1. Two chopper switches simultaneously ON and all other phases freewheeling; during this interval, the load current will be rising towards the peak I_p.
2. Only one $(2-1)$ chopper switch ON and all other phases freewheeling; during this interval, the load current will be falling towards the valley magnitude I_v.

For $2/m<D<3/m$, similar considerations show that there will again be two modes of conduction:

1. Three chopper switches ON, when the load current will be rising towards the peak I_p.
2. Only two $(3-1)$ chopper switches ON, when the load current will be falling towards the valley magnitude I_v.

Proceeding in this manner, we can generalize as follows, for any value of D. For $(p-1)/m<D<p/m$, where p is an integer less than or equal to m, there will be two conducting modes:

1. p chopper switches ON, and the current rising towards the peak I_p.
2. Only $p-1$ chopper switches ON, and the load current falling towards the valley magnitude I_v.

We shall now write the circuit equations for mode 1 when p chopper phases are ON and the other $m-p$ phases are freewheeling. Referring to Fig.

4.28(a), we shall assume that the conducting phases are $1, \ldots, p$ and the freewheeling phases are $p + 1, \ldots, m$. The filter L_1–C_1 will be assumed to be ideal, so that the input voltage V_1 to the chopper will be constant. We have

$$R_2 i_1 + L_2 \frac{di_1}{dt} + R_s i_s + L_s \frac{di_s}{dt} = V_1 \qquad (4.39)(1)$$

$$R_2 i_2 + L_2 \frac{di_2}{dt} + R_s i_s + L_s \frac{di_s}{dt} = V_1 \qquad (4.39)(2)$$

$$\vdots \qquad\qquad \vdots \qquad\qquad \vdots$$

$$R_2 i_p + L_2 \frac{di_p}{dt} + R_s i_s + L_s \frac{di_s}{dt} = V_1 \qquad (4.39)(p)$$

$$R_2 i_{p+1} + L_2 \frac{di_{p+1}}{dt} + R_s i_s + L_s \frac{di_s}{dt} = 0 \qquad (4.39)(p+1)$$

$$\vdots \qquad\qquad \vdots \qquad\qquad \vdots$$

$$R_2 i_m + L_2 \frac{di_m}{dt} + R_s i_s + L_s \frac{di_s}{dt} = 0 \qquad (4.39)(m)$$

Adding (4.39)(1)—(m), we get

$$R' i_s + L' \frac{di_s}{dt} = V' \qquad (4.40)$$

where

$$R' = R_2 + mR_s \qquad (4.41)$$

$$L' = L_2 + mL_s \qquad (4.42)$$

$$V' = pV_1 \qquad (4.43)$$

Equation (4.40) is valid during the rising part of the output current, when p chopper switches are ON, as we stated earlier. This period lasts from the instant of turn ON of the chopper switch p to the instant of turn OFF of chopper switch 1. Denoting this growth period of the output current by T_r, we have

$$T_r = DT - \frac{p-1}{m} T = T\left(D - \frac{p-1}{m}\right) \qquad (4.44)$$

Taking the reference zero for time at the instant of commencement of the growth period, that is, at the instant of turn ON of chopper switch p, the initial condition for (4.40) will be

$$i_s = I_v \quad \text{at } t = 0 \qquad (4.45)$$

On this basis, the solution of (4.40) will be

$$i_s = I_v e^{-t/\tau} + \frac{V'}{R'}(1 - e^{-t/\tau})$$

(4.46)

where

$$\tau = L'/R'$$

(4.47)

The peak current I_p occurs at $t = T_r$. Therefore

$$I_p = I_v e^{-T_r/\tau} + \frac{V'}{R'}(1 - e^{-T_r/\tau})$$

(4.48)

where T_r is given by (4.44).

Decay period. During the decay period of i_s, only $p - 1$ chopper switches will be ON, and the other phases will be freewheeling. This period will last from the turn OFF of switch 1 to the turn ON of switch $p + 1$. Denoting this interval by T_f, we have

$$T_f = \frac{p}{m} T - DT = T\left(\frac{p}{m} - D\right)$$

(4.49)

We can write down the loop equations for the decay period in a similar manner to what we did for the growth period. The only difference here will be that, out of the m-simultaneous equations, only $p - 1$ will have V_1 on the right-hand side. For the rest, the right-hand side will be zero. By adding all m equations, in a manner identical to what we did for the growth period, we can write a single equation for i_s as

$$R'i_s + L'\frac{di_s}{dt} = V''$$

(4.50)

where R' and L' will be the same as defined earlier by (4.41) and (4.42), and V'' will be

$$V'' = (p - 1)V_1$$

(4.51)

Taking a new reference zero for time at the commencement of the decay period, the initial condition for (4.50) will be

$$i_s = I_p \quad \text{at } t = 0$$

(4.52)

On this basis, the solution (4.50) will give the expression for the current during the decay period as:

$$i_s = I_p e^{-t/\tau} + \frac{V''}{R'}(1 - e^{-t/\tau})$$

(4.53)

The current falls to the valley magnitude at $t = T_f$. Therefore

$$I_v = I_p e^{-T_f/\tau} + \frac{V''}{R'}(1 - e^{-T_f/\tau}) \tag{4.54}$$

Equations (4.48) and (4.54) are two simultaneous equations in I_p and I_v, and their solution gives

$$I_p = \frac{V_1}{R'}\left(p - \frac{e^{-T_f/\tau} - e^{-T/m\tau}}{1 - e^{-T/m\tau}}\right) \tag{4.55}$$

$$I_v = \frac{V_1}{R'}\left(p - \frac{1 - e^{-T_f/\tau}}{1 - e^{-T/m\tau}}\right) \tag{4.56}$$

In the above equations, p is an integer less than m, depending on the duty cycle, and has to be determined using the relationship stated earlier and rewritten below:

$$\frac{p-1}{m} < D < \frac{p}{m} \tag{4.57}$$

T/m is the period of the output current ripple, which will be

$$T/m = T_r + T_f \tag{4.58}$$

DC components of the output voltage and current. For this, we shall refer to Fig. 4.28(a) and remember that V_s is zero. All the chopper phases are supplying current in parallel to the load. The voltage step down ratio for each individual chopper phase is D, a result that we derived for the single-phase chopper studied previously. Since in Fig. 4.28 we have shown a resistance R_2 in each individual chopper phase, there will be a DC voltage drop in it. If we denote the DC component of the load current by I_d, the DC component of each individual chopper current will be I_d/m, since the current is shared equally by the m identical chopper phases. Therefore the DC component of the output voltage at the load terminals P–N will be, for each phase,

$$V_{PN} = DV_1 - R_2 I_d/m \tag{4.59}$$

$$= I_d R_s \tag{4.60}$$

From (4.59) and (4.60), we get I_d and V_{PN} as

$$I_d = \frac{DV_1}{R_s + R_2/m} \tag{4.61}$$

$$V_{PN} = \frac{DV_1 R_s}{R_s + R_2/m} \tag{4.62}$$

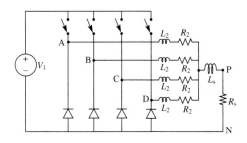

FIGURE 4.30
Four-phase chopper of Example 4.18.

Illustrative example 4.18. The four-phase chopper of Fig. 4.30 is working at a frequency of 250 Hz and a duty cycle of 70%. Other data are $R_2 = 0.4\,\Omega$, $L_2 = 2\,\text{mH}$, $R_s = 2.4\,\Omega$, $L_s = 4.5\,\text{mH}$ and input voltage 200 V. Determine the peak-to-peak ripple in the output current and the frequency of this ripple.

Solution. The chopper period $T = 1/f = 4\,\text{ms}$, the number of phases m 4 and the duty cycle $D = 0.7$. Therefore, from (4.57), $p = 3$. From (4.44), the growth period of the ripple current $T_r = 4(0.7 - 0.5) = 0.8\,\text{ms}$. From (4.49), the decay period of the ripple current $T_f = 4(0.75 - 0.7) = 0.2\,\text{ms}$. So the total ripple period $T_r + T_f = 1\,\text{ms}$. Therefore the ripple frequency is 1000 Hz, that is, four times the chopper frequency.

From (4.41) and (4.42),

$$R' = 0.4 + 4 \times 2.4 = 10\,\Omega$$
$$L' = 2 + 4 \times 4.5 = 20\,\text{mH}$$

The time constant $\tau = 20/10 = 2\,\text{ms}$. Numerical substitution into (4.55) and (4.56) gives the following values for the peak and valley currents:

$$I_p = 56.76\,\text{A}$$
$$I_v = 55.16\,\text{A}$$

Therefore the peak-to-peak ripple current is $56.76 - 55.16 = \mathbf{1.6\,A}$.

Illustrative example 4.19 In the previous example determine
(a) the DC component of the output voltage across the load resistor R_s;
(b) the DC component of the load current;
(c) the peak-to-peak ripple current in the load as a percentage of the DC current.

Solution.
(a) The DC component of the voltage across R_s (V_{PN}) is found by numerical substitution into (4.62). This gives

$$V_{PN} = \frac{0.7 \times 200 \times 2.4}{2.4 + 0.1}$$

$$= \mathbf{134.4\,V}$$

(b) The DC component of the load current is found by numerical substitution into (4.61). This gives

$$I_d = \frac{0.7 \times 200}{2.4 + 0.1} = \mathbf{56\,A}$$

(c) The peak-to-peak ripple in the current output, determined in the previous example, when expressed as a percentage of the above DC current, becomes

load current peak-to-peak ripple $= 1.6/56 = 0.0286 = \textbf{2.86\%}$

4.5.3 Load with Back e.m.f.

In applications such as speed control of DC motors, there will be a back e.m.f. in the load circuit, as shown by V_s in Fig. 4.28. Such an operation of the multiphase chopper can be analyzed by a simple extension of our analysis without back e.m.f., assuming continuous current flow in each chopper phase. The m loop equations 4.39(1)–(m) can be rewritten with an additional term on the right-hand side equal to $-V_s$. The new equations can be solved in a very similar manner to what we did earlier, and the corresponding relationships derived. This procedure is left as an exercise to the reader. We give below the expressions for the peak and valley currents thus obtained:

$$I_p = -\frac{mV_s}{R'} + \frac{V_1}{R'}\left(p - \frac{e^{-T_t/\tau} - e^{-T/m\tau}}{1 - e^{-T/m\tau}}\right) \tag{4.63}$$

$$I_v = -\frac{mV_s}{R'} + \frac{V_1}{R'}\left(p - \frac{1 - e^{-T_t/\tau}}{1 - e^{-T/m\tau}}\right) \tag{4.64}$$

Limiting condition for the onset of discontinuous conduction mode. The limiting condition for the commencement of the discontinuous load current flow can be obtained by equating I_v to zero in (4.64). This gives

$$\frac{V_s}{V_1} = \frac{1}{m}\left(p - \frac{1 - e^{-T_t/\tau}}{1 - e^{-T/m\tau}}\right) \tag{4.65}$$

Illustrative example 4.20. In Example 4.18, assume that the load is a DC motor with a back e.m.f. of 100 V. All other data remain the same.
(a) Determine the peak and valley currents and the peak-to-peak ripple in the motor current.
(b) If the speed of the motor goes up, with a corresponding increase in the back e.m.f., determine the value of the back e.m.f. at which discontinuous conduction will commence, all other data remaining the same.

Solution.
(a) Substitution of the given numbers into (4.63) and (4.64) give

$$I_p = \textbf{16.76 A}$$

$$I_v = \textbf{15.16 A}$$

This gives the peak-to-peak ripple current as

$$I_{pp} = \textbf{1.6 A}$$

Notice that the peak-to-peak ripple current is the same as in Example 4.18, and its value has not changed because of V_s.

(b) The value of V_s at which discontinuous conduction will commence is obtained by numerical substitution into (4.65). This gives

$$V_s = \mathbf{137.9 \ V}$$

4.5.4 Voltage Step Up Mode and Two-Quadrant Operation of Multiphase Choppers

The circuit configuration of the voltage step up multiphase chopper is obtained in the same manner as for the single-phase chopper, namely by interchanging the positions of the controlled switching elements and the power diodes. This is shown in Fig. 4.31. In this figure, the input side, which is the low voltage side, is shown on the right. The input source is V_s. Such an arrangement can be used for the regenerative braking of the DC motor in a speed control application of the chopper. In such a case, the power recovered by braking will be fed back into the DC source shown on the left side, labeled V_1. The $L–C$ filter on the high voltage side is also included in Fig. 4.31.

The analysis of the multphase chopper in the voltage step up mode can be done by extending the corresponding analysis for the single-phase chopper in a manner similar to what we did for the step down mode. This is left as an exercise for the reader.

Two-quadrant multiphase choppers enable both voltage step up and voltage step down modes of operation. The changeover between voltage step up and voltage step down modes can be implemented either by using mechanical changeover schemes similar to what we described for the single-phase two-quadrant chopper, or statically by channeling the switching control pulses to the appropriate power semiconductor switches. With mechanical changeover, the number of power switching devices needed for the chopper will be half of what will be needed for static changeover. This will be a consideration in a very large power application, such as a chopper controlled electric train. Static changeover schemes are faster and more convenient, and

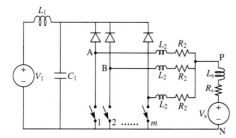

FIGURE 4.31
Multiphase voltage step up chopper.

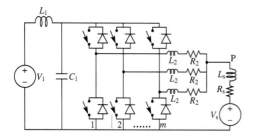

FIGURE 4.32
Multiphase two-quadrant chopper with static changeover between operating modes.

are to be preferred for low power applications, when the cost of a few extra switching devices may not be significant. Such a scheme is shown in Fig. 4.32.

4.5.5 Use of Coupled Reactors in Multiphase Choppers

In the multiphase chopper configuration that we have described so far, each phase has an individual reactor on the low voltage side, which we have labeled L_2 in our figures. The unavoidable ohmic resistance of this coil has been labeled R_2. In addition to the individual phase reactor L_2, we also included a common reactor in the load circuit, labeled L_s, for additional smoothing. Typically, these reactors are iron-cored and have also to carry a large DC current without magnetic saturation. For these reasons, they contribute significantly to the total weight of the equipment. One technique to reduce the overall weight is to use coupled reactors. The individual reactors labeled L_2 in the figures are also magnetically coupled to each other by winding them on a common core structure. The manners in which this is done in biphase and three phase choppers are shown respectively in Figs. 4.33(a) and (b).

In Fig. 4.33(a), both phase coils are wound on a single core. Examination of the direction of winding of the two sections of the coil will show that the DC ampere-turns of the two sections will cancel each other. In this way, the DC core saturation is elminated, and this should contribute to the reduction in the

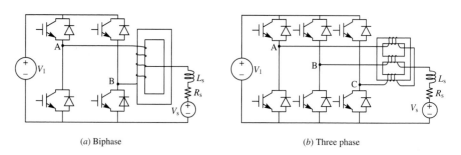

(a) Biphase (b) Three phase

FIGURE 4.33
Multiphase choppers with coupled reactors.

size of the reactor. Similar statements are applicable to the three-limb reactor shown in Fig. 4.33(*b*) for a three-phase chopper. In addition to the coupled reactor, a common reactor labeled L_s is also shown in each of the circuits in Fig. 4.33. It may also be desirable to use such a reactor because the coupled reactor may not be effective in the same way for reducing all the ripple harmonics.

4.6 THYRISTOR CHOPPERS

4.6.1 Need for Force Commutation

For the proper operation of the chopper circuits, or any power electronic circuit, the controlled switching elements have to be turned ON and turned OFF at the required instants of time. In most power semiconductor switching devices, this is done by providing the appropriate switching control signal input at the control terminal of the device. For example, an IGBT is turned ON by a positive voltage on its gate terminal and turned OFF by a zero volt signal at the same terminal. The thyristor, which is an important power switching device, is a singular exception to this rule. The gate terminal is useful only for the turn ON switching. It is ineffective for implementing the turn OFF switching. The practical method employed to turn OFF a thyristor that is already ON in a DC circuit is to impress a reverse voltage across its power terminals, namely, the anode and the cathode. In this way, by "forcing" the anode to go negative, at least temporarily, it can be turned OFF or "commutated." (Commutation implies turn OFF switching in the terminology of Power Electronics.) The circuit that provides this force commutation impulse is in the power circuit, connected between the anode and cathode of the thyristor. Since the force commutation circuit is part of the power circuit of the converter, its operation will to some extent depend on the rest of the power circuit. Force commutation circuits are necessary when thyristors are employed as switching elements on a DC side in power electronic converters. For a thyristor converter, such as a rectifier for conversion from AC to DC, the reversals of the AC line voltage present will serve to reverse-bias and turn OFF the thyristors, and no force commutation circuits are normally necessary. The turn OFF switching in such converters takes place by "line commutation." In practice, line commutation is employed in AC/DC thyristor converters, and their switching frequencies are the relatively low power system frequencies, such as 50 or 60 Hz. For these, it is economical to use a category of thyristors, with longer turn OFF switching times, often classified as line commutated types. Another category of thyristors, with short turn OFF times, often called "fast turn OFF" type, has to be chosen for force commutated thyristor converters such as choppers.

Several types of force commutation circuits have been developed, each with its relative merits and demerits. For the force commutated thyristor chopper that we are going to describe here, we have selected a practical circuit whose operation is relatively simple. The method of analysis that we shall present can generally be used with appropriate changes for other force

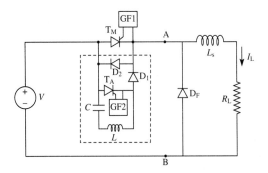

FIGURE 4.34
Force commutated thyristor chopper.

commutation circuits also. Some more force commutation circuits are included in the end of chapter exercises.

4.6.2 Operation and Analysis of a Force Commutated Thyristor Chopper

Figure 4.34 shows the circuit of a force commutated thyristor chopper for voltage step down conversion. The input is the DC voltage source labeled V. The load is the resistor labeled R_L. L_s is the smoothing inductance on the load side. An input filter is omitted for the sake of clarity. The power switching elements of the chopper are the freewheeling diode labeled D_F and the controlled switching element, which is the main thyristor labeled T_M. Turn ON switching of the main thyristor is achieved by the circuit block labeled GF1, which supplies the gate firing pulses to the gate of T_M. The circuit block shown boxed inside the broken lines is the force commutation circuit, which serves to turn OFF the main thyristor T_M. The force commutation circuit shown has five circuit elements. Two of these are the diodes labeled D_1 and D_2. Their role will be understood from the description that follows. Two other elements of the force commutation circuit are the "commutating capacitor" labeled C and the "commutating inductor" labeled L. L and C constitute an oscillatory circuit, whose oscillations serve to achieve the commutation of the main thyristor T_M in the manner to be described below. T_A is an auxiliary thyristor, for which the gate turn ON pulses are provided by the circuit block labeled GF_2. The turn OFF switching (commutation) of the main thyristor is initiated by a gate pulse on the gate of the auxiliary thyristor T_A, through the action of its gate firing circuit GF_2. Then follows an extremely fast sequence of circuit changes, usually lasting only some microseconds, culminating in the commutation of T_M. We shall now go through these circuit changes, to give a clear understanding of the force commutation process. For this purpose, we shall assume that the chopper has been operating long enough to justify the assumption of repetitive conditions. We shall also assume that the load smoothing inductance L_s is large and therefore the load current I_L can be assumed constant during the commutation sequence.

Examination of the complete circuit shows that there are several static switches—two thyristors and three diodes. Every time a device switches, the circuit configuration changes, and when one cycle of switching is completed, we return to the first configuration and the next cycle commences.

We shall commence from the configuration when T_M is OFF and so also are T_A, D_1 and D_2. Under this condition, the source V is disconnected from the load, and therefore the load current will be freewheeling through the diode D_F. The resulting circuit is shown in Fig. 4.35(a). During this interval, the

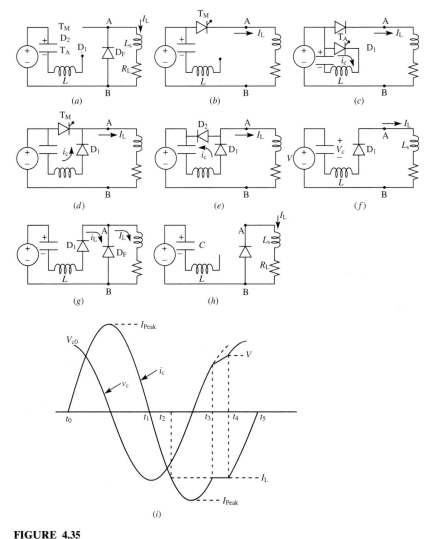

(a) (b) (c)

(d) (e) (f)

(g) (h)

(i)

FIGURE 4.35
Sequential changes in the circuit configuration during the progress of force commutation and the associated voltage and current waveforms of the commutating capacitor.

commutating capacitor will be charged with a voltage V_{c0} of polarity as indicated in (a). V_{c0} cannot be less than the source voltage V, because otherwise D_1 will be forward-biased and stay ON until this condition is met.

The next switching occurs when we fire the main thyristor T_M. The DC source voltage appears across the load, and D_F gets turned OFF. The load current I_L now flows through T_M. The new circuit configuration is shown in Fig. 4.35(b).

Now, when we want to turn OFF the main thyristor, we initiate the commutation sequence by firing the auxiliary thyristor T_A. The resulting circuit configuration is shown in Fig. 4.35(c). The commutating capacitor C, which had an initial voltage V_{c0}, is now short-circuited through the commutating inductor L. This causes an oscillatory current i_c in the L–C loop, which, neglecting resistance of the loop, obeys the following loop equation:

$$\frac{1}{C}\int i_c \, dt + L\frac{di}{dt} = 0 \tag{4.66}$$

The initial conditions are

$$i_c = 0 \quad \text{and} \quad v_c = V_{c0} \quad \text{at } t = 0$$

The solution of this is an oscillatory current i_c given by

$$i_c = I_{\text{peak}} \sin \beta t \tag{4.67}$$

where the oscillatory radian frequency is given by

$$\beta = \sqrt{\frac{1}{LC}} \tag{4.68}$$

and the peak value of the oscillatory current by

$$I_{\text{peak}} = V_{c0}\sqrt{\frac{C}{L}} \tag{4.69}$$

If the commutation is to be successful I_{peak} must be considerably larger than the load current I_L. We shall therefore assume that this condition is satisfied. On this basis, the waveforms of the capacitor voltage v_c and the current i_c are as sketched in Fig. 4.45(i). The first half-cycle of the oscillatory current flows through T_A.

When the second half-cycle of the oscillatory current commences, T_A turns OFF, because it cannot conduct a reverse current. This causes D_1 to turn ON, and the second half cycle of the oscillatory current progresses through D_1 and T_M. The new circuit configuration will be as shown in Fig. 4.35(d). Notice that T_M is now allowing the load current to flow in the forward direction through it and the oscillatory current in the reverse direction. This is possible as long as the resultant current through it is in the forward direction. The resultant current becomes zero when the magnitude of the reverse oscillatory current becomes equal to the load current I_L. This happens at the instant labeled t_2 in (i).

The main thyristor T_M is commutated at t_2, because the resultant current through it tends to reverse as the oscillatory current continues to increase in magnitude. Now the diode D_2, which is in antiparallel with T_M, provides the path for the continued flow of both i_c and I_L. Therefore the new circuit configuration will be as shown in Fig. 4.35(e).

During the conduction period of D_2, the main thyristor remains reverse-biased by the small forward voltage drop of D_2. The diode D_2 permits the flow of both the load current and the oscillatory current through it only as long as the resultant is a forward current for it. The oscillatory current reaches the reverse peak, after which its magnitude decreases. During the decrease, when its magnitude falls to that of I_L, the resultant current through D_2 becmes zero. This happens at the instant labeled t_3 in (i).

After t_3, the current tends to reverse through D_2, and therefore D_2 is commutated. The new circuit configuration will be as shown in Fig. 4.35(f). Here, we see that there is only one circuit loop with L and L_s in series. Since L_s is assumed to be large, the current $i_c = I_L$ remains constant. This therefore is an interval during which the capacitor has a constant charging current equal to I_L. The capacitor voltage therefore linearly increases from the instant t_3 as shown in (i).

During the constant current charging interval, the voltage occurring across the load is $V - v_c$. Therefore, as the capacitor voltage increases, the load voltage V_{AB} decreases and becomes zero when $v_c = V$. After this, V_{AB} tends to reverse, and therefore the freewheeling diode D_F turns ON. This happens at the instant labelled t_4 in (i).

The new circuit configuration will be as shown in Fig. 4.35(g). Inspection of (g) shows that there are two circuit loops, both of which are completed through the static switch D_F. On the right side is the freewheeling loop with a constant current I_L. On the left side is a loop consisting of the source V, the commutating capacitor C and the commutating inductor L. This second loop is an oscillatory circuit, consisting of V, L and C in series, whose loop equation, neglecting resistance, is

$$\frac{1}{C}\int i_c \, dt + L\frac{di_c}{dt} = V \tag{4.70}$$

The initial conditions are that at $t = 0$ [corresponding to the instant t_4 in (i)], $i_c = I_L$ and

$$\frac{1}{C}\int i_c \, dt = V$$

The solution of this equation is

$$i_c = I_L \cos \beta t \tag{4.71}$$

Because the charging current is continuing in the same direction, the capacitor voltage keeps increasing, and it gets overcharged above V. This continues until $i_c = 0$, which will be at $t = \pi/2\beta$ according to the above equation. The capacitor over-voltage above V will be given by

$$v_{c(over\text{-}voltage)} = \frac{1}{C} \int_0^{\pi/2\beta} I_L \cos \beta t \, dt = \frac{I_L}{\beta C}$$

$$= I_L \sqrt{\frac{L}{C}} \qquad (4.72)$$

At the instant t_5, the charging current i_c becomes zero and tends to reverse. Therefore D_1 turns OFF. The new circuit configuration will be as shown in Fig. 4.35(h). This is exactly the same as our starting circuit of (a). Therefore one switching sequence is completed at t_5. The circuit as in (h) = (a) persists until T_M is turned ON again.

Waveform of load voltage. The voltage V_{AB} is constant at V from the instant at which T_M is turned ON till D_2 turns OFF at t_3 [Fig. 4.35(i)]. From t_3 to t_4, which is the linear charging interval of the capacitor, V_{AB} falls linearly to zero. It stays at zero till T_M is turned ON again in the next converter switching cycle.

Overcharging of commutating capacitor. We have neglected the resistance in the L–C circuit and the forward voltage drop of D_1 in determining the capacitor over-voltage. Therefore our expression for this voltage is approximate. However, we notice that this voltage is load-dependent. A simple circuit modification will eliminate the over-voltage. The modified circuit, shown in Fig. 4.36, includes a resistance r that has a large value. It provides a discharge path for the over-voltage until the capacitor voltage falls to V. Since r is large, it does not otherwise significantly affect the functioning of the force commutation circuit.

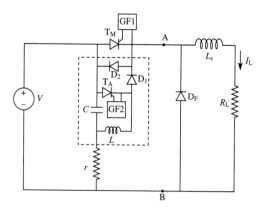

FIGURE 4.36
Modified force commutation circuit.

4.6.3 Dependence of the L and C Values of the Force Commutation Circuits on the Turn OFF Time of the Main Thyristor

During the force commutation sequence described above, the forward current through the main thyristor first becomes zero. After this has happened, the main thyristor is reverse-biased for a short time. If the commutation is to succeed, this short time for which the main thyristor is reverse-biased should not be less than the specified "turn OFF time" of this thyristor. The turn OFF time of a thyristor is a parameter specified in its data sheet. It is defined as the minimum time for which the thyristor has to be held in reverse bias, after the forward current through it has become zero, to ensure that subsequent application of forward voltage will not cause it to turn ON without the application of a turn ON gate pulse. In the following analysis, we shall derive the minimum values of the commutating capacitance and the commutating inductance necessary to provide the minimum turn OFF time to the main thyristor and thus ensure successful commutation. This will also highlight the need to select fast turn OFF thyristors for force commutated thyristor converters, to minimize the size of the commutating circuit elements, irrespective of whether the repetitive switching frequency of the converter is low or high.

Reference should be made to Figs. 4.35(i) and 4.37. In the latter, we have highlighted the portion up to $t = t_3$ in the negative segment of the capacitor current waveform of Fig. 4.35. The period during which the main thyristor is reverse-biased is the interval t_2 to t_3 in both the figures. During this interval, the diode D_2 is conducting and the main thyristor is reverse-biased by a voltage equal to the forward voltage drop of this diode.

During the interval t_2 to t_3, the magnitude of the oscillatory capacitor current exceeds the DC load current I_L. In a typical design, the L and C values of the commutation circuit are chosen so that the oscillatory peak current is twice the maximum current to be commutated. In the modified commutation circuit of Fig. 4.36, the capacitor voltage at the commencement of the commutation sequence will always be equal to V. Therefore, from (4.69), the peak oscillatory current will be given by

$$I_{peak} = V \sqrt{\frac{C}{L}} \tag{4.73}$$

We shall follow the norm of limiting the maximum current to be commutated

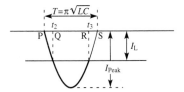

FIGURE 4.37
Turn OFF time of main thyristor.

at 50% of I_{peak}, and shall denote the maximum current to be commutated by I_L. On this basis, the above equation becomes

$$2I_L = V\sqrt{C/L} \tag{4.74}$$

The half-period of the oscillatory current, which is labeled T in Fig. 4.37, is given by

$$T = \pi/\beta = \pi\sqrt{LC} \tag{4.75}$$

where the oscillatory radian frequency β is given by (4.68).

Referring to Fig. 4.37, for $I_{\text{peak}} = 2I_L$, t_2 will correspond to an angular interval 30° from the commencement of the negative half-period to the current. Also, the instant t_3 will at $180° - 30° = 150°$. Therefore the turn OFF time t_{off} available to the main thyristor, which is from t_2 to t_3, will be given by $\frac{2}{3}T$. Therefore

$$t_{\text{off}} = \tfrac{2}{3}\pi\sqrt{LC} \tag{4.76}$$

Equations (4.74) and (4.76) may be treated as two simultaneous equations from which the values of L and C can be determined. Their solution gives

$$L = \frac{3}{4\pi}\frac{V}{I_L}t_{\text{off}} \tag{4.77}$$

$$C = \frac{3}{\pi}\frac{I_L}{V}t_{\text{off}} \tag{4.78}$$

The above equations show that the L and C values are proportional to the turn OFF time to be provided to the main thyristor.

Illustrative example 4.21. In the chopper shown in Fig. 4.38, the main thyristor has a turn OFF time equal to 20 μs. The maximum load current to be commutated

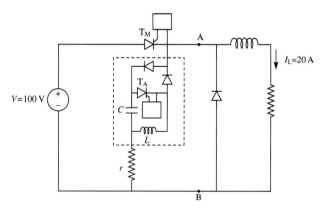

FIGURE 4.38
Chopper of Example 4.21.

is 20 A. Determine the minimum values necessary for the commutating capacitor and the commutating inductor, on the basis of the oscillatory peak being equal to twice the maximum current to be commutated. What is the duration of an oscillatory period of the force commutation circuit?

Solution. On the basis of $I_{peak} = 2I_L$, we can use (4.77) and (4.78) to determine minimum values of L and C. We have $V = 100$ V, $I_L = 20$ A and $t_{off} = 2 \times 10^{-5}$ s. Numerical substitution into (4.77) and (4.78) gives

$$L = \frac{3}{4\pi} \frac{100}{20} \times 2 \times 10^{-5} = 23.9 \ \mu\text{H}$$

$$C = \frac{3}{\pi} \frac{20}{100} \times 2 \times 10^{-5} = 3.8 \ \mu\text{F}$$

One oscillatory period is $2\pi\sqrt{23.9 \times 3.9} = 60 \ \mu\text{s}$.

4.7 SWITCHING CONTROL CIRCUITS FOR CHOPPER CONVERTERS

4.7.1 Functional Requirements of the Switching Control Circuit

A chopper converter, like other power electronic converters, has a power circuit section and a switching control circuit section. The ultimate purpose of the switching control circuit section in a power electronic converter is to provide the necessary inputs to the control terminal of every power semiconductor switching element of the converter, so that it switches ON and OFF with the correct timing, to enable the converter to operate in the required manner. Referring to Fig. 4.39, which applies generally to a typical power electronic

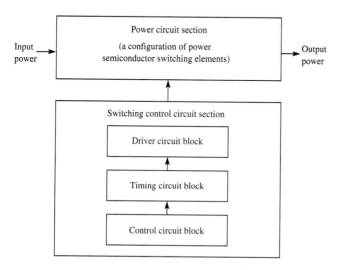

FIGURE 4.39
Functional blocks of a typical power electronic converter.

converter, the power circuit section consists of a configuration of power semiconductor switching elements through which power from its input terminals is delivered to a load circuit connected to its output terminals. Filter circuits, which may also exist on the imput and output sides, are not included in Fig. 4.39, since these are external to the power switching section. The power circuit section ordinarily has to handle relatively large voltages and currents.

The switching control section, on the other hand, normally works from low voltage power supplies, and generally consists of integrated circuits and other low power electronic components, usually assembled on printed circuit boards. In Fig. 4.39 we have shown three functional circuit blocks within the switching control section. These are labeled as the control block, the timing block and the driver block.

The timing block actually generates the timing pulses, according to the required switching pattern of the converter. For example, in the case of a chopper, it should produce the pulses of the required frequency and duration needed for the controlled chopper switch (or switches if the chopper is multiphase). But if the chopper is used for a specific application, it will be necessary to vary the duty cycle to satisfy the needs of the application. For example, if the chopper is used for the speed control of a DC motor, we may use a closed-loop control circuit to maintain the speed at the wanted value. The closed-loop controller will sense the actual speed and compare it with the reference input to it. There may also be other inputs to the closed loop controller, such as the motor current, because the controller may have to limit the motor current within safe limits. All of these are included in what we have grouped as the control circuit block. This block processes all the input information into it and decides the duty cycle at which the chopper should work at a given instant of time. It provides an input voltage to the timing circuit block to vary the widths of the timing pulses, and so achieve this duty cycle. What we want to point out here is that the controller block basically depends on the specific application of the chopper. Since this chapter is not intended to cover the applications of the chopper, we shall not give further consideration to the controller block here.

The output pulses of the timing block, in general, may not be suitable for being directly applied to the control terminals of the power semiconductor switching elements. The three main reasons for this are as follows.

1. The power capability of the timing pulses may be insufficient.
2. The nature of the timing pulses may not suit the particular type of power switching element. For example, if the switching element is a GTO, what it needs to turn ON is a positive current pulse, of short duration only, into its gate terminal. An IGBT, on the other hand, requires a positive voltage input for the entire duration of the ON time.
3. There is usually a need to provide electrical isolation between the timing circuit (in fact, the whole switching control section) and the power circuit.

The driver circuit block serves to meet these requirements. The driver circuits employed depend, in general, on the type of power semiconductor switch employed. For these reasons, we shall limit further treatment in the present chapter to the timing circuit block only.

4.7.2 Generation of Timing Pulses for a Single-Phase Chopper

The block schematic of a circuit scheme for generating the timing pulses for a single phase chopper is shown in Fig. 4.40(a). The related waveforms are shown in Fig. 4.40(b). The switching frequency of the chopper is set by programming the circuit block labeled "clock oscillator" to output pulses at this frequency. If there is a need to vary the chopper frequency then we may use a voltage controlled oscillator for our clock and adjust the frequency as required, by adjusting the reference voltage to it, as is shown in the figure. The output of the clock generator is used to synchronize the frequency of the ramp generator. The ramp voltage output is compared with an adjustable control

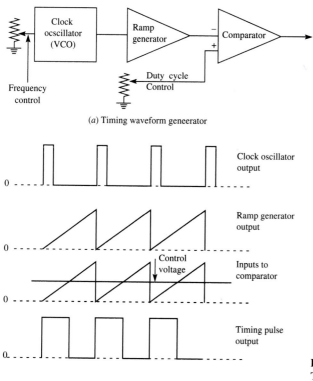

(a) Timing waveform geneerator

(b) Waveforms

FIGURE 4.40
Timing pulse generation for a single-phase chopper.

voltage using a comparator chip. The comparator output serves as the timing pulses for the chopper. The ON period of the chopper will be the pulse width of the timing pulses from the comparator.

The output pulse width, and therefore the duty cycle of the chopper, could be varied by varying the control voltage input to the comparator, as shown in the figure. The operation of the circuit is further made evident by the relevant waveforms sketched in Fig. 4.40(*b*).

If the power semiconductor switch of the chopper is a latching device such as a GTO, what is mainly needed to turn it ON will be a pulse of short duration at each leading edge of the timing pulse train. For turn OFF switching, it will need another pulse at each trailing edge of the timing pulse train. These can be obtained by using two monostable multivibrator chips—one triggered by the leading edges and the other triggered by the trailing edges. In the case of a force commutated thyristor chopper, which employs a force commutation circuit such as described in Section 4.6.2, the monostable outputs at the leading edges can be used to time the gate firing input of the main thyristor and the monostable outputs at the trailing edges can be used to time the gate firing of the auxiliary thyristor.

4.7.3 Generation of Timing Pulses for Multiphase Choppers

A multiphase chopper will require one timing waveform for each phase. A scheme that achieves this, and ensures the exact phase shift between phases, is illustrated in Fig. 4.41 for a three-phase chopper. This is a simple extension of the circuit of Fig. 4.40. For a multiphase chopper with m phases, we shall use a

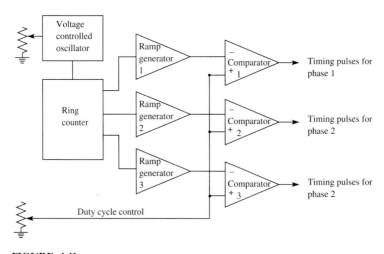

FIGURE 4.41
A circuit scheme for generating timing pulses for a three-phase chopper.

clock oscillator frequency equal to m times the chopper frequency. We can then divide this frequency by m using an m-stage ring counter. Each of the m outputs of the ring counter serves as the clock for each individual phase, which can be used in exactly the same manner as in Fig. 4.40. The control voltage input to all the m comparators will be common, so that the duty cycles of all the choppers can be simultaneously adjusted by the same control voltage.

PROBLEMS

4.1 An electric train in a 3000 V DC railway system uses chopper control with two selectable discrete switching frequencies, the higher of which is 5 kHz. During starting and at very low speeds, it is necessary to reduce the motor voltage to as low as 30 V. The power semiconductor switching device has a minimum effective ON time of 50 μs. Determine
    (a) the highest possible value for the lower of the two discrete frequencies, which is to be used as starting and low speeds,
    (b) the minimum possible motor voltage, if a frequency change is not made at low speeds.

4.2 A battery powered electric car uses a DC motor drive, controlled by an IGBT chopper working at 5 kHz. The battery voltage is 240 V. The minimum effective OFF time of the chopper switch is 20 μs. Determine the highest voltage that the chopper can deliver to the motor.

4.3. The data relating to the chopper circuit of Fig. P4.3 are $V = 120$ V, $L = 4$ mH, $R = 20\,\Omega$, frequency 1 kHz and duty cycle 70%. Determine
    (a) the DC load current;
    (b) the peak-to-peak ripple in the load current.

4.4. The chopper in Fig. P4.4 is operating at 1 kHz. Other data are $V = 240$ V, $L = 10$ mH, $R = 10\,\Omega$ and duty cycle 60%.
    (a) Determine the DC component of the load current and the peak-to-peak ripple in the load current.
    (b) By how much will the above values change if the frequency is increased to 2 kHz other data remaining the same.
    (c) What will be the change in the values determined in (a) if the frequency is unchanged but the inductance value is increased to 20 mH, other data remaining the same.

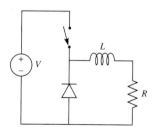

FIGURE P4.3

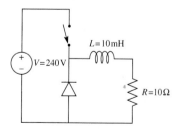

FIGURE P4.4

4.5. The data for the chopper circuit of Fig. P4.5 are $V = 500$ V, $R = 25\,\Omega$ and $f = 2$ kHz, while the duty cycle is variable. The maximum value of the peak-to-peak ripple current is to be limited to 1 A.

(a) Determine the required value of the smoothing inductance L.

(b) The actual value of the inductor used is 40% of what has been determined in (a). The frequency is raised to satisfy the same ripple limits. What is the lowest frequency that will ensure this?

4.6. A chopper is used to charge a 60 V battery from a 240 V DC supply. The internal resistance of the battery is $0.5\,\Omega$. The chopper frequency is 1 kHz. Determine the value of the inductance necessary to limit the peak-to-peak ripple in the battery to 2 A. The DC charging current is to be 40 A.

4.7. A battery powered, chopper controlled electric car is cruising at a constant speed, and the data are motor voltage 80 V, battery voltage 120 V, chopper frequency 2 kHz, duty cycle 80%, motor resistance $0.5\,\Omega$ and total load circuit inductance $200\,\mu$H. Determine

(a) the DC current drawn by the motor;

(b) the peak-to-peak motor current ripple.

4.8. An industrial DC motor drive is controlled by a chopper operating at a fixed frequency of 2 kHz. The DC bus voltage is 250 V. The motor resistance is $0.5\,\Omega$ and the total load circuit inductance is 0.25 mH. At a particular speed, the motor voltage is 150 V. Determine the minimum duty cycle at this speed below which discontinuous conduction will commence in the motor.

4.9. Data relating to the chopper circuit of Fig. P4.9 are $V = 400$ V, chopper frequency 2 kHz, duty cycle 60% and load current (assumed to be ripple-free because of large smoothing inductance) 20 A. Assume that the filter inductance L_1 is also very large and that the input current I_1 may be considered ripple-free.

(a) Determine the waveform of the current i_c in the filter capacitor. Sketch the waveform, indicating relevant magnitudes and durations.

(b) Determine a suitable value for the filter capacitor C that will limit the peak-to-peak ripple voltage at the chopper to 4 V.

4.10. A voltage step down chopper operating at 2 kHz and at a duty cycle of 0.7 supplies a load current of 120 A, which can be assumed to be constant and ripple-free because of large inductive smoothing. The DC bus voltage is 450 V. Determine the maximum and minimum voltage occurring across the filter capacitor C if its value is $100\,\mu$F.

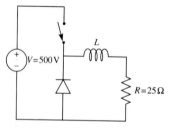

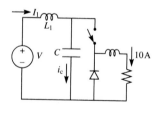

FIGURE P4.5   **FIGURE P4.9**

FIGURE P4.12

4.11. A 5 kHz voltage step down chopper uses an IGBT as the controlled switching element. It is to work from a 200 V DC bus, at a maximum duty cycle of 80%, to deliver a maximum load current on the output side equal to 120 A. Determine the theoretical voltage and current ratings needed for the IGBT and the freewheeling diode. Give the r.m.s. average and repetitive peak current ratings separately for each. Assume ideal smoothing on the output side and a perfect input filter.

4.12. The chopper in Fig. P4.12 is operating in the voltage step up mode, drawing a current of 10 A DC from a 100 V DC source. The duty cycle is 40% and $R = 1\,\Omega$. Determine the output voltage on the high voltage side. Assume ideal elements and continuous current flow in the DC source.

4.13. A battery powered car uses a 240 V battery, and the driving motor is controlled by a fixed frequency chopper operating at 5 kHz. While going down a slope, the car is being regeneratively braked. The motor voltage is 70 V. The motor current is 20 A. The resistance of the motor circuit is 2 Ω. Assume ideal filtering on both the input and output sides of the chopper. Calculate:
(a) the duty cycle of the chopper;
(b) the current being fed back to the battery.

4.14. A chopper controlled, battery powered electric vehicle is being regeneratively braked. The recharging current flowing into the battery is 12 A. Other related data are battery voltage 150 V, chopper frequency 2 kHz, duty cycle 0.4 and motor resistance 0.5 Ω. Determine the motor back e.m.f. and the motor current under these conditions. Assume ideal filters and continuous motor current flow.

4.15. The driving motor of a battery powered electric car has a resitance of 0.8 Ω. The total inductance in the motor circuit is 0.4 mH, the battery voltage is 220 V, the chopper frequency is 2 kHz and the motor back e.m.f. is 120 V. Determine the minimum duty cycle below which discontinuous current flow will commence.

4.16. In Problem 4.15, determine the DC braking current in the motor at the limiting value of the duty cycle.

4.17. A battery powered car is cruising down a gradient while the chopper is operating in the voltage step up mode, recharging the battery. The motor circuit has a resistance of 0.6 Ω and a total inductance of 150 μH. The chopper frequency is 1 kHz and the duty cycle is 80%. The battery voltage is 160 V. Determine the maximum value of the motor back e.m.f. above which discontinuous conduction will commence. Calculate the DC motor current for the limiting condition.

4.18. In Problem 4.17, determine the new value of the limiting motor back e.m.f. if
(a) the chopper frequency is increased to 2.5 kHz, all other data remaining the same;
(b) the inductance is increased to 300 μH without any change in the chopper frequency, all other data remaining the same.

4.19. A three-phase chopper is operating at a duty cycle of 80% and supplying power into a resistance load from a 250 V DC bus. The smoothing inductance of each phase coil is 2 mH. The resistance of each phase coil is 1 Ω. The load resistance and inductance are respectively $3\,\Omega$ and $6\,\text{mH}$. The chopper frequency is $500\,\text{Hz}$. Determine the following:

(a) the DC load current;

(b) the peak-to-peak ripple in the load current and the ripple frequency.

4.20. In a $1500\,\text{V}$ DC electric railway system, a train has a $1\,\text{kHz}$ four-phase chopper controlling a driving motor. The motor resistance and inductance are $0.05\,\Omega$ and $0.01\,\text{mH}$. Additionally, each chopper phase has an independent smoothing inductance of $0.4\,\text{mH}$ whose resistance is $0.02\,\Omega$. The chopper duty cycle of each phase is 90%. Determine the limiting value of the motor back e.m.f. above which discontinuous conduction will commence in the motor.

4.21. A five-phase chopper is feeding a DC load. The chopper frequency is $200\,\text{Hz}$. The current flow is continuous for each phase of the chopper. Determine

(a) the minimum number of chopper switches that will be simultaneously ON at any instant of time in a switching cycle;

(b) the maximum number of switches that will be simultaneously ON in any instant;

for each of the following values of the chopper duty cycle D: 10%; 35%; 45%; 70%; 90%.

For each of the above duty cycles, also determine the duration of the time interval when the output current is

(c) rising;

(d) falling.

4.22. A driving motor in a $1500\,\text{V}$ DC electric train is controlled by a thyristor chopper using the force commutation circuit of Fig. 4.36. The maximum current to be commutated is $300\,\text{A}$. the specified turn OFF time of the thyristor is $25\,\mu\text{s}$. Determine suitable values for the commutating capacitor and the commutating inductor. State the basis on which the values were determined.

What will be the duration of an oscillatory half-period of the commutating circuit? Determine the peak amplitude of the current through the auxiliary thyristor.

CHAPTER
5

INVERTERS

5.1 INTRODUCTION—FUNCTIONS AND FEATURES OF INVERTERS

An inverter converts from a DC input into an AC output statically, that is, without any rotating machines or mechanical switches. The power circuit configuration of an inverter consists of semiconductor power devices that function as static switches, that is, switches without moving contacts. The inverter also has a switching control circuit that provides the necessary pulses to turn ON and turn OFF each static switching element with the correct timing and sequence. These switches are repetitively operated in such a way that the DC source at the input terminals of the inverter appears as AC at its output terminals.

When an inverter is required for a practical power conversion or power control application, the following features of the AC output need consideration.

The AC frequency. In static inverters, the AC frequency is precisely adjustable by adjustment of the switching frequency of the power switching elements. This is usually determined by the frequency of a "clock oscillator" in the switching control section of the inverter.

Magnitude of AC voltage. There is usually provision for adjustment of the voltage in one of two ways.

1. The voltage may be varied by varying the DC input voltage to the inverter. In this case, the adjustment is outside the inverter and is independent of the inverter switching.
2. The alternative way of AC voltage variation is within the inverter by a technique called pulse with modulation (PWM). The PWM strategies will be presented later in this chapter.

Waveform of the AC voltage. The sinusoidal waveform is usually the most desirable for many applications. However, the AC output of practical inverters will inevitably have a certain amount of harmonic content. There are two practical ways by which the harmonic content can be brought down to a low value or at least within acceptable limits. One method is to use a filter circuit on the output side of the inverter. The filter circuit will, of course, have to handle the large power output from the inverter. The second scheme employs a pulse width modulation strategy that will change the harmonic content in the output voltage in such a way that the filtering needed will be minimal or zero depending on the type of application. This second technique is based on a suitably designed switching strategy within the inverter. Several such strategies have been developed. In one technique, known as sinusoidal pulse width modulation (SPWM), the residual harmonic content in the output voltage will be at high frequencies, and a filter circuit, if at all required, will be small. The permissible harmonic content in the voltage is dependent on the type of load on the AC side. In this context, it may be stated that the waveforms of the AC voltage and the AC current will be normally different. For example, if the load is an AC motor that is highly inductive, the harmonic content in the current will be considerably less than the harmonic content in the voltage applied to the motor. This is because the motor inductance itself acts as a filter element, considerably attenuating the higher-order harmonic currents. The sinusoidal pulse width modulation technique will be presented later in this chapter.

The phase number and the phase sequence. Inverters are usually designed to provide either single- or three-phase output. Larger industrial applications, such as AC motor drives, require three-phase AC. Three-phase inverters will have a larger number of switching elements, and the switching control circuit will be organized in such a way as to provide the three-phase AC at its three output terminals. For reversible motor drive applications, the phase sequence of the three-phase supply fed to the motor needs to be reversed, to reverse its direction of rotation. In the case of an inverter fed motor, there is no need to use a mechanical reversing switch to reverse the phase sequence of the voltage applied to the motor. The phase sequence can be reversed by reversing the sequence of switching of the power semiconductor switches of the inverter. This can be achieved electronically inside the inverter.

5.2 INVERTER APPLICATIONS

An inverter is an efficient and convenient source of AC power, whether it be a fixed frequency application, or one that requires variable frequency. As such, they find wide use in industrial equipment. To highlight the role of inverters in power electronic technology and to indicate the performance requirements that they may have to meet, we shall briefly describe the place of the inverter in a typical power conversion scheme. We shall also briefly overview two important practical applications here, which will be treated in detail in later chapters dealing with applications, namely,

1. adjustable speed AC motor drives;

2. uninterruptible power supplies (UPS).

5.2.1 DC Link Inverters

Electric power is usually available as AC at a fixed frequency, such as 60 or 50 Hz. For some applications, we may need a frequency that is not the same as the available supply frequency, or we may need an adjustable frequency. The AC we need may be three-phase, when the available power is single-phase. All these requirements can be met by the use of static inverters. A block schematic of the conversion scheme is shown in Fig. 5.1.

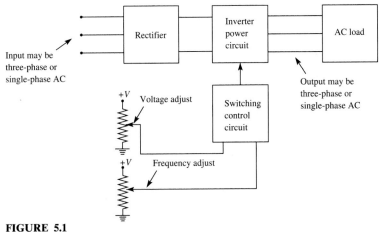

FIGURE 5.1
The DC link inverter.

In Fig. 5.1 the input AC, which may be single- or three-phase, is first converted into DC by a rectifier circuit. This may be an uncontrolled rectifier using power diodes, or a controlled rectifier using thyristors. In the latter case, the DC voltage will be adjustable. The DC output of the rectifier serves as the input to the inverter. An intermediate filter circuit will be desirable to minimize the ripple in the DC link voltage, which will inevitably be present after the rectification. The inverter may be designed to provide either three- or single-phase output as required. The adjustment of the AC output voltage, if required, can be made outside the inverter by varying the DC link voltage. Alternatively, this adjustment can be made by pulse width modulation within the inverter, if the inverter switching control circuit is designed on this basis. In the scheme shown in Fig. 5.1, the voltage adjustment is done internally by PWM within the inverter. The manner in which this is done will be explained subsequently in this chapter. For the present, we may state that this can be done by varying a DC control voltage applied at a certain location in the switching control circuit.

Similarly, for adjustment of AC frequency, we provide another adjustable DC control voltage at a separate location in the switching control circuit. This is usually the control voltage of a voltage controlled oscillator chip, which determines the "clock" frequency of the inverter switching control circuit. In the scheme shown in the figure, the frequency and the voltage are shown as adjustable by two independent control voltages, and the manner of control shown is by "open loop," wherein each is manually set by an appropriate value of the respective control voltage.

5.2.2 Adjustable Speed AC Drives

A major application of static inverters is the speed control of AC motors. The speed of an AC motor of the induction type or of the synchronous type is determined by the frequency of the AC. Therefore the speed could be conveniently adjusted by supplying the motor from an inverter of adjustable frequency. At the same time, it must be stated that the level of magnetic flux density in the motor is determined by both the frequency and the magnitude of the AC voltage applied to the motor. The flux density is approximately proportional to the reciprocal of the frequency. On the other hand, the flux density increases with the voltage—approximately proportionally. A low flux level results in poor torque characteristics, whereas a high flux density is associated with magnetic saturation of the steel and increased magnetizing current and losses in the motor. Therefore it is necessary to maintain the optimum flux density in the motor, at all speeds, when the speed is varied by varying the AC frequency. What this means in a practical speed control scheme is that it is necessary to vary the inverter voltage also, when the frequency is varied to implement a speed change. One way of doing this will be to make the reference voltage for voltage adjustment in Fig. 5.1 a function of the reference voltage for frequency adjustment—approximately proportional to it. This aspect will be discussed in more detail when we deal with speed control of AC motors in Chapter 8.

If the source of power is DC, as is the case in a battery powered road vehicle, the rectifier shown in Fig. 5.1 is not needed. The present trend is to use AC motors for battery powered cars, even though the power available is DC. This is because the three-phase AC motor is sturdier, lighter and requires relatively less maintenance, because it has no rubbing contacts (brushes) since there is no need to make electrical connections to the rotor. The so-called "brushless DC motor," sometimes used for vehicle drives, is, in fact, a three-phase synchronous AC motor fed through a static inverter. This type of motor has permanent magnets on the rotor, and therefore there is no need to provide brushes to feed magnetizing current to the rotor. The other type of motor that is popular for this application is the three-phase induction-type motor, which also does not have brushes since the rotor currents are induced by electromagnetic induction from the stator side. Present-day inverters are highly efficient and the control strategies are easy to implement. Therefore inverter fed AC motors are considered to be better alternatives to the earlier concept of using DC motors for battery powered road vehicles.

The statements we have made regarding the use of AC motors for battery powered cars are also, by and large, valid for electric trains supplied from DC. The majority of rapid transit systems for city and suburban traffic are supplied by DC power, because they were originally designed to use DC motors. Therefore, here again, the trend now is to use inverter fed AC motors for future designs, the inverters being fed from the existing DC supply.

5.2.3 Uninterruptible Power Supplies (UPS)

It is rare that an AC power distribution system is absolutely reliable. There are occasions when, owing to circumstances beyond control, interruptions ("power supply blackouts") occur. Sometimes the problem may not be a total failure of power, but a deterioration in its quality, such as a decrease in voltage, making it unsuitable for sensitive equipment ("power supply brownout"). Such a blackout or brownout may not be permissible for certain types of equipment, computers for instance, where loss of data can occur irretrievably. Uninterruptible power supplies (UPS) are made to solve this problem, for use with such equipment. The equipment receives its AC power from the UPS. The AC power provided by the UPS is from an inverter, which reliably maintains the correct voltage and frequency. This inverter works from an internal DC bus in the UPS. When the power lines are healthy, this DC bus is fed through a rectifier circuit within the UPS. But the DC bus also has a floating battery. The battery automatically feeds the AC during power supply failures, so that the inverter continues to function without interruption, thus maintaining the power supply to the equipment. There is also a charging circuit for the batteries, which charges the batteries when the power lines are healthy, thereby ensuring that the batteries are always ready to provide power to the inverter.

In the scheme described above, the power supply to the equipment is always through the inverter. But there is an alternative scheme in which the inverter is called upon to provide power only when power lines fail. In this arrangement, the power supply to the equipment is normally directly from the power lines when these are healthy. When a blackout or brownout occurs, this is automatically detected by a sensing circuit, and the power supply to the equipment is instantly transferred to the inverter. The changeover of the connections is implemented using a static bypass switch, which consists of semiconductor power devices that operate so fast that the interruption is not felt by the equipment. When the power lines regain their healthy condition, the equipment is switched back in a similar manner.

5.3   TYPES OF INVERTERS

Static inverters may be classified into one of the following categories, on the basis of the type of AC output.

1. voltage source inverters;
2. current source inverters;
3. current regulated inverters (hysteresis-type);
4. phase controlled inverters.

Of these, the phase controlled inverter does not generate an independent AC. It only serves to feed power from a DC source into an existing AC source. It is in fact a phase controlled rectifier operating in the "inversion" mode, that is, with the direction of power flow reversed. This type of inverters is dealt with in detail in Chapter 2.

The present chapter will deal almost exclusively with the first category, namely, the voltage source inverter. But we shall briefly state here the basis of the classification given above.

5.3.1   Voltage Source Inverters

This is the most commonly used type of inverter. The AC that it provides on the output side functions as a voltage source. The input is from a DC voltage source. The input DC voltage may be from the rectified output of an AC power supply, in which case it is called a "DC link" inverter. Alternatively, the input DC may be from an independent source such as a battery.

5.3.2 Current Source Inverter

On the output side, this type of inverter functions as an AC current source. This type is also a DC link inverter. But, in this case, the DC link functions like a DC current source. A block schematic of the current source inverter is shown in Fig. 5.2. The DC link is provided by a phase controlled thyristor rectifier bridge. However, the phase control is made to function in such a way that the DC link current is maintained constant and equal to a reference value provided to the controller. The actual DC current is sensed by means of a current sensing circuit and compared against the reference current. Any error detected between the reference current and the sensed current is made to operate the phase control circuit in such a way that this error is cancelled. In the scheme shown in the figure, the current sensing is shown on the AC side of the rectifier, whereas it is the DC current that needs to be sensed. But the current on the AC side of the rectifier bridge is a measure of the DC current. It is more convenient to sense the AC side current through current transformers, because they provide electrical isolation. Therefore, in practice, it is usual in current source inverters to use the rectified output of the current transformers as the feedback voltage proportional to the DC current. The figure also shows a large inductor in series with the DC link. This is an essential requirement to minimize fluctuations in the DC link current. The reason is that the switchings

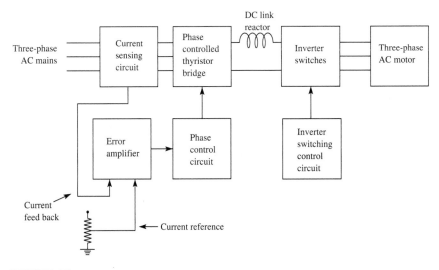

FIGURE 5.2
Block schematic of a current source inverter for a three-phase AC motor drive.

in the phase controlled rectifier bridge take place only at discrete intervals, and these intervals are relatively large and are dictated by the frequency of the AC mains. The DC link reactor serves to minimize current variations during the interval between successive switchings (commutations) in the phase controlled thyristor bridge. In this way, the DC link is made to function as a better current source.

The switching elements in the inverter block in Fig. 5.2 only serve to channel the DC link current to the output terminals of the inverter in such a way that the output terminals function as a three-phase current source. This is an adjustable current source, the adjustment being done by changing the reference current input to the phase controlled rectifier bridge.

Current source inverters of this type have proved to be very useful for the speed control of three-phase AC motors. The motor is automatically protected against over-currents even if it stalls owing to excessive load torque. There is torque even in the stalled condition, so that it can be reversed and released if there is jamming. The motor can be regeneratively braked by making it go into the generating mode. This is possible because the operating mode of the phase controlled rectifier can change into phase controlled inversion, without reversal of the DC current, thereby making it possible to feed the braking power back into the AC supply lines. Current source inverters will be described in more detail in Chapter 8.

5.3.3 Current Regulated Inverters (Hysteresis-Type)

Inverters of this category are increasingly becoming popular for the speed control of AC motors. The input DC is the same as in conventional voltage source inverters. This could be a DC voltage link provided by a rectifier or an independent DC source such as a battery. The internal switching of the inverter is controlled in such a way that, at the output terminals, it behaves as an AC current source. The waveform of the current required is provided to the switching control circuit as a reference waveform. There is a current sensing circuit that senses the actual value of the current at every instant. This sensed value of the prevailing current is compared against the value as dictated by the reference waveform, and the switching inside the inverter is altered as necessary to correct any error. In this way, the output current waveform is made to conform, as accurately as possible, to the input reference waveform. This operating principle may be further explained by reference to the block schematic of Fig. 5.3(a), where the inverter circuit block is fed from a fixed DC voltage source. The current sensor on the output side provides a feedback signal voltage proportional to the value of the instantaneous output current. The actual instantaneous current required is provided as a reference waveform.

The switching control circuit has a module that compares the required instantaneous value of the output current and the actual instantaneous value sensed by the current sensor. If the error, by which we mean the difference

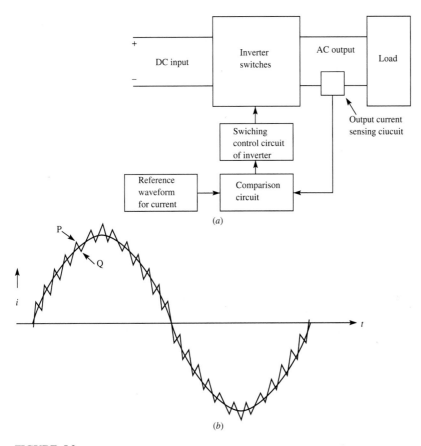

FIGURE 5.3
(*a*) Block schematic of current regulated inverter. (*b*) Hysteresis-type current control used in current regulated inverters.

between the required value and the actual value of the current, exceeds a preset limit then a switching is made in the inverter in such a way that the error decreases. Figure 5.3(*b*) shows a typical reference waveform and also the actual waveform of the output for an inverter of this type. To illustrate, consider, for example, an instant at which the actual current is indicated by the point P in Fig. 5.3(*b*). At this instant, the actual current value is higher than the required value as shown by the reference waveform by the finite preset limit. Therefore the comparison module causes the inverter to switch in such a way as to cause the current to decrease. Therefore the current will start decreasing. The decrease will continue, and at the instant corresponding to Q, the actual current will be lower than the required value again by a finite preset limit. Therefore, at this instant, the comparison circuit will cause the inverter to switch again in such a way as to cause the current to increase. This manner of switching back and forth will keep on happening, as a result of which the actual current waveform will be confined between two narrowly spaced envelopes, one of which is above the reference waveform by a small preset value and the other below the reference also by a small preset limit. The two preset limits of error are generally referred to as hysteresis limits. The controller itself is referred to as a hysteresis-type controller. By making the hysteresis limits narrow, it is possible to make the actual current waveform conform more closely to the reference. But the repetitive switching frequency of the inverter will increase. The lowest possible hysteresis limits are determined by the highest repetitive switching frequency capability of the power semiconductor switching element used in the inverter. A more detailed treatment of this type of inverter will be presented in Chapter 8.

5.4 THE HALF-BRIDGE INVERTER

There are two circuit topologies commonly used for inverters. They are (a) the half-bridge and (b) the full bridge. For certain low power applications, the half-bridge may suffice. But the full bridge is more convenient for adjustment of the output voltage by pulse width modulation techniques. The half-bridge, however, is the basic building block of the full bridge, and therefore we shall describe the half-bridge first.

5.4.1 The Half-Bridge Inverter—Circuit Configuration and Switching

Figure 5.4(a) shows the power circuit configuration of the half-bridge. The half-bridge inverter has two controlled static switching elements, which are labeled S_1 and S_2, each of which has an antiparallel diode. It is evident from

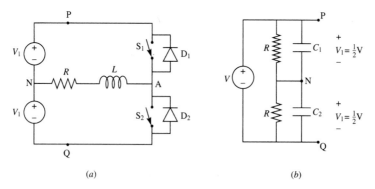

(a) (b)

FIGURE 5.4
Half-bridge inverter.

the presence of the diodes that the switching devices S_1 and S_2 need not have the capability to block reverse voltages. If the actual device used has reverse voltage blocking capability, that capability is not utilized in the inverter circuit topology. If the switching element is a power MOSFET, there may not be a need to use the antiparallel diodes, because the device structure has an internal antiparallel diode. Manufacturers of power semiconductor devices now supply switching modules in a single package, consisting of the antiparallel diode and the controlled switching element, which may be a power transistor, IGBT or some other device. Such switching modules are also available in multiple units, a single package consisting of two, four or six switching elements, each with its antiparallel diode. The necessary insulation is also provided internally and between the modules and the casing, so that the entire package can be mounted on a single heat sink. Such modules are very convenient for use in inverter topology.

As shown in Fig. 5.4, the input DC to the half-bridge has to be a split power supply. This means that, besides the positive terminal (labeled P in the figure) and the negative terminal (labeled Q), the midpotential terminal, labeled N, must also be available. Therefore the DC input to the inverter is shown as two identical voltage sources, each labeled V_1 connected in series at N. If the midpotential terminal of the DC source is not available, the requirement can be met with the arrangement shown in Fig. 5.4(*b*). Two identical capacitors, each labeled C_1, are shown connected in series across the DC source. The junction point N of the two capacitors may be used as the DC source midpoint. Two large and equal resistors, labeled R, may be connected as shown, to ensure correct voltage division. They also enable the capacitors to discharge when the DC supply is switched off. If the capacitors are large, the voltage will be constant across each, and for analysis purposes we can replace them by the DC voltage sources as shown in Fig. 5.4(*a*). Because of the DC voltage on them, electrolytic capacitors can be used, and these are inexpensive and available in large microfarad values in compact sizes. The DC input terminals of the inverter are P and Q. The AC output terminals are the terminal A, which is the junction of the two controlled switches, and N, which is the DC midpotential terminal. The load shown in the figure consists of a resistance R and inductance L in series. Since the majority of industrial loads are inductive in nature, we shall analyze the inverter operation for such a load, which we shall represent by a series R–L circuit as in the figure.

To operate the inverter to provide an AC output of frequency f, the switches S_1 and S_2 are turned ON and turned OFF alternately, each switch being kept ON for one half-period of the AC, while the other is kept OFF. It is important to ensure that both the switches are never ON simultaneously at any time. If that happens, it is equivalent to a short-circuit across the DC input, resulting in excessive current and possible damage to the switching elements. For this reason, it is customary in practical inverters to provide a "dead time" after the turn OFF of one switch and the turn ON of the other. The switching control circuit of the inverter is not shown in Fig. 5.4 for the sake of clarity. It

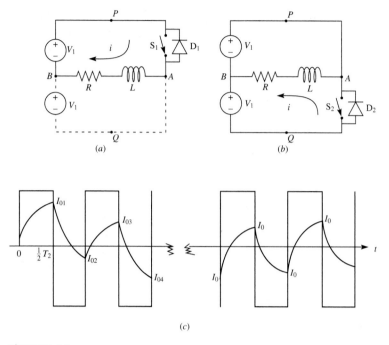

FIGURE 5.5
Output voltage and current waveforms of the half-bridge inverter.

is, however, assumed, that the switching control citcuit is present and that by means of it the switches S_1 and S_2 are operated in the manner stated above.

5.4.2 Analysis of the Half-Bridge Inverter with Inductive Load

Let us assume zero initial current in L and that the switch S_1 is turned ON at $t = 0$. The loop equation will be (see Fig. 5.5(a)]

$$L\frac{di}{dt} + Ri = V_1 \tag{5.1}$$

the initial conditions being $i = 0$ at $t = 0$. The solution is

$$i = \frac{V_1}{R}(1 - e^{-t/\tau}) \tag{5.2}$$

where the load circuit time constant $\tau = L/R$. The exponential growth of current as given by (5.2) is sketched in Fig. 5.5(c), commencing from $t = 0$. For an output frequency f, one cycle time will be

$$T = 1/f$$

At $t = \frac{1}{2}T$, the first half-period of the AC ends and the circuit configuration changes, because S_1 is turned OFF. At this instant, the current will be, according to (5.2),

$$I_{01} = \frac{V_1}{R}(1 - e^{-T/2\tau}) \tag{5.3}$$

The second half-period commences at $t = \frac{1}{2}T$. During the second half-period, S_2 is ON instead of S_1. Let us consider now what will happen if, after turning OFF S_1, we do not instantly turn ON S_2, but delay this by a short "dead time." When we turn OFF S_1, the current cannot fall to zero instantly, because of the inductance L, which has stored energy due to I_{01}. Therefore, when S_1 is turned OFF, the induced e.m.f. due to the inductance L will cause the diode D_2 to become forward-biased and turn ON. In this manner, the terminal A will get connected to the terminal Q instantly. Therefore it will not make any difference to the voltage across the load, or the current through it, if we delay the turn ON of S_2, as long as this delay does not exceed the time it takes for the current to fall to zero. Therefore the circuit configuration for the second half-period will be as shown by Fig. 5.5(b), whether or not we provide a short dead time. We shall take a new reference zero for t, at the commencement of the second half-period, for the loop equation and the initial conditions, which are

$$\left. \begin{array}{l} L\dfrac{di}{dt} + Ri = -V_1 \\[2mm] i = I_{01} \quad \text{at } t = 0 \end{array} \right\} \tag{5.4}$$

The solution of this and similar equations may be written as the sum of two parts: a "growth" part giving the growth of current for zero initial conditions and a "decay" part corresponding to the decay of the initial current. If written in this manner, the solution becomes

$$i = -\frac{V_1}{R}(1 - e^{-t/\tau}) + I_{01}e^{-t/\tau} \tag{5.5}$$

The current during the second half-period, as given by (5.5), is sketched in Fig. 5.5(c). We notice from this that, at the end of the second half-period, the current has a negative peak, labeled I_{02} in the figure. Therefore, when we solve the loop equation for the third half-period, the positive peak that we shall get at the end of this half-period, which is labeled I_{03} in the figure, will be lower in magnitude than I_{01}. In this way, as the switching progresses, the positive peaks will become less and less and the negative peaks will become more and more. This happens during the initial "transient" or "current build up" period. After several cycles of switching, the difference in the current waveform between successive cycles become negligible. We may then take it that the transient current build up period is over and repetitive conditions have been reached. When repetitive conditions prevail, symmetry considerations show that the magnitudes of the positive and negative current peaks are equal. The current and voltage waveforms, when repetitive conditions prevail, are sketched on the right-hand side of Fig. 5.5(c). This is an extension of the same waveforms on the left of the same figure corresponding to start up conditions. Notice that the voltage waveform is repetitive from the very beginning, whereas the current waveform needs several cycles of operation to attain repetitive conditions. We shall denote the magnitude of the repetitive peak current by I_0 and assume repetitive conditions for the following analysis.

We shall use the term "switching block" for the combination of a controlled switch and its antiparallel diode. When a switching block is in the ON state, only one of the constituent switching elements will be ON, depending on the direction of current through it. But, irrespective of which element is ON, the voltage across the block will be ideally zero.

Taking $t = 0$ at the instant of turn ON of switching block 1 in Fig. 5.5(a), the loop equation for this half-cycle will be

$$\left. \begin{array}{c} L\dfrac{di}{dt} + Ri = V_1 \\[2mm] i = -I_0 \quad \text{at } t = 0 \end{array} \right\} \tag{5.6}$$

The solution of this is

$$i = -I_0 e^{-t/\tau} + \frac{V_1}{R}(1 - e^{-t/\tau})$$

At $t = \frac{1}{2}T$, from symmetry considerations, $i = I_0$. Therefore

$$I_0 = -I_0 e^{-T/2\tau} + \frac{V_1}{R}(1 - e^{-T/2\tau})$$

This gives I_0 and the expression for i as

$$I_0 = \frac{V_1}{R}\frac{1 - e^{-T/2\tau}}{1 + e^{-T/2\tau}} \tag{5.7}$$

$$i = -\frac{V_1}{R}\frac{1 - e^{-T/2\tau}}{1 + e^{-T/2\tau}}e^{-t/\tau} + \frac{V_1}{R}(1 - e^{-t/\tau}) \quad \text{for } 0 \leqslant t \leqslant \frac{1}{2}T \tag{5.8}$$

During the second (negative) half-period, the waveform of the current will be a repetition of the positive half-period, with the sign reversed. Therefore the expression for the current will be given by (5.8), with the sign reversed and the zero of t treated as the instant of commencement of the negative half period, that is, when switching block 2 turns ON.

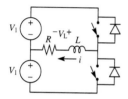

FIGURE 5.6
The half-bridge of Example 5.1.

Illustrative example 5.1. For the half-bridge inverter shown in Fig. 5.6, the AC frequency is 1 kHz, $R = 10\,\Omega$, $L = 10\,\text{mH}$ and $V_1 = 100\,\text{V}$. Assume repetitive conditions.

(a) Obtain the expression for the current for one half-period of the AC.

(b) Sketch the waveforms of the output voltage and current and indicate therein (1) the peak magnitudes of each and (2) the time displacement between the zero-crossings of voltage and current. Express this in angular measure, treating one full period as 360°.

Solution. We have load time constant $\tau = L/R = 10^{-3}\,\text{s}$ and time period of inverter $T = 1/f = 10^{-3}\,\text{s}$, so that

$$\text{magnitude of peak current } I_0 = \frac{V_1}{R}\frac{1 - e^{-T/2\tau}}{1 + e^{-T/2\tau}} = \frac{100}{10}\frac{1 - e^{-0.5}}{1 + e^{-0.5}}$$

$$= 2.45\,\text{A} \tag{5.9}$$

The expression for i during the positive voltage half-period will be

$$i = -2.45e^{-1000t} + 10(1 - e^{-1000t})$$

$$= 10 - 12.45e^{-1000t} \tag{5.10}$$

The same waveform is repeated during the second half-period (negative voltage half-period), with reversal of sign. The waveforms are sketched in Fig. 5.7. The instant of zero-crossing of the current during the positive voltage half-period is determined by equating (5.10) to zero. This gives the time displacement between voltage and current zero-crossings as $t = 0.219\,\text{ms}$. This time displacement is the interval between instants such as t_1 and t_2 in Fig. 5.7. To express this in angular measure, we take one full time period of the AC, that is 1 ms in this example, as equal to 360°. This gives the displacement angle as $\theta = 78.8°$.

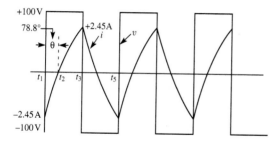

FIGURE 5.7
Waveforms of voltage and current.

5.4.3   Commutation Sequence of the Switching Elements in the Half-Bridge Inverter with Inductive Load

Figure 5.8(a) shows one period of the AC output voltage and current waveforms. This corresponds to repetitive operating conditions. The power circuit configuration is shown in Fig. 5.8(b). The power circuit has four switching elements: S_1, D_1, S_2 and D_2. Each of them turns ON and turns OFF once in an AC period. The turn OFF switching of one element and the turn ON switching of another element, by which a transfer of current takes place from one element to another, is called "commutation" in power electronics terminology. This term is also used to describe the turn OFF switching only, whether or not there is a transfer of current taking place to another element. The instants at which the four commutations take place in a half-bridge inverter are the instants of zero-crossings of the voltage and current waveforms. These instants are labeled t with an appropriate suffix in Fig. 5.8(a). Our purpose here is to describe the manner in which each of the four power switching elements go through the commutation sequence when the inverter is supplying an inductive load as in Fig. 5.8(b).

Referring to Fig. 5.8(a), during the interval t_1 to t_2, the load voltage v_L is positive and the load current i is negative. Therefore the conducting device is D_1. At t_2, the diode D_1 is commutated and the current shifts to S_1. From t_2 to t_3, S_1 is conducting. At t_3, S_1 is turned OFF (commutated) by the action of the switching control circuit, and the current transfers to D_2. At t_4, D_2 is commutated, and S_2 takes over conduction from t_4 to t_5. At t_5, one AC period is completed, and the next sequence begins by the commutation of S_2 and the current shifting to D_1.

We notice from an examination of the commutation sequence that the commutations at the zero-crossings of the voltage waveform, namely t_1 and t_3, are implemented directly by the action of the switching control circuit. For example, the commutation at t_1 takes place when the switching control circuit turns OFF the switch S_2. But the situation is not the same as regards the other two commutations, which occur at the zero-crossing instants of the current waveform, namely t_2 and t_4. The instant at which the current goes through

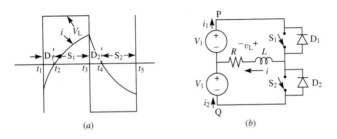

(a)   (b)

FIGURE 5.8
Commutation sequence in the half-bridge inverter.

zero depends on the load circuit time constant, which can be different for different loads. Considering, for example, the instant t_2, which is the positive-going zero-crossing instant of the current, the switching control circuit should be so designed that the control signal for the turn ON switching of S_1 must be present at this instant, although this instant itself is variable and dependent on the load circuit time constant. The practical way in which this can be done is to commence the turn ON switching signal of S_1 after a short "dead time" after the instant t_1, even though the switch itself is not yet ready to turn ON. The signal should continue to be present when the switch is able to turn ON, that is, at the zero-crossing of the current. The variable nature of the instant of zero-crossing of the current needs specially to be borne in mind when the switching device is of the type that only needs a short current pulse to turn ON. An example is the GTO, which is commonly used in high power inverter circuits. The GTO is turned ON by a current pulse on its gate terminal of very short duration. If the turn ON gate pulse used is of short duration, which arrives and disappears before the GTO is ready to turn ON, then it will not turn ON at all, and the inverter will not operate in the required manner.

5.4.4 Current Waveform on the DC Side

During the positive voltage half-period of the AC, the current is supplied from the top section of the split DC power supply and during the negative voltage half-period it is from the bottom section. The waveform of the current in each DC section of the input can be obtained from the AC current waveform of Fig. 5.8(a). These are sketched in Figs. 5.9(a) and (b), which are respectively the currents supplied from the top and bottom sections of the split power supply. Inspection of these waveforms shows that each of these input currents has a ripple component besides the DC component. In fact, the amplitude of the ripple current is so large in this case that the resultant current is actually negative during intervals of time such as t_1 to t_2. An input side filter will be necessary to reduce DC source current ripple to low values. A large enough filter will be able to limit the DC side ripple current within acceptable limits and also eliminate the necessity for the DC source to handle reverse currents.

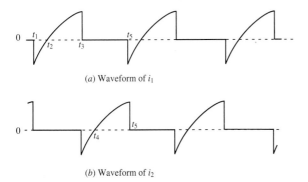

(a) Waveform of i_1

(b) Waveform of i_2

FIGURE 5.9
DC side current waveforms.

We are making this statement, because a DC bus supplied from a rectifier and not having a filter will not be able to handle a reverse current at any instant of time. We shall delay further treatment of input side filter circuits till we describe the more popular full-bridge inverter topology.

5.5 ADJUSTMENT OF AC FREQUENCY AND AC VOLTAGE

For most practical applications of inverters, there will be need to adjust the AC frequency and the AC voltage. Frequency is easily adjusted by variation of the frequency of the clock oscillator of the switching control circuit. Therefore this aspect will not require further consideration here.

Voltage adjustment can be made by varying the input DC voltage to the inverter. In such a case, the voltage control is external to the inverter and is independent of the switching within the inverter configuration. The technique used for implementing voltage control within an inverter is known as pulse width modulation (PWM). The half-bridge inverter topology does not lend itself to easy adjustment of voltage by PWM. For this, the full-bridge topology, which we shall present later in this chapter, will be preferable. However, we shall explain the nature of the output waveform with PWM and how the output r.m.s. value becomes adjustable.

A typical voltage waveform of a PWM inverter is shown in Fig. 5.10. We may treat the interval from t_1 to t_4 as one (the positive) half-period of the AC voltage and t_4 to t_7 as the negative half-period. During the positive half-period, the output voltage consists of a single pulse of amplitude V_1 and of duration t_2 to t_3, which is shorter than the total half-period.

We shall define the pulse duty cycle D as

$$D = \frac{\text{actual duration of the pulse in a half-period } (t_2 \text{ to } t_3)}{\text{duration of one half-period } (t_1 \text{ to } t_4)}$$

Therefore the mean square value of the AC output voltage will be given by

$$V_{ac}^2 = DV_1^2$$

Therefore we get the r.m.s. value of the AC output voltage as

$$V = \sqrt{D}V_1 \qquad (5.11)$$

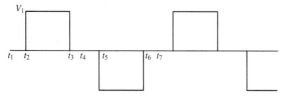

FIGURE 5.10
PWM technique for output voltage control of static inverters.

Illustrative example 5.2. A PWM inverter is designed to give its rated AC output voltage at a duty cycle of 0.7. What is the new duty cycle needed to increase the output voltage to 110% of the rated value?

Solution. We have

$$V_{\text{rated}} = \sqrt{0.7}V_1 \tag{5.12}$$

where V_1 is the pulse amplitude. We want

$$1.1V_{\text{rated}} = \sqrt{D}V_1 \tag{5.13}$$

From (5.12) and (5.13) by division,

$$1.1 = \frac{\sqrt{D}}{\sqrt{0.7}}$$

which gives $D = 0.847$, that is, the required duty cycle is 84.7%.

IMPLEMENTATION OF PWM. Referring to Fig. 5.10, if PWM is to be implemented in an inverter according to this pattern, it is necessary to achieve the following. During the intervals t_1 to t_2, the voltage across the load has to be maintained at zero. From t_2 to t_3, it has to be constant at V_1. Then, from t_3 to t_4, it again has to be zero. Similar statements apply for the next half-period, with due consideration for voltage polarity.

Now the question is whether it is possible to implement this pattern of voltage by switching control within the half-bridge topology. The answer is yes if the load is a pure resistance, but no if the load is inductive. The reasons are stated below.

Resistance load. Referring to the circuit topology [Fig. 5.11(a)], if the load is a pure resistance only, the voltage across it can be made zero by keeping both the switching blocks (1 and 2) OFF, as required from t_1 to t_2 and from t_3 to t_4. With switching block 1 ON, the voltage will be V_1, as required for the interval t_2 to t_3. The zero-voltage condition is possible because the current will instantly fall to zero when both the switches S_1 and S_2 are turned OFF.

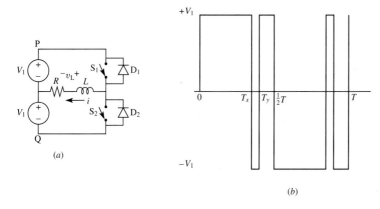

FIGURE 5.11
Negative voltage excursions during switching in the half-bridge inverter.

Inductive load. With an inductive load, such as that shown in Fig. 5.11(a), the current will not instantly fall to zero when the associated controlled switching element is turned OFF, but finds an alternative path through an uncontrolled switch (diode). For example, referring to Figs. 5.11(a) and (b), at the instant labeled T_x, the current in the load is positive, with S_1 conducting. If we turn OFF S_1 at this instant, the current cannot instantly fall to zero, because of the stored energy in the inductance. The induced voltage in the inductance will cause the current to shift from S_1 to D_2. As a result, the voltage output will make a negative excursion to $-V_1$, as shown in (b). If we now turn ON S_1 again at T_y as shown, the negative excursion may last for the entire duration from T_x to T_y if the current does not die down to zero before that. In any case, a zero voltage level is not entirely possible with S_1 turned OFF. Also, if the negative excursion lasts for the entire duration of T_x to T_y, as shown in the figure, the mean square value, and therefore the r.m.s. value, of the output voltage will not be affected. (the r.m.s. value of the fundamental sinusoidal component of the voltage may change, and so also may the composition of harmonics in the output voltage).

We therefore conclude that, with the half-bridge topology, it is not always possible to implement PWM in the manner indicated by Fig. 5.10. We shall see that with the full-bridge topology, the difficulty can be overcome by a suitable choice of switching scheme.

5.6 OUTPUT WAVEFORM CONSIDERATIONS

The ideal output waveform required for most applications of inverters is a sinusoid. But the waveforms we have described so far are rectangular in nature. This means that the output voltage has, besides the fundamental sinusoidal component, a significant content of harmonics. It is possible to reduce the harmonic content and make the output closer to a sine wave, within acceptable limits, by using a filter circuit on the output side of the inverter. But it should be remembered that such a filter is also a power circuit and has to handle the voltage and current for which the inverter is designed, which are usually large, for power electronic applications. It will add to the size, weight and cost of the equipment and also cause added power loss, thereby adversely affecting the overall efficiency. Therefore it will be advisable to shape the output waveform by switching in such a way that the filter requirement, if needed, is minimal. The filter requirement actually depends on the permissible limits of voltage distortion for the load fed from the inverter. For example, if the load is an AC motor, the inductance of the motor circuit performs the role of a filter element to some extent, and serves to reduce the harmonic content in the current waveform to a lower level than that in the voltage waveform.

There are basically two techniques used to shape the output voltage waveform. One is to shape the voltage waveform by switching in such a way as to selectively eliminate or minimize certain undesirable harmonics. The

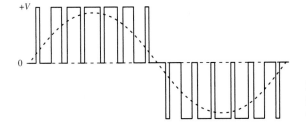

FIGURE 5.12
Sinusoidal pulse width modulation (SPWM).

second method is to shape the voltage waveform in such a way that the spectrum of harmonic frequencies is totally shifted in the direction of high frequency. In this case, the filter elements, if required, will be considerably smaller, because the harmonics to be eliminated will all be at high frequencies. The second method is more general and often favored for many applications. We shall briefly present a preview of it here, and describe the manner of its implementation more fully later, after we describe the full-bridge topology. The technique is called sinusoidal pulse width modulation (SPWM).

5.6.1 Sinusoidal Pulse Width Modulation (SPWM)

The shaping of the output voltage waveform is generally achieved by having multiple pulses in each half-period of the AC waveform. SPWM is a particular type of multiple-pulse PWM. A typical voltage waveform of a SPWM inverter is sketched in Fig. 5.12.

In each half-period, the pulse width is maximum in the middle. From the center, the pulse widths decrease as cosine function towards either side. In the SPWM technique, voltage control is implemented by varying the widths of all the pulses, at the same time maintaining the cosine relationship. In an SPWM waveform, the total harmonic content is still very significant. But all the harmonic frequencies get shifted towards the high-frequency direction of the frequency spectrum. The order of the harmonics in the SPWM waveform depends on the number of pulses per half-cycle employed. If the pulse repetition frequency employed is made higher, the harmonic frequencies also become higher. The lower frequency harmonics become insignificant. If SPWM is implemented in an inverter with a large number of pulses per half-cycle, the harmonic frequencies will be so large that for many applications, such as motor speed control, no separate filter may be needed on the output side. If a clean sine wave is required for the output voltage, this can be achieved by a filter consisting of only small values of L and C.

5.6.2 SPWM Using the Half-Bridge Topology

The SPWM waveform of Fig. 5.12 has zero-voltage intervals between successive pulses. We have seen, that in the half-bridge topology, with a typical inductive load, this is not possible. Therefore, it is generally not possible to

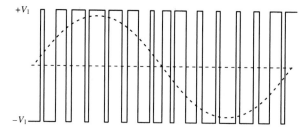

FIGURE 5.13
SPWM waveform for a half-bridge inverter (with reverse voltage excursions).

implement SPWM in the half-bridge inverter in the manner shown by Fig. 5.12. This does not mean that SPWM cannot be implemented using the half-bridge. SPWM can be implemented in the half-bridge also, but with negative voltage excursions. A typical SPWM waveform with both positive and negative pulses is shown in Fig. 5.13.

If we look at one half-cycle of this waveform, say the positive one, the positive pulses will be widest in the middle and decrease in width in both directions from the middle. On the other hand, the negative pulses will be narrowest in the middle and increase in width in either direction from the middle. This type of SPWM waveform also has the advantages, stated earlier, for the waveform without reverse voltage excursions (Fig. 5.12), from the filtering point of view. The actual switching control strategy for the implementation of SPWM, in the half- and full-bridge inverters, will be considered later, after we present the full-bridge topology.

5.7 THE FULL-BRIDGE CONFIGURATION

Figure 5.14(a) shows the full-bridge topology. The full bridge has four

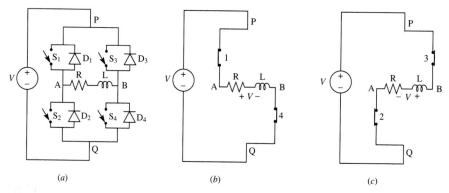

(a) (b) (c)

FIGURE 5.14
Single-phase full-bridge inverter.

"switching blocks," each consisting of a controlled switch and its antiparallel diode. The full bridge can be operated with or without pulse width modulation. For operation without PWM, the sequence of switching will be as follows.

5.7.1 Operation without PWM

The switches S_1 and S_4 are turned ON and kept ON for one half-period of the AC [Fig. 5.14(b)]. We may call this the positive voltage half-period. Next, S_1 and S_4 are turned OFF and the switches S_2 and S_3 are kept ON [Fig. 5.14(c)] for the duration of the negative voltage half-period. The output voltage and current waveforms can be derived by following the same procedure we did for the half-bridge inverter.

An inductive load represented by series R and L elements is shown in Fig. 5.14. When S_1 and S_4 are ON (or, to be more exact, when switching blocks 1 and 4 are ON) and the other two are OFF, the circuit will be as shown in Fig. 5.14(b). The loop equation will be

$$L\frac{di}{dt} + Ri = V \tag{5.14}$$

Similarly, when switching blocks 2 and 3 are ON and the other two are OFF [Fig. 5.14(c)], the loop equation will be

$$L\frac{di}{dt} + Ri = -V \tag{5.15}$$

These two equations are identical to those [(5.1) and (5.4)] that we wrote earlier for the half-bridge inverter, except for the fact that for the latter we used the voltage V_1, which is the voltage of one section of the split power supply. The full-bridge does not need a split power supply, and in (5.14) and (5.15), the voltage V is the full DC voltage supply to the bridge. Just as we saw for the half-bridge, for the full bridge also there will be an initial current build up period, after which we may assume repetitive conditions. The expressions for the current will be the same as we got earlier for the half-bridge, with V_1 replaced by the full-bridge voltage V. Therefore, when repetitive conditions can be assumed, the expression for the load current i will be, by reference to (5.8),

$$i = -\frac{V}{R}\frac{1-e^{-T/2\tau}}{1+e^{-T/2\tau}}e^{-t/\tau} + \frac{V}{R}(1-e^{-t/\tau}) \tag{5.16}$$

where $\tau = L/R$. For the negative voltage half-period, the expression for i will

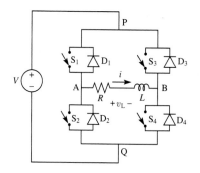

FIGURE 5.15
The full bridge of Example 5.3.

be the same, with the sign reversed for V and the reference zero for t as the instant of commencement of this half-period.

Illustrative example 5.3. The following data relate to the full-bridge inverter shown in Fig. 5.15. It is operated without PWM: $R = 20\,\Omega$, $L = 0.2\,$H, $V = 400\,$V and frequency $f = 100\,$Hz. Assume repetitive conditions.

(a) Obtain a numerical expression for the current in one half-cycle of the AC output.

(b) Sketch the waveforms of the output current and voltage. Determine the instants at which the zero-crossings occur, commencing from the positive-going zero-crossing of the voltage.

(c) State the conduction sequence of all eight power semiconductor switching elements and the conduction angle of each element.

Solution.

(a) The expression for the current during the positive voltage half-cycle is given by (5.16), which, on substitution of numbers, becomes

$$i = -\frac{400}{20}\frac{1 - e^{-0.01/2 \times 0.01}}{1 + e^{-0.01/2 \times 0.01}}e^{-t/0.01} + \frac{400}{20}(1 - e^{-t/0.01})$$

$$= 20 - 24.9e^{-100t} \tag{5.17}$$

(b) The zero-crossing instant is determined by equating (5.17) to zero. This gives $t = 2.19$ ms.

One full period of the AC output is equal to $1/f = 10$ ms. Taking this as $360°$ in angular measure, the angular delay of the zero-crossing of current with respect to the zero-crossing of voltage is

$$\frac{2.19}{10} \times 360° = 78.84°$$

The waveforms of current and voltage are sketched in Fig. 5.16. The magnitudes of the voltage peaks are 400 V and those of the current peaks are $I_{p0} = 4.9$ A.

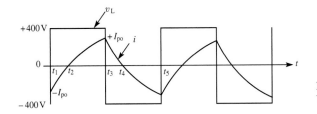

FIGURE 5.16
Voltage and current waveforms.

(c) The zero-crossing instants of both the waveforms are labeled as shown in Fig. 5.16. By reference to these waveforms, the conduction sequence may be described as follows:

Interval	Conduction angle	Conducting devices
t_1 to t_2	78.84°	D_1, D_4
t_2 to t_3	101.16°	S_1, S_4
t_3 to t_4	78.84°	D_2, D_3
t_4 to t_5	101.84°	S_2, S_3

5.7.2 The Half-Bridge Topology Viewed as the "Building Block" of the Full Bridge

It is helpful for certain types of analysis, to look upon the full-bridge topology as a superposition of two half-bridge topologies. This may be explained by reference to Fig. 5.17(a), (b) and (c).

Figure 5.17(a) shows the full-bridge configuration. In Fig. 5.17(b), we have replaced the DC input voltage V by an equivalent split DC supply consisting of $\frac{1}{2}V$ and $\frac{1}{2}V$ in series. We did this to get the hypothetical DC source midpoint N_S. Also, we have replaced the AC load $R-L$ by equal loads of half the value, each consisting of $\frac{1}{2}R$ and $\frac{1}{2}L$ in series. This gives us the hypothetical load midpoint, which is labeled N_L in the figure. N_L is the midpotential point between A and B at every instant of time, irrespective of whether the current is constant or varying. This is evident from the following:

$$V_{AB} = Ri + L\frac{di}{dt} \tag{5.18}$$

$$V_{AN_L} = \frac{1}{2}Ri + \frac{1}{2}L\frac{di}{dt}$$

$$= \frac{1}{2}V_{AB} \tag{5.19}$$

Equation (5.19) shows that the source midpoint N_S and the load midpoint N_L are at the same potential at every instant of time, irrespective of whether the switching blocks that are ON are 1 and 4 or 2 and 3. Therefore we can connect N_L to N_S without disturbing the operation. Such a connection is shown in Fig. 5.17(c). This single connection is shown as two separate connections between the same two points N_L and N_S to bring out the fact that each connection identifies a half-bridge topology. Figure 5.17(c), which shows a superposition

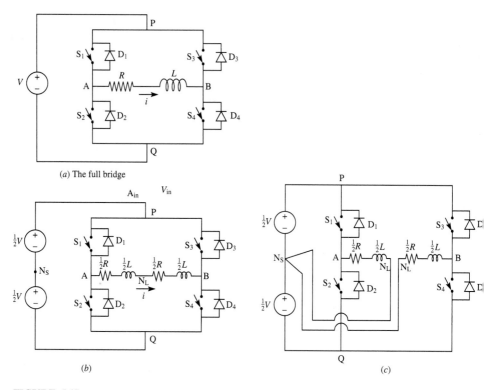

FIGURE 5.17
A full-bridge shown as a superposition of two half-bridges.

of two half-bridge inverters, is exactly the same and functions in the same manner as the full bridge shown in Fig. 5.17(a).

5.7.3 The Single-Phase Full-Bridge Topology Viewed as a Biphase Midpoint Topology

Referring to Fig. 5.17(c), which is the twin half-bridge equivalent of the full-bridge, let us look at the AC output voltages at the points A and B with respect to the source midpoint N_S (which is the same as N_L).

At every instant, as seen from (5.19),

$$V_{AN_S} = \tfrac{1}{2}V_{AB}$$
$$V_{BN_S} = -\tfrac{1}{2}V_{AB}$$

This means that the AC potential at A with respect to the source midpoint N_S is exactly equal in magnitude to the AC potential at B with respect to the same point, but reversed in sign. In AC, where we use angular measure to express time on the basis of 360° for one full period of the AC, the reversal in sign between two AC voltages is the same as a phase difference of 180 degrees between them. Therefore we can look upon the two half-bridges of Fig. 5.17(c) as giving us two AC voltages differing in phase by 180°, each with respect to the midpoint of the DC supply. The "leg" A of the inverter consisting of switching blocks 1 and 2 constitutes the A phase. Similarly, the leg B consisting of the switching blocks 3 and 4 constitutes the B phase. Using this approach, we can look upon each leg of the inverter as providing one output phase of the AC. The three-phase bridge inverter then becomes one more extension of this approach. To obtain a three-phase output, we provide one more leg to the inverter topology, consisting of two more switching blocks, which may be labeled as switching blocks 5 and 6 across the DC supply terminals P and Q in the same way as 1, 2 for the A phase and 3, 4 for the B phase. The terminal C linking 5 and 6 becomes the output terminal of the third phase C, in exactly the same manner as the output terminals A and B of the A and B phases. To obtain the phase difference of 120°, the switching pattern of each half-bridge is implemented with a successive phase difference of 120°, that is, one-third of the AC time period. This approach is quite general, and can be extended to implement multiphase inverter bridge topologies with more than one output phase, by providing the appropriate additional pairs of switching blocks and correspondingly modifying the time phase difference between the switching of successive phase legs. However, the three-phase bridge is the one that is most widely used in large power applications, and therefore we shall present it in detail later—considering both the power circuit and the timing circuit of the static switching elements.

Voltage between the load neutral terminal N_L and DC source midpoint N_S in multiphase bridge topologies. The single-phase bridge configuration of Fig. 5.17(a) may be termed a "biphase" bridge if we want to be consistent with the above approach of treating each leg of the inverter as providing one output phase. We have already shown above that in this biphase topology, the source midpoint N_S and the load midpoint N_L are at the same potential at every instant of time, and therefore it makes no difference whether or not these two are connected together. In multiphase bridge topologies having three or more legs, one way of connecting the phase loads is the "Y" connection in which the phase loads have a common point, usually called the neutral terminal, which we shall label N. In such cases, if we decide to operate the bridge with the load neutral isolated from the source midpoint, we may not always be justified in assuming that these two terminals are at the same potential at every instant of time. This is because the division of voltage on the load side may not be equal if more than two switching blocks are simultaneously ON at any instant. In fact, we shall show later that in the case of the three-phase bridge there will be an AC voltage between the load neutral and the source midpoint, at a frequency three times the inverter frequency. Even with such a limitation, the approach of treating a multiphase bridge as an assembly of several individual half-bridge topologies is useful for a better understanding of both the power circuit and the timing circuits of the switches. We shall therefore use this approach where it is justified and helpful.

5.8   CONTROL OF THE AC OUTPUT VOLTAGE—PULSE WIDTH MODULATION

With the single-phase full-bridge inverter, it is possible to implement the PWM output voltage waveform, such as shown in Fig. 5.18(b) and also earlier in Fig. 5.10. We showed earlier that it is not always possible to implement this type of PWM waveform with the single-phase half-bridge topology. We shall now describe the timing sequence of the static switching elements of the full bridge, to achieve the PWM voltage waveform shown in Fig. 5.18(b). We shall also describe a practical circuit scheme for providing the required timing signals for the switches.

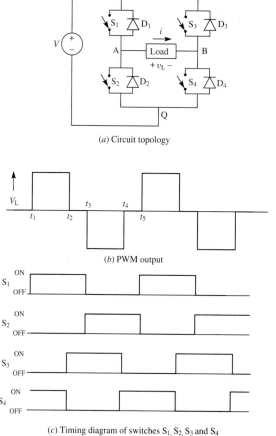

(a) Circuit topology

(b) PWM output

(c) Timing diagram of switches S_1, S_2, S_3 and S_4

FIGURE 5.18
Switching strategy for implementing PWM.

5.8.1 Sequence of Switching

The inverter circuit and the required voltage waveform are shown in Fig. 5.18(a) and (b) for ready reference. Inspection of the circuit, together with the waveform, gives us the timing requirements of the four static switching elements S_1, S_2, S_3 and S_4. For this purpose, let us look at one full time period of the AC output, starting from the instant labeled t_1 to the instant labeled t_5. It is evident that from t_1 to t_2, switches S_1 and S_4 should be ON, to give us the positive voltage. Similarly, from t_3 to t_4, switches S_2 and S_3 should be ON, to give us the negative output voltage. However, there are two intervals in a cycle, namely from t_2 to t_3 and from t_4 to t_5, during which the output voltage is to be zero. These are the "zero voltage freewheeling" intervals. During these freewheeling intervals, we should ensure that the current flow in the load is possible in either direction, because the actual direction of current will depend on the nature of the load. The inverter should be capable of handling any type of load—resistive, inductive or capacitive. Therefore, during the freewheeling intervals, two switches should be simultaneously ON, which may be the two on the top side of each leg (positive side or the P side) or the bottom side of the legs (negative side or Q side). In this way, we shall ensure that the output voltage is zero and that current flow is possible in either direction in the load.

We therefore have a choice between freewheeling on the P side and freewheeling on the Q side. As there are two freewheeling intervals in a cycle (t_2 to t_3 and t_4 to t_5), we have four switching patterns from which to choose, namely

1. both freewheelings on the P side;
2. both freewheelings on the Q side;
3. the t_2 to t_3 freewheeling on the Q side and the t_4 to t_5 freewheeling on the P side;
4. the t_2 to t_3 freewheeling on the P side and the t_4 to t_5 freewheeling on the Q side.

In Fig. 5.18(c), we have drawn the timing diagrams for the four switches based on alternative 4. The timing diagrams for each of the other three alternatives are left as exercises for the reader (see Problems 5.4–5.6). It will be found that alternatives 3 and 4 will give equal ON and OFF timing requirement for every switching block. If either of the alternatives 1 or 2 is chosen, it will be found that all four switching blocks will not have the same ON time. Also, for the same switching block, the ON and OFF times will be different, and will depend on the duty cycle of the output voltage. It should also be pointed out here that the timing diagram of Fig. 5.18(c) shows only the required timing signal for each switch. The switch may not conduct during the entire duration of the ON signal. The conduction is actually shared between the switch S and the antiparallel diode D, depending on the direction of the current.

5.8.2 Generation of the Timing Pulses

Figure 5.19 shows a block schematic of a circuit scheme for the generation of timing pulses to suit the switching pattern shown in Fig. 5.18(c). The signal waveforms at different locations in this timing circuit are also shown in the same figure, against the corresponding functional circuit block. It may be stated here that a circuit designer may also employ alternative schemes to achieve the same timing functions. We have given the scheme of Fig. 5.19 primarily to provide greater insight into the working of the inverter, rather than as a unique design for the switching control circuit. A brief description of the circuit of Fig. 5.19 is given below.

The circuit block labeled "clock generator" serves to set the frequency of the AC output. This could be a voltage-to-frequency converter IC that provides a repetitive pulse train whose frequency can be adjusted by the input voltage at one of its terminals. This adjustment is shown as the inverter frequency control in the figure. The pulse repetition frequency of the inverter has to be set at twice the inverter AC frequency in the scheme of Fig. 5.19.

The output of the clock generator is fed to a ramp generator, which may

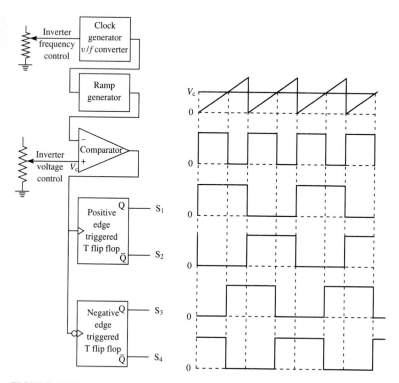

FIGURE 5.19

A scheme for the generation of timing pulses to inplement PWM according to the switching pattern of Fig. 5.18.

be an op-amp circuit. This circuit block outputs a ramp waveform, which is synchronized with the clock pulses. The ramp voltage is compared with an adjustable DC voltage in a voltage comparator IC chip. The adjustable DC input voltage to the comparator is the one that determines the duty cycle of the AC output voltage. Therefore this adjustment shown in Fig. 5.19 is the duty cycle adjustment and therefore the adjustment for the AC output voltage of the inverter. The comparator gives a pulse width modulated output, as shown in the figure, the pulse width being adjustable by means of the control voltage V_c.

The PWM output of the comparator is at twice the required inverter frequency. This pulse train is used to trigger two edge triggered toggle (T) flip flops—one at the positive edges and the other at the negative edges. The total of four outputs from the two T flip flops (the direct and the complementary outputs from each flip flop) constitute the required timing pulses for the four static switches S_1, S_2, S_3 and S_4. This may be verified by reference to Fig. 5.18(c).

The timing pulses so obtained may not be suitable for directly operating the four static switches. The type of drive signal required for a switch will depend on the nature of the latter, depending on whether it is a power bipolar transistor, or a power MOSFET or a GTO or an IGBT or some other device. But the timing signals shown in Fig. 5.19 may be used to time the drive circuits, which should be designed to suit the type of power switching element actually chosen for the inverter.

A circuit refinement not shown in Fig. 5.19 is the provision of a "dead time" between two switches on the same leg of the inverter. The need for dead time may be explained as follows. An examination of the timing waveform shows that, on any single leg of the inverter, only one switch will be ON at any instant of time, the other being OFF. It is important to ensure that both switches on an inverter leg are never simultaneously ON at any instant of time. Such an occurrence is a direct short-circuit across the DC input to the inverter—a major fault condition, which is described as a "shoot through" fault. As a precaution against such a shoot through fault, which may occur owing to a finite delay time in the turn OFF switching of the outgoing switch, it is advisable to provide a short, but finite, dead time between the turn OFF of the outgoing switch and the turn ON of the incoming switch, on the same leg of the inverter. One way of providing this is to delay the rising edge of each of the timing pulse waveform in Fig. 5.19 by the required dead time. A circuit scheme for this is not included in Fig. 5.19, but is left as an exercise for the reader (see Problem 5.7).

5.8.3 Waveform of the Output Current in the PWM Inverter

The waveform of the output current of a PWM inverter can be determined by proceeding on the same lines as we did earlier for the inverter without PWM.

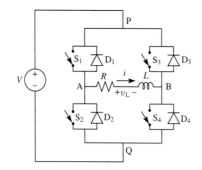

(a) Inverter bridge feeding an R–L load

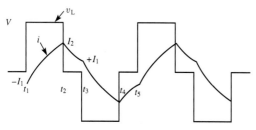

(b) Output voltage and current waveforms

FIGURE 5.20
Current and voltage waveforms with PWM operation.

In the case of the PWM inverter, there is additionally a zero voltage freewheeling interval in each half-period, and therefore an additional discontinuity in the current waveform in each half-period. We shall illustrate this by deriving the current waveform for the inverter bridge of Fig. 5.20(a) feeding an R–L load on the AC side. We shall assume that the inverter has been ON for sufficient number of switching cycles to justify the assumption of repetitive conditions. The AC frequency $(1/T)$ is labeled f and the duty cycle D.

Referring to Fig. 5.20(b), in the interval t_1 to t_2, when the load voltage has a positive magnitude V, the loop equation is

$$L\frac{di}{dt} + Ri = V \tag{5.20}$$

We shall take the reference zero of time at the instant t_1 and the initial condition as [see Fig. 5.20(b)]

$$i = -I_1 \quad \text{at } t = 0 \tag{5.21}$$

The solution with the stated inital condition is

$$i = -I_1 e^{-t/\tau} + \frac{V}{R}(1 - e^{-t/\tau}) \tag{5.22}$$

where $\tau = L/R$ is the load time constant.

At the instant labeled t_2, $t = \frac{1}{2}DT$. Therefore the peak current, labeled I_2 in Fig. 5.20(b), will be

$$I_2 = -I_1 e^{-DT/2\tau} + \frac{V}{R}(1 - e^{-DT/2\tau}) \tag{5.23}$$

The freewheeling begins at the instant t_2, and during this interval the loop equation becomes

$$L\frac{di}{dt} + Ri = 0 \tag{5.24}$$

We shall take a new reference zero for time t for this equation at the instant t_2. The initial condition for this equation may then be stated as

$$i = I_2 \quad \text{at} \quad t = 0 \tag{5.25}$$

The solution with the stated initial current will be

$$\begin{aligned} I &= I_2 e^{-t/\tau} \\ &= \left[-I_1 e^{-DT/2\tau} + \frac{V}{R}(1 - e^{-DT/2\tau}) \right] e^{-t/\tau} \end{aligned} \tag{5.26}$$

with $t = \frac{1}{2}(1 - D)T$ at the instant t_3. At this instant, we have completed one half-period of the AC cycle. Therefore, considering the fact that repetitive conditions prevail, and the symmetry of the current waveform, at this instant, $i = I_1$. Therefore

$$I_1 = \left[-I_1 e^{-DT/2\tau} + \frac{V}{R}(1 - e^{-DT/2\tau}) \right] e^{-(1-D)T/2\tau} \tag{5.27}$$

From this we get I_1 as

$$I_1 = \frac{V}{R} \frac{e^{-(1-D)T/2\tau} - e^{-T/2\tau}}{1 + e^{-T/2\tau}} \tag{5.28}$$

We may summarize our analysis by recalling the equations for the current waveform in the following manner.

The magnitude of the current at the instants t_1 and t_3, which are instants of discontinuity in the waveform and also the instants at which freewheeling is completed, is given by (5.28).

The magnitude of the peaks at the instants t_2 and t_4, which are the other two instants of discontinuity and the instants at which freewheeling commences, is given by (5.23), where the value of I_1 to be used is obtained from (5.28).

The expression for the current during the interval t_1 to t_2 is given by (5.22). The reference zero for t in this expression is the commencement of this interval, namely t_1.

The expression for the current during the interval t_2 to t_3 is given by either of the two expressions in (5.26). The zero reference for t in this equation is the commencement of this interval, namely t_2.

During the second half-period, the current waveform is a repetition of the first half-period, with the sign reversed. The nature of the current waveform is sketched in Fig. 5.20(b).

Illustrative example 5.4. The PWM inverter of Fig. 5.20 operates with duty cycle 60%, $f = 100$ Hz, $V = 200$ V, $R = 20\ \Omega$ and $L = 0.1$ H.
(a) Determine the value of the current at each discontinuity in its waveform and obtain mathematical expressions for it in each interval.
(b) Determine the zero-crossing instants of the current. Express these in angular measure on the basis of one AC time period being 360°.
(c) Sketch the current and voltage waveforms. Indicate the instant at which each discontinuity occurs in the current waveform and also each zero-crossing instant.
(d) List the conduction sequence and the conduction angle of each power semiconductor switching element, including the diodes.

Solution.
(a) The time period $T = 1/f = 0.01$ s and the time constant $\tau = L/R = 0.005$ s, $D = 0.6$, $V = 200$ V and $R = 20\ \Omega$. Substitution of these numbers into (5.28) gives

$$I_1 = 2.211\ \text{A} \tag{5.29}$$

Substituting this value for I_1 into (5.22) gives the following expression for i in the interval t_1 to t_2, with t measured from t_1:

$$i = 10 - 12.211e^{-200t} \tag{5.30}$$

The positive peak I_2 is obtained by substituting numbers into (5.23), which gives

$$I_2 = 3.299\ \text{A} \tag{5.31}$$

Therefore, during the interval t_2 to t_3, substitution of numbers into (5.26) gives

$$i = 3.299e^{-200t} \tag{5.32}$$

t being measured from t_2. As a check on our calculations, we can find i at t_3 by putting $t = \frac{1}{2}(1 - D)T$ in (5.32). The magnitude of i thus obtained matches with the value obtained above in (5.29).

For the next half-period, we get the expressions for the current by reversing the sign of the corresponding expressions obtained above, and measuring t from the commencement of each interval. This procedure gives the following. From t_3 to t_4, with t measured from t_3

$$i = -10 + 12.211e^{-200t} \tag{5.33}$$

From t_4 to t_5, with t measured from t_4,

$$i = -3.299e^{-200t} \tag{5.34}$$

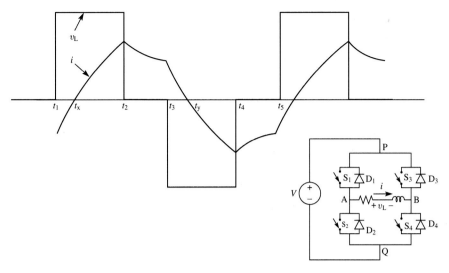

FIGURE 5.21
Conduction sequence and conduction angles of power semiconductor switching elements.

(b) The instant of positive-going zero-crossing is obtained by putting $i = 0$ in (5.30). This gives $t = 1$ ms. Since one period is 10 ms and is equated to 360° in angular measure,

$$t = \tfrac{1}{10} \times 360° = 36° \text{ measured from } t_1$$

This instant is labeled t_x in Fig. 5.21.

The negative-going zero-crossing occurs with the same delay from the instant t_3, and is labeled t_y in Fig. 5.21.

(c) On the basis of the above results, the waveform of the output current is as sketched in Fig. 5.21.

(d) The conducting devices at any instant can be identified in the power circuit topology by looking at the sign of the voltage and the sign of the current in the waveforms at that instant. This procedure gives the following. From t_1 to t_x, the conducting devices are D_1 and D_4. Their conduction angles during this interval are 36°, as determined above.

From t_x to t_2; the conducting devices are S_1 and S_4. Their conduction angles are $D \times 180° - 36° = 108° - 36° = 72°$.

From t_2 to t_3, the conducting devices are S_1 and D_3 on the basis of freewheeling on the P side. Their conduction angles during this interval are $(1 - D) \times 180° = 72°$.

From t_3 to t_y, the conducting devices are D_2 and D_3. Their conduction angles during this interval are 36°.

From t_y to t_4, the conducting devices are S_2 and S_3. Their conduction angles during this interval are 72°.

From t_4 to t_5, the conducting devices are S_2 and D_4 on the basis of freewheeling on the Q side. Their conduction angles during this interval are 72°.

Total conduction angles of each device, on the basis of the above results, are

D_1	36°
D_2	36°
D_3	108°
D_4	108°
S_1	144°
S_2	144°
S_3	72°
S_4	72°

It should be pointed out here that the total conduction angle of each "switching block" (S and D) is 180°. During this 180° interval, the conduction is shared between S and D according to the direction of the current through the "switching block."

5.8.4 Voltage Harmonics—Fourier Components

The ideal voltage waveform desired from an inverter is usually a sinusoid. The actual waveforms that we have considered so far depart considerably from this ideal. The harmonic components of the actual waveform can be determined by the standard procedure of Fourier analysis. We shall illustrate this for the general case of a PWM waveform with a duty cycle D. The results are also valid for a waveform without PWM if we put the duty cycle $D = 1$.

The waveform is sketched in Fig. 5.22 with $\theta = 2\pi f_1 t$ along the horizontal axis. Also, we have taken $\theta = 0$ at the middle of the positive voltage half-period for convenience. We may express the output AC voltage mathematically as

$$v = \begin{cases} V & \text{for } 0 < \theta < \tfrac{1}{2}D\pi \\ 0 & \text{for } \tfrac{1}{2}D\pi < \theta < \pi - \tfrac{1}{2}D\pi \\ -V & \text{for } \pi - \tfrac{1}{2}D\pi < 0 < \pi \end{cases} \qquad (5.35)$$

Also, the waveform satisfies the symmetry condition, which may be stated as

$$v_\theta = -v_{\theta-\pi} \qquad (5.36)$$

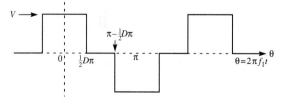

FIGURE 5.22
Harmonic analysis of PWM voltage waveform.

Fourier analysis of this waveform yields the fundamental as well as the individual harmonic amplitudes. This is left as an exercise for the reader (Problems 5.10–5.13). The results are summarized below.

1. There will be no even-order harmonics.
2. The amplitude (in V) of each frequency component is given by

$$A_n = \frac{4V}{n\pi} \sin \tfrac{1}{2}Dn\pi \tag{5.37}$$

3. The amplitude of the fundamental is obtained by putting $n = 1$ in (5.37). This gives

$$A_1 = \frac{4V}{\pi} \sin \tfrac{1}{2}D\pi \tag{5.38}$$

Therefore the r.m.s. value of the fundamental voltage will be:

$$V_1 = \frac{2\sqrt{2}V}{\pi} \sin \tfrac{1}{2}D\pi \tag{5.39}$$

The r.m.s. value of the total harmonic components will be given by the following relationship:

$$V_{rms}^2 = V_1^2 + V_H^2 \tag{5.40}$$

where

$$V_{rms} = \sqrt{D}V \tag{5.41}$$

is the r.m.s. value of the PWM voltage. Therefore the r.m.s. total harmonic component will be:

$$V_H = \sqrt{DV^2 - \frac{8V^2}{\pi^2} \sin^2 \tfrac{1}{2}D\pi} \tag{5.42}$$

Illustrative example 5.5. The DC input to a single-phase full-bridge PWM inverter is 100 V.
(a) Determine the r.m.s. values of the fundamental and the two lowest-order harmonics for a duty cycle of 100%. Also determine the r.m.s. total harmonic component.
(b) Repeat (a) for a duty cycle of 70%.
(c) Determine the duty cycle that will eliminate (1) the third harmonic and (2) the fifth harmonic.

Solution.
(a) The r.m.s. fundamental voltage V_1 is obtained by substitution into (5.39). This gives

$$V_1 = 90 \text{ V}$$

The r.m.s. third harmonic is obtained by first determining the peak using (5.37) for $n = 3$ and then dividing by $\sqrt{2}$. This gives

$$V_3 = 30 \text{ V}$$

The same procedure with $n = 5$ gives the r.m.s. value of the next higher-order harmonic, which is the fifth:

$$V_5 = 18 \text{ V}$$

The r.m.s. total harmonic component is obtained from (5.41) and (5.42). This gives

$$V_H = \sqrt{100^2 - 90^2} = 43.5 \text{ V}$$

(b) For $D = 0.7$, (5.39) gives the r.m.s. fundamental voltage as

$$V_1 = 80.2 \text{ V}$$

Using (5.37) for the peak value of the third harmonic, we get the r.m.s. value as

$$V_3 = 4.7 \text{ V}$$

A similar procedure for $n = 5$ gives the r.m.s. fifth harmonic as

$$V_5 = 12.7 \text{ V}$$

The r.m.s. value of the PWM voltage for a 70% duty cycle is, from (5.41),

$$V_{rms} = 83.7 \text{ V}$$

Using this value in (5.40) gives the r.m.s. total harmonic content as

$$V_H = 23.8 \text{ V}$$

(c) The duty cycle to totally eliminate the third harmonic will be obtained by making $A_n = 0$ in (5.37) for $n = 3$. This gives

$$\tfrac{3}{2} D\pi = \pi, \ 2\pi, \ 3\pi, \ \ldots$$

From this, the highest duty cycle is obtained as $\tfrac{2}{3} = 66.7\%$.

A similar procedure for $n = 5$ gives the highest duty cycle to eliminate the fifth harmonic as 80%.

5.8.5 Current Waveform—Superposition of Fundamental and Harmonics

The output current waveform may be considered as a superposition of the fundamental frequency component and the different harmonics. Each of these can be calculated separately. This is illustrated by the following example.

Illustrative example 5.6. The single-phase bridge inverter of Example 5.4 is operating at 60 Hz. The AC load consists of a resistance $R = 5\,\Omega$ in series with an inductance $L = 10$ mH.

(a) For a duty cycle of 100%, determine the r.m.s. values of (1) the fundamental AC output current and (2) the next two lowest-order harmonic currents.

(b) Repeat (a) for a duty cycle of 70%.

Solution. The load impedance at the fundamental frequency and at the two lowest-order harmonic frequencies are as follows: At $f_1 = 60$ Hz,

$$Z_1 = 5 + j2 \times \pi \times 60 \times 10 \times 10^{-3}$$
$$= 5 + j3.77 \ \Omega$$
$$|Z_1| = 6.26 \ \Omega$$

Similarly, at the third harmonic frequency $f_3 = 180$ Hz,

$$Z_3 = 5 + j11.31 \ \Omega$$
$$|Z_3| = 12.37 \ \Omega$$

And at the fifth harmonic frequency $f_5 = 300$ Hz,

$$Z_5 = 5 + j18.85 \ \Omega$$
$$|Z_5| = 19.50 \ \Omega$$

We shall use the r.m.s. values of the fundamental and harmonic voltages determined in Example 5.5.

(*a*) The fundamental r.m.s. current component $I_1 = V_1/Z_1$. This gives

$$I_1 = 90/6.26 = 14.38 \ A$$

Similar procedures for I_3 and I_5 give

$$I_3 = 30/12.37 \ A = 2.43 \ A$$
$$I_5 = 18/19.50 = 0.92 \ A$$

(*b*)
$$I_1 = 80.2/6.26 = 12.81 \ A$$
$$I_3 = 4.7/12.37 = 0.38 \ A$$
$$I_5 = 12.7/19.50 = 0.65 \ A$$

5.9 SHAPING OF OUTPUT VOLTAGE— SINUSOIDAL PULSE WIDTH MODULATION (SPWM)

The concept underlying SPWM is to build up the total waveform of the AC output voltage by means of multiple pulses, with the pulse widths distributed sinusoidally, as shown earlier in Fig. 5.12. We shall describe one practical way in which this can be done. There are alternative ways, but the differences, generally, are in the details and not in the basic approach. The implementation strategy that we describe here is chosen primarily from the point of view of highlighting the essential aspects of the technique.

In SPWM, it is always desirable to use a large number of pulses to synthesize the AC output waveform. For example, if we use 100 individual pulses to synthesize one half-cycle of the AC, the pulse repetition frequency for a 60 Hz output will be 12 kHz. This is well within the range of realistic frequencies actually used in practical equipment. A higher pulse repetition frequency serves to shift the harmonic components in the output towards the high frequency direction in the frequency spectrum of the output voltage. If it is intended to use an output filter to further refine the waveform to achieve a very pure sinusoid then the size of the filter elements will be significantly reduced if the harmonic frequencies are high. However, in our presentation of the SPWM technique, we have chosen a relatively small number of pulses per half-cycle. Otherwise, in the graphical description of the output waveform, there will be too large a number of closely spaced very narrow pulses, because of which the physical relationships that we need to highlight relating to pulse widths will be obscured.

5.9.1 SPWM with Reverse Voltage Excursions

We have pointed out already that there are two types of SPWM waveforms—those without and those with reverse voltage excursions. These were illustrated by Figs 5.12 and 5.13. Further, we had shown earlier that with the half-bridge topology it is not always possible to avoid reverse voltage excursions. But, with the half-bridge, it is always possible to implement SPWM with reverse voltage excursions. This can be confirmed by referring to Fig. 5.23(a). If S_1 is ON and S_2 is OFF, the output voltage has the positive peak. The negative peak is achieved by keeping S_1 OFF and S_2 ON.

An SPWM waveform with reverse voltage excursions is shown in Fig. 5.23(b). The time interval AB in the figure is for one half-period only. This half-period chosen corresponds to the positive voltage half-period of the

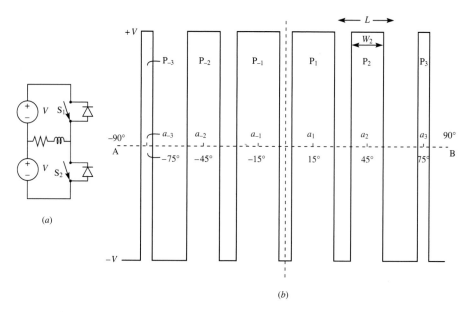

(a)

(b)

FIGURE 5.23
Sinusoidal PWM for a half-bridge.

corresponding sinusoid. We confine our description to one half-period only, because the negative voltage half-period is implemented in an exactly similar manner, with the voltage polarities reversed.

In the SPWM waveform of Fig. 5.23(b), we have divided the half-period AB into s equally spaced intervals, the duration of each interval being $180°/s$. We have chosen $s = 6$ in Fig. 5.23, so that the duration of each interval is $30°$. We shall, for convenience, measure angular intervals from the middle of the half-period, and so our half-period is from $-90°$ to $+90°$. The waveform is built up with s pulses, each centered at the middle of its interval. Because we chose to describe SPWM with reverse voltage excursions, each pulse rises from $-V$ to $+V$ on its rising edge and falls from $+V$ to $-V$ on its trailing edge. Because we chose to have an even number, 6, for s, there are six pulses, which we have labeled in pairs as P_{-1}, P_1, P_{-2}, P_2 and P_{-3}, P_3. The pulses are centered at angular instants labeled $a_{-1} = -15°$, $a_1 = 15°$, $a_{-2} = -45°$, $a_2 = 45°$, $a_{-3} = -75°$ and $a_3 = 75°$.

Let us look at any one interval, such as that centered about $a_2 = 45°$. The total duration of the interval is $180°/s = L$, as labeled in the figure. Let the duration of the pulse be W_2 for this interval. Therefore, during this interval, we have a positive voltage of duration W_2 and negative voltage of total duration $L - W_2$. The average value of the voltage during this interval is therefore

$$\frac{1}{L}[W_2 - (L - W_2)]V = \frac{2W_2 - L}{L}V \tag{5.43}$$

In the SPWM technique we are presenting here, our objective is to make the average value for the voltage during any interval proportional to the cosine of the angle about which the interval is centered. Therefore, to achieve this objective, we must have a way of determining the pulse width W_2 in (5.43) such that

$$\frac{2W_2 - L}{L} = M \cos a_2 \tag{5.44}$$

or, in general,

$$\frac{2W_n - L}{L} = M \cos a_n \tag{5.45}$$

A strategy for realising the switching instants of the inverter so that the pulse width in each interval is in conformity with (5.45) will be the key to the practical implementation of SPWM to achieve our objective.

5.9.2 The Sine Triangle Comparison Method

This is the classical method for the implementation of sinusoidal pulse width modulation. In this, we use two voltage waveforms: (1) a sinusoidal voltage of the same frequency as the inverter, which we shall call the "reference" voltage, and (2) a high frequency voltage, which has a triangular waveform, which we

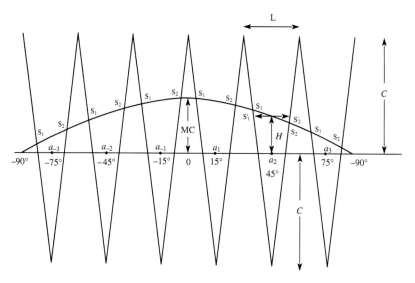

FIGURE 5.24
Implementation of sinusoidal pulse width modulation by sine triangle comparison.

shall call the "carrier" voltage. These are shown in Fig. 5.24 for one half-period of the inverter. To conform to our SPWM waveform of Fig. 5.23, where we had six pulses per half-period, we have chosen a carrier frequency equal to 12 times the reference frequency. We have also positioned our carrier waveform so that the positive peak of the reference coincides with a positive peak of the carrier. Therefore we have six full periods of the carrier in one half-period of the reference. We shall count one full period of the carrier between its successive positive peaks. In this way, each full period of the carrier thus defined becomes the interval L that we assigned for each pulse in Fig. 5.23. To be general, we shall call the pulse in this interval P_n and the pulse width W_n. For our specific example, n happens to be 2. Also, we shall label the middle instant of this interval a_n.

The triangular carrier waveform has a fixed amplitude. The amplitude of the reference sine wave is usually made adjustable. We shall define a modulation index M as follows:

$$M = \frac{\text{amplitude of reference wave}}{\text{amplitude of carrier}}$$

The variation of the reference wave amplitude, keeping the carrier amplitude fixed, is the usual means employed to adjust the modulation index. We shall see presently that adjustment of the modulation index gives us a convenient way of adjusting the AC output voltage of the inverter. Since the inverter output frequency is the same as the reference sine wave, the inverter output frequency is adjustable by adjustment of the reference wave frequency.

If the amplitude (in V) of the carrier is C, for a modulation index M, the reference wave may be expressed as

$$v_{\text{ref}} = MC \cos \theta \qquad (5.46)$$

We have chosen a modulation index of 50% in drawing the reference wave in Fig. 5.24, and so its amplitude is 50% of the carrier amplitude. The waveforms of the carrier and the reference are used to obtain the timing waveform to operate the static switches S_1 and S_2 (see Fig. 5.23(a)) in such a way as to get the SPWM output from the inverter. When S_1 is ON and S_2 is OFF, we get the positive output pulse. When S_1 is OFF and S_2 is ON, we get the negative pulse. To achieve an SPWM waveform with reverse voltage excursions, one (and only one) of the two switches must be ON at any given instant of time. To explain the role of the reference voltage waveform and that of the carrier voltage waveform in deciding which of these two switches is to be ON at a given instant of time, let us arbitrarily choose any one of the pulse intervals, say that centered at $a_n = a_2$ (a_2 being equal to 45° in Fig. 5.23), corresponding to the pulse P_2 in the same figure. Referring to Fig. 5.24, at the instant a_2, the reference wave has a magnitude $MC \cos a_2$, which is the height of the reference wave at this instant. This height is labeled H in Fig. 5.24. We have drawn a horizontal line at this height to intersect the triangle wave at the instants labeled S_1' and S_2'. From the following procedure, we shall see that the horizontal interval from S_1' to S_2' is the required pulse width to satisfy (5.45). Referring to Fig. 5.24, by considering similar triangles,

$$\frac{S_1'S_2'}{L} = \frac{H + C}{2C} \qquad (5.47)$$

Therefore

$$\frac{2S_1'S_2' - L}{L} = \frac{2(H + C) - 2C}{2C}$$

$$= \frac{H}{C} = M \cos a_2 \qquad (5.48)$$

Comparison of (5.44) [which is a specific case of (5.45)] with (5.48) shows that the interval $S_1'S_2'$ is the exact interval W_2 equal to the required pulse width centered at the instant a_2. Therefore, to get the exact pulse according to our SPWM strategy in any interval, the thing to do is to draw a horizontal line through the reference wave at the instant at which the pulse is centered. The intersection of this horizontal line with the triangular carrier waveform will give the exact switching instants S_1' and S_2' of the static switches S_1 and S_2. The switch S_2 should be turned OFF and S_1 turned ON at the instant S_1' at which reference wave is above the carrier. The reverse operation should take place at the instant S_2' when the reference waveform is below the carrier.

The intersection instants of the carrier and the reference waveforms are also indicated in Fig. 5.24. These instants are labeled S_1 and S_2. Inspection shows that the difference between S_1S_2 and $S_1'S_2'$ is very little. In the practical implementation of SPWM, it is often more convenient to implement the switchings at the instant of intersection of the carrier waveform and the reference waveform. The small errors in pulse width and in the centering of the pulse are considered as worthwhile prices to pay, because of the resulting circuit simplicity. For actual implementation, we therefore compare the triangular carrier waveform and the sine wave reference waveform by means of an analog comparator chip in such a way that the comparator output goes high when the reference wave goes higher than the carrier, and vice versa. In this way, we get a comparator output that is a replica of the SPWM output of the inverter. This comparator output is used as the timing wave for the switching of the static switches S_1 and S_2 of the inverter. For varying the output frequency, we vary the frequency of the reference wave. For varying the magnitude of the output voltage, we vary the amplitude of the reference wave, thereby varying the modulation index. Our derivation of (5.44) and (5.45) shows that the modulation index determines the average value of the output voltage the inverter, averaged over a pulse period. By "pulse period," we mean the time interval allocated for a pulse, which is one time period of the carrier between its two successive positive peaks. Therefore a change in the modulation index results in a change, by the same factor, in output voltage of the inverter. A practical circuit scheme for implementing SPWM on the above stated lines is described next.

5.9.3   Implementation of the SPWM by Sine Triangle Comparison

A block schematic of a scheme for implementing SPWM is shown in Fig. 5.25. Circuit block 1 is the reference sine wave generator. It has provision for independent adjustment of the frequency and the amplitude. These two adjustments are shown as manual potentiometer adjustments of the corresponding input voltage at the appropriate terminals of the waveform generator.

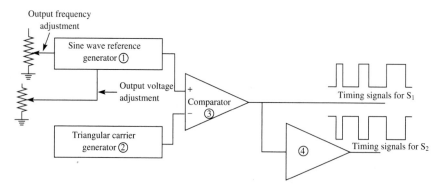

FIGURE 5.25

Block schematic of circuit scheme for implementing SPWM by sine triangle comparison.

In Fig. 5.27 we give a more detailed schematic of this circuit block, which we shall describe later. The circuit block labeled 2 is the circuit block for generating the triangular carrier waveform. A more detailed description of this is also given later, along with Fig. 5.26. Circuit block 3 is the analog comparator chip that compares the two waveforms. Notice that we have connected the reference voltage to the noninverting input terminal of the comparator. Therefore the output of the comparator will go high when the reference wave is more positive than the carrier. Therefore, in conformity with the switching pattern described earlier with reference to Fig. 5.24, the rising edges of the comparator output pulses will be the ON switching instants of the static switch S_1 and also the OFF switching instants of the static switch S_2. The falling edges of the comparator output are the instants when the reference voltage becomes less positive than the carrier, and these will be the instants for the OFF switching of S_1 and the ON switching of S_2. Later, when we extend the sine triangle comparison technique of SPWM to full-bridge topologies, both single- and three-phase, we shall follow the same convention—that is, the rising edges of a comparator output pulse will correspond to the ON switching instant of the top switch of an inverter leg and the OFF switching of the bottom switch on the same leg, and vice versa. We may state here a convention that is often followed in the literature and that we will also follow. This is the practice of using odd numbers to label the top switches of the legs of the bridge and even numbers to label the switches on the lower sides of the legs. Therefore, when the comparators are connected as shown in Fig. 5.25, the leading edges of the comparator output will determine the OFF switching instant of an even numbered switch and the ON switching of the odd numbered switch on the same leg. The trailing edges will determine the OFF switching instant of the odd numbered switch and the ON switching of the even numbered switch. In Fig. 5.25, we have also included a circuit module, labeled 4, that will invert the output of the comparator and thereby provide the timing pulses for our even numbered switch, which is switch S_2.

We may repeat here a statement made earlier, because of its great practical importance. This is the need to ensure, as a safety measure against "shoot through fault," that, before a switch on one side of the inverter leg comes ON, the other switch should have turned OFF. We usually provide a dead time for this purpose—a short delay time after the turning OFF of the outgoing switch and the turning ON of the incoming switch on the same leg of the inverter. The circuit detail for providing the dead time is not included in Fig. 5.25. But the purpose can be achieved by introducing a time delay between the rising edge of the timing waveform for a switch and the actual turn ON instant of this switch.

Another detail that we have left out in Fig. 5.25 is the synchronization between the reference waveform and the carrier waveform. In our description of SPWM with reference to Fig. 5.24, we had used a carrier frequency that is an integral multiple of the reference frequency. Therefore we had an integral number of carrier time periods in one time period of the reference. Also, we

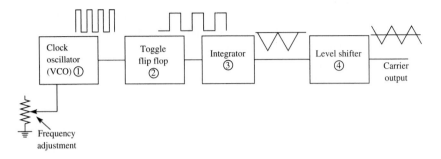

FIGURE 5.26
Block schematic of a circuit scheme for generation of the triangular carrier wave.

had the two waveforms synchronized in such a way that the peaks of the carrier occurred at the zero-crossing of the reference. It will need more circuit complexity if the carrier and reference waveforms are to be synchronized. We have treated the waveform error in the inverter output, due to the absence of synchronization between the carrier and the reference, as negligible.

Generation of the carrier waveform. A circuit scheme for the generation of the triangular carrier voltage waveform is shown in Fig. 5.26. The principle employed here is that the integration of a square wave results in a triangular wave. The circuit block labeled 1 is a voltage controlled oscillator chip. It is programmed to generate a pulse train whose frequency is twice the desired carrier frequency. The carrier frequency adjustment is shown in Fig. 5.26 as a manual potentiometer adjustment of the control voltage input of the VCO. In general, the VCO output may not be a symmetrical (that is, of exactly 50% duty cycle) square wave. Therefore we use an edge triggered toggle flip flop, labeled 2 in Fig. 5.26, which toggles, say, at each positive edge of the output of the VCO. The output of the toggle flip flop will therefore be a square wave of exactly 50% duty cycle and of frequency half of the VCO output frequency. The output square wave of the toggle flip flop is integrated, using an operational amplifier as an integrator (circuit block 3) (Any DC component in the square wave should be blocked by a capacitor, before integration). The output of the integrator may not have the correct zero level. Therefore we get the correct zero level by the level shifter, labeled 4, in the output of the integrator. The output of the level shifter will serve as the carrier.

Generation of the reference sine wave. A circuit scheme for the generation of the reference sine wave is shown in Fig. 5.27. This scheme allows independent adjustment of the frequency and amplitude of the sine wave. It employs a digital technique, wherein the waveform is digitally stored in a programmable read only memory. The circuit block labeled 1 is a voltage controlled oscillator. The control voltage input to this chip is shown in Fig. 5.27 as manually

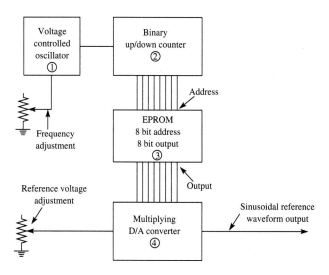

FIGURE 5.27
Digital circuit scheme for the generation of the sinusoidal reference voltage for implementation of SPWM.

adjustable. This is the adjustment for the frequency of the sine wave. The circuit block labeled 2 is described as a binary Up/Down counter. It may actually consist of several ICs. Its functions is to, first Up count the output pulses of the VCO, starting from count zero. As soon as the count reaches a preset binary number, it starts Down counting. As soon as the count becomes zero, it again Up counts, and so on. The counter used in Fig. 5.27 has an 8 bit output. The signal for initiating the down counting may be generated automatically by using an 8 bit binary magnitude comparator. This comparator compares the output of the binary 8 bit counter, which we shall call number A, with a fixed preset binary number, which we shall call number B. When $A = B$, the binary magnitude comparator chip outputs a signal at its $A = B$ output terminal. This output pulse can be routed to the binary 8 bit counter to initiate the down counting in every counting cycle. The 8 bit output of the binary counter serves as the address inputs to an erasable programmable read only memory (EPROM) chip (block 3 in Fig. 5.27). Each address location of the EPROM is allocated to an angular instant of the reference sine wave. The sine of this angle is stored at this address location as an 8 bit binary number. Therefore, as the output of the binary counter changes, successive locations of the memory are sequentially addressed, and 8 bit numbers corresponding to the sines of successive angles appear at the output of the EPROM. The 8 bit binary output of the EPROM is converted into an analog voltage using a D/A converter chip. The D/A converter used in Fig. 5.27 is a multiplying type. In this type of D/A converter chip, it is possible to introduce a scaling factor in the analog output by adjustment of a reference DC voltage input provided to the chip. This adjustment constitutes the magnitude control of the reference sine wave. Because of the discrete steps in which the voltage is outputted from the D/A converter, this analog output will actually be a stepped wave closely approximated to a sine wave. We have used 8 bit resolution in the scheme described in Fig. 5.27. Use of higher numbers of binary bits for the counter, the EPROM and the D/A converter will further improve the resolution, if that is needed.

The analog output of the D/A converter in Fig. 5.27 will be of one polarity only. We can load the EPROM memory in such a way that this gives us one-half, say, the positive half-period, of the sine wave signal. To obtain the negative half-period, we shall have to reverse the polarity of the analog output of the DA converter during the next Up/Down counting cycle of the binary counter. Therefore we have to have an additional circuit block that will reverse the polarity of the output voltage of the D/A converter after each Up/Down counting cycle. This circuit block is not included in Fig. 5.27 for the sake of clarity. This is, however, given as Problem 5.15.

5.9.4 SPWM in a Full-Bridge Inverter

We are using the approach, explained earlier, of treating the full bridge as a superposition of two half-bridge topologies. Therefore, referring to the full-bridge topology shown in Fig. 5.28(a), the static switches S_1 and S_2 constitute one half-bridge, which produces an AC output at the terminal A with respect to the load midpoint (which may be real or hypothetical). Similarly, the static switches S_3 and S_4 constitute the second half-bridge, which produces another AC voltage at the terminal B with respect to the same load midpoint, the output at B differing in phase by exactly 180°. Therefore, to implement sinusoidal pulse width modulation, we treat each half-bridge separately. We use the same carrier waveform. But we provide separate reference waveforms for each half-bridge. These two reference waveforms will have the same amplitude, but will differ in phase by 180°. One way of doing this is to obtain a second reference by simply reversing the polarity of the first by using an operational amplifier as an inverting amplifier with unity gain. To the block schematic for sine triangle comparison given in Fig. 5.25, we shall add one more comparator and associated circuits to provide timing pulses for the second leg of the inverter consisting of S_3 and S_4. The modified scheme for the full-bridge is shown in Fig. 5.28(b).

Output voltage waveform of the full bridge. The sinusoidally pulse width modulated output waveform of the full bridge can be obtained from the switching sequence of the four static switches of the inverter resulting from the sine triangle comparison. This is illustrated by Fig. 5.29, where we have drawn the carrier waveform that is common to both the legs of the inverter and the two sinusoidal reference waveforms—one for each leg of the inverter. The reference waveforms are phase-shifted by 180°. These have been drawn for

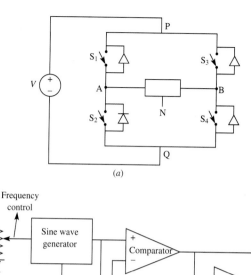

(a)

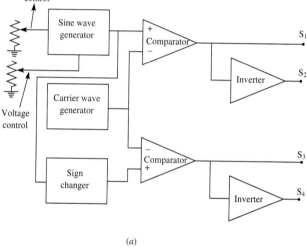

(a)

FIGURE 5.28
(a) Full bridge. (b) Generation of timing signals for S_1, S_2, S_3 and S_4 to implement SPWM.

one half-period of the inverter, which is the positive voltage half-period. A modulation index of approximately 50% has been used in drawing the reference wave, so that its amplitude is approximately 50% of that of the carrier. The ON switching instants of each static switch are marked in the figure, based on the sine triangle intersection, as explained earlier for the half-bridge. It is understood that the OFF switching of the other switch on the same inverter leg occurs ideally at the same instant.

The output voltage of the inverter can be drawn by taking into consideration the status of each of the four switches at each instant of time. After ascertaining the status of the inverter switches at any instant, the inverter output voltage at that instant can be found from the following reasoning, which can be verified by inspection of the configuration of the switches in Fig. 5.28(a).

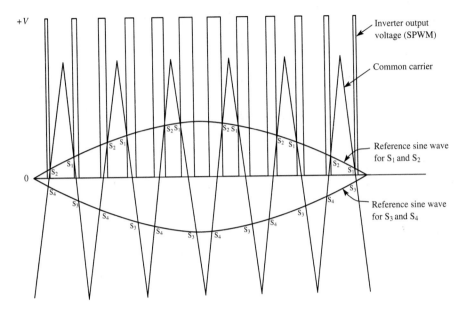

FIGURE 5.29
Single-phase full-bridge inverter SPWM.

1. Two switches of the inverter will be ON at any instant of time. These two switches will be on the two separate legs of the inverter bridge.
2. If the two switches that are ON, are both on the top of the inverter legs, that is, if they are both odd numbered, then freewheeling is taking place on the top (positive side) in the inverter. The output voltage of the inverter is zero.
3. If the two switches that are ON are both on the lower side of the legs, that is, if both are even numbered, then, also, freewheeling is taking place, but on the lower (negative) side. In this case as well, the output voltage is zero.
4. If the two switches that are ON are S_1, which is on the top side of leg A, and S_4, which is on the lower side of leg B, then the output voltage is equal to the DC input voltage V to the inverter.
5. If the two switches that are ON are S_2, which is on the lower side of leg A, and S_3, which is on the upper side of leg B, then the output voltage of the inverter is $-V$.

The output waveform of the inverter for one half-period is drawn in Fig. 5.29, based on the above reasoning. Examination of the output waveform leads us to the following important conclusions.

1. The inverter output pulses for the full bridge are unipolar in nature. That means that for the full bridge, all pulses are of one polarity only. Since we have drawn the output for the positive half-period only of the inverter, all

output pulses are positive during this half-period. This result is in contrast to the case of the half bridge, where we found that our sine triangle intersection technique resulted in an inverter output with reverse voltage excursions. The half-bridge is not very commonly used in practical equipment. We began our presentation of the SPWM using the half-bridge topology, primarily as a means of developing the underlying principles of this SPWM strategy in a logical sequence and in a way that can be easily extended to the single phase full bridge and later to the three phase full bridge topologies.

2. The number of output pulses in one half period of the inverter is seen to be 12 in Fig. 2.59. There are only 6 carrier voltage periods during this interval. Therefore we are getting a pulse repetition frequency equal to twice the carrier frequency.

A close inspection of the manner in which the two reference waveforms intersect with the carrier waveform will provide the reader with the reasons for the unipolar nature of the inverter output pulses and the higher pulse repetition frequency, both of which are advantages in comparison with the half-bridge topology.

5.10 THE THREE-PHASE INVERTER

The three-phase inverter derives its importance from the fact that most AC industrial equipment, such as motors, are designed to work from three-phase AC. Static three-phase inverters have become popular as an efficient means of providing adjustable three-phase AC, and therefore we treat them separately in this section.

5.10.1 Circuit Configuration and Switching Sequence

The most commonly used inverter circuit topology is the three-phase full bridge configuration, which is shown in Fig. 5.30. There are six controlled

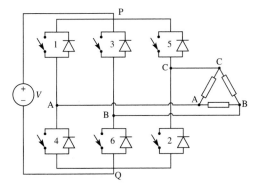

FIGURE 5.30
Three-phase bridge inverter.

static switches, labeled $S_1, \ldots, S_6$, each with its own antiparallel diode. Each switch with its antiparallel diode constitutes a "switching block." There are three "legs" for the inverter, each consisting of a switching block on the top (positive) side and a switching block on the lower (negative) side. In Fig. 5.30 we have followed the usual practise of numbering the switching blocks of a three phase bridge in the order in which the switches are turned ON. This may be verified by reference to the timing waveforms of Fig. 5.31. The AC output terminals of the inverter are A, B and C, each of which is the link connecting the upper switching block and the lower switching block on an inverter leg. The input DC terminals are labeled P (positive) and Q (negative). In Fig. 5.30, we have shown a three-phase Δ-connected load across the AC output terminals of the inverter.

Switching sequence. The three-phase full bridge of Fig. 5.30 is, in fact, a superposition of three single-phase half-bridge configurations. For example, the leg A, consisting of switching blocks 1 and 2, constitutes the A phase half-bridge. Similar statements apply for the B phase and C phase half-bridges. The switches of each half-bridge are operated in a manner exactly similar to what we have described earlier for a single-phase half-bridge. For example, on the leg A, the switch S_1 is ON and S_4 is OFF for one half-period of the AC. Then S_1 is turned OFF and S_4 turned ON for the next half-period. The switches of leg B are operated in the same manner, except that a phase delay of 120°, that is, one-third of an AC period, is maintained with respect to the switching instants of the corresponding switches of the leg A. Similarly, the operations of the leg C switches are delayed by 120° with respect to the leg B switches. This switching pattern for the six switches is shown graphically by the six timing diagrams of the switches on the top in Fig. 5.31. In each of these, the shaded interval is the ON time of the switch.

Output voltage waveform of the inverter. The output voltage waveforms of the inverter can be ascertained on the basis of the following considerations.

When a top side switch of an inverter leg is ON, the output terminal of that leg gets connected to the positive terminal of the DC input. Therefore this output terminal is at the same potential as the positive DC terminal. If we use the DC source midpoint (which may be real or hypothetical) as our zero-potential reference then the positive DC source terminal will be at $+\frac{1}{2}V$ and the negative DC source terminal will be at $-\frac{1}{2}V$. It follows from this that when a top switch of an inverter phase leg is ON, the output terminal of that phase will be at $+\frac{1}{2}V$. Similarly, when the lower switch of an inverter leg is ON, the output terminal on that leg will be at $-\frac{1}{2}V$. Therefore, looking at the timing diagram of each switch of an inverter leg, we are able to draw the waveform of the potential of the output terminal of that leg, with respect to the DC source midpoint. This potential should be a square wave with a positive peak value equal to $+\frac{1}{2}V$ and a negative peak equal to $-\frac{1}{2}V$. On this basis, the potentials of the three output terminals A, B and C have been drawn in Fig. 5.31. From

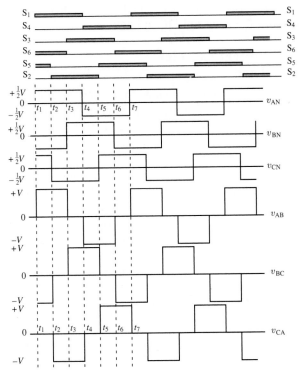

FIGURE 5.31
Timing diagrams of the switches and the resulting voltage waveforms.

these three waveforms we can derive the three output "line voltages" of our three-phase inverter, based on the following approach.

The output line voltages of the three-phase inverter. A line voltage is, by definition, the voltage between two output terminals. For example, the line voltage v_{AB} is the voltage of terminal A with respect to the terminal B. Since we already have the potentials of the terminals A and B with respect to a common reference, which is the DC source midpoint, we can determine the voltage v_{AB} at any instant of time by subtracting the potential of B from the potential of A at that instant. In this way, by taking the terminal voltages of A, B and C in pairs and subtracting one from the other in the manner stated, we can derive the three line voltage waveforms v_{AB}, v_{BC} and V_{CA}. The three line voltages derived in this manner are also sketched in Fig. 5.31.

It will be noticed that the line voltage waveform is a square wave with zero-voltage steps between the positive and negative intervals. Such a waveform is often referred to in the literature as a "quasi-square wave." The three line voltage waveforms are identical in shape. Each line voltage is successively displaced in phase by 120° with respect to the previous one. There are four steps in each waveform—two of which are zero-voltage steps, one a positive voltage step equal to the full DC voltage and the other a negative voltage step, also equal in magnitude to the full DC voltage.

5.10.2   Waveforms of Phase Current and Line Current With a Δ-Connected Load

The load in Fig. 5.30 is Δ-connected, which means that each one of the three-phase loads is connected across a pair of output terminals. Therefore the voltage appearing across a phase load is the same as the corresponding line voltage. Therefore, to determine the current in a load phase, we can consider that phase independently of the other two. In this manner, it is possible to get the current waveform for each phase. If the load is a balanced one, by which we mean that the three load impedances are identical, the three-phase current waveforms will also be identical, but displaced in phase by 120°. By the "line current," we mean the current in the connecting line from an output terminal on an inverter leg to the corresponding "corner" of the delta. The three lines can be seen in Fig. 5.30 as AA, BB and CC. The current in a line at any instant can be seen to be the difference between the two phase currents at the corner of the delta that is fed by that line. On this basis, the three line currents can also be determined by taking the difference between the corresponding phase currents. We shall illustrate these procedures by a numerical example.

> **Illustrative example 5.7.** The three phase bridge inverter of Fig. 5.30 is operating at 100 Hz from a DC supply of 300 V. Each of the three Δ-connected impedances consists of a resistor of 15 Ω in series with an inductor of 75 mH.
> (a) Obtain expressions for all the segments of the phase current waveform, for one phase.
> (b) Sketch the waveform for all the three phase currents and indicate the magnitude at each discontinuity and the instants at which they occur.
>
> **Solution.** Since we know the voltage across each phase load, this being the line voltage output of the inverter, we shall consider a phase load separately, to determine the current in it. We have drawn the phase load Z_{AB} and the voltage v_{AB} across it in Fig. 5.32(a). The waveform of the voltage v_{AB} is shown in Fig. 5.32(b), which is in conformity with what we derived in Fig. 5.31.

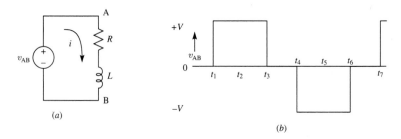

FIGURE 5.32
The circuit of phase AB and the waveform of phase AB voltage.

Interval t_1 **to** t_3. The loop equation for the circuit during the interval t_1 to t_3 may be written as

$$L\frac{di}{dt} + Ri = V \tag{5.49}$$

The initial condition may be stated as

$$i = I_1 \quad \text{at} \quad t = 0$$

(We shall later determine I_1 using the assumption of repetitive conditions.) The solution with the stated initial condition is

$$i = I_1 e^{-t/\tau} + \frac{V}{R}(1 - e^{-t/\tau}) \tag{5.50}$$

where the load time constant $\tau = L/R$. At the instant t_3, $t = \frac{1}{3}T$, where T is the AC time period. The current at t_3, which we shall label I_3, will be, from (5.50),

$$I_3 = I_1 e^{-T/3\tau} + \frac{V}{R}(1 - e^{-T/3\tau}) \tag{5.51}$$

Interval t_3 **to** t_4**.** During the interval t_3 to t_4, the voltage is zero, and therefore the loop equation has to be modified as

$$L\frac{di}{dt} + Ri = 0 \tag{5.52}$$

Taking a new zero reference for time t at t_3, the initial condition for this equation may be stated as

$$i = I_3 \quad \text{at} \quad t = 0$$

The solution of (5.52) with the stated initial condition is

$$i = I_3 e^{-t/\tau} \tag{5.53}$$

$$= \left[I_1 e^{-T/3\tau} + \frac{V}{R}(1 - e^{-T/3\tau}) \right] e^{-t/\tau} \tag{5.54}$$

At the instant t_4, $t = \frac{1}{6}T$. Therefore, the current at t_4, which we shall label as I_4, will be, from (5.53) and (5.54),

$$I_4 = I_3 e^{-T/6} \tag{5.55}$$

$$= \left[I_1 e^{-T/3\tau} + \frac{V}{R}(1 - e^{-T/3\tau}) \right] e^{-T/6\tau} \tag{5.56}$$

When repetitive conditions prevail, to satisfy waveform symmetry,

$$I_4 = -I_1$$

Therefore, replacing I_4 by $-I_1$ in (5.56) and solving for I_1, we get the value of I_1 as

$$I_1 = -\frac{V}{R}\frac{e^{-T/6\tau} - e^{-T/2\tau}}{1 + e^{-T/2\tau}} \tag{5.57}$$

Equation (5.57) gives the value of the current at instant t_1 which is one of the four instants at which discontinuities occur in the waveform. The value I_4 at t_4 is given by the same equation with the sign reversed. The other two discontinuities are at t_3 and t_6. The value I_3 at t_3 is obtained from (5.51), into which the value of I_1 has to be substituted from (5.57). This gives

$$I_3 = \frac{V}{R}\frac{1 - e^{-T/3\tau}}{1 + e^{-T/2\tau}} \tag{5.58}$$

At the instant t_6, the current I_6 has the same magnitude as I_3, but with the sign reversed.

We have now determined the current value at every discontinuity in the waveform. We shall now substitute, in the equations, the numerical data given, namely $V = 300$ V, $R = 15\,\Omega$, $L/R = 5$ ms and $T = 1/f = 10$ ms. The resulting values of the current are

$$I_1 = -5.10 \text{ A}$$

$$I_3 = 7.11 \text{ A}$$

$$I_4 = 5.10 \text{ A}$$

$$I_6 = -7.11 \text{ A}$$

Substituting these values into the appropriate equations, we can now numerically express the value of the current in each segment, between successive discontinuities in the waveform, starting from the instant t_1. It is important to remember that for each of the expressions, the zero of time $(t = 0)$ is the instant of commencement of the interval. It is therefore different for each expression.

Interval t_1 to t_3 (with t measured from t_1). From (5.50),

$$i = -5.1e^{-200t} + 20(1 - e^{-200t})$$

This simplifies to

$$i = 20 - 25.1e^{-200t} \tag{5.59}$$

Interval t_3 to t_4 (with t measured from t_3). From (5.53),

$$i = 7.11e^{-200t} \tag{5.60}$$

Interval t_4 to t_6 (with t measured from t_4). By a sign change in (5.59),

$$i = -20 + 25.1e^{-200t} \tag{5.61}$$

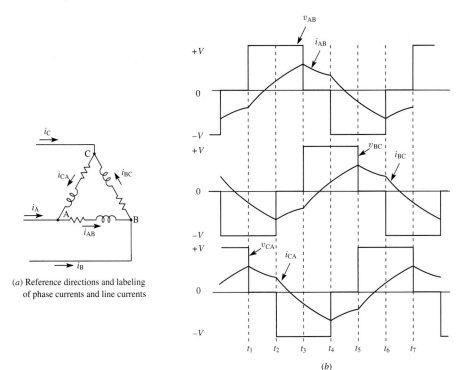

(a) Reference directions and labeling
of phase currents and line currents

(b)

FIGURE 5.33
Phase current waveforms for the Δ-connected load.

Interval t_6 to t_7 (with t measured from t_4). By a sign change in (5.60),

$$i = -7.11e^{-200t} \tag{5.62}$$

The waveform of the current over a complete period is plotted in Fig. 5.33(b) for the phase AB. The reference directions for the current in all the phases and their labeling are shown in Fig. 5.33(a).

The waveform of the current in the phase BC, labeled i_{BC}, is identical to that of i_{AB} except that it is delayed by 120°. However, the mathematical expression for each segment of the waveform is identical to what we have obtained for the corresponding segment of i_{AB}. This is because in our derivation of the expressions, we have always taken the reference zero for time as the instant of commencement of the segment. Similar statements are applicable to the waveform of the current i_{CA} in the phase CA. All the three phase current waveforms are sketched in Fig. 5.33(b).

Line currents. The waveforms of each of the three line currents for the case of the Δ-connected load can be obtained by taking the difference of the currents in the two phases that are linked to that line at their common Δ link. This is illustrated by the following numerical example, for the inverter circuit that we considered in the previous example.

Illustrative example 5.8. For the three-phase bridge inverter feeding a Δ-connected load that was considered in Example 5.7
(a) obtain mathematical expressions for each segment of the line current i_A [for the labeling of currents, see Fig. 5.33(a)];
(b) sketch the waveforms of all the three line currents.

Solution.
(a) The line current i_A is obtained by taking the difference of the phase currents at the corner of the delta fed by the line. Therefore, referring to Fig. 5.33(a),

$$i_A = i_{AB} - i_{CA} \tag{5.63}$$

To obtain the mathematical expressions for the line current, we can use those obtained in Example 5.7 for the phase currents. There are four discontinuities in each phase current during one complete time period of the AC, as can be seen from the phase current waveforms sketched in Fig. 5.33(b). Therefore there are four separate expressions for the segments of the waveform between successive discontinuities, and we have used a separate reference zero for time t for expressing each segment. Two of the discontinuities in the phase current i_{AB} occur at the same instant as two in the phase current i_{CA}. But the other two instants are not coincident. Therefore there will be six discontinuities in the line current waveform, and so six segments, each of duration 60°. To obtain the mathematical expression for each of the line current segments by taking the difference of the phase currents, we have to modify the expression for the phase current segments, where necessary, to make the reference zero for t the same for each pair of expressions. This procedure is shown in tabular form in Table 5.1.

We have divided the full AC period into six intervals, each of duration 60°. These intervals are identified in column 1 of the table, in conformity with

TABLE 5.1

(1) Interval	(2) Phase current i_{AB}	(3) Phase current i_{CA}	(4) Line current $i_A = i_{AB} - i_{CA}$	(5) i_A at the start of the interval
t_1 to t_2	$20 - 25.1e^{-200t}$	$7.11e^{-200t}$	$20 - 32.21e^{-200t}$	-12.21 at t_1
t_2 to t_3	$20 - 17.98e^{-200t}$	$-20 + 25.1e^{-200t}$	$40 - 43.08e^{-200t}$	-3.08 at t_2
t_3 to t_4	$7.11e^{-200t}$	$-20 + 17.98e^{-200t}$	$20 - 10.87e^{-200t}$	$+9.13$ at t_3
t_4 to t_5	$-20 + 25.1e^{-200t}$	$-7.11e^{-200t}$	$-20 + 32.21e^{-200t}$	$+12.21$ at t_4
t_5 to t_6	$-20 + 17.98e^{-200t}$	$20 - 25.1e^{-200t}$	$-40 + 43.08e^{-200t}$	$+3.08$ at t_5
t_6 to t_7	$-7.11e^{-200t}$	$20 - 17.98e^{-200t}$	$-20 + 10.87e^{-200t}$	-9.13 at t_6

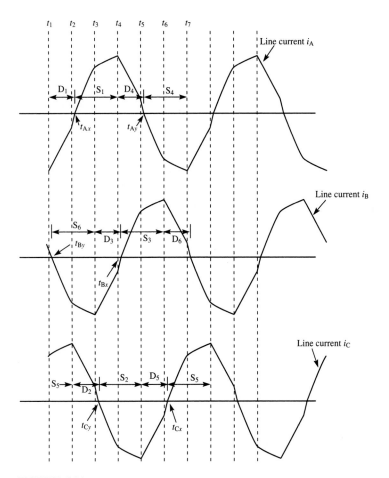

FIGURE 5.34
Line current waveforms of a three-phase Δ-connected load fed from a three-phase inverter bridge.

the labeling in Fig. 5.33 and Fig. 5.34. In column 2, we have given the expressions for the phase current i_{AB} for each of the six intervals. These are obtained from the expressions obtained in Example 5.7, with the zero of t shifted, wherever necessary, to correspond to the commencement of the interval. We have repeated the procedure of column 2 in column 3 for the phase current i_{CA}. We then get the line current i_A by subtracting the expression in column 3 from that in column 2. This is done in column 4. We have also listed the value of the line current at each discontinuity in column 5. These values have been obtained by putting $t = 0$ in the corresponding expression in column 4.

(b) The waveform of the line current i_A is sketched in Fig. 5.34, based on the results listed in Table 5.1. The waveform of the line current i_B is identical to that of i_A, but delayed by 120°. Similarly, the waveform of i_C is obtained by a further delay of 120°. These two waveforms are also sketched in Fig. 5.34 on this basis.

Zero-crossing of the line current. The zero-crossings of the line current are the instants at which the current shifts between the static switch and its antiparallel diode in the same switching block. These instants can be ascertained by equating the appropriate expressions for the line current to zero. Commutations within a switching block occur at these instants. Commutations between switching blocks occur at the zero-crossings of the voltage, which are, in fact determined by the timing diagrams of the switches. This is illustrated by the following example.

Illustrative example 5.9. Consider the inverter-fed Δ-connected load of Example 5.7.

(a) determine the zero-crossing instants of the line current and indicate them on the line current waveforms.

(b) Use the zero-crossings of the voltage and of the current to identify the conducting devices at each instant. On this basis, mark the conducting intervals of the individual switching devices, on all the legs of the inverter, on the waveforms of Fig. 5.34.

(c) Show the sequential changes in the circuit configuration, as the switchings of the individual switching blocks occur, in a complete AC period.

Solution.

(a) Inspection of the Table 5.1 and the waveform in Fig. 5.34 shows that for the line current i_A, the positive slope zero-crossing occurs in the interval t_2 to t_3 and the negative slope zero crossing in the interval t_5 to t_6. The positive slope zero-crossing can be determined by equating the corresponding expression in column 4 of Table 5.1 to zero:

$$40 - 43.08e^{-200t} = 0 \tag{5.64}$$

This gives $t = 0.371$ ms. Since one complete period is 10 ms ($f = 100$ Hz), the interval in angular measure will be

$$\frac{0.371}{10} \times 360° = 13.4°$$

In a similar manner, or from waveform symmetry considerations, the negative slope zero-crossing will be found to occur after the same delay (13.4°), from the instant t_5. These zero-crossing instants are indicated in Fig. 5.34 by the labels t_{Ax} and t_{Ay}.

The delay angle determined above is used to locate the zero-crossings of the other two line currents. These are marked on the waveforms in Fig. 5.34.

(b) On each inverter leg, the conducting switching block in any instant is determined by the timing diagram in Fig. 5.31. The conducting device within a switching block will be determined by the direction of current at that instant. On this basis, the conducting interval of each individual device is marked in Fig. 5.34 for all the legs of the inverter.

(c) At the instant of switching of each switching block in a cycle, there is a change in the circuit configuration. Therefore, looking at the state of the switching blocks, which is determined by the timing diagram of the switches as shown in Fig. 5.31, we conclude that there are six intervals, each of 60° duration, during which there are different circuit configurations. What the inverter basically does is to change the configuration of the circuit, consisting of the DC voltage source and the Δ-connected load, sequentially, six times in a complete period of the AC, by the switching action of the devices. Each of the six configurations can be drawn by looking at the state of the switching block in the appropriate 60° interval. On this basis, the six circuit configurations are drawn in Fig. 5.35. The interval corresponding to each configuration is also indicated in the figure, starting from the instant labeled t_1 in Fig. 5.34. In Fig.

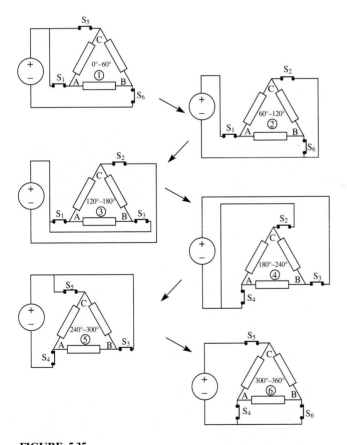

FIGURE 5.35
Sequential changes in the power circuit configuration resulting from the switching of the inverter switching blocks.

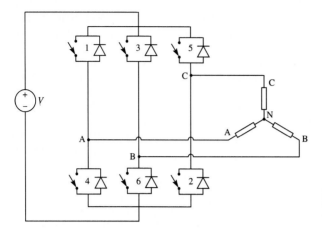

FIGURE 5.36
Three-phase inverter feeding a
Y-connected load.

5.35, the switches that are shown ON in each configuration should be interpreted as the switching blocks to which they belong. The actual device that is conducting in a switching block will depend on the direction of current through the switching block and is not indicated in the figure.

5.10.3 Three-Phase Inverter Feeding a Y-Connected Load

Figure 5.36 shows a Y-connected load fed from a three-phase inverter bridge. When the load impedances are connected in this manner, each load impedence comes in series with a line. However, since the inverter, which is connected to the load terminals A, B and C, is external to the load, the internal switchings of the inverter only change the interconnection from the DC source to the load terminals A, B and C. The switching pattern and timings in the inverter are the same, irrespective of whether the load is Δ- or Y-connected. What this means to us, in the context of the Y-connected load, is that here, also, we get six different circuit configurations as we explained for the Δ-connected load, each configuration lasting for 60° in a complete period. Furthermore, each of these configurations and the sequence in which they occur will be the same as we obtained for the case of the Δ-connected load, if we replace the Δ connection of the load by the Y connection. On the basis of this reasoning, we have drawn the six circuit configurations that occur sequentially in an AC period in Fig. 5.37. Each of these is the same as the corresponding circuit in Fig. 5.35, with the Δ replaced by Y.

In the Y connection, we have a neutral terminal, labeled N in Figs. 5.36 and 5.37. The voltage occurring across any load phase is the voltage between its line terminal and the neutral terminal. Therefore, to find the voltage across a load phase in each of the 60° intervals during which the circuit configuration for that interval in Fig. 5.37 is valid, we have to determine the manner in which the total DC voltage V is shared between the load impedances in that

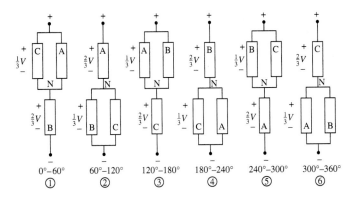

FIGURE 5.37
Three-phase inverter with Y-connected load—sequential changes in the circuit configuration in a switching cycle.

configuration. In this manner, we can determine the phase voltage waveform, as explained below.

Phase voltage waveform for a Y-connected load. In each of the six circuit configurations shown in Fig. 5.37, we notice that two of the load impedances are in parallel. This parallel combination is in series with the third impedance. The DC voltage V is applied across this series–parallel circuit. We shall only consider the case of a balanced three-phase load, which means that the three impedances are identical. If the load is purely resistive, it follows that the voltage across the parallel combination will be $\frac{1}{3}V$ and that across the third resistance will be $\frac{2}{3}V$. The following reasoning shows that this manner of voltage division is true for typical inductive loads also.

Three identical series R–L circuits are shown in Fig. 5.38, with two of them connected in parallel and the third in series with the parallel combination.

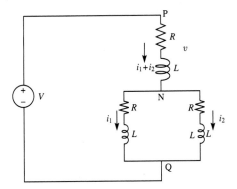

FIGURE 5.38
Instantaneous voltage division between impedances in a Y-connected inductive load.

The DC voltage V comes across this series parallel circuit. Using the labeling of currents and terminals in Fig. 5.38, we may express the voltages to neutral as follows:

$$v_{NQ} = Ri_1 + L\frac{di_1}{dt} \tag{5.65}$$

$$= Ri_2 + L\frac{di_2}{dt} \tag{5.66}$$

and, from these two equations, by addition,

$$2v_{NQ} = R(i_1 + i_2) + L\frac{d}{dt}(i_1 + i_2)$$

$$= v_{PN}$$

giving

$$v_{PN} = \tfrac{2}{3}V$$

$$v_{NQ} = \tfrac{1}{3}V$$

Therefore we conclude that the voltage division is the same as for a resistive load.

If we use the neutral terminal N as the reference to express the voltage of each phase, then the polarity of the voltage of a load phase will be negative when it appears below N in Fig. 5.37 and positive when it appears above N. The waveform of each phase voltage is drawn on this basis in Fig. 5.39. This will be a six-step waveform, in conformity with the six circuit configurations in Fig. 5.37. The middle step in each half-period will have a magnitude equal to $\tfrac{2}{3}V$ and the two steps on either side will have magnitudes equal to $\tfrac{1}{3}V$, in agreement with the manner of voltage division determined above.

Neutral terminal oscillations. In drawing the waveform of each phase voltage in Fig. 5.39, we have used the neutral terminal of the load as the reference point. However, we show below that the load neutral potential actually fluctuates with respect to the DC supply terminals. The manner of this fluctuation is determined below. For this purpose, we shall use the midpotential point of the DC supply, which may be real or hypothetical, as our reference. We shall label the midpotential point of the DC supply as N_S.

Referring to Fig. 5.38, the potential of the load neutral N with respect to the hypothetical DC source midpoint N_S is seen to be:

$$v_{NN_S} = \tfrac{1}{3}V - \tfrac{1}{2}V = -\tfrac{1}{6}V$$

When the parallel load branch is on the upper side (positive side) of N, v_{NN_S} will be equal to $+\tfrac{1}{6}V$. The parallel load branch shifts between the positive and negative sides every 60° in Fig. 5.37. Therefore the neutral potential oscillates as shown in Fig. 5.40. The frequency of oscillation is the same as the

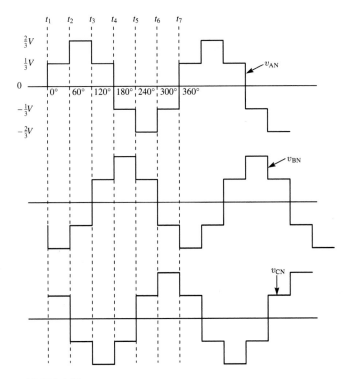

FIGURE 5.39
Phase voltage waveforms in a Y-connected load.

third-harmonic frequency of the AC output of the inverter. The amplitude of the oscillation is one-sixth of the DC input voltage to the inverter.

Current waveforms for the Y-connected load. For the Y-connected load, the line current and the phase current are the same. We have already obtained the waveform of the phase voltage in Fig. 5.39. Therefore it becomes a straightforward procedure to obtain the phase current waveform for any given load impedance. There are six steps in the voltage waveform in an AC period, that is, three steps in a half-period. We can write the voltage equation for each interval and solve for the current, assuming an initial unknown value of the current at the beginning of the first step (for the typical case of an inductive

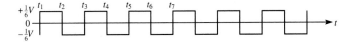

FIGURE 5.40
Oscillations of the neutral potential in a Y-connected load fed from a three-phase bridge inverter.

load) as I_1. On the assumption that repetitive conditions prevail, the final current at the end of the third interval must be equal to the negative of I_1, because of the waveform symmetry. We can therefore use this condition to determine I_1. The solutions obtained for the first half-cycle will apply to the second half-cycle with the signs reversed. This procedure is illustrated by the following numerical example.

Illustrative example 5.10. A three-phase bridge inverter is operating at 100 Hz. The DC input voltage is 300 V. The load is Y-connected, each phase consisting of a resistance $R = 5\,\Omega$ in series with an inductance $L = 25\,\text{mH}$. Determine the current waveforms.

Solution. Referring to the phase voltage waveform derived in Fig. 5.39, we shall solve for the current in the three 60° intervals that constitute one half-period of the AC, commencing from the interval t_1 to t_2. In solving for the current, we shall use a separate reference zero for time for each interval, which will be the instant of commencement of that interval. Repetitive conditions will be assumed.

Interval t_1 to t_2. The voltage during this interval is $\frac{1}{3}V$. Therefore

$$L\frac{di}{dt} + Ri = \frac{1}{3}V \tag{5.67}$$

We denote the initial current at $t = 0$ by I_1. We shall later determine I_1 using the condition for waveform symmetry. The solution of (5.67) with the stated initial condition is

$$i = \frac{1}{3}\frac{V}{R}(1 - e^{-t/\tau}) + I_1 e^{-t/\tau} \tag{5.68}$$

where the load time constant $\tau = L/R$. At the instant t_2, $t = \frac{1}{6}T$, where $T = 1/f$ is the AC period, and

$$I_2 = \frac{1}{3}\frac{V}{R}(1 - e^{-T/6\tau}) + I_1 e^{-T/6\tau} \tag{5.69}$$

Interval t_2 to t_3. During this interval, the voltage is $\frac{2}{3}V$. Therefore

$$L\frac{di}{dt} + Ri = \frac{2}{3}V \tag{5.70}$$

The initial condition is

$$i = I_2 \quad \text{at } t = 0$$

The solution of (5.70) with this initial condition is

$$i = \frac{2}{3}\frac{V}{R}(1 - e^{-t/\tau}) + I_2 e^{-t/\tau} \tag{5.71}$$

At t_3, $t = \frac{1}{6}T$, and

$$I_3 = \frac{2}{3}\frac{V}{R}(1 - e^{-T/6\tau}) + I_2 e^{-T/6\tau} \tag{5.72}$$

Substitution of (5.69) for I_2 into (5.72) gives

$$I_3 = \frac{2}{3}\frac{V}{R}(1 - e^{-T/6\tau}) + \left[\frac{1}{3}\frac{V}{R}(1 - e^{-T/6\tau}) + I_1 e^{-T/6\tau}\right]e^{-T/6\tau}$$

$$= \frac{1}{3}\frac{V}{R}(2 - e^{-T/6\tau} - e^{-T/3\tau}) + I_1 e^{-T/3\tau} \tag{5.73}$$

Interval t_3 to t_4. The voltage during this interval is $\frac{1}{3}V$. Therefore

$$L\frac{di}{dt} + Ri = \frac{1}{3}V \tag{5.74}$$

The initial condition is

$$i = I_3 \quad \text{at } t = 0$$

The solution with this initial condition is

$$i = \frac{1}{3}\frac{V}{R}(1 - e^{-t/\tau}) + I_3 e^{-t/\tau} \tag{5.75}$$

At t_4, $t = \frac{1}{6}T$, and

$$I_4 = \frac{1}{3}\frac{V}{R}(1 - e^{-T/6\tau}) + I_3 e^{-T/6\tau} \tag{5.76}$$

We now substitute (5.73) for I_3 into (5.76), giving

$$I_4 = \frac{1}{3}\frac{V}{R}(1 + e^{-T/6\tau} - e^{-T/3\tau} - e^{-T/2\tau}) + I_1 e^{-T/2\tau} \tag{5.77}$$

To satisfy the waveform symmetry condition, $I_4 = -I_1$. Therefore, replacing I_4 by $-I_1$ in (5.77) and solving for I_1, we get

$$I_1 = -\frac{V}{3R}\frac{1}{1 + e^{-T/2\tau}}(1 + e^{-T/6\tau} - e^{-T/3\tau} - e^{-T/2\tau}) \tag{5.78}$$

We may now substitute numbers from the given data to determine I_1 using (5.78). We have $V = 300\,\text{V}$, $R = 5\,\Omega$, $\tau = L/R = 5\,\text{ms}$ and $T = 1/f = 10\,\text{ms}$, and so $I_1 = -12.21\,\text{A}$.

This value of I_1 may be substituted into (5.69) to calculate the value of I_2. This gives

$$I_2 = -3.08\,\text{A}$$

We may use this value of I_2 to determine I_3 by substitution into (5.72). This gives

$$I_3 = 9.13\,\text{A}$$

We have now determined the value of the current at each of the three discontinuities in the half cycle. We can use these values to express the equation for the three segments of the current waveform in the half-cycle using respectively (5.68) for the interval t_1 to t_2, (5.71) for the interval t_2 to t_3, and (5.75) for the interval t_3 to t_4. The same three expressions with the signs reversed will apply to the corresponding segments of the current waveform during the second half-period. The six expressions for the six segments of the current waveform thus obtained are as follows:

$$t_1 \text{ to } t_2 \qquad i = 20 - 32.21e^{-200t} \qquad (5.79)$$

$$t_2 \text{ to } t_3 \qquad i = 40 - 43.08e^{-200t} \qquad (5.80)$$

$$t_3 \text{ to } t_4 \qquad i = 20 - 10.87e^{-200t} \qquad (5.81)$$

$$t_4 \text{ to } t_5 \qquad i = -20 + 32.21e^{-200t} \qquad (5.82)$$

$$t_5 \text{ to } t_6 \qquad i = -40 + 43.08e^{-200t} \qquad (5.83)$$

$$t_6 \text{ to } t_7 \qquad i = -20 + 10.87e^{-200t} \qquad (5.84)$$

Inspection of (5.79)–(5.84) shows that these equations are the same as those listed in column 4 of Table 5.1, which we got earlier for the line currents for the case of the Δ-connected load in Example 5.7. This was to be expected, because the impedance parameters we chose for the Y-connected load are exactly one-third of the values we had for the Δ-connected load in Examples 5.7 and 5.8. The Y–Δ transformation relationship in basic circuit theory states that a Δ connection of three equal impedances is equivalent to a Y connection of three other equal impedances of one-third the value. We chose the impedance values that satisfy this condition and also the same DC voltage and frequency for the inverters, to enable an additional check on our calculations.

Since the expression for each segment of the waveform is the same as that for the line current waveform in Example 5.7, the current waveforms for the present example are identical to Fig. 5.34. For this reason, they are not separately plotted here.

5.10.4 Voltage Control and Waveform Shaping in Three-Phase Inverters—Sinusoidal Pulse Width Modulation

Adjustment of the AC output voltage can always be done by adjustment of the input DC voltage. For instance, we can use a phase controlled rectifier to provide an adjustable DC input to the inverter, in the case of a DC link inverter, in which the DC link is obtained by rectification from the AC power mains. Another scheme commonly employed is to use a DC/DC converter (chopper converter) of adjustable voltage ratio on the input side of the inverter. Since these schemes are external to the inverter, we shall not consider them further here. The technique of sinusoidal pulse width modulation, which we described for the single-phase inverter, can also be used in three-phase inverters. This technique will enable both voltage adjustment and waveform shaping within the inverter. We shall describe here the manner of implementation of SPWM in three-phase bridge inverters.

In describing SPWM in the single-phase full bridge inverter, we used the approach of treating the full bridge as a combination of two half-bridges. Each half-bridge was viewed as providing an AC output with respect to the midpotential point. In the case of the single-phase half-bridge inverter, these two AC voltages were phase displaced by 180°. Therefore, to implement

SPWM by sine triangle comparison, we used two sine wave references that were phase-displaced by 180° for comparison with the carrier wave—one for each leg of the inverter. We now extend the same approach to the three-phase bridge and treat it as a combination of three half-bridges in which each half-bridge provides an AC voltage, the phase shift between successive AC outputs being 120°. Therefore, to implement SPWM by sine triangle comparison, we need to use three reference sine waves—one for each leg of the inverter—with a phase shift of 120° between successive sine waves.

Figure 5.41 shows the block schematic of the scheme used to obtain the timing instants of the six static switching elements of the inverter. The six timing voltage outputs shown in the figure are numbered to correspond to the numbering of the six static switches in the inverter topology, which is also shown in the figure. The provision for a dead time between switchings is not included in Fig. 5.41. This can be incorporated by providing a delay in the turn ON switching of each incoming switch by introducing a corresponding delay with respect to the rising edge of the timing pulse output for its turn ON switching. The reader may compare Figs. 5.25, 5.28 and 5.41, and will notice that the last two are only extensions of the same sine triangle comparison strategy presented in the first for the half-bridge, and extended in the second for the full bridge with two legs, and then to the three-phase bridge with three legs. In Fig. 5.41, the reference voltage generator is shown as a single circuit block, providing all three reference wave outputs. The three sine waves are to have a mutual phase difference of 120°, but the same amplitude and frequency. The frequency adjustment is shown as common for all the three sine waves, and so also is the magnitude adjustment.

A scheme for the generation of the three reference sine waves is shown in Fig. 5.42. This is an adaptation of the digital scheme that we described earlier for the single-phase half-bridge in Fig. 5.27, and the same description applies. The addition is that now we have three EPROMs, all of which are addressed by the output of the same binary counter. The 120° phase shift between the three sine waves is achieved by introducing a corresponding shift in the memory locations, in which the same sine wave value is stored, in the three EPROMs. A common adjustable reference voltage is used for all three multiplying D/A converters. This makes the amplitude of the sine waves adjustable, but the same, for all the three outputs. As in Fig. 5.27, the outputs shown in Fig. 5.42 are for one half-period only. The polarity reversal circuit for generating the second half-periods of the sine waves is not included in this figure.

5.11 INPUT CURRENT RIPPLE—USE OF AN INPUT FILTER

In our treatment of inverters, we have assumed that the DC input voltage to the inverter is constant and ripple-free. But the input current will have ripple. For the half-bridge inverter, we did obtain the waveform of the input current,

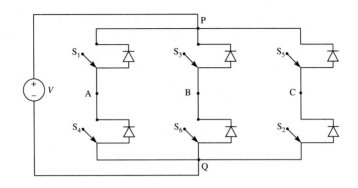

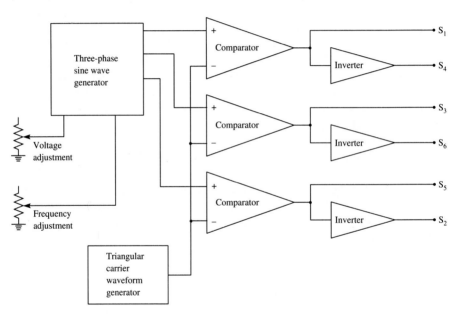

FIGURE 5.41
Implementation of SPWM using sine triangle comparison in a three-phase inverter.

in Fig. 5.9, as supplied from each section of the twin DC power supply. For full-bridge inverters, there is only one DC power source. The waveform of the current supplied by this source can be derived from the waveform of the output AC current by looking at the switching blocks of the inverter through which this current flows. We shall illustrate this procedure by the numerical examples that follow, for

1. the single-phase full-bridge inverter without PWM;

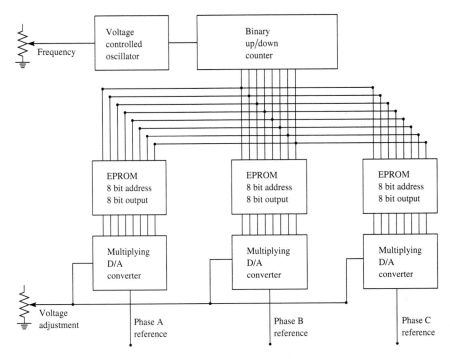

FIGURE 5.42
Generation of three-phase sine wave references for implementing SPWM.

2. the single-phase full-bridge inverter with PWM;
3. the three-phase quasi-square-wave inverter.

5.11.1 Single-Phase Full-Bridge Inverter—Input Current Waveform

Illustrative example 5.11. For the full-bridge inverter of Example 5.3,
(*a*) obtain the waveform of the input current;
(*b*) determine the value of the DC component of the input current, by averaging the mathematical expression for it.

Solution.
(*a*) In the solution of Example 5.3, the waveform of the output current was obtained and shown in Fig. 5.16. Referring to this waveform and the configuration of the switching blocks given in Fig. 5.15 for the inverter, we find that during the interval t_1 to t_3, the current flow is through switching blocks 1 and 4. Therefore the input current is the same as the output current i. Similarly, during the interval t_3 to t_5, the current flow is through switching blocks 2 and 3. Therefore the input current is the negative of the output current. In this manner, the input current waveform can be derived from the

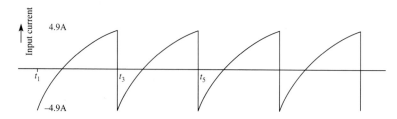

FIGURE 5.43
Waveform of input current in a full-bridge inverter without PWM.

waveform of the output current. The waveform thus obtained is plotted in Fig. 5.43.

(*b*) From the waveform of the input current, we notice that it is repetitive and that the repetitive frequency is twice the AC output frequency. The input current has a DC component, which can be determined by averaging over one repetitive period. For this purpose, we recall the mathematical expression given by (5.17) in the solution of Example 5.3. This is valid for each repetitive period of the input current:

$$i = 20 - 24.9e^{-100t}$$

One repetitive period for the input current is 0.005 s (for $f = 100$ Hz). Therefore the DC component of the input current will be:

$$I_{DC} = \frac{1}{0.005} \int_0^{0.005} (20 - 24.9e^{-100t}) \, dt$$

$$= 0.405 \text{ A} \tag{5.85}$$

Illustrative example 5.12. For the PWM inverter of Example 5.4,
(*a*) obtain the waveform of the input current;
(*b*) determine the DC component of the input current by averaging the appropriate mathematical expression.

Solution.

(*a*) The waveform of the AC output current was obtained in Fig. 5.21. Referring to this figure and the power circuit topology shown along with it, we find the following.

During the interval t_1 to t_2 the conduction is through switching blocks 1 and 4. Therefore, during this interval, the input current is the same as the output current *i*.

During the interval t_2 to t_3, the output current is freewheeling through the switching blocks on the top of the two inverter legs. The switching blocks on the bottom of both the legs of the inverter are in the OFF state. Therefore the input current is zero.

During the interval t_3 to t_4, the current flow is through switching blocks 2 and 3. Therefore the input current to the inverter is the negative of the output current.

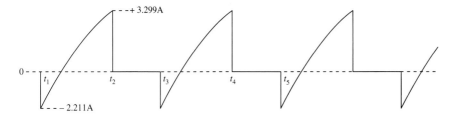

FIGURE 5.44
Input current waveform for full-bridge PWM inverter.

During the interval t_4 to t_5, the input current is zero as in the case of the interval t_2 to t_3. The input current waveform is derived from the output current waveform on the basis of the above conditions, and is given in Fig. 5.44.

(b) From the above waveform, we notice that the repetitive frequency is twice the inverter frequency. For the given $f = 100$ Hz, the repetitive period is 0.005 s. During this interval, current flow is only for 0.003 s. The expression for the current is given by (5.30) (see the solution of Example 5.4):

$$i = 10 - 12.211e^{-200t}$$

The DC component of the input current is therefore

$$I_{\text{DC}} = \frac{1}{0.005} \int_0^{0.003} (10 - 12.211e^{-200t})\, dt$$

$$= 0.49 \text{ A} \tag{5.86}$$

5.11.2 Three-Phase Quasi-Square-Wave Inverter—Input Current Waveform

Illustrative example 5.13. For the three-phase quasi-square-wave inverter considered in Example 5.10,
(a) obtain the waveform of the input current;
(b) determine the ripple frequency of the input current;
(c) determine the DC component of the input current by averaging the mathematical expression for it.

Solution.
(a) The load in this case is Y-connected, and the waveforms of the line currents, which are also the same as the phase currents, are obtained in Fig. 5.34. Each of these waveforms has six segments in a complete period. We can look at each of these six intervals and ascertain the current path by identifying the switching blocks through which the current flow takes place. A simple way of doing this is to look at the six changes in the power circuit configuration that takes place in one period. We have obtained these six configurations in Fig.

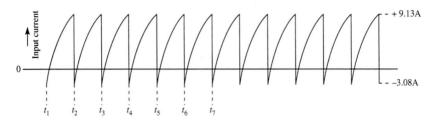

FIGURE 5.45
Input current waveform for three-phase quasi-square-wave inverter.

5.37, each of which is valid for one 60° interval. We can identify the input current in terms of the output currents as follows by referring to Fig. 5.37:

t_1 to t_2	(0–60°)	phase current B reversed
t_2 to t_3	(60°–120°)	phase current A
t_3 to t_4	(120°–180°)	phase current C reversed
t_4 to t_5	(180°–240°)	phase current B
t_5 to t_6	(240°–300°)	phase current A reversed
t_6 to t_7	(300°–360°)	phase current C

Based on the above list, we can pick up the appropriate phase current segment from the waveforms in Fig. 5.34 for each interval, and reverse the sign if it is so indicated for that segment in the list above, to obtain the input current waveform over one complete AC period. The waveform so obtained is drawn in Fig. 5.45.

(b) Inspection of the above waveform shows that the repetitive frequency is six times the inverter frequency. Since the inverter frequency is 100 Hz, the input ripple current frequency is 600 Hz.

(c) The mathematical expressions for each segment of the current for the A phase are listed in (5.79)–(5.84) in the solution of Example 5.10. Furthermore, each of these expressions is valid for the phases B and C, for intervals successively delayed by 120°. On this basis, the correct expression for the input current can be picked up, for each interval, from the list of expressions obtained in Example 5.10. When we do so, we find that the same expression holds for the input current in each 60° interval, and that this expression is (5.80):

$$i = 40 - 43.08e^{-200t}$$

The time interval for which this is valid is one-sixth of the AC period, which, for the inverter frequency of 100 Hz, is $\frac{1}{6} \times \frac{1}{100} = \frac{1}{600}$ s. Therefore the DC component of the input current is obtained as follows, by averaging the above expression over this interval:

$$I_{DC} = 600 \int_0^{1/600} (40 - 43.08e^{-200t})\, dt$$

$$= 3.36 \text{ A} \tag{5.87}$$

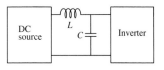

FIGURE 5.46
Use of an input filter.

Use of an input filter. We see from the input current waveforms of Figs 5.43, 5.44 and 5.45 that the input current on the DC side of the inverter has, in addition to the DC component, a large ripple. In each of the above examples, the ripple current amplitude is so large that there are intervals in a ripple period when the actual current is negative. If the DC is obtained by rectification of an AC supply, a negative current is not possible, because the diodes or thyristors, as the case may be, that perform the rectification will not permit reverse current flow. Even if the DC source is such that a reverse current is possible, it may still be not desirable to allow a large ripple, for reasons of increased power loss, or because of the electromagnetic disturbances resulting from high frequency currents of large magnitude and steep wavefronts. These problems can be overcome by the use of a filter circuit between the DC source and the inverter. The filter circuit consists of an inductor L and a capacitor C, connected as shown in Fig. 5.46. It should be noted here that L and C are part of the power circuit and must be capable of handling the typically large values of the inverter current and voltage, without magnetic saturation and voltage breakdown.

If the value of the inductance L is large, the ripple current in the DC source will be negligible. A large value of the capacitor C will make the voltage ripple at the input terminals of the inverter negligible, enabling the inverter to operate in the manner we have described. In the following numerical example we shall assume an ideal filter with large values of L and C.

Illustrative example 5.14. The filter circuit of Fig. 5.46 is used on the input side of the three-phase quasi-square-wave inverter considered in Example 5.13. Assume that L and C values are large. Determine
(*a*) the current supplied from the DC source;
(*b*) the peak-to-peak ripple current in the capacitor.

Solution.
(*a*) For a large value of L, the ripple component of the current in the DC source will be negligible. Therefore the DC source current will be the DC component only of the inverter input current. We determined this in Example 5.13
$I_{DC} = 3.36$ A.
(*b*) The ripple current will still flow in the inverter. But this will be supplied by the capacitor. The input current to the inverter obtained in Fig. 5.45 shows that the fluctuation is between -3.08 and 9.13 A. Therefore the peak-to-peak ripple current is 12.21 A.

5.12 INVERTER OPERATION WITH REVERSE POWER FLOW

In the normal operation of an inverter, the direction of power flow is from the DC input to the AC output. But there is an important practical application in which an inverter, which normally supplies power from the DC side to the AC side, will be called upon, from time to time, to transfer power in the reverse direction. This happens in the important applications area of inverter-fed adjustable speed AC motor drives. Take for example the case of a battery powered electric car, driven by an inverter-fed three-phase AC motor. Three-phase AC motors fed through inverters are more favored at the present time than chopper-fed DC motors for this application. During normal driving, the inverter supplies power to the AC motor. However, during braking, the motor is made to function as a genarator, which is driven by the stored kinetic energy of the vehicle, or by gravity if the braking is done while going down a slope. When this is done, the inverter has to deliver power from the AC side to the DC side, which will serve to recharge the battery. In this section, we intend to show that such a mode of operation of the inverter is quite feasible.

For the steady flow of power in the reverse direction, there has to be a source of power on the AC side, which, in the case of regenerative electric braking of the car, is the AC motor now functioning as a generator. Since our purpose here is only to show that the inverter can steadily transfer power from the AC side to the DC side, we shall formulate a rather idealized situation as shown in Fig. 5.47, in which we have a single-phase inverter that has on its output side an opposing AC voltage source of higher magnitude, in series with an $R-L$ circuit. We shall determine the nature of the current on the AC and the DC sides. The procedure is best illustrated by the following numerical example.

Illustrative example 5.15. In the inverter circuit of Fig. 5.47, V_2 is an AC voltage whose waveform is a square wave of amplitude 300 V. The phase of V_2 is the same as that of the output voltage v_{AB} of the inverter. The inverter frequency

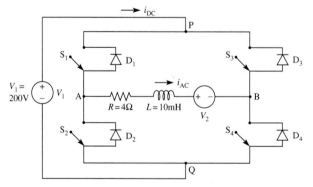

FIGURE 5.47
Inverter operation with reverse power flow.

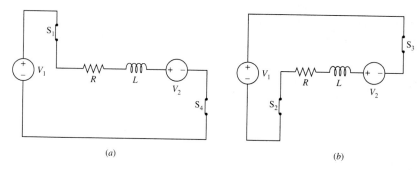

FIGURE 5.48
Power circuit configuration in each half-cycle.

$f = 100\,\text{Hz}$, and the remaining data are given in the figure. Assume repetitive conditions. Determine the following:
(a) the waveform of the current i_{AC} on the AC side of the inverter;
(b) the waveform of the current i_{DC} on the DC side of the inverter;
(c) the magnitude and direction of the DC component of the current on the DC side; therefore state the direction of power flow through the inverter.

Solution. The circuit of configuration when the inverter switching blocks 1 and 4 are ON is shown in Fig. 5.48(a). We shall label the beginning and end of this half-period as t_1 and t_2. Similarly, when switching blocks 2 and 3 are ON, the circuit configuration will be as shown in Fig. 5.48(b). This will correspond to the second half-cycle t_2 to t_3.

The loop equation during t_1 to t_2 will be

$$L\frac{di}{dt} + Ri = V_1 - V_2$$

We shall state the initial condition as

$$i = I_0 \quad \text{at } t = 0$$

The solution of this is

$$i = \frac{V_1 - V_2}{R}(1 - e^{-t/\tau}) + I_0 e^{-t/\tau} \tag{5.88}$$

where $\tau = L/R$.

At t_2, $t = \frac{1}{2}T$ (where $T = 1/f$ is one full period). From waveform symmetry, the current at t_2 must be equal to $-I_0$. Therefore

$$-I_0 = \frac{V_1 - V_2}{R}(1 - e^{-T/2\tau}) + I_0 e^{-T/2\tau} \tag{5.89}$$

From this, we get I_0 as

$$I_0 = -\frac{V_1 - V_2}{R}\frac{1 - e^{-T/2\tau}}{1 + e^{-T/2\tau}} \tag{5.90}$$

During this interval, $V_2 = +300$ V. Numerical substitution of this, and the other data, into (5.90) gives

$$I_0 = 19.04 \text{ A}$$

Substitution of this value of I_0 into (5.88) gives the expression for the current during the half-period t_1 to t_2 as

$$i = -25 + 44.04e^{-400t} \tag{5.91}$$

The above equation expresses both the AC side current i_{AC} and the DC side current i_{DC} during this half-period. This is because the switching blocks through which the current flows, in this half-period are 1 and 4. Therefore both currents have the same reference polarity.

The expression for the current for the second half-period can be obtained either by continuing the analysis in the above manner or by consideration of waveform symmetry. To satisfy waveform symmetry, the expression for the current should be the same as (5.91) with the sign reversed, the new reference zero of time being t_2, the commencement of the second half-period. Therefore the expression for the current during the second half-period t_2 to t_3 is

$$i = 25 - 44.04e^{-400t}, \quad \text{with } t \text{ measured from } t_2 \tag{5.92}$$

Equation (5.92) applies only to the AC side current. The reference polarity for the current on the DC side is opposite to that of the current on the AC side, because the current flow during this half-period is through switching blocks 2 and 3. For this reason, the expression for the current on the DC side is the same as (5.91) during this half-period also.

(a) The waveform of the current on the AC side, i_{AC}, is sketched in Fig. 5.49(a) on the basis of (5.91) and (5.92). The magnitude of the peaks is $I_0 = 19.04$ A.

(b) The waveform of the current on the DC side, i_{DC}, is sketched in Fig. 5.49(b) on the basis of (5.91). The ripple frequency is twice the AC frequency. The

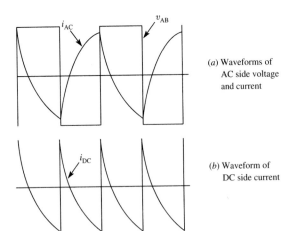

(a) Waveforms of AC side voltage and current

(b) Waveform of DC side current

FIGURE 5.49
Waveforms for inverter operation with reverse power flow.

maximum value, is +19.04 A and the minimum value is −19.04 A. Therefore the peak-to-peak ripple current is 38.08 A. The DC component is calculated below in the answer to (c).

(c) The DC component is obtained by averaging the expression for i_{DC} given by (5.91) over one ripple period. Thus

$$I_{DC} = \frac{1}{0.005} \int_0^{0.005} (-25 + 44.04e^{-400t}) \, dt$$

$$= -5.96 \text{ A}$$

This current has a negative value. Therefore the resultant flow of power is from the AC side into the DC source.

The ability of the inverter to operate with reverse power flow, makes it possible to move between driving and braking in adjustable speed AC motor drives. It is, however possible only if the DC side is capable of absorbing the returned power.

PROBLEMS

5.1 For the half-bridge inverter of Fig. P5.1, the data are as follows: $V_1 = 200$ V, $R = 25 \, \Omega$, $L = 10$ mH and $f = 2$ kHz. Assume repetitive conditions.
(a) Obtain an expression for the current for one half-period of the AC.
(b) Sketch the waveforms of the output voltage and output current, indicating the related magnitudes and time durations.

5.2 A pulse width modulated square wave inverter works from a 120 V DC input. The inverter frequency is 2 kHz. It is operating at a duty cycle of 80%. Determine
(a) the r.m.s. value of the AC output voltage;
(b) the r.m.s. value of the fundamental frequency component of the output AC voltage;
(c) the r.m.s. value of the total harmonic component of the output voltage.
If the voltage is to be reduced to 50% of the result in (a), what should be the new duty cycle?

5.3. The bridge inverter of Fig. P5.3 is operated without PWM. The data are as

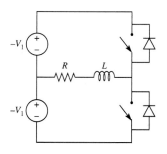

FIGURE P5.1

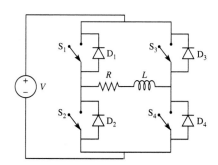

FIGURE P5.3

follows: $V = 200\,\text{V}$, $f = 1\,\text{kHz}$, $R = 25\,\Omega$ and $L = 20\,\text{mH}$. Assume repetitive conditions

(*a*) Obtain an expression for the current in one half-cycle of the AC output.

(*b*) Sketch the waveform of the AC output current and voltage.

(*c*) Determine the delay of the zero-crossing of the current with respect to the zero-crossing of the voltage. Express this in angular measure.

(*d*) State the conduction sequence of all eight power semiconductor switching elements of the inverter.

5.4. The controlled switches in the full-bridge inverter of Fig. P5.4 are labeled S_1, S_2, S_3 and S_4. The load on the AC side is inductive. Give a timing diagram for the four controlled switches on the same lines as Fig. 5.18(*c*), but on the basis of a freewheeling pattern with both freewheelings t_2 to t_3 and t_4 to t_5 on the P side.

5.5. Repeat Problem 5.4 on the basis of both freewheeling on the Q side.

5.6. Repeat Problem 5.4 on the basis of a freewheeling pattern with the t_2 to t_3 freewheeling on the Q side and the t_4 to t_5 freewheeling on the P side.

5.7. In the circuit scheme given in Fig. 5.19, there is no provision for a "dead time" between the turn OFF switching and the turn ON switching of the switches on one leg. Give a modified circuit scheme by adding more functional blocks to Fig. 5.19 so that a programmable dead time can be provided.

5.8. The bridge inverter of Fig. P5.8 operates with PWM. The data are as follows: $V = 120\,\text{V}$, $R = 5\,\Omega$, $L = 10\,\text{mH}$, $f = 50\,\text{Hz}$ and duty cycle 70%.

(*a*) Obtain mathematical expressions for each segment of the current waveform in the AC load.

(*b*) Determine the magnitude of the current at each discontinuity in the waveform and also the instants of the zero-crossings of the current. Sketch the waveforms of the current and the voltage.

(*c*) List the conduction sequence of each power semiconductor switch, giving the conduction periods in angular measure, on the basis of one AC period being equal to 360°.

5.9. Use the waveform of the load current determined in Problem 5.8 and the conduction periods of the devices to obtain the waveform of the current drawn from the DC source. Sketch the waveform of the DC source current.

5.10. A full-bridge single-phase inverter operates without PWM. The DC source voltage is 150 V and the frequency 60 Hz. The AC load consists of a resistance of

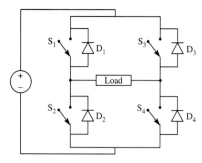

FIGURE P5.4

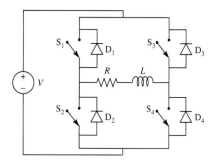

FIGURE P5.8

37.7 Ω in series with an inductance of 0.1 H. Determine the r.m.s. value of the fundamental frequency components of the voltage and of the current. What is the phase angle between the fundamental frequency components of the voltage and current.

5.11. A full-bridge PWM inverter operates with a duty cycle of 80% from a 200 V DC supply. Determine
(*a*) the r.m.s. value of the AC output voltage;
(*b*) the r.m.s. value of the fundamental frequency component of the AC voltage;
(*c*) the r.m.s. total harmonic content in the output voltage.

5.12. A full-bridge PWM inverter operates from a 120 V DC bus with a duty cycle of 80%.
(*a*) Determine the r.m.s. values of the two lowest-order harmonics in the voltage output.
(*b*) Determine the duty cycle that will totally eliminate the third harmonic. What will be the r.m.s. values of the total output voltage and the fundamental frequency voltage at this duty cycle.

5.13. A 60 Hz PWM inverter operates with a duty cycle of $\frac{2}{3}$ from a 300 V DC supply. The AC load consists of a resistance of 30 Ω in series with an inductance of 15 mH. Determine the frequencies and the r.m.s. values of the two lowest-order harmonics in the AC load current.

5.14. In the circuit scheme given in Fig. 5.27, the EPROM used has an 8 bit address. Determine the angular resolution possible in the sine wave output with this number of bits. How can the resolution be further improved?

5.15. The circuit scheme for the generation of the reference sine wave given in Fig. 5.27 does not provide for the negative part of the sine wave. Give a modified circuit using additional circuit blocks for the generation of the negative half-periods of the reference sine wave.

5.16. A 60 Hz full-bridge inverter uses sinusoidal pulse width modulation based on sine–triangle intersection. If the triangular carrier frequency is 18 kHz, what will be the number of pulses per half-period in the output waveform. With 90% modulation index, what will be the width of the longest pulse?

5.17. A simple scheme for the generation of timing pulses for a three-phase inverter without pulse width modulation is given in Fig. P5.17. In this, a clock oscillator

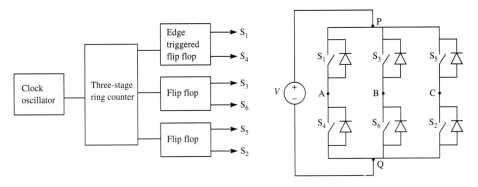

FIGURE P5.17

output is fed to a ring counter to divide the frequency into three and provide three identical channels—one for each phase. The positive edge triggered flip flop is used to give a symmetrical pulse of exactly 50% duty cycle. The two complementary outputs of the flip flop are used to time the switches in each phase leg of the inverter.

(a) In this scheme, determine the frequency of the clock for an inverter frequency of 60 Hz.

(b) Introduce additional circuit blocks in the output of each flip flop to delay the rising edge by a programmable "dead time."

(c) Introduce additional logic elements to implement reversal of the "phase sequence" of the AC output by interchanging the timing pulses to any two legs of the inverter. Such reversals of phase sequence are needed in practice to reverse the direction of a three-phase motor supplied from the inverter. The reversing should be implemented by changing the logic level of an input in the additional logic included for phase sequence reversal.

5.18. A three-phase quasi-square-wave inverter is operating from a 150 V DC source at 60 Hz. The AC load consists of three identical impedances connected in delta, each consisting of a resistance of 6 Ω in series with an inductance of 30 mH.

(a) Determine the r.m.s. line voltage.

(b) Obtain mathematical expressions for each segment of the waveform for the current in one phase and sketch the waveform, indicating relevant magnitudes and intervals.

5.19. For the inverter of Problem 5.18, obtain expressions for each segment of the line current waveform and sketch the line current waveforms for all the three phases, indicating relevant magnitudes and intervals. Identify the conducting periods of each of the 12 power semiconductor devices in the inverter.

5.20. A three-phase quasi-square-wave inverter has a balanced three-phase load on the AC side, which is Y-connected. Each load phase consists of a resistance of 2 Ω in series with an inductance of 10 mH. The inverter frequency is 60 Hz and the DC source voltage 150 V.

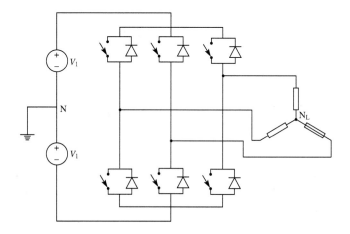

FIGURE P5.22

(*a*) Sketch the waveform of the voltage of each load phase, indicating relevant magnitudes and time durations.

(*b*) Obtain mathematical expressions for the different segments of the waveform of the current in one phase and sketch the current waveform, indicating magnitudes at the boundaries of each segment.

5.21. For the inverter of Problem 5.20, assume that the DC source is ideal.

(*a*) Determine the waveform of the current drawn from the DC side.

(*b*) Determine the frequency of the ripple current on the DC side.

(*c*) Sketch the waveform of the DC side current, indicating relevant magnitudes and time durations.

5.22. The three-phase quasi-square-wave inverter of Fig. P5.22 is supplied from a split DC supply with the midpoint labled N grounded. The three-phase load is balanced. $V_1 = 100$ V and $f = 60$ Hz. Sketch the waveform of the voltage between the load neutral N_L and ground. What is the frequency of this voltage and its amplitude?

CHAPTER

6

POWER
SUPPLY
SYSTEMS

6.1   INTRODUCTION

The power supply unit is an essential circuit block in practically all electronic equipment—computers, instruments or some other signal processing equipment. Such equipment generally works from the AC power mains. The power supply unit is the power interface between the AC mains and the rest of the functional circuits of the equipment. The functional circuits of the equipment usually need power at one or more fixed DC voltages, which have to be maintained within close limits to ensure reliable working of the equipment. The power supply unit has to meet certain terminal specifications, such as the output voltage or voltages, the permissible limits of output voltage variation (regulation) when the load current or AC mains voltage vary, ripple voltage component, automatic current limiting under fault or overload conditions etc. As long as the specified terminal conditions are met, the electrical design of the power supply unit is, by and large, independent of the design of the rest of the equipment. The design and manufacture of power supply systems for electronic equipment has therefore evolved as an independent and competitive industrial technology. Many manufacturers of electronic equipment find it economical to buy power supply units and incorporate them in their equipment rather than design and manufacture their own.

376

The class of power supply systems most widely used in electronic equipment use the switching technique, and are known as switch mode power supplies (SMPS). They are also called switching power supplies or switching regulators. The alternative to an SMPS is a "linear regulator." These are good for low power applications, but are uneconomical and inefficient when the power requirement is high. We shall begin this chapter with a study of linear regulators and highlight their differences from SMPS. We shall then deal with the different configurations of SMPS and their operation.

In this chapter, we shall also present uninterruptible power supply (UPS) systems. A UPS is usually an external interphase between the AC mains and the equipment for which it provides the electric power. A UPS has to be provided with an additional input from a rechargeable battery. When the power mains is healthy, the UPS ensures that the battery is fully charged and that power is provided to the equipment. If there happens to be a mains "blackout," that is, a total interruption of mains power, or a mains "brown-out," that is, a voltage or frequency change outside the limits for the reliable functioning of the equipment, the UPS automatically ensures continued reliable supply of power to the equipment using the battery source. Therefore the equipment will continue to operate, without being sensitive to the power supply disturbance, for as long as there is battery power available. In many situations, all that may be needed is to make sure that the equipment is shut down in the proper sequence, without irreparable loss of data. In such a case, the battery capacity needed will be small.

6.2 LINEAR REGULATORS

Linear regulators are manufactured as integrated circuit chips, which can be programmed to give the required fixed voltage output by the use of a few external components. They are also available to provide the commonly used fixed voltages such as $+12$ V, -12 V, $+15$ V, -15 V, $+5$ V etc. without the need for external programming resistors. Even fixed voltage regulator chips can be programmed to provide regulated voltages higher than their specified output, by using external resistors in the manner that we shall describe in this section. The output current capabilities of IC voltage regulators are usually limited because of internal power dissipation constraints. However, by using them along with external power transistors, in the manner that we shall describe, they can still be used for higher current ratings. IC regulator chips incorporate internal circuitry to automatically limit the output current to a safe value in the event of an accidental short-circuit across their output terminals. They recover and function again normally, under healthy conditions of output.

6.2.1 Working Principle

Figure 6.1(a) shows the basic elements of a linear voltage regulator, available as an integrated circuit module, omitting some of the details that are not

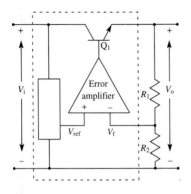

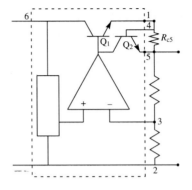

(a) Basic elements of a linear voltage regulator

(b) Automatic current limiting

FIGURE 6.1
Linear voltage regulator.

immediately relevant, for the sake of clarity. The internal functional circuit blocks are shown inside the box of broken lines. The "raw" or "unregulated" DC input to the regulator is labelled V_i in the figure. V_i is usually obtained from the AC mains through a "front end rectifier" circuit, consisting of a transformer to provide electrical isolation and the required voltage level, a rectifier to convert to DC and a filter capacitor to minimize the ripple component in the DC voltage. The front end rectifier circuit is the same whether the regulator is the linear type or the switching type, and will be considered in detail when we take up the study of switching regulators. The regulated DC output is labeled V_o in the figure. The purpose of the regulator is to maintain V_o at a constant value when the input voltage V_i or the output current or both vary. This is achieved by automatically adjusting the voltage drop across the "series pass transistor" labeled Q_1 in the figure. For this, the base drive of Q_1 is provided by the output of an operational amplifier, labeled "Error amplifier." The regulator chip has an internal circuit to generate an accurate reference voltage labeled V_{ref} in the figure. The second input to the error amplifier, which is labeled V_f (the feedback voltage), is a predetermined fraction of the output voltage V_o. Neglecting the input current of the error amplifier,

$$V_f = \frac{R_2}{R_1 + R_2} V_o \tag{6.1}$$

The error amplifier amplifies the error between the reference input and the feedback voltage, and varies the base drive to the series pass transistor, and thereby the voltage drop across it, in such a way that the error voltage is made negligibly small. Therefore, ideally

$$V_o = \frac{R_1 + R_2}{R_2} V_{ref} \tag{6.2}$$

We can therefore choose suitable values for the external programming resistors R_1 and R_2 to achieve the desired output voltage from the regulator.

For the linear regulator to work satisfactorily, there should be a minimum voltage differential across the series pass transistor, which is in effect the differential between the unregulated input voltage and the regulated output voltage. This minimum differential voltage is specified in the data sheet of the regulator chip, and it could be typically 3 V or lower. This specified minimum differential voltage, in effect, specifies the lowest value to which the unregulated input can swing, under the extreme conditions of mains voltage dip or load current, without affecting the satisfactory working of the regulator. In other words, the maximum voltage for which the output can be programmed is lower than the anticipated lowest value of the unregulated voltage that can occur under all conditions of operation, by this specified minimum differential voltage. A large differential voltage will cause excessive power dissipation in the series pass transistor, and so there is also an upper limit for it.

6.2.2 Automatic Current Limiting

Voltage regulators usually have an automatic current limiting feature, which will limit the current to a safe value, in the event of an external short circuit. Figure 6.1(b) shows how this feature can be incorporated in the regulator of Fig. 6.1(a). The regulator has two input terminals for current sensing, which are labeled 4 and 5 in Fig. 6.1(b). The current is sensed by sensing the voltage across the current sense resistor labeled R_{cs}. When the current, and therefore the current sense voltage, exceed a threshold level, the transistor labeled Q_2 starts conducting, thereby bypassing and reducing the base drive current of the series pass transistor Q_1. The current limit is determined by the value chosen for the external programming resistor R_{cs}. The regulated output voltage is taken across the terminals labeled 5 and 2 in Fig. 6.1(b). Since the output voltage sensing is also done across these two terminals, the voltage regulation is not affected by the voltage drop across the current sensing resistor.

6.2.3 Use of an External Series Pass Transistor

A programmable linear IC regulator may have a low current capability, which may be insufficient for a particular power application. It is usually possible to achieve a higher current capability by the additon of an external series pass transistor of high current rating. Figure 6.2 shows a circuit scheme for this purpose, using the regulator of Fig. 6.1.

The external series pass transistor is labeled Q_3 in Fig. 6.2. This transistor receives its base drive current from the internal series pass transistor of the IC module. Q_3 should be chosen to provide the required load current capability. Automatic current limiting is achieved in a similar manner as in Fig. 6.1(b).

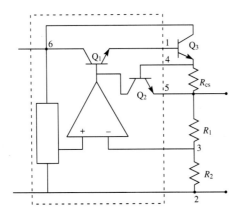

FIGURE 6.2
Use of external series pass transistor.

Illustrative example 6.1. The internal reference voltage of a linear IC regulator chip is 3 V. The current sense threshold is 0.6 V. It is used with an external series pass transistor to operate up to a current limit of 2 A. The minimum differential voltage needed for the external series pass transistor is 3.5 V. The maximum collector power dissipation permissible in the external series pass transistor is 40 W. The regulated output required is 12 V. Determine

(a) suitable values for the output voltage programming resistors;
(b) the value of the current limit resistor;
(c) the maximum and minimum permissible limits of the unregulated DC input, based on the stated limits of the external series pass transistor.

Solution

(a) To choose the values of the output programming resistors, we use (6.2). Substitution of numerical data into this gives

$$12 = \frac{3(R_1 + R_2)}{R_2}$$

From this,

$$R_1 = 3R_2$$

We have several choices possible that will satisfy the above relationship. Maximum power dissipation in the resistors will be one of the constraints. We arbitrarily choose 20 kΩ for R_2. This gives the value for R_1 as 60 kΩ.

(b) The value of the current sensing resistor R_{cs} should be such that the threshold voltage is reached at the limiting value of current. This gives

$$R_{cs} = \frac{0.6}{2} = 0.3 \ \Omega$$

(c) The voltage differential V_D across the external series pass transistor is given by

$$V_D = V_i - (V_o + V_{cs})$$

where V_{cs} is the voltage across the current sensing resistor. Substituing the limiting values into the above equation, the minimum value of V_i is

$$3.5 + 12 + 0.6 = 16.1 \text{ V}$$

The collector power dissipation in the external series pass transistor is given by

$$P = V_D I$$

where I is the output current. Substituting the limiting values into the above relationship, we get the maximum permissible differential voltage across the external series pass transistor as 20 V. Therefore the upper limit to which the input voltage can swing without adversely affecting the operation of the regulator is

$$V_i = 20 + 12 + 0.6 = 32.6 \text{ V}$$

6.2.4 Three-Pin Regulators

Integrated circuit linear voltage regulators are also made for the commonly needed fixed output voltages by incorporating all the programming resistors and the series pass transistor of Fig. 6.2 internally on the same chip. Such regulators are convenient to use, since they have only three external terminals, labeled I, G and O, which are respectively the input, the ground and output pins. Such an IC can be connected in the manner shown in Fig. 6.3(a) to give a regulated output from an unregulated input.

Even a fixed voltage three-pin regulator can be programmed to give a higher regulated voltage than its fixed value, by using it with external programming resistors in the manner shown in Fig. 6.3(b). In choosing the values of R_1 and R_2 in (b), it is necessary to take into account the current through the terminal G, as shown below in Example 6.2. The value of this current is usually specified in the data sheet of the device. By making one of the two programing resistors adjustable, the regulated output voltage can be made adjustable.

6.2.5 Ripple Rejection

The unregulated DC voltage input to a voltage regulator usually has a superimposed ripple voltage content, because this voltage is usually obtained by the rectification of an AC voltage. The regulator chip responds to the instantaneous changes in the input voltage, resulting from the ripple, and acts fast enough to minimize corresponding changes on the output side. In this manner, the regulator chip serves to reduce the ripple on the output side to a

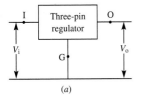

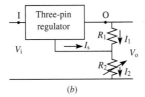

FIGURE 6.3
Three-terminal fixed voltage regulator.

large extent. The ripple rejection capability is usually stated in decibels, in the data sheet of the regulator chip.

Illustrative example 6.2. The fixed voltage of the three-pin regulator of Fig. 6.3 is 5 V. The ground pin current listed in its data sheet is 75 μA. It is to be used to provide a regulated voltage, whose value is adjustable from 5 to 25 V.

(a) If the value of the programming resistor R_1 is to be 40 kΩ, determine the range of values of R_2.

(b) The specified ripple rejection of the regulator is 60 dB. For a peak-to-peak 0.2 V ripple on the input, determine the output peak ripple, when the output is set for 5 V.

Solution

(a) The regulator maintains a constant voltage equal to 5 V across R_1. Therefore in Fig. 6.3(b)

$$I_1 = 5/40 \text{ mA} = 125 \ \mu\text{A}$$

Since the current through the ground pin of the regulator chip is 75 μA in the direction shown in Fig. 6.3(b), this makes the current I_1 through R_2

$$I_2 = 75 + 125 = 200 \ \mu\text{A}$$

The maximum voltage across R_2 occurs at the highest voltage setting of the power supply, and this is equal to $25 - 5 = 20$ V. Therefore

$$R_2 = 20/0.2 = 100 \text{ k}\Omega$$

Therefore, by choosing a 100 kΩ adjustable resistor for R_2, the power supply can be made adjustable from 5 to 25 V, provided the range of input voltage will conform to the required minimum differential voltage across the series pass transistor and the maximum permissible power dissipation in it.

(b) The ripple rejection in dB is

$$20 \log \frac{\text{input peak-to-peak ripple voltage}}{\text{output peak-to-peak ripple voltage}} = 60$$

Therefore the ripple voltage ratio is 10^3. This give a peak-to-peak output ripple of

$$0.2/1000 = 0.2 \text{ mV}$$

6.2.6 Transient Response—Use of a Capacitor on the Output Side; Reverse Diode for Protection

When a relatively large load current is suddenly turned ON on the output side, there will be a momentary "dip" in the output voltage because of the finite time it takes for the control loop of the regulator to respond and correspondingly readjust the voltage across the series pass transistor. Similarly, there is a transient overshoot of output voltage when the load current, or a substantial part of it, is suddenly switched OFF, for the same reason. Power supplies often have to meet stringent specifications regarding the transient response. These

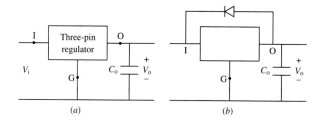

FIGURE 6.4
Capacitor on output side for improved transient performance.

(a)　　　　(b)

requirements are usually specified in terms of the maximum permissible overshoot and the maximum time within which the unit has to settle down to the normal value. One way of improving the transient performance of a regulated power supply is to provide a capacitor on the output side, as shown labeled C_o in Fig. 6.4(a). This capacitor helps to hold the output voltage, while the closed-loop control circuit of the regulator readjusts to the sudden load changes.

The use of a large value of C_o can be the cause of damage to the regulator, in case an accidental short circuit or large decrease in voltage occurs on the input side of the regulator. In such a situation, the capacitor C_o will maintain the output voltage level for a short time, whereas the input side voltage is zero or near zero. As a result, a large reverse voltage will appear across the series pass transistor of the regulator. The series pass transistor is a bipolar device whose reverse voltage withstand capability is very little. The method used to overcome this problem is to use a diode in reverse across the output and input terminals of the regulator, as shown in Fig. 6.4(b).

6.2.7　Power Loss and Efficiency

Linear regulators give good performance under both steady and transient conditions. Their main disadvantage is the large voltage drop in the series element. This element also has to carry the full load current. Therefore there is a significant dissipation of power in the regulator, resulting in poor efficiency and the need to handle problems of heat dissipation and temperature rise. The following example highlights this disadvantage.

Illustrative example 6.3. A 5 V three-pin linear regulator has a rated output of 10 A. The upper limit of variation of the raw input voltage is 12 V. Determine
(a) the maximum power dissipation in the regulator;
(b) the conversion efficiency under this condition.

Solution. At the highest value of the input voltage, the differential voltage across the regulator is $12 - 5 = 7$ V.
(a) For full load power output under this condition, the internal power loss is
$7 \times 10 = 70$ W.
(b) The output power under this condition is $5 \times 10 = 50$ W. Therefore the conversion efficiency is $50/(50 + 70) = 41.7\%$.

6.2.8 Linear Regulator versus Switching Regulator

The high internal power dissipation and poor efficiency may be acceptable trade-offs for the good performance of the linear regulator, when the power requirements are small. But at higher power levels, these limitations become almost prohibitive. The choice then always falls on the switching type of regulator.

A switch mode power supply uses one or more power semiconductor switching elements, such as a power MOSFET, to repetitively switch ON and OFF at a high switching frequency, and achieve output voltage regulation by pulse width modulation. When an ideal switch is used as the controlling element in an electric circuit, there is no power dissipation in the switch, either in its ON state or in its OFF state. Therefore the switching mode of control is, in principle, an efficient means of control. However, practical power semiconductor devices have finite internal power dissipation when used as switches. But, even so, the power dissipation when the device is used as a switch is much smaller than when it is used as a linear series regulating element. Therefore switch mode power supplies are always the choice for high power requirements. We now begin our study of this important class of power supplies.

6.3 FUNCTIONAL CIRCUIT BLOCKS OF AN "OFF LINE" SWITCHER

"OFF LINE" switcher is technical slang for a switch mode power supply (SMPS) that works off a standard AC power supply line which is usually either the 115 V, 60 Hz system or the 230 V, 50 Hz system, depending on the country. The functional circuit blocks that constitute a typical power supply unit of this kind are shown in Fig. 6.5. In this figure, we have shown each major block boxed within broken lines and labeled alphabetically. The constituents of each major block are shown inside it, boxed in firm lines and labeled numerically.

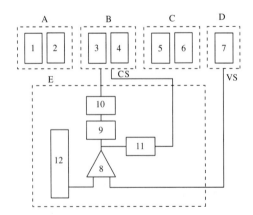

FIGURE 6.5
Functional circuit blocks of "OFF LINE" switcher.

In this section, we shall give a brief description of the functions of each block. The controller block is usually in the form of an integrated circuit module. There are a number of different types of ICs presently available for use as control elements in SMPS. We shall explain the main functions to be performed by the controller. We shall also explain the differences between two alternative types of closed-loop control strategies generally used by means of the controller chip, namely the voltage mode of control and the current mode of control.

6.3.1 Circuit Block "A"—The Front End
Rectifier and Filter

The purpose of this block is to provide the raw DC voltage from the input AC. The main circuit block "A" consists of the rectifier labeled 1 and the filter labeled 2. The rectifier is usually a diode bridge, available as a single four-terminal module. Normally, the rectification is direct, without the use of a transformer. The raw DC voltage value will depend on which of the two systems (115 or 230 V) is the input AC. There is, however, a simple and often used means by which the same unit can be used with 115 or 230 V input by the simple change of a link connection. We shall describe this later, when we treat this circuit block in more detail, in Section 6.4.

The block labeled 2 is a filter used to bring down the ripple content in the rectified voltage to a level acceptable to the next circuit block (B). Block 2 is actually a bank of electrolytic capacitors. This bank of capacitors will also retain the voltage for a short time in the event of a momentary interruption in the AC input. The capacitor value can be suitably chosen to make the output of the SMPS insensitive to power supply interruptions of several milliseconds.

6.3.2 Block "B"—The Switching Converter

This block converts the unregulated DC into high frequency pulse width modulated AC of the required voltage level, by repetitive switching. Switching frequencies may be as high as several hundred kilohertz. The switching is achieved by the block labeled 3, which is a configuration of power semiconductor switches. The most popular switching device at the present time for SMPS is the power MOSFET, because of its very high switching frequency capability. Bipolar power transistors and IGBTs are alternative devices, which have higher power handling capabilities, that can be used for switching, but their frequency capabilities are lower than those possible with power MOSFETs.

Circuit block 4 is the high frequency transformer. This serves to provide the required voltage level on the output side. It also serves to provide electrical isolation of the regulated output from the input AC bus. The core material of this transformer is always a suitable grade of ferrite, because of the high frequency of operation.

There are several different switching circuit topologies suitable for the switching converter block of an SMPS. We shall make a detailed study of these converter topologies later in this chapter.

6.3.3   Circuit Block "C"—Output Rectifier and Filter

The input to this circuit block is the high frequency pulse width modulated AC from the secondary of the ferrite power transformer of the previous block (4). Circuit block 5 is the rectifier that converts this AC to DC. The frequency capability of these rectifying diodes have to be very high. The diodes used are very often the Schottky barrier type. Block 6 is the output filter. This usually consists of a high frequency ferrite core inductor and a bank of electrolytic capacitors.

6.3.4   Voltage and Current Sensing

Since it is the output voltage that has to be automatically regulated, it is necessary for the controller to sense the output voltage. The sensing block is shown labeled D in Fig. 6.5. In this block, the voltage sensing circuit is shown labeled 7. For the practical operation of the SMPS, it is also necessary to sense the current. However, the current sensing circuit is not shown in block D for the following reason.

The need for current sensing in a SMPS arises for one or both of the following reasons. The first is to provide a current limiting feature that will automatically limit the maximum output current to a predetermined value, in the event of overloading or short circuit, on the output side. For implementing the current limit feature, it is necessary for the controller to sense the current. In many designs of SMPS, the current limiting is implemented by "pulse-by-pulse current limiting." This means that the maximum current that can occur is limited in every switching cycle, thereby automatically limiting the maximum current that can occur in the output also. From the point of view of the switching elements, this is obviously a safer and better way to implement current limiting.

The second reason for sensing the current arises, when the type of closed-loop control used for the SMPS is "current mode control." The differences between current mode and voltage mode types of control will be explained presently by reference to Fig. 6.6. But, at this point, we may state that the current mode type of control also calls for the sensing of the instantaneous current in every switching cycle.

Because of the above reasons, it is now common practice to sense the instantaneous current in the switching converter itself, in the switching block labeled "B" in Fig. 6.5, rather than sense the output current. A convenient way of doing this is to insert a low value resistor in series on the primary side of the high frequency transformer, and sense the voltage drop across it. In Fig. 6.5, the current sense line is labeled CS and the voltage sense line VS.

6.3.5 Circuit Block "E"—The Closed-Loop Controller

The primary function of the controller block labeled E in Fig. 6.5 is to sense the output voltage, compare it with an accurate internal reference, amplify any error between the two, and implement pulse width modulation of the switching elements in such a way as to negate the error. Controllers for SMPS are now available from several manufacturers, as integrated circuit modules. In fact there are several types of ICs available, giving the designer a choice of control techniques. Block "E" in Fig. 6.5 shows two basic functions of such an IC, which are output voltage regulation and pulse-by-pulse current limiting. These ICs also provide additional useful features such as over-voltage and under-voltage lock-out, facility for programming the switching frequency, etc.

The controller chip requires power supply for its operation. In some designs this voltage is supplied from a separate small power supply unit. Alternatively, there are ICs available that can start the operation of the SMPS using control power from the unregulated DC of the circuit block "A" through a resistor network to reduce the voltage to the low level needed to power the chip. After the switching converter starts operating, the power supply to the controller will automatically switch over to a DC from the switching converter, by means of a diode changeover circuit.

The regulator chip has an internal circuit (12) that generates an accurate and stable reference voltage. The circuit module 8 is an operational amplifier that compares the sensed output voltage against this internal reference voltage and amplifies any error. The output of this error amplifier adjusts the timing pulses generated in the timing circuit labeled 9 in the figure. The timing instants provided by the circuit module 9 determine the switching instants of the switching elements in the converter (circuit block 3). In this manner, the error amplifier implements automatic regulation of the output voltage by pulse width modulation.

The timing pulse output from the circuit module 9 usually does not have the power capability to drive the power switches. The driver block labeled 10, serves as an interphase between the timing circuit and the power switching devices to provide the required drive power. When the switching elements are large, as is the case when the power rating of the SMPS is large, the internal driver circuit (labeled 10 in Fig. 6.5) within the controller chip may be insufficient to provide the required power. It may then be necessary to have an external driver circuit to further boost the drive power.

The circuit module labeled 11 implements automatic pulse-by-pulse current limiting. The required limiting value of the current has to be set externally, as one of the inputs to the block 11. The other input is the sensed value of the instantaneous current. The circuit block 11 compares the two. As long as the sensed value of the current is below the set limiting current value, the circuit block 11 is inactive and does not interfere with the switching. But if the sensed value of the instantaneous current becomes equal to the set limit, the output from 11 will override the output of the voltage error amplifier, and initiate the switching of the power switching element in such a way that no further increase of current is possible in that switching cycle.

6.3.6 Current Mode Control

The control scheme illustrated by Fig. 6.5 is the voltage mode of control. There is basically only one control loop which compares the output voltage against a reference and automatically adjusts the switching to negate the voltage error. The sensing of the current is done only for the purpose of limiting the current if a situation should arise, where the current exceeds a preset limit. The current sense circuit therefore essentially performs a protection function in the scheme of Fig. 6.5.

The principle underlying the current mode type of control is illustrated by Fig. 6.6. In this figure, the voltage error amplifier is labeled VEA. The inputs to the voltage error amplifier are the same as in the case of Fig. 6.5. But the output of the voltage error amplifier does not directly operate on the timing circuit. Instead, it provides the reference input to a second controller. This second controller is labeled CC in Fig. 6.6, to signify "current controller". The second input to the current controller is the sensed instantaneous current from the power switching circuit, provided by the current sense line labeled CS. The current controller compares the sensed instantaneous current, against the current reference provided to it by the voltage error amplifier, and varies the switching instants in such a way as to correct the current error. The timing circuit block 9 and the driver block 10 perform the same functions as in Fig. 6.5. In this way, the current controller makes the current in each switching

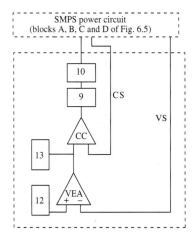

FIGURE 6.6
Current mode control.

period equal to the value set by the voltage error amplifier. Pulse-by-pulse current limiting feature can also be incorporated in the scheme of Fig. 6.6. This is done by the clamping circuit labeled 13, which sets the current limit by preventing the output terminal of VEA from exceeding this set value of the current limit. In this way, the maximum current reference provided to the current controller will be the preset current limit value.

Current mode control has advantages in some designs of SMPS. One of the benefits is that it allows satisfactory parallel operation of two or more identical regulator circuits. Current mode control makes the proper sharing of current between several identical units working in parallel automatic. Therefore modular design becomes easily possible, by means of which the required number of units can be assembled in parallel to achieve high output current ratings. Several types of control ICs for SMPS are now available for easy implementation of current mode control.

6.4   THE FRONT END RECTIFIER

We now take up the study of the power circuit blocks of a switch mode power supply. We shall first consider the front end rectifier, which is a circuit block common to all types of "OFF LINE" power supplies. The front end rectifier converts the input AC into DC. It also has a filter element to limit the ripple component in the unregulated DC to the level necessary for the satisfactory operation of the next circuit block, which is the switching converter.

6.4.1   Bridge Rectifier and Filter Capacitor

Figure 6.7(a) shows the rectifier and filter. A diode bridge rectifier module is normally used for rectification. This is followed by a bank of electrolytic

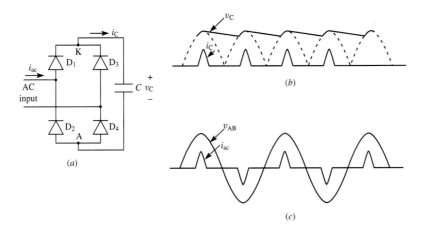

FIGURE 6.7
Front end rectifier and filter.

capacitors for filtering. In each AC half-cycle the capacitor gets charged to the peak magnitude of the AC voltage. When the instantaneous value of the AC voltage magnitude starts falling, the capacitor still retains the charge, and its voltage stays near the peak. Therefore the bridge diodes get reverse-biased and turn OFF. The load current, from then on, is supplied by the capacitor. The diodes remain OFF till the rising magnitude of the AC input voltage during the next half-cycle begins to exceed the capacitor voltage. Therefore the voltage across the capacitor, which is the same as the output voltage of the front end rectifier, will have the waveform shown in Fig. 6.7(b). The capacitor charging current waveform is also shown in the same figure. This will be pulses of short duration near the peaks of the AC input voltage. The waveform of the AC line current is shown in Fig. 6.7(c). This again will be in the form of pulses, whose path will alternate between D_1–D_4 and D_3–D_2. Reference to Fig. 6.7(b) shows that the output voltage has a ripple component.

The peak-to-peak ripple can be kept within acceptable limits by choosing a large enough value for the filter capacitor. A large value of C also gives the power supply unit the ability to ride over short interruptions of AC mains power. This is illustrated by the following example.

Illustrative example 6.4. The AC input to the front end rectifier of Fig. 6.7 is 120 V at 60 Hz. Assume that the diodes are ideal.

(a) What is the value of the raw DC input to the regulator when the current drawn by it is negligible?

(b) The capacitor C has a value of $10\,000\,\mu$F. The DC current drawn by the regulator is 2 A. The switching converter will work satisfactorily down to 20 V below the value found in (a). Determine the maximum duration of a power supply interruption which the unit can ride over. How many cycles of the input AC will this be?

Solution

(a) The peak magnitude of the unregulated DC voltage will be equal to the peak magnitude of the input AC voltage, neglecting the voltage drop across the bridge diodes. This will be

$$V_{DC} = \sqrt{2} \times 120 = 169.7 \text{ V}$$

(b) Let t denote the maximum permissible duration of a power supply failure in seconds. The loss of charge from the filter capacitor during t will be $2t$ C. The corresponding drop in the raw DC voltage fed to the switching converter will

$$V = 2t/C \text{ V},$$

Equating this to the specified maximum permissible voltage drop of 20 V,

$$t = 20C/2 = 0.1 \text{ s, or } 100 \text{ ms}$$

For the 60 Hz AC frequency, this will be equal to 6 cycles.

6.4.2 Inrush Current Limiting

The large values of the filter capacitor needed to limit the ripple and to provide
the ability to ride over short interruptions of the power mains is not a problem,
because electrolytic capacitors, which are available in compact sizes, can be
used. Their polarized nature (unidirectional voltage requirement) does not
pose a problem, because the resultant voltage is undirectional (DC). But the
large value of C required creates a problem when the unit is initially switched
ON. Initially, the capacitor is uncharged. A large uncharged capacitor is initially
like a short-circuit across the supply, when the supply voltage is switched ON,
until the capacitor builds up sufficient voltage. This results in a large initial
"inrush current," which can disturb the supply bus voltage, blow the line fuses,
or damage the bridge diodes. Therefore, for SMPS units of large rating, some
form of inrush current limiting is invariably used. Two techniques for this are
described below.

(*a*) **Use of a negative temperature coefficient (NTC) resistor.** An NTC
resistor is a nonlinear resistor, whose ohmic value decreases rapidly from the
value at room temperature to a low value when its temperature rises, as will
happen when it carries an electric current of sufficient magnitude. Such a
resistor can be used in the manner shown in Fig. 6.8(*a*) for the inrush current
limiting of a SMPS. When a NTC resistor is used for inrush current limiting, it
is not advisable to turn ON the equipment immediately after it has been turned
OFF, because the resistor may need time to cool and regain its higher ohmic
value, and thereby its current limiting capability.

(*b*) **Use of a temporary current limiting series resistor.** Figure 6.8(*b*) shows
an alternative scheme in which a resistor (labeled R in the figure) is inserted in

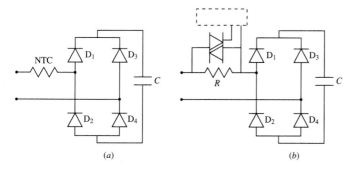

(*a*) (*b*)

FIGURE 6.8
Inrush current limiting.

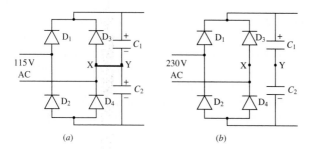

FIGURE 6.9
Dual voltage capability.

series initially to limit the inrush current. After the converter has begun functioning, this resistor is bypassed by switching ON a parallel connected triac. The triac is maintained in the ON state as long as the converter is operating. For this purpose, the trigger input to the gate of the triac may be provided from an auxiliary winding of the converter transformer.

6.4.3 Dual Voltage Operating Capability

A practice often followed by manufacturers of power supplies is to design the front end rectifier in such a way that the equipment is usable from either of the two standard voltage systems, which are (1) the 50 Hz system of nominal voltage 230 V or near 230 V and (2) the 60 Hz system of nominal voltage of half this value. The selection can be done without the use of a transformer, just by changing one link connection. The way in which this facility is provided, is illustrated by Figs. 6.9(a) and (b).

The filter capacitor now consists of two capacitors labeled C_1 and C_2, of equal value. For 115 V operation, a link connection is made between an AC input line ("X" in Fig. 6.9) and the junction of C_1 and C_2 ("Y" in Fig. 6.9). During the positive half-cycle of the input AC, the capacitor C_2 does not receive any charge, but the capacitor C_1 receives charge through diode D_1. As a result, C_1 gets charged to the peak value of the AC voltage, which is $\sqrt{2} \times 115 = 162.6$ V. During the negative half-cycle of the input AC voltage, the capacitor C_1 does not receive any charge, but C_2 gets charged to the peak value of the AC input voltage through the diode D_2. This therefore is a voltage doubler circuit, which provides a total output voltage of $2 \times \sqrt{2} \times 115 = 325.2$ V, that is, twice the peak value of the input AC voltage.

For operating from a 230 V AC input, the link connection between X and Y is removed, and the circuit will be as shown in Fig. 6.9(b). In this case, the two equal series-connected capacitors C_1 and C_2 are equivalent to a single capacitor of half the value, and the four diodes function as a bridge rectifier, giving a DC output equal to $\sqrt{2} \times 230 = 325.2$ V. It should be noted here that if the SMPS is designed for dual voltage operation at nominal voltages of 115 and 230 V, the switching converter should be designed for a nominal raw voltage of $\sqrt{2} \times 230$ V.

6.5 MINIMIZATION OF INPUT CURRENT HARMONICS

Reference to Fig. 6.7(c) shows that the current flow in the AC input lines is in the form of relatively narrow pulses. This is far from the ideal sinusoidal wave shape. The fundamental component only of this distorted current wave contributes to useful power flow into the equipment. All the harmonic components only lower the input power factor by increasing the r.m.s. value of the total line current. The power factor, by definition, is the ratio of real power to the product of the r.m.s. values of voltage and current. These harmonic currents also have another serious and undesirable effect on the power supply bus voltage. The harmonic currents have to flow through the feeder lines that bring power to the local bus. This will cause voltage drops at harmonic frequencies in the impedances of lines and the AC sources from which the local bus receives power. Therefore the local AC bus voltage gets distorted. If there are only a few pieces of electronic equipment that receive power from an AC bus of large capacity, the degree of voltage distortion that they can create in the AC bus, will not be significant. Therefore this problem has not received the attention it deserves, until very recently. Recent years have seen a proliferation of the use of computers and other electronic equipment that work from the AC mains. The pollution they cause to the power supply environment, because of the harmonic currents created by them in the system, has now become a cause of serious concern, primarily because of the adverse effect on the reliable operation of all equipment that receives power from the same power supply bus. For this reason, studies relating to "Power quality" are now gaining importance. These studies involve systematic assessments of the adverse effects on the AC supply systems of harmonics generated by the power supply units of electronic equipment, and ways of preserving the quality of the power supply environment. The International Electro-technical Commission has prepared standards (IEC Document No. 555-2), that set quantitative limits to the harmonics that an equipment should be permitted to spill into the power supply bus. The power supply unit is the interphase between an electronic equipment and power system bus. To conform to IEC 555-2 and similar standards, it becomes necessary to add additional circuitry to the front end rectifier, to limit the harmonics introduced into the power bus.

The ideal waveform of the input current is a sinusoidal one, in phase with the sinusoidal voltage waveform. If this could be achieved, the resulting value of the input "power factor" will have the highest possible value, which is unity. Minimization of harmonics helps towards the achievement of a high input power factor. There are two practical approaches to achieve this, which are listed below.

(*a*) The use of an input filter. A passive filter circuit is inserted between the AC input terminals and the front end rectifier. The purpose of such a filter is not to prevent the creation of current harmonics by the front end rectifier, but to confine the path of the harmonics so that the amount spilling over into the supply lines is minimal.

The main disadvantage of this approach is that, large inductance and capacitor values will be needed for the filter circuit, because of the relatively low fundamental frequency, which is either 60 or 50 Hz. The filter elements also have to handle the relatively large currents and voltages needed. Therefore the filter will add significantly to the dimensions and weight of the equipment.

(*b*) High power factor preregulator. In this technique, an additional circuit block, called a "preregulator," is introduced immediately after the diode bridge and before the capacitor as shown in Fig. 6.10(*b*). The preregulator is actually a DC-to-DC voltage step up chopper converter, which is operated in

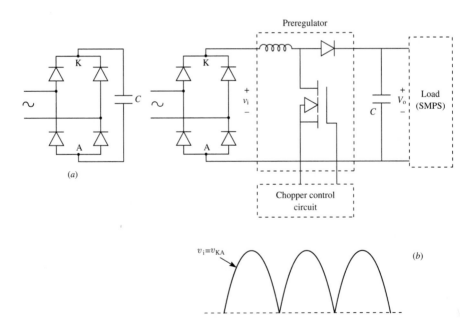

FIGURE 6.10
High power factor preregulator.

such a way that the current drawn by it through the diode bridge has the required waveform that, ideally, will result in a power factor of unity. This technique employs an active switching circuit, in contrast to the use of passive filter elements. We shall present below the basic features of this relatively new technique for minimizing harmonics and achieving a high input power factor.

6.5.1 High Power Factor Preregulator—Power Circuit and Working Principle

The working principle of the high power factor preregulator can be better understood if we first examine more closely the reason why the current flow is in pulses in the circuit without the preregulator.

The circuit without the preregulator is recalled for ready reference in Fig. 6.10(*a*). In this circuit, the voltage across the capacitor *C* is always very close to the peak value of the AC input voltage. The diode pair, through which charge flows into the capacitor in a particular half-cycle of the AC voltage waveform, remains reverse-biased during the major part of the half-cycle. They are forward-biased only during a very short interval when the magnitude of the AC input voltage exceeds the capacitor voltage. In other words, the bridge has to supply current into a circuit, which will accept current only when the instantaneous magnitudes of a rectified sine wave are close to the peak magnitude.

The preregulator is, in fact, a voltage step up chopper. The power circuit configuration of the preregulator is shown inside the box of broken lines in Fig. 6.10(*b*). This is the same as that of the voltage step up chopper studied in Chapter 4. We have shown there that the DC voltage step up ratio of the chopper is given by $V_o/V_i = 1/(1 - D)$, assuming ideal elements and continuous current flow in the inductor *L*. Therefore, for a fixed value V_o of the output voltage, the input voltage V_i will be given by

$$V_i = (1 - D)V_o \tag{6.3}$$

In a practical preregulator, the chopper is operated at a high switching frequency, typically around 100 kHz. The duty cycle *D* of the chopper is continuously varied during each half-cycle of the low frequency AC input (which in practice is around 50 or 60 Hz) in such a way that the instantaneous value of V_i is just below the rectified value of the AC input. The chopper is operated in this manner, with a chosen fixed value of V_o that is higher than the peak value of the rectified AC voltage. The waveform of the rectified AC voltage consists of positive halves of sine waves, sketched as V_{KA} in Fig. 6.10(*b*). This is the input waveform to the chopper, labeled V_i. The chopper steps up this voltage to the high fixed value V_o at its output. Even though the input voltage V_i is continuously changing from zero to the peak and back again to zero in every half-cycle, the chopper maintains the output at the fixed high value V_o by automatic variation of the duty cycle, and thereby the step up voltage ratio. Thereby the diode bridge is enabled to feed current into the preregulator during the entire duration of each half-cycle.

We have now found a way of making the current flow continuously from the diode bridge, and our next objective is to ensure that the waveform of this current is also a rectified half sine wave, just like that of the voltage. Only by ensuring this can we eliminate the current harmonics and achieve unit power factor.

Illustrative example 6.5. A switch mode power supply is designed with a high power factor preregulator. The output voltage of the preregulator is 400 V. Determine the range of variation of the chopper duty cycle, if the unit is to work satisfactorily from
(a) an AC bus of nominal voltage 120 V and range of variation from 90 to 130 V;
(b) an AC bus of nominal 230 V and range of variation from 185 to 265 V;
(c) either of the above AC systems.

Solution. The relationship between the input voltage V_i and the output voltage V_o of the chopper is given by (6.3) (derived in Chapter 4). This gives the duty cycle as,

$$D = 1 - \frac{V_i}{V_o}$$

(a) For an AC bus voltage at the nominal value of 120 r.m.s., the instantaneous value of the rectified voltage sine wave, which is the input voltage to the chopper, will vary continuously between $\sqrt{2} \times 120$ V and zero in every half-cycle of the AC. On substituting these numbers into the above relationship, we get, for $V_o = 400$ V, the range of variation of the duty cycle in every half-cycle as

$$57.6\% < D < 100\%.$$

When the AC input voltage has its lowest value of 90 V, the same procedure gives

$$68.2\% < D < 100\%$$

When the AC input has its highest value of 130 V,

$$54.0\% < D < 100\%$$

From the above results, we conclude that the range of variation of the duty cycle when the voltage of the AC bus fluctuates within the low and high limits will be, in a half-period of the AC,

$$54.0\% < D < 100\%$$

(b) If the unit is to be designed for working from an the AC bus voltage of 230 V, which can fluctuate between 185 and 265 V, we follow the same procedure as in (*a*). This gives the limits of variation of the duty cycle of the chopper, within a half-cycle of the AC, as

$$6.3\% < D < 100\%$$

(*c*) The above results show that the preregulator can be designed to work from both voltage systems, and can handle the limits of voltage fluctuation specified for both systems. The range of variation of the chopper duty cycle will then be, in a half-cycle of the AC

$$6.3\% < D < 100\%$$

6.5.2 The Switching Control Circuit of the Preregulator

The main objective to be achieved by the control circuit of the preregulator is to operate the chopper in such a way that the output voltage has a fixed value V_o and that the waveform of the input current is a rectified half sine wave similar to that of the input voltage V_i. Control circuit modules that achieve these functions are now available as IC chips. An example is the IC type UC3854 from Unitrode Integrated Circuits Corporation. Although the different IC's have differences in their features and circuitry, they all achieve the main objectives stated above. To be specific, we shall base our description on one of these ICs, and we choose the UC3854. The underlying principle of the technique is to control the instantaneous value of the input current to the chopper by continuously adjusting the duty cycle automatically by closed-loop control in such a way that the input current waveform conforms to a reference waveform. To achieve this objective by closed-loop control, it is necessary to take the following steps.

1. Using input signals from the circuit, the required reference waveform of the current should be formulated. In other words, the required instantaneous value of the current should be determined.
2. The actual value of the current at each instant should be sensed. This should be compared against the required value. Any error between the required instantaneous current and the actual sensed value of the current should be made minimal. This is done by causing the current error signal to automatically adjust the duty cycle of the pulse width modulated chopper switch.

The second of the above two stated functions is nothing new. It is a feature available in all ICs for pulse width modulation of SMPS, which provide for current mode control and pulse-by-pulse current limiting. Such IC's are

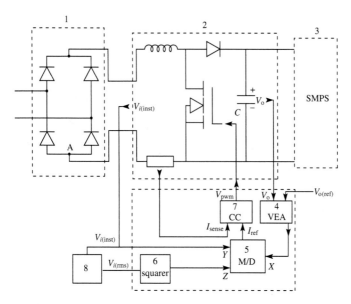

FIGURE 6.11
High power factor preregulator.

now produced by several manufacturers. Therefore, in the following description, our main focus will be on the manner of generating the reference waveform of the input current to the chopper.

The major functional circuit blocks of the IC controller chip are shown in the simplified circuit of Fig. 6.11. The power circuit blocks, labeled 1, 2 and 3, are respectively the diode bridge, the chopper and the load on the output side, which is the switch mode power supply. The main functional circuit blocks of the IC chip are shown, labeled 4, 5, 6 and 7, inside the box of broken lines. These are respectively the voltage error amplifier (VEA), the multiplier/divider (M/D), the squarer (SQ) and the current controller (CC).

The inputs to the IC chip are the following. These inputs will be processed by the IC in the manner to be described subsequently.

1. The output voltage V_o across the capacitor C. Actually, a scaled-down value will be used, by means of a resistor voltage divider that is not shown in Fig. 6.11.

2. The instantaneous value of the input voltage to the chopper, labeled $V_{i(inst)}$. This is the same as the rectified output from the diode bridge, and will have the waveform of a rectified sine wave, with a peak magnitude equal to the peak value of the AC input voltage. A scaled-down value will actually be used, employing a resistor voltage divider.

3. The r.m.s. value of the AC input voltage. This value is the same as the r.m.s. value of the rectified output V_i of the diode bridge. Since the r.m.s. value, the average value and the peak value of the rectified sine wave are all related through constant numerical factors, and the input has to be scaled in any case, it is convenient to utilize a diode circuit to rectify a fraction of this voltage and use a filter circuit to obtain either the peak or the average value from V_i. This circuit block is shown labeled 8 in the figure. What this circuit block provides at the IC terminal is a voltage proportional to the r.m.s. value of the AC input voltage to the diode bridge. This circuit block is external to the IC. We have labeled the output of this circuit block $V_{i(r.m.s.)}$.

4. The actual value of the input current to the chopper. This may be obtained by sensing the voltage drop across a noninductive resistor of low value. The value of the current sensing resistor will determine the scaling factor of the sensed current. The current sense input is labeled I_{sense} in the figure.

The main output from the IC is the switching control signal that turns ON and turns OFF the chopper switch. The chopper switch is invariably a power MOSFET because of the very high switching frequency. This output is labeled V_{pwm}. It may sometimes be necessary to introduce another "driver stage" between V_{pwm} and the gate of the power MOSFET, to ensure fast switching, if the gate capacitance is large, calling for a high transient charging current capability beyond what is directly possible from V_{pwm}.

Besides the above outputs and inputs, there are other terminal connections for the IC, which are not very relevant to our present description. These include the timing capacitor and timing resistor connections, which determine the switching frequency of the chopper, the power supply connections for the IC, and other terminal connections for supervisory and protection functions such as shutting down and peak current limiting.

Generation of the reference waveform of the input current. The reference waveform of the input current to the chopper should satisfy the following two conditions.

1. Its value, averaged over a period of time, should match the power requirement of the load.

2. Its instantaneous value should correspond to a rectified sine wave in phase with the waveform of V_i.

If the load current supplied by the preregulator is zero, the chopper will ideally need zero input current also. The output capacitor C will then remain charged at a fixed voltage, which we shall label as $V_{o(ref)}$. If now the load draws power, the current drawn by it will be practically proportional to the power consumed by it, because the voltage has practically a fixed value. As a result, the capacitor will lose charge and an error voltage will begin to appear between $V_{o(ref)}$ and the actual value of V_o. This error voltage is a measure of the current required for the load to meet its power needs. Referring now to Fig. 6.11, circuit block 4, which is the voltage error amplifier, has, on its input side, scaled-down values of $V_{o(ref)}$ as defined above and the actual sensed value of V_o. On the basis of what we stated above, the output of the error amplifier, which is the amplified error voltage $V_{o(ref)} - V_o$, may be treated as a measure of the output current required from the chopper. On the AC input side, the power is determined by the r.m.s. value of the input current if the AC input voltage is constant. Therefore the output of block 4 will be a measure of the r.m.s. AC current required, assuming that the AC input voltage has a constant value. Since the r.m.s. and peak currents are related by the constant numerical factor $\sqrt{2}$ for a sine wave, we can also treat the output of the voltage error amplifier as a measure of the peak AC current demanded. Accordingly, we will treat the output of block 4, which we have labeled X in Fig. 6.11, as the demanded peak current from the AC input.

To get the *instantaneous* value of the demanded current, X has to be multiplied by $\sin \theta$. Circuit block 5 does the multiplication. The $\sin \theta$ input to the multiplier is the input labeled V_i, which is a rectified half sine wave. This input to block 5 is labeled Y in the figure.

If the input AC voltage is constant, the product XY should be sufficient to provide the instantaneous current reference. However, the input AC voltage is liable to vary. We must have a means of modifying the product XY for variations in the input AC voltage. When the current and voltage waveforms are both sine waves that are in phase, what the AC source sees is a resistor. In such a resistor, the power consumed is proportional to the square of the r.m.s. AC voltage. The power on the AC side is actually the product of V_{rms} and I_{rms}. With ideal elements, the input power should match the output power. Therefore, for the same power, the demanded current should decrease inversely with the r.m.s. AC voltage when the latter changes. If the r.m.s. AC voltage increases, the input Y also increases proportionately. Therefore it is necessary to divide by the square of the r.m.s. voltage to get the new reference value for the current. So we have a third input to circuit block 5. This third input, which we have labeled Z, is proportional to the square of the r.m.s. AC input voltage. The signal voltage Z is obtained from another circuit block, labeled 6, which we call the squarer. The input to the squarer is the r.m.s. AC input voltage signal, derived as explained before.

Therefore circuit block 5 does both the multiplication and division. We shall call it the multiplier/divider (M/D). Its output is proportional to the instantaneous current reference labeled I_{ref}. I_{ref} is one of the two inputs to circuit block 7, which is the pulse width modulator. We shall call circuit block 7 the current controller. It generates the gate drive signals for the power MOSFET, which functions as the voltage step up chopper. The PWM frequency is programmable by means of a timing capacitor and a timing resistor to be connected externally to the designated terminals of the IC chip. The PWM frequency is typically of the order of 100 kHz. The second input to the current controller is the actual instantaneous value of the current, obtained typically across a noninductive current sensing resistor. This signal is labeled I_{sense} in the figure. This is the feedback current signal. The current controller compares this feedback signal against the reference signal labeled I_{ref}, and adjusts the duty cycle of the chopper in such a way that the difference between the reference and feedback is made negligible. In this way, the actual current is made to conform to the reference, which is a half sine wave in phase with the voltage and satisfies the power requirements of the switch mode power supply.

The use of high power factor pre-regulators is now becoming more widespread because of the increased awareness of the adverse effects of current harmonics on the electrical power supply environment.

6.6 SMPS CONVERTER CIRCUIT TOPOLOGIES—THE "BUCK" OR "FORWARD" CONVERTER

There are several circuit topologies that have evolved and are used in switch mode power supplies (SMPS). We shall classify them into the following groups, based on their operating principle:

1. the "buck" converter—also known as the forward converter;
2. the "boost" converter—also known as the "fly back" converter;
3. the "buck–boost" version of the "fly back" converter;
4. the Ćuk converter;
5. resonant converters;
6. the half-bridge and full bridge configurations.

In this section, we shall present the buck converter (forward converter).

6.6.1 Circuit Configuration and Working Principle

This circuit can be viewed as having evolved from the basic voltage step down chopper studied in Chapter 4. Its configuration and working principle are best

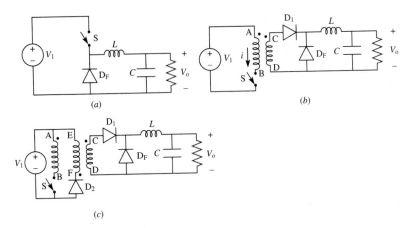

FIGURE 6.12
Evolution of the buck converter topology.

understood if we start from the basic voltage step down chopper and proceed step by step to the final converter topology. For this, we may use Fig. 6.12. Figure 6.12(a) shows the basic voltage step down chopper with an input DC voltage labeled V_1. If the controlled switch S is being repetitively operated with a duty cycle D, the output DC voltage under ideal assumptions is given by $V_o = DV_1$. L is the filter inductor and D_F is the freewheeling diode. The capacitor C also contributes to the filtering. The output voltage is regulated by pulse width modulation, that is, by the control of the duty cycle of the switch, which is defined as the fraction of the switching period for which it is ON. The output voltage can only be lower than the input voltage V_1 because the duty cycle can only be less than 1—hence the name "buck" converter. The transfer of energy from the input side to the output side occurs when the switch S is ON. If the ON switching is viewed as the "forward" operation of the switch then the energy transfer takes place during the forward operation—hence the name "forward" converter.

The circuit of Fig. 6.12(a) does not provide electrical isolation between input and output, which is often a requirement. Therefore we shall try modifying it by introducing a transformer as shown in Fig. 6.12(b). The "forward" feature of the circuit is preserved because of the diode D_1, which will allow flow of power into the load only when the switch S is turned ON. This will be evident if we consider the transformer terminal polarities indicated by the "dots" near the terminals A and C in the figure. Although (b) is functionally similar to (a) and provides electrical isolation, it is, from a practical point of view, an unworkable circuit. The reason for this may be explained as follows. When the switch S is turned ON, the voltage V_1 is applied across the primary terminals, with terminal A positive. This causes the secondary terminal C to go positive with respect to D, causing a flow of current into the output side through D_1. At the end of the ON period of the switch, there will be a current i_1 in the primary coil of the transformer and a flux ϕ in the core of the transformer. Therefore there will be a certain amount of energy stored in the magnetic field of the transformer. When the switch S is opened, this field collapses to zero. The circuit (b) does not provide an electrical channel for the flow of this energy. When S is opened, terminal B goes positive on the primary side, and terminal D goes positive on the secondary side. Current flow is blocked on both sides, because of the open switch on the primary side and the diode D_1 on the secondary side. Because of this, an excessive voltage will be induced in the transformer windings when S is opened, which can be destructively large for the components, especially the static switch and the diode D_1. Therefore we arrive at the practical version of the buck converter, by introducing one more modification, as shown in Fig. 6.12(c). This modification consists in providing one more winding on the converter transformer. The terminals of this winding are labeled E and F in (c). This winding is connected across the DC source, with the diode labeled D_2 in series with it. The diode polarity is such that the DC source cannot drive a current through this winding. Power flow is possible only from the winding to the DC source, and not in the opposite direction. When the switch S is closed, the terminal F becomes positive with respect to the terminal E according to the winding polarity shown by the dots in (c). This causes the diode D_2 to be reverse-biased, the reverse voltage being equal to the DC source voltage plus the induced voltage in the coil EF. When the switch S is opened, the voltage polarity in EF will be reversed; that is, terminal F will go negative with respect to terminal E. The induced voltage in EF will exceed the source voltage and will forward-bias the diode D_2, causing the flow of current and power into the DC source. In this way, the energy stored in the magnetic field of the transformer is fed back into the DC source. For this reason, the winding EF is sometimes called the "feedback winding" and the diode D_2 the "feedback diode." As long as the feedback is taking place, that is, during the ON period of the diode D_2, the voltage across the feedback winding is clamped to V_1. This lasts until the energy transfer from the magnetic field of the transformer is completed and the current in diode D_2 falls to zero. The following numerical example illustrates how the voltage waveforms across the transformer and across the static switch can be determined.

Illustrative example 6.6. The input voltage V_1 for the forward converter shown in Fig. 6.13 is 160 V (obtained from the front end rectifier and filter section). The

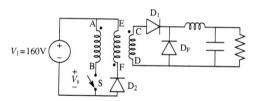

FIGURE 6.13
Circuit for Example 6.1.

converter transformer has 120 turns on the primary winding AB and 120 turns on the feedback winding EF. The switching frequency is 50 kHz and the duty cycle 25%. Determine the following:

(a) the waveform of the voltage across the primary winding AB;
(b) the waveform of the voltage across the feedback winding EF;
(c) the maximum blocking voltages of the static switching element.

Solution

(a) When the switch S is ON, $V_{AB} = V_1 = 160$ V. When the S is turned OFF, the diode D_2 turns ON and clamps the voltage across EF to V_1, that is, 160 V. The voltage polarity will be terminal E positive with respect to terminal F.

Therefore, by transformer action, the voltage induced in the winding AB is $V_{FE} N_{AB}/N_{EF}$, where N_{AB} and N_{EF} are the numbers of turns for the two windings. Therefore

$$V_{AB} = -V_1 \times \frac{120}{120} = -160 \text{ V}$$

This negative voltage will last as long as the feedback diode is ON. The ON period T_F of the feedback diode is to be determined.

The waveform of the voltage across AB is sketched in Fig. 6.14. The positive pulse lasts for T_{ON}, which is determined by the duty cycle of the switch:

$$T_{ON} = 0.25 \times \text{one switching period} = 0.25 \times \frac{1}{50 \times 10^3} = 5 \text{ } \mu s$$

We can determine the interval T_F from the following reasoning. The waveform of the voltage V_{AB} across the primary cannot have a DC component. If there is a DC component, the current in AB will build up from cycle to cycle indefinitely. This implies that the stored energy is not fully returned to the source through the feedback diode in each cycle of switching. We determine T_F on the assumption that the circuit is working stably, such as to make the DC component in the voltage across the transformer zero. Therefore in the waveform of Fig. 6.14, the area under the positive pulse must be equal to the area under the negative pulse:

$$T_{ON} V_1 = T_F V_1$$

This gives $T_F = T_{ON} = 5 \text{ } \mu s$.

(b) Since the ratio of turns between AB and EF is unity, the waveform of V_{EF} is the same that of V_{AB}, but reversed in sign.

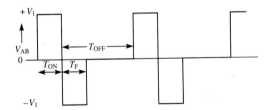

FIGURE 6.14
Waveform of transformer primary voltage.

(c) If we look at the primary circuit loop, we find that the voltage labeled V_s across the switch is

$$V_s = V_1 - V_{AB}$$

On the basis of this relationship, the voltages occurring across the switch in one switching period may be listed as follows:

$$V_s = 0 \qquad \text{during } T_{ON}$$
$$V_s = +320 \qquad \text{during } T_F$$
$$V_s = +160 \qquad \text{for the rest of the swtiching period}$$

The maximum blocking voltage is 320 V forward blocking. There is no reverse blocking voltage requirement.

Illustrative example 6.7. Repeat the previous example with the number of turns in AB changed to (a) 150 and (b) 60.

Solution
(a) With $N_{AB} = 150$ and $N_{EF} = 120$, the voltage across AB during the interval T_F becomes

$$V_{AB} = -\frac{150}{120} V_1 = -\frac{150}{120} \times 160 = -200 \text{ V}$$

T_F will be determined by equating the positive and negative sides of the waveform of V_{AB}. Therefore

$$160 T_{ON} = 200 T_F$$

$$T_F = \frac{160}{200} \times 5 = 4 \ \mu s$$

The waveform of V_{AB} is sketched in Fig. 6.15(a).

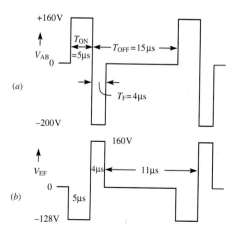

(a)

(b)

FIGURE 6.15
Waveforms of voltages across the primary and feedback windings.

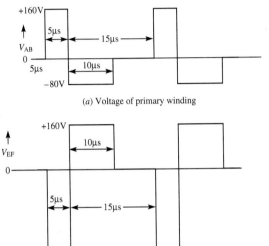

(a) Voltage of primary winding

(b) Voltage of feedback winding

FIGURE 6.16
Waveforems of the voltages across the primary and feedback windings.

The waveform of V_{EF} will be reversed in sign and the magnitude decreases according to the turns ratio, which is now 120/150. The waveform of V_{EF} is sketched Fig. 6.15(b).

Proceeding in the same manner as in the previous example, we find that the maximum forward voltage to be blocked by the static switch is $160 - (-200) = 360$ V.

(b) $N_{AB} = 60$ turns and $N_{EF} = 120$. In this case, the ratio $N_{AB}/N_{EF} = 60/120 = 0.5$. Therefore the voltage V_{AB} during the feedback becomes $-0.5 \times 160 = -80$ V and T_F becomes $160 \times 5/80 = 10\ \mu$s.

The waveform of the voltage across the primary is sketched in Fig. 6.16(a) on this basis. The waveform of the voltage across EF is obtained by dividing V_{AB} by 0.5 and reversing the polarity. This waveform is sketched in Fig. 6.16(b).

Proceeding on the same lines as in the previous cases, we get the maximum forward voltage to be blocked by the static switch as $V_{s(max)} = 160 - (-80) = 240$ V.

6.6.2 Duty Cycle Constraint on the Isolated Buck Converter

We have stated that the DC component of the transformer primary voltage has to be zero for stable continuous operation of the buck converter. This means that the area under the positive part of the waveform should be equal to the area under the negative part. This means that the static switch should be in the OFF state for sufficient time to complete the feedback of stored energy through the feedback circuit. The T_{OFF} of the static switch should be longer than T_F.

This condition therefore puts a limit to the maximum duty cycle permissible for an isolated buck converter. The upper limit of the duty cycle depends on the time needed for the feedback of energy. It is evident from the results of the two previous illustrative examples that this depends on the turns ratio between the primary and the feedback windings of the converter transformer. The upper limiting values of the duty cycle for the cases considered in the two previous examples are worked out in the next example.

It is important to ensure that the limiting value of the duty cycle is not exceeded during the operation of the converter. It may also be stated that the limit exists because we use a transformer for isolation. If the primitive buck converter topology is employed without an isolating transformer then the constraint on duty cycle for this reason does not exist.

The turns ratio between the primary and the main secondary winding does not introduce a constraint on the duty cycle from the operational stability point of view. This ratio will depend on the output voltage requirement of the converter and the range of variation of the output required.

Illustrative example 6.8. Determine the maximum permissible duty cycle values for the turns ratios considered in the two previous illustrative examples.

Solution. For a turns ratio $N_{AB}/N_{EF} = 1$, the feedback time $T_F = T_{ON}$ of the switch. This is the minimum permissible value of the OFF time of the switch. For the limiting case, $T_{OFF} = T_{ON}$. The maximum permissible duty cycle is 50%.

For $N_{AB}/N_{EF} = 1.25$, the minimum value of T_{OFF} for the static switch is given by the feedback time T_F, which is

$$T_F = T_{ON}/1.25 = 0.8 T_{ON}$$

On this basis, the limiting duty cycle becomes

$$D = \frac{T_{ON}}{T_{ON} + T_{OFF}} = \frac{T_{ON}}{T_{ON} + 0.8 T_{ON}} = 0.55$$

Therefore the maximum permissible duty cycle is 55%.

For $N_{AB}/N_{EF} = 0.5$, minimum value of $T_{OFF} = T_F = T_{ON}/0.5 = 2 T_{ON}$ and limiting value of

$$D = \frac{T_{ON}}{T_{ON} + T_{OFF}} = \frac{1}{1 + 2} = 0.33$$

The maximum permissible duty cycle is 33%.

6.6.3 Use of Two Static Switches to Eliminate the Feedback Winding

In the isolated buck converter topology, the feedback winding can be eliminated by making the primary winding itself perform the feedback

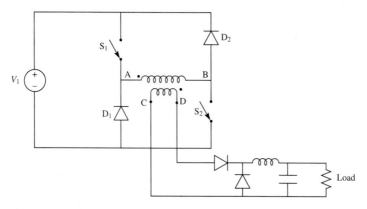

FIGURE 6.17
Isolated buck converter without a separate feedback winding.

function by using an additional static switch and an additional feedback diode. The modified circuit topology is shown in Fig. 6.17. In this circuit, the static switches are turned ON and turned OFF simultaneously. When both are turned OFF, the feedback of energy occurs through the two feedback diodes. The circuit functions in the same manner as a single-switch buck converter with a turns ratio of unity between the primary and feedback windings. The same duty cycle constraint applies.

Illustrative example 6.9. In the two-switch buck converter of Fig. 6.17, $V_1 =$ 160 V. Determine
(a) the maximum permissible duty cycle;
(b) the peak forward voltage across each switch;
(c) the number of secondary turns on the transformer to give an output voltage of 8 V DC, given that the number of primary turns is 80.

Solution
(a) The maximum permissible duty cycle is the same as for the isolated single-switch topology, for which the turns ratio between the primary winding and the feedback winding is unity. Therefore

$$D_{max} = 50\%$$

(b) The peak forward voltage occurs across the switch S_1 when D_1 is conducting and across S_2 when D_2 is conducting. In each case, it is equal to V_1. Therefore

$$V_{Fmax} = 160 \text{ V}$$

(c) We shall consider the case of maximum duty cycle of 50%. At this duty cycle, the voltage on the primary of the transformer will be a square wave of amplitude $\pm160\,$V. If a is the transformer voltage ratio, defined as secondary voltage/primary voltage, the secondary side of the transformer will give a square wave of amplitude $= \pm a \times 160\,$V. But because of the diodes on the secondary side, the voltage appearing at the output will have a duty cycle of 50% and amplitude $+a \times 160\,$V. This will have a DC component $= a \times 160/2$. Because of the series inductance in the output filtering circuit, which will absorb the AC components, the output voltage will be this DC component. Equating this to the required value of $8\,$V, we get

$$\frac{a \times 160}{2} = 8 \Rightarrow a = \frac{1}{10}$$

Therefore $a = \frac{1}{10}$. The required value of the secondary turns is $80/10 = 8$ turns. This is based on ideal elements and the maximum 50% duty cycle.

6.7 THE "BOOST" CONVERTER AND THE "BUCK–BOOST" CONVERTERS— THE "FLYBACK" MODE

The "boost" converter and also the "buck–boost" converter topologies employ the so called "flyback" mode of operation. They may be viewed as having evolved from the basic voltage step up chopper studied in Chapter 4. We shall follow this approach, starting from the basic voltage step up chopper and proceeding step by step till we arrive finally at the "buck–boost" configuration used in practical switch mode power supplies. This approach will make the essential characteristics of the buck–boost topology better understood.

Figure 6.18(a) shows the conventional voltage step up chopper circuit. In this circuit, assuming continuous current flow in V_1 and L and ideal elements, the waveform of the voltage labeled V_s across the switch is as shown in Fig. 6.18(b). The amplitude will be zero when the switch is ON and equal to V_o

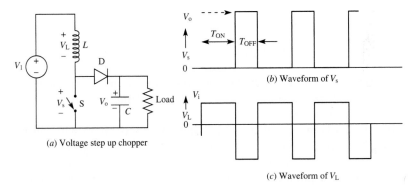

(a) Voltage step up chopper

(b) Waveform of V_s

(c) Waveform of V_L

FIGURE 6.18
Voltage step up chopper. Waveforms of V_s and V_L.

when it is OFF, because the diode D will be conducting when the switch is OFF. The input voltage V_1 will be equal to the DC component of V_s, the AC component being absorbed across the inductance L. Therefore

$$V_1 = V_o(T_{OFF}/T) = V_o(1 - D)$$

This gives

$$V_o = \frac{V_1}{1 - D}$$

the relationship that we obtained earlier for the voltage step up chopper. Therefore the output voltage V_o is always greater than the input voltage V_1, since D is always less than unity. Because of this voltage step up feature, we call this a boost converter. There is no transfer of energy from the source to the output side when the switch is ON, which we call the "forward" operation of the switch. Energy flows from the input side to the output side when the switch is turned OFF, that is, when it is made to "flyback"—hence the name flyback converter.

Figure 6.18(c) shows the waveform of the voltage across the inductor L, which is labeled V_L. It is this inductor which receives energy from the input source when the switch S is ON, and then, together with assistance from the voltage source V_1, pumps energy into the output side, during the flyback of the switch. During the T_{ON} of the switch, this voltage V_L has a positive magnitude equal to the source voltage V_1. During T_{OFF}, since we are assuming continuous conduction and the diode D will be ON,

$$V_L = V_1 - V_o$$

This will be a negative voltage, since V_o is greater in magnitude than V_1. The positive area under V_L during T_{ON} is given by

$$V_o(1 - D)T_{ON} = V_o(1 - D)DT$$

The negative area during T_{OFF} is

$$(V_1 - V_o)T_{OFF} = [V_o(1 - D) - V_o]T_{OFF} = -DV_oT_{OFF} = -V_oD(1 - D)T$$

Therefore we find that the area under the positive-voltage region is the same as that area under the negative-voltage region. This confirms that the voltage across L has no DC component.

If the switch S is kept OFF all the time (duty cycle $= 0$) there will still be an output voltage, because the capacitor will receive charge through D and get charged to the voltage V_1. When the switch S is turned ON and turned OFF, additional charge will be pumped into the capacitor, thus making the output voltage greater than the input V_1. This is another way of describing the voltage boost behavior of the circuit.

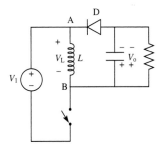

FIGURE 6.19.
Buck–boost topology without isolation.

We shall now make a simple modification of the circuit. This consists in transferring the connections from the output side from across the switch to across the inductor L, as shown in Fig. 6.19. The voltage across AB, labeled V_L in the figure, during T_{ON}, is equal to V_1. During T_{OFF}, assuming continuous conduction of the diode D, the voltage $V_L = -V_o$. Equating the positive and the negative volt-seconds,

$$V_1 T_{ON} = V_o T_{OFF} \tag{6.4}$$

This gives

$$V_o = V_1 \frac{T_{ON}}{T_{OFF}} = V_1 \frac{D}{1-D}$$

This shows that the output can be less or more than the input, depending on the duty cycle.

For $D < 0.5$, it is a buck converter.

For $D > 0.5$, it is a boost converter.

In both cases, it is a flyback converter, but there is no isolation between input and output.

6.7.1 The Isolated Buck–Boost Converter

The isolated version of the buck–boost converter is obtained by having a secondary coil for the winding AB of Fig. 6.19. This winding is labeled EF in Fig. 6.20, which shows the modified circuit.

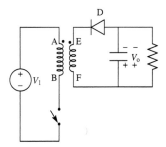

FIGURE 6.20.
The isolated buck–boost converter.

If the turns ratio a is defined as

$$a = \frac{N_{AB}}{N_{EF}}$$

then

$$V_o = \frac{1}{a} V_1 \frac{D}{1-D}$$

The following points may be stated in relation to the isolated buck–boost converter topology of Fig. 6.20. The basic distinction between the "forward" converter and the flyback circuit of Fig. 6.20 is that the energy transfer from the input side to the output side takes place during the ON period of the switch in the forward converter, while in the case of the flyback circuit, the energy transfer takes place during the OFF period of the switch (flyback period).

In the isolated flyback circuit of Fig. 6.20, the transformer serves both for isolation and as the inductor for energy storage. Therefore it is more appropriate to call it a "transformer–inductor" or "transformer choke," as is usually done.

Forward blocking voltage of the switch S. During flyback, the voltage across the primary winding AB is equal to $V_o(N_{AB}/N_{EF})$, where N_{AB}/N_{EF} is the primary to secondary turns ratio. Therefore the peak repetitive forward blocking voltage of the switch is

$$V_1 + V_o \frac{N_{AB}}{N_{EF}} = V_1 + V_1 \frac{D}{1-D} = V_1 \frac{1}{1-D}$$

This relationship assumes 100% magnetic coupling between the primary and the secondary. In actual cases, there will be some leakage reactance, which will create additional voltage stress on the switch during the flyback period. It is therefore important to use a properly designed snubber circuit to protect the switching transistor. It is also possible to limit the blocking voltage on the switch by using two transistors in the manner to be described.

Illustrative example 6.10. For the isolated buck–boost converter of Fig. 6.20, the output voltage is to be 35 V at a duty cycle of 30%. The DC input is obtained from a front end rectifier without voltage doubling, fed from 115 V AC. Determine the peak forward blocking voltage of the switching element.

Solution. The input voltage from the front end rectifier is

$$V_1 = 115 \times \sqrt{2} \times 162.6 \text{ V}$$

The peak forward blocking voltage is

$$V_1 \frac{1}{1-0.3} = 232.3 \text{ V}$$

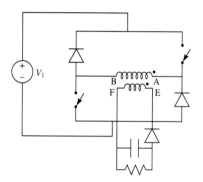

FIGURE 6.21
Dual switch topology of the isolated buck–boost flyback converter.

6.7.2 Dual Switch Topology of the Isolated Buck–Boost Converter

The maximum blocking voltage requirement of the switching element can be limited to V_1 by modifying the circuit in the manner shown in Fig. 6.21. This circuit requires an additional switching element and an additional diode. There are similarities between this circuit and the two-switch version of the forward converter given in Fig. 6.17. But the fundamental difference can be seen if we look at the transformer terminal polarity indicated by the dots. This shows that the circuit of Fig. 6.21 is a flyback converter, as distinct from Fig. 6.17, which is a forward converter.

> **Illustrative example 6.11.** For the dual switch isolated flyback converter of Fig. 6.21, the data are as follows:
>
> $$V_1 = 160 \text{ V}, \quad N_{AB} = 100 \text{ turns}; \quad N_{EF} = 20 \text{ turns}, \quad \text{duty cycle} = 40\%$$
>
> Determine the output voltage V_o.
>
> *Solution.* The turns ratio
>
> $$a = \frac{N_{AB}}{N_{EF}} = \frac{100}{20} = 5$$
>
> $$V_o = V_1 \frac{1}{a} \frac{D}{1-D} = 160 \times \frac{1}{5} \times \frac{0.4}{1-0.4} = 21.3 \text{ V}$$

6.8 THE ĆUK CONVERTER

We have seen that in the forward (buck) converter, the energy transfer from the input to the output side occurs when the static switch is in the ON state. In the flyback converter (buck–boost), this transfer takes place when the static switch is turned OFF. In both cases, the energy transfer is not continuous. We overcome this limitation by providing adequate filtering. The filter consists of energy storage elements such as an inductor or capacitor or both, which serve as reservoirs of energy and ensure that the flow of energy into the load is continuous and ripple-free.

In contrast to the above, in the converter developed by Ćuk, energy transfer from the input side to the output occurs both during the ON time and the OFF time of the static switch. Ideally, the Ćuk converter can function with zero ripple. In this section, we shall describe the Ćuk converter topology and its manner of operation, starting from a simple circuit configuration and proceeding step by step to arrive at the final topology that incorporates the essential features of the converter.

6.8.1 Energy Transfer through an Intermediate Capacitor

In the circuit shown in Fig. 6.22, when the switch S is opened, the capacitor C_1 is charged through the diode D. The capacitor receives the charge and energy from the source V_1 and also from L_1, if L_1 has stored energy prior to the opening of the switch S.

When S is closed, the capacitor C_1 comes across the diode, reverse-biasing it. The capacitor is now connected directly across the output side of the converter, and energy is transferred from it to the output. L_2 and C_2 are the output filter inductance and filter capacitor respectively.

For this circuit, assuming ideal elements, we can derive an expression for the output voltage by first deriving an expression for the voltage V_{C_1} across the capacitor C_1. We shall assume that the switch is operating at a fixed frequency f and duty cycle D. We shall also assume that the capacitor C_1 is very large, so that V_{C_1} can be assumed to be constant with negligible ripple.

When S is ON, the voltage across L_1 is equal to V_1:

$$V_{L_1} = V_1 \quad \text{during } T_{ON}$$

When S is OFF, the diode D will be conducting. We shall assume that the diode is ON for the entire OFF period of the switch. Therefore

$$V_{L_1} = -(V_{C_1} - V_1) \quad \text{during } T_{OFF}$$

For steady-state continuous operation, the waveform of V_{L_1} cannot have a DC component. Otherwise, an indefinite current build up would occur in L_1.

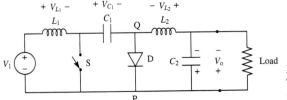

FIGURE 6.22
Two-step energy transfer through intermediate capacitor C_1.

Therefore we equate the positive and negative volt-seconds areas under the waveform:

$$V_1 T_{ON} = (V_{C_1} - V_1) T_{OFF}$$

This gives

$$V_{C_1} = \frac{V_1}{1-D} \tag{6.5}$$

Voltage across PQ (across the diode). The voltage V_{PQ} across the diode will be equal to V_{C_1} when the switch S is ON and zero when it is OFF. Therefore the DC component of this voltage will be given by

$$V_{PQ(DC\,component)} = DV_{C_1}$$

With ideal elements, the DC component of the voltage V_{PQ} will be the same as the output voltage V_o across the capacitor C_2. We shall assume C_2 to be large enough so that the ripple voltage is negligible. Therefore

$$V_o = DV_{C_1} = \frac{D}{1-D} V_1 \tag{6.6}$$

This result shows that the converter is a buck–boost converter: buck for duty cycle values less than 0.5 and boost for duty cycle values greater than 0.5.

Illustrative example 6.12. The data for the converter of Fig. 6.23 are as follows: $V_1 = 160$ V and frequency 25 kHz. For a duty cycle of 25%,
(a) determine V_{C_1} and V_o;
(b) the peak forward voltage that the switch S has to block;
(c) sketch the waveforms of V_{L_1} and V_{L_2}.

Solution

(a)
$$V_{C_1} = \frac{V_1}{1-D} = \frac{160}{1-0.25} = 213.3 \text{ V}$$

$$V_o = V_1 \frac{D}{1-D} = 160 \times \frac{0.25}{1-0.25} = 53.3 \text{ V}$$

(b) The voltage across the switch is

$$V_s = \begin{cases} 0 & \text{when S is ON (during } T_{ON}) \\ V_{C_1} & \text{when S is OFF (during } T_{OFF}) \end{cases}$$

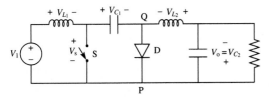

FIGURE 6.23
Circuit of Example 6.11.

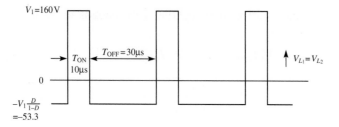

FIGURE 6.24
Waveforms of voltages V_{L_1} and V_{L_2}.

The peak forward voltage that the switch has to block is $V_{C_1} = 213.3$ V.
(c) During T_{ON}, $V_{L_1} = V_1 = 160$ V. During T_{OFF},

$$V_{L_1} = V_1 - V_{C_1} = -(V_{C_1} - V_1) = -\left(\frac{V_1}{1-D} - V_1\right) = -\frac{D}{1-D}V_1$$

$$= -\frac{0.25}{1-0.25} \times 160 = -53.3 \text{ V}$$

The switching period is

$$\frac{1}{25 \times 10^3} = 40 \ \mu s$$

$$T_{ON} = DT = 0.25 \times 40 = 10 \ \mu s$$

$$T_{OFF} = (1-D)T = 0.75 \times 40 = 30 \ \mu s$$

The waveform of V_{L_1} is sketched in Fig. 6.24 on the basis of the above.
During T_{ON},

$$V_{L_2} = V_{C_1} - V_o = V_1\frac{1}{1-D} - V_1\frac{D}{1-D}$$

$$= V_1 = 160 \text{ V}$$

During T_{OFF},

$$V_{L_2} = -V_o \qquad \text{(because diode } D \text{ is ON)}$$

$$= -\frac{D}{1-D}V_1 = -53.3 \text{ V}$$

Therefore V_{L_2} has the same waveform of V_{L_1}, which is sketched in Fig. 6.24.
We also notice from the related expressions that the equality of V_{L_1} and V_{L_2} is
valid for all values of the duty cycle.

The above example highlights an important aspect of the Ćuk converter,
which we shall arrive at by further modification of the circuit of Fig. 6.22. The
output voltage V_o is the DC component of the voltage across PQ in Fig. 6.22.
The purpose of the inductor L_2 is to absorb the AC components in the voltage
across PQ. If L_2 is large enough, practically the entire ripple component will

be absorbed across it, and the resulting ripple current will be very small. With a large value for the filter capacitor C_2, the output ripple voltage resulting from the flow of ripple current in the inductor will also be made small, so that the output voltage will be practically ripple-free. These statements indicate the need to use a large filter to get a ripple-free output.

The Ćuk topology in principle is designed to eliminate the need for the filter elements such as L_2 and C_2. For this, we modify the circuit of Fig. 6.22 in such a way that an additional voltage is introduced in the output circuit, which exactly cancels the ripple voltage that would otherwise be present. We notice from the above example that the ripple voltage absorbed across L_2 has exactly the same waveform as the voltage across L_1. Therefore, by winding a second coil with identical number of turns as L_1 on the same magnetic core, we can get a voltage that can balance the output ripple and provide a resultant output voltage that is ripple-free. This is illustrated by Fig. 6.25. In Fig. 6.25(a) we have introduced on the output side an AC voltage source that is identical to V_{L_1}, with the terminal polarity as marked. This AC voltage is obtained by having a second coil wound on the same magnetic core of L_1 and with the same number of turns as L_1. Ideally, with 100% magnetic coupling, the voltage induced in the second coil should be the same as V_{L_1}. We shall show presently that, with such an arrangement, no filter is required on the output side. Therefore we have shown this filter in (a) as a circuit block that is nonessential, and we have eliminated this block in Fig. 6.25(b). We shall now look at the voltage appearing across the terminals labeled R and S in (b), during the ON time T_{ON} and the OFF time T_{OFF} of the switch S.

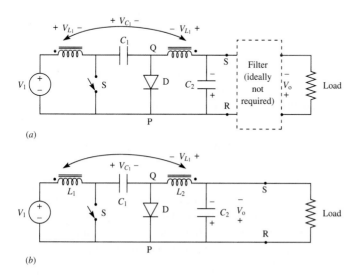

(a)

(b)

FIGURE 6.25
Basic Ćuk converter topology without isolation.

During T_{ON},

$$V_{RS} = V_{C_1} - V_{L_2} = V_{C_1} - V_{L_1}$$

$$= V_1 \frac{1}{1-D} - V_1$$

$$= V_1 \frac{D}{1-D}$$

During T_{OFF}, assuming that D is ON for the entire duration of T_{OFF},

$$V_{RS} = -V_{L_2} = -V_{L_1}$$

$$= -\left(-\frac{D}{1-D}\right)V_1$$

$$= V_1 \frac{D}{1-D}$$

The expressions above for V_{RS} show that it has the same value during both T_{ON} and T_{OFF} equal to $V_1(D/(1-D))$. Therefore the voltage has zero ripple, and ideally no filter is needed. In the nonideal real situation, a certain amount of filtering will be needed. For this, we have included the capacitor labeled C_2 in (b).

Figure 6.25(b) shows the basic Ćuk topology without isolation between input side and output side. In this, L_1 and L_2 are coupled inductors wound on a common magnetic core. The input–output voltage relationship for this basic Ćuk topology, as derived above, will be

$$V_o = V_1 \frac{D}{1-D} \tag{6.7}$$

From this, we see that this converter belongs to the "buck–boost" category: "buck" for duty cycle values less than 0.5 and "boost" for $D > 0.5$. It cannot strictly be classified either as a "forward" converter or as a "flyback" converter, because energy transfer occurs from the input side to the output side both during T_{ON} and during T_{OFF} of the switch S.

6.8.2 The Isolated Version of the Ćuk Converter

Electrical isolation is invariably a requirement to be met by regulated power supplies. The basic Ćuk converter topology can be further modified to provide this feature. To explain the functioning of the isolated version, we shall start with the nonisolated topology of Fig. 6.25(b) and proceed step by step, introducing a hypothetical circuit change at each step until we finally arrive at the isolated circuit. For this we shall use Fig. 6.26(a)–(e).

(a) shows the basic nonisolated Ćuk converter topology, and is the same as Fig. 6.25(b).

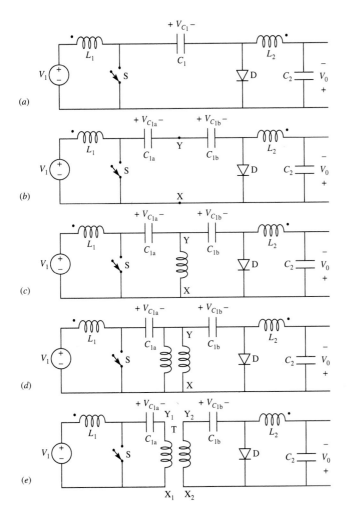

FIGURE 6.26
Step-by-step modifications of the nonisolated Ćuk topology to arrive at the isolated version.

In (b) we replace C_1 by two capacitors C_{1a} and C_{1b} in series in such a way that the series combination has the same value as C_1. Therefore the two capacitors should satisfy the following condition (there will be several combinations of values that will satisfy this condition):

$$\frac{C_{1a}C_{1b}}{C_{1a} + C_{1b}} = C_1 \tag{6.8}$$

For a given operating duty cycle, we shall choose C_{1a} and C_{1b} in such a way that the resultant voltage appearing across the series link connection labeled Y in the figure and the terminal labeled X has no DC component. For this, notice that, during T_{ON} of the switch S,

$$V_{XY} = V_{C_{1a}}$$

During T_{OFF}, $V_{XY} = -V_{C_{1b}}$, since diode D is in conduction. For zero DC component, the positive and negative volt-seconds must be equal. Therefore

$$V_{C_{1a}}T_{ON} = V_{C_{1b}}T_{OFF} \Rightarrow V_{C_{1a}}D = V_{C_{1b}}(1-D)$$

$$V_{C_{1b}} = \frac{D}{1-D}V_{C_{1a}} \tag{6.9}$$

From the above relationships, we also get

$$V_{C_{1a}} + V_{C_{1b}} = V_{C_1} = \frac{V_1}{1-D}$$

$$V_{C_{1a}} = V_1$$

$$V_{C_{1b}} = V_1\frac{D}{1-D}$$

Therefore, since

$$\frac{V_{C1a}}{V_{C_{1b}}} = \frac{C_{1b}}{C_{1a}}$$

we get

$$\frac{C_{1b}}{C_{1a}} = \frac{1-D}{D} \tag{6.10}$$

This shows that we can in fact choose the capacitances C_{1a} and C_{1b} to satisfy the stated condition that V_{XY} has no DC component. The converter will therefore operate exactly as in (a).

Our next hypothetical circuit change is shown by (c). This consists in connecting a coil of very large inductance across XY. Since the voltage V_{XY} has no DC component, and since the inductance is very large, the current drawn by it is negligible. For these reasons, the operation of the circuit is not affected by this modification.

The next hypothetical modification is shown by (d). This consists in connecting one more coil of very large inductance identical to the first across the same terminals XY. By the same token as before, this change also does not affect the circuit operation.

(d) shows that both the coils have the same voltage across it. Therefore we can have both the coils wound on the same magnetic core with the same number of turns. This is done in (e). Since the voltages across the coils are identical, it will not make any difference if the parallel connecting links between the two are removed. This is also shown in (e). The two coils now become the primary and secondary of a $1:1$ isolating transformer. The voltages across X_1Y_1 and across X_2Y_2 are the same, each being equal to the voltage V_{XY} of (b)–(d). Therefore the circuit of (e) functions in the same manner as the previous ones, and at the same time provides total electrical isolation between the input and the output.

In the above derivation, we assumed that the values of C_{1a} and C_{1b} are chosen such as to satisfy the stated relationships. We did this only to ensure that the voltage across XY in (b) is a pure AC with no DC component *before* we connect the inductance L, so that there is no change in the operation of the circuit when we make the hypothetical change from (b) to (c). But, if we consider repetitive operating conditions, we can show that any values of C_{1a} and C_{1b} will suffice, provided these values are large enough to justify the assumption the voltages $V_{C_{1a}}$ and $V_{C_{1b}}$ can be assumed to be constant and ripple-free. The voltages across C_{1a} and C_{1b} will automatically adjust to satisfy the condition that the DC voltage across the coil is zero when repetitive conditions prevail. This will be so irrespective of the actual values of the capacitances. This may be shown as follows. During T_{ON} of the switch,

$$V_{C_{1a}} = V_{XY}$$

During T_{OFF} of the switch,

$$V_{C_{1b}} = -V_{XY} \qquad \text{(because diode D is ON)}$$

Assuming that stable repetitive operation prevails, V_{XY} cannot have a DC component, and we shall therefore equate the positive and negative volt-seconds using the above relationships. Therefore

$$V_{C_{1a}} T_{ON} = V_{C_{1b}} T_{OFF}$$

$$\frac{V_{C_{1a}}}{V_{C_{1b}}} = \frac{T_{OFF}}{T_{ON}} = \frac{1-D}{D} \tag{6.11}$$

We see that (6.11) is the same relationship as given previously in (6.9).

In the above derivations, we have assumed that steady-state repetitive conditions prevail. This assumption needs to be particularly emphasized in the present case. For example, our theory is not valid for start-up conditions. If the switch has been OFF for a long time, the primary side capacitor C_{1a} will be charged to a voltage equal to V_1, whereas the secondary side capacitor will be uncharged. It will take several cycles of operation before C_{1b} gets charged to the correct repetitive value of its voltage. During this time, a large value of voltage equal to V_1 is being injected into the output circuit through the coupled reactors L_1–L_2. The injected voltage polarity (refer to the terminal dots of L_1 and L_2) is such as to make the output voltage of the converter reversed in sign during the start up. This reversal of output voltage polarity during the start up is a limitation of this circuit. But this can be overcome by incorporating further refinements to the circuit. But, since our purpose has been to explain the distinguishing feature, which is the zero output voltage ripple property, we shall conclude our description of this topology at this point.

6.8.3 Integrated Magnetics

In the isolated Ćuk converter circuit of Fig. 6.26(e), there are two separate magnetic elements:

1. the coupled reactor L_1–L_2 coils wound on a common magnetic core.
2. the isolating transformer T consisting of two windings on another common core.

The coupled reactor and the isolating transformer are two independent magnetic elements. The core of the coupled reactor that constitutes its magnetic circuit is distinct and separate from the core that constitutes the magnetic circuit of the isolating transformer. There is no magnetic coupling between the coupled reactor coils and the windings of the transformer. Even so, it is possible to implement the magnetic curcuits of the coupled reactor and the isolating transformer using a single magnetic core, at the same time making sure that there is no magnetic coupling between the two and that they function independently. We are familiar with integrated electric circuits in which more than one functional electric circuit, such as operational amplifiers or logic gates, that work independently of each other can be fabricated on the same silicon wafer. By analogy, the term "integrated magnetics" has been used to describe the technique of fabricating more than one functional magnetic circuits by using a single magnetic core structure. A manner in which the coupled reactor L_1–L_2 and the isolating transformer T of the Ćuk converter can be integrated on a single magnetic core structure is shown in Fig. 6.27. In this figure the coils AB and CD each constitute one half of the primary

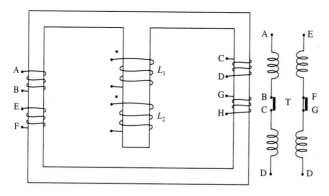

FIGURE 6.27
Integration of the magnetics of the coupled reactor and the isolating transformer.

winding of the isolating transformer T. When they are connected as shown in "series," the terminals A–D constitute the primary terminals of the transformer. Careful attention whould be given to the directions in which the coil sections AB and CD are wound on the core. When AB and CD are connected as shown and carry current, the magnetomotive forces (ampere-turns) of AB and of CD add in the outer magnetic circuit consisting of the two outside vertical legs and the top and bottom horizontal limbs. But the ampere-turns of AB acting on the middle limb cancel the ampere-turns of CD acting on the middle limb. Therefore there is no magnetic coupling from the primary of the isolating transformer to the coils on the middle leg. EF and GH constitute the two halves of the secondary of the isolating transformer. The same consideration as we made for the primary shows that there is no magnetic coupling from the secondary of the isolating transformer to the coils on the middle leg.

The coils L_1 and L_2 of the coupled reactor are wound on the middle leg. When a coil on the middle leg carries current, the resulting flux links with the coils AB and CD in such a way that the induced voltage in the winding AB due the rate of change of this flux cancels the induced voltage in the winding CD when AB and CD are series-connected as shown. Therefore there is no coupling from the coils on the middle leg to the primary winding of the transformer. Similar considerations show that there is no coupling from these coils to the secondary circuit of the transformer.

Notice the relative terminal polarities of the coils L_1 and L_2 as shown by the dots. In the actual converter circuit, coil L_1 carries a DC current equal to the input DC current from the source V_1 [see Fig. 6.26(e)]. The coil L_2 carries the DC load current. The directions of these currents are such [see the terminal dots in Fig. 6.26(e) for L_1 and L_2] that the DC ampere-turns are additive in the magnetic circuit of these two coils. The total DC ampere-turns of L_1 and L_2 can cause magnetic saturation of the core. To avoid this, an air gap is provided on the middle leg, as shown in Fig. 6.27.

6.9 RESONANT CONVERTERS

The choice of a higher switching frequency in a switch mode power supply greatly helps to make the unit smaller and more compact by reducing the size of the filter elements. Therefore, in the design of switch mode power supplies, the aim is to use as high a switching frequency as possible. Currently frequencies in the region of 100–200 kHz are quite common. The main difficulty in the way of increasing the frequency to much larger values is the increased switching power loss in the switching element. At each turn ON and turn OFF switching, a certain amount of energy dissipation occurs in the switching element, whose average is what we call the switching power loss. The switching power loss in the device is directly proportional to the switching frequency. Therefore, as the switching frequency is increased, there will be a proportional increase in the power dissipation in the switching element, causing higher temperature rise and lowering the overall efficiency of power conversion. If the switching power loss can be eliminated or minimized, it will be possible to use higher switching frequencies, and at the same time achieve higher power conversion efficiencies. Resonant converter topologies serve this purpose in the following manner.

The power dissipation in the switch during a switching transition occurs because, during the transition, the voltage across the switch and the current through it have nonzero finite values. The power dissipation at an instant during the transition is the product of the instantaneous current and voltage. If either the current or the voltage is zero, or has a low value during the switching transition, the power dissipation will be zero or low. Operating the switch when the voltage across it is zero is called zero-voltage switching (ZVS). Operating the switch when the current through it is zero is called zero-current switching (ZCS). With either ZVS or ZCS type of switching, the switching power loss will be zero. Resonant converter topologies make such switching possible. If a series L–C circuit is made to resonate, the current and the voltage automatically go through zeros without the need for a switch. The resonant converter topologies implement either ZVS or ZCS by taking advantage of this feature. In this way, the switching frequencies can be increased without increasing the power loss, and overall power conversion efficiencies can be improved. For these reasons, there is a great deal of interest in resonant converter topologies and their implementation at the present time, together with expectation that smaller and more efficient switch mode power supplies can be realized by this technique. In this section, we intend to introduce the technique of power conversion by means of the resonant mode of switching, by considering simple circuits that illustrate the principle. A detailed coverage of the various circuit topologies is not proposed.

6.9.1   The Half-Bridge, Series-Loaded, Series Resonant Converter

The circuit of this converter is shown in Fig. 6.28. It is a half-bridge configuration needing a dual power supply consisting of two identical DC voltages V in (a). If a split power supply is not available, this can be implemented by using two large equal capacitors across a single power supply, to provide the DC source midpoint as shown in (b). The resonant circuit is a series circuit consisting of the inductance L and the capacitance C. The load is shown as a series-connected resistor R. Actually, the load will be on the secondary side of an isolating transformer as shown in (b). R is the reflected resistance of the load as seen from the primary side of this transformer. The leakage inductance of the transformer, if any, will be included in the value L of the resonating inductor. On the secondary side of the isolating transformer, the induced voltage will not have a DC component. The DC is obtained by rectification using diodes as shown. There will also be a filter circuit to minimize output ripple voltage. This is not shown in (b), to make the treatment simple. There are two controlled switching devices, which typically

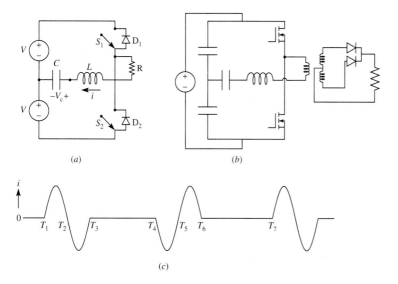

FIGURE 6.28
Half-bridge resonant converter.

may be power MOSFETs. Each switching element has an antiparallel diode. In the case of power MOSFETs, this diode may be the "body diode" of the MOSFET. With other devices, a separate antiparallel diode will be needed for the operation to be described. It is also possible to operate the circuit without the antiparallel diode, but such an operation will be different from what we propose to describe here.

When the switch S_1 is turned ON for the first time at the instant labeled T_1 in Fig. 6.28(c), the voltage V of the top section of the split power supply will be applied across the series resonant circuit and the load resistor R. The latter should be small enough to make the circuit underdamped, so that the current is oscillatory. Therefore the current will be a damped sinusoid. The first half-cycle of this current will flow through the switch S_1. This current will charge the capacitor C to the maximum positive value of the voltage, labeled V_C. The current becomes zero at the end of the first half-cycle, when the capacitor voltage is maximum. This instant is labeled T_2 in (c). This maximum will be close to twice the magnitude of V. The second half-cycle of the oscillatory current will start immediately, causing the capacitor to discharge. This second-half-cycle is possible because of the presence of the diode D_1, which provides the path for the flow of this current. At any instant during the flow of this reverse current, the voltage across the switch S_1 is negligible, because the diode D is conducting. Therefore, at any time during this interval T_2–T_3, it can be turned OFF without switching power loss. If this is done, it will be a case of zero-voltage switching. The first switching of S_1 at T_1 was a case of zero-current switching, because the current was initially zero and has to rise according to the rate determined by the resonant circuit, after the switching is initiated. The turn ON switching of the diode is also similarly a case of zero-current switching. The diode turns OFF at T_3 at the end of the second half-cycle, when the current becomes zero—a case of zero-current switching. After the second half-cycle of the oscillatory current, the capacitor will be discharged and no further current flow is possible.

The next current flow is initiated at T_4 when the switch S_2 is turned ON. The behavior of the circuit from the instant labeled T_4 is similar to what we have described from the instant T_1. It may be noted that the direction of current in the resonant circuit for the first half-cycle after S_2 is switched ON is the same as the direction in the second half-cycle after S_1 was switched ON. This is also shown in Fig. 6.28(c). One switching period is completed after the second half-cycle of current flow through D_2 and we arrive at the instant labeled T_7, which is the instant at which we again turn ON S_1. Notice that in this mode of operation, the switching period is longer then the resonant period, one switching period being the interval from T_1 to T_7. In one switching period, we have created four half-cycles of oscillatory current in the load, which are made unidirectional by rectification on the output side of the transformer. There are also two time gaps, with no current flow, from T_3 to T_4 and from T_6 to T_7. These time gaps exist because the switching frequency is below the resonant frequency. They decrease as the switching frequency is increased. Therefore, if the switching frequency is increased, the number of output pulses occurring in the load per second will proportionately increase. This is the means employed for varying the voltage conversion ratio of the converter. This type of operation is called the discontinuous current mode of operation. In this mode, the output control is achieved by varying the switching frequency. It is possible to operate a resonant converter in the continuous current mode by making the switching frequency higher than the resonant frequency. However, we shall not consider such an operation here.

The sequence of operation of the resonant half-bridge converter that we have described above may be verified by the following mathematical analysis. Figure 6.29 shows the active part of the circuit of Fig. 6.28(a) when the switch S_1 is closed for the first time at T_1.

We shall take T_1 as the reference zero of time and write the loop

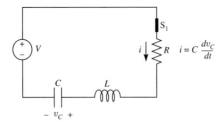

FIGURE 6.29
Resonant Circuit with S_1 closed.

equation as follows:

$$L\frac{di}{dt} + Ri + v_C = V \tag{6.12}$$

since $i = C\, dv_C/dt$, (6.12) may be rewritten as

$$LC\frac{d^2v_C}{dt} + RC\frac{dv_C}{dt} + v_C = V$$

that is,

$$\frac{d^2v_C}{dt^2} + \frac{R}{L}\frac{dv_C}{dt} + \frac{1}{LC}v_C = \frac{1}{LC}V \tag{6.13}$$

The particular integral in the solution of this differential equation is

$$v_C = V$$

For writing the complementary function of the solution, we use the the roots of the auxiliary equation, which are

$$s_1, s_2 = -\frac{R}{2L} \pm \sqrt{\frac{R^2}{4L^2} - \frac{1}{LC}} \tag{6.14}$$

For resonant converter operation, the circuit must be underdamped. Therefore

$$\frac{R^2}{4L^2} < \frac{1}{LC} \tag{6.15}$$

The roots s_1, s_2 of the characteristic equation may be written as

$$s_1, s_2 = -\frac{R}{2L} \pm j\beta$$

where $\tag{6.16}$

$$\beta = \sqrt{\omega_0^2 - \frac{R^2}{4L^2}}$$

and

$$\omega_0 = \frac{1}{\sqrt{LC}}, \qquad \text{the undamped resonant angular frequency}$$

The complementary function part of the solution is a damped sinusoid, which may be written as

$$v_C = e^{-(R/2L)t}(A\cos\beta t + B\sin\beta t) \tag{6.17}$$

where A and B are constants to be determined by initial conditions in the complete solution.

The complete solution, which is the sum of the particular integral and the complementary function, may be written as

$$v_C = e^{-(R/2L)t}(A \cos \beta t + B \sin \beta t) + V \tag{6.18}$$

The two initial conditions may be stated as

$$v_C = 0 \quad \text{at } t = 0$$
$$i = 0 \quad \text{at } t = 0$$

Substitution of the first initial condition gives

$$A = -V \tag{6.19}$$

To use the second initial condition, we obtain the expression for the current by differentiation of (6.18):

$$i = C\frac{dv_C}{dt}$$

$$= C\left[e^{-(R/2L)t}(-A\beta \sin \beta t + B\beta \cos \beta t) - \frac{R}{2L}e^{-(R/2L)t}(A \cos \beta t + B \sin \beta t)\right]$$

$$= Ce^{-(R/2L)t}\left[\left(B\beta - \frac{AR}{2L}\right)\cos \beta t - \left(\frac{BR}{2L} + A\beta\right)\sin \beta t\right] \tag{6.20}$$

Since $i = 0$ at $t = 0$, we get from (6.20)

$$B\beta = \frac{AR}{2L} = -\frac{R}{2L}V$$

which gives

$$B = -\frac{R}{2\beta L}V$$

Substitution for A and B in (6.18) gives the solution of the differential equation as

$$v_C = V - Ve^{-(R/2L)t}\left(\cos \beta t + \frac{R}{2L\beta}\sin \beta t\right) \tag{6.21}$$

Similar substitution in (6.20) gives the expression for the current as

$$i = CVe^{-(R/2L)t}\left(\frac{R^2}{4L^2\beta} + \beta\right)\sin \beta t$$

On simplification, this gives

$$i = \frac{CV\omega_0^2}{\beta}e^{-(R/2L)t}\sin \beta t \tag{6.22}$$

Equation (6.22) shows that the current is a damped sinusoid, as stated in our earlier description.

The amplitude of the current is

$$I = \frac{CV\omega_0^2}{\beta} \tag{6.23}$$

The duration of the half-period is given by

$$\beta t = \pi \Rightarrow t = \frac{\pi}{\beta}$$

6.9.2 Parallel-Loaded Half-Bridge Resonant Converter Topology

In the configuration of Fig. 6.28, the load is shown connected in series with the resonant circuit. An alternative scheme is shown in Fig. 6.30. In this, the effective load resistance, which is the reflected resistance of the load as seen from the primary side of the isolating transformer, is connected across the resonant capacitor. This is still a series resonant circuit, because the capacitor section and the resonating inductor are in series. It could be described as a series resonant parallel-fed half-bridge topology. This circuit functions in a manner similar to the series-fed circuit. We shall derive the differential equation for the voltage across the capacitor and show that it is very similar to the equation for the series-fed circuit that we derived in (6.13).

When the static switch S_1 is turned ON, the active part of the circuit is as shown in Fig. 6.31. In this circuit,

$$i_C = C\frac{dv_C}{dt}, \quad i_R = \frac{v_C}{R}$$

and

$$i = i_C + i_R = C\frac{dv_C}{dt} + \frac{v_C}{R}$$

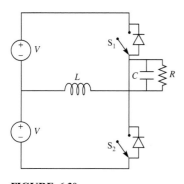

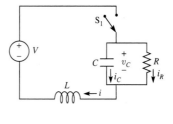

FIGURE 6.30
Parallel-fed half-bridge series resonant converter.

FIGURE 6.31
Parallel-fed converter operation.

The loop equation may be written as

$$L\frac{di}{dt} + v_C = V$$

$$LC\frac{d^2v_C}{dt^2} + \frac{L}{R}\frac{dv_C}{dt} + v_C = V$$

$$\frac{d^2v_C}{dt^2} + \frac{1}{CR}\frac{dv_C}{dt} + \frac{1}{LC}v_C = \frac{1}{LC}V$$

This equation is the same as (6.13) for the series-fed circuit if the time constant L/R in (6.13) is replaced by the time constant CR. The operation of the circuit can be described on similar lines as for the series-fed circuit.

6.10   HALF-BRIDGE AND FULL-BRIDGE INVERTER TOPOLOGIES FOR SMPS

The half-bridge inverter and the full-bridge inverter, which were studied in Chapter 5, are very often the choice for switch mode power supplies when the power ratings are large, typicallly in the region of 1 kW or higher. The half-bridge topology was described in Section 5.4 and the full-bridge topology in Section 5.7. Their operation is similar when used for SMPS. When used for SMPS, the load will consist of the high frequency isolation transformer, with the rectifier and filter circuits on the secondary side. The secondary side circuit will be the same for the full bridge and the half-bridge. These are shown in Fig. 6.32(a) and (b) respectively. The voltage control aspects are also discussed in Chapter 5. Therefore we shall not deal further with these configurations here.

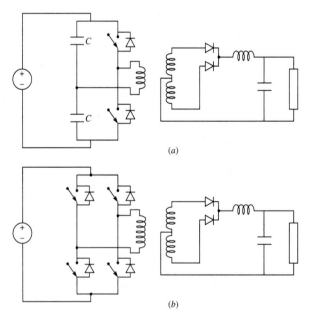

(a)

(b)

FIGURE 6.32
(a) Half-bridge and (b) full-bridge inverter topologies for SMPS.

6.11 CONTROLLERS FOR SWITCH MODE POWER SUPPLIES

Integrated circuit chips that simplify the design and fabrication of switch mode power supplies are available from several manufacturers. These chips are designed to provide the switching signals for the converter circuit, and are generally adaptable for the different topologies that we have considered. For the circuit details and specifications of these ICs, reference will have to be made to the respective data sheets. Our purpose here is to give a brief introduction. There are several types available, with differences in detail. Our description will be confined to some basic functions that they perform and the manner in which they do them.

Basically, these ICs provide pulse width modulated (PWM) switching signals at a programmable frequency. The pulse outputs from the ICs are generally adequate to drive the gates of power MOSFETS, which are the most commonly used switching elements in switch mode power supplies, which operate at high switching frequencies. Figure 6.33 shows a block schematic illustrating one way in which this may be done. This scheme is based on current mode control.

In Fig. 6.33, the circuit block labeled 1 is the clock oscillator that determines the switching fequency. This frequency is determined by the values of an external timing capacitor C_t and an external timing resistor R_t, which are to be connected to the designated terminals of the IC. In this way, the switching frequency is made programmable. The data sheet of the IC will indicate how to choose the values of the timing components to obtain the desired frequency.

The clock oscillator generates a train of pulses at the designed frequency. These pulses are fed to the set input S of the set–reset flip-flop, which is

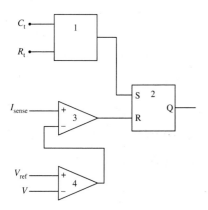

FIGURE 6.33
Generation of PWM switching signals in an IC controller for SMPS.

shown as circuit block 2. The Q output of the S–R is used as the switching signal for the controlled static switch of the SMPS topology. Therefore, in our scheme, the switch will be turned ON at the arrival of the clock pulse from the clock oscillator. The output from the S–R flip-flop will remain high until a reset pulse arrives at its reset input labeled R. This reset signal to the flip-flop comes from the circuit block labeled 3, which is a comparator circuit. The input to the noninverting input terminal of the comparator is shown as the actual sensed value of the current, usually obtained by taking the voltage drop across a current sensing resistor in the SMPs circuit topology. The second input to the comparator, which is shown as the inverting terminal, is the current reference. Therefore the comparator output will go high when the actual sensed value of the current exceeds the current reference. When the comparator output goes high, the flip-flop will be reset and the output from it will terminate. In this manner, the switch in the SMPs topology will be turned ON at the beginning of each switching cycle at the arrival of the clock pulse, and will remain ON until the current, at the location where it is sensed, rises to the reference value provided to the comparator. The switch will remain OFF until the beginning of the next switching period.

The reference to the current comparator is provided by an operational amplifier, labeled as circuit block 4. This is the voltage error amplifier of the controller. The noninverting input to the voltage error amplifier is the reference voltage, which will be a fixed scaled down value of the output voltage to be provided by the SMPS. The second input to the voltage error amplifier is a similarly scaled value of the actual prevailing voltage output of the SMPS, obtained through a voltage sensing circuit. The output of the voltage error amplifier serves as the reference input to the current comparator. Therefore there is an outer voltage control loop and an inner current control loop for the controller. This type of closed-loop control of a SMPS in which an outer voltage control loop provides the reference input to an inner current control loop is called "current mode control." This enables automatic pulse-by-pulse current limiting. One advantage of the current mode control is that such a scheme makes it possible to design modular units that can work in parallel without the danger of individual units getting overloaded through unequal current sharing between parallel units. At present, several PWM controller ICs for SMPs incorporate the facility for current mode control. The sensing circuits for voltage and current generally need to provide electrical isolation, because the output of the SMPS is usually isolated electrically from the input side. Such a requirement will occur if the controller is operated from the primary side of the converter transformer. The current sensing generally can be made from the primary side. But the voltage sensing has to be done from the output side itself. There are functional ICs that can be used for providing this isolation. We shall not discuss this aspect further here.

For providing the fixed voltage reference to the voltage error amplifier, a stable fixed reference voltage is internally generated by the IC controller itself and made available at a designated terminal of it. This voltage has to be

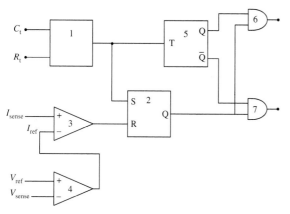

FIGURE 6.34

Generation of complementary switching signals.

properly scaled, using a resistance voltage divider, before being connected to the reference input terminal of the voltage error amplifier.

Complementary outputs. The scheme described above produces only one output. Some circuit topologies may require two complementary outputs, exactly 180° apart. Figure 6.34 shows the manner in which this is usually done inside the controller chip. The circuit blocks 1–4 in Fig. 6.34 are identical to the circuit blocks with the same labels in Fig. 6.33. The additional blocks are labeled 5, 6 and 7. Block 5 is a toggle flip-flop that toggles at each clock pulse. Therefore its Q and $\bar{Q}$ outputs go high after successive clock pulses. When Q is high, the output from the S–R flip-flop (block 2) is gated through the AND gate labeled block 6. When $\bar{Q}$ is high, it is gated through the AND gate labeled 7. In this manner, the output from the S–R flip-flop is gated alternately through the gates labeled 6 and 7. Notice that this scheme effectively makes the switching frequency of the converter half, if we treat the two outputs from 6 and 7 as part of the same switching cycle of the converter. This should be kept in mind when choosing the timing capacitor and resistor values for programming the clock oscillator frequency.

We may also state here that the output from the logic gates or from the S-R flip-flop may not have the necessary capability to drive a power MOSFET. For fast switching of the power MOSFET, it is necessary to quickly charge the relatively large gate input capacitance, including the "Miller" capacitance, as was explained in Section 1.6.6, This will need a relatively large transient current capability. Therefore an interface circuit is provided between the outputs of the logic gates or the flip-flop that will provide the necessary current capability. This is usually in the form of a "totem pole" transistor pair for each signal. This is not included in Figs 6.33 and 6.34.

In addition to the basic features described above, the PWM controller chips for SMPS also provide additional features for supervisory and protective functions. Typically, these may include an automatic shut down when an overvoltage or undervoltage is sensed on the input side. There may be a "soft start" feature, which means the pulse width will slowly increase from a small value when initially started. Full details of such features will be available on the data sheet of the particular IC module.

6.12 UNINTERRUPTIBLE POWER SUPPLY SYSTEMS (UPS)

An uninterruptible power supply is used as a power interface for an electrical or electronic equipment to ensure its reliable and uninterrupted operation in the face of possible power blackouts or power "brownouts," that is, total interruption of mains power, or when the power supply quality such as voltage or frequency deteriorate to levels outside what is permissible for reliable working of the equipment. Investment in an UPS may be justified when the failure of an equipment, because of loss of power supply, or bad power supply, can have serious consequences, such as irreparable loss of valuable data in a computer or failure of a critical system. The UPS ensures continued supply of clean power even when power supply blackouts or brownouts occur. For this, the UPS relies on a battery source. The length of time for which the UPS is able to maintain power to the equipment in the event of a power failure depends on the capacity of the battery. Many commercial designs of UPS come with an internal sealed rechargeable battery unit. There are other designs where the user can connect additional batteries to increase the time duration. The battery unit gets discharged while the UPS is drawing power from it. However, the UPS keeps the batteries charged when the power lines are healthy.

Figure 6.35 shows an example of the functional circuit blocks of an UPS system. The static power conversion blocks in an UPS are actually what we have already studied earlier in this and previous chapters. They are put

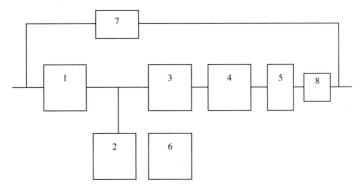

FIGURE 6.35
Organization of circuit blocks in a typical UPS.

together, and their operations are coordinated by the control circuit block in such a way as to achieve the purpose of the equipment, which is to provide a clean power output continuously, irrespective of whether or not the AC mains power lines are healthy.

The circuit block labeled 1 is the rectifier unit. The input to it is from the mains AC supply. It is similar to the front end rectifier unit in a SMPS. As in the case of switch mode power supplies, the recent trend is to incorporate a high power factor preregulator to minimize the line harmonics generated. This is because specifications limiting the maximum harmonics that are allowed to be generated in the AC network are likely to be enforced in the future, on the lines of the International Electrotechnical Commission's Document IEC 555-2.

The circuit block labeled 2 is the rechargeable battery unit. This will be charged from the DC output of the unit 1 when the lines are healthy. It will be floating across the DC output bus of the rectifier, and will provide power to the circuit block 3 in the event of a power supply failure. The battery voltage in this arrangement will have to be large to match the output of the front end rectifier. We shall describe a modified arrangement later, in which the battery need be only of a lower voltage and has a separate charger unit.

The circuit block labeled 3 is the inverter. The modern trend is to use a sinusoidal pulse width modulated inverter. This type of inverter was studied in Chapter 5. Switching frequencies for sinusoidal pulse width modulation may be in the neighborhood of 20 kHz or more. If a square wave inverter with single-pulse pulse width modulation is employed, a significantly larger filter will be needed to extract the sine wave and eliminate the harmonics. Therefore such a scheme is not favored. With sinusoidal PWM and a high switching frequency, a clean sine wave can be achieved with a very small filter.

Circuit block 4 is the filter unit. This is a low-pass LC filter. As stated above, the L and C values will be low if a high switching frequency is used for sinusoidal PWM. A consequent benefit will be low internal voltage drop and low power dissipation in this circuit block.

Circuit block 5 is a power transformer. This provides electrical isolation. Its turns ratio can be chosen to take care of the voltage output of the previous circuit block. The transformer may or may not be required, depending on the nature of the load, and whether or not electrical isolation is required. This transformer, if included, is rated for the low power line frequency.

Circuit block 6 is the control circuit block. This block provides all the switching control signals. It is generally desirable to synchronize the inverter switching to the AC line frequency, so that both voltages may have the same frequency and phase. Therefore the AC mains constitutes one of the inputs to this block. The other inputs to this block are: the sensed values of the output voltage and output current of the UPS. The voltage is needed for automatic closed-loop stabilization of the output voltage. The sensed value of the current is used for protection, or for providing a current limiting feature. It can also be used for ensuring proper current sharing when several modular units are required to operate in parallel to achieve a larger current rating.

The circuit block labeled 7 is a static AC switch that can be used to bypass the UPS and supply the load directly from the mains. In some systems, the power is supplied directly from the mains as long as the supply lines are healthy. The static bypass switch is used to transfer the load to the inverter whenever a blackout or brownout is detected by a line monitoring system. The transfer switching is a duty that can be assigned to the control circuit block 6. Another static switch is also shown, labeled 8. To transfer load to the mains, we turn ON the static switch 7 and turn OFF 8. The reverse set of operations will transfer from the mains to the inverter. Switches 7 and 8 are power semiconductor AC switches for fast transfer. But a parallel mechanical switch is also sometimes used with each to eliminate the voltage drop and power dissipation in the power semiconductor switch after it has been turned ON.

Use of a lower battery voltage. In the scheme described above, the battery floats on the DC voltage bus of the front end rectifier. This calls for a large battery voltage. It is often more convenient to use a battery with lower number of cells when the power capacity of the UPS is not large. Figure 6.36 shows a block schematic of such a scheme. Only the modifications to the scheme of Fig. 6.35 are shown.

In addition to the battery unit labeled 2 in Fig. 6.35, two more additional blocks are added between the battery unit and the DC bus of the rectifier. These are the circuit blocks labeled 2A and 2B.

The circuit block 2A is the battery charger. It is basically a DC-to-DC voltage step down converter (chopper) through which the low voltage battery is charged from the high voltage DC bus.

The circuit block 2B is a voltage step up chopper. When a power blackout or brownout is detected, this converter is brought into operation. It steps up the battery voltage to that of the DC bus and supplies power to the inverter. The rest of the scheme is as described with reference to Fig. 6.35.

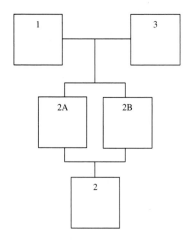

FIGURE 6.36
Additional circuit blocks for using a lower battery voltage.

PROBLEMS

6.1. The reference voltage for a linear voltage regulator chip is 4 V. The current limit threshold is 0.6 V. This IC is to be used for a voltage regulator that will provide a regulated output that is adjustable in the range 5–20 V. The current limit is to be 3 A. Give suitable values for the external programming resistors for voltage adjustment and for current limit.

6.2. An IC regulator chip is to be used with an external series pass transistor to give a maximum current capability of 2.5 A as shown in Fig. 6.2. The regulated output is to be 12 V. The minimum differential voltage needed for the external series pass transistor is 2 V. The maximum power dissipation permissible in the external series pass transistor is 20 W. Determine the maximum permissible voltage variation in the unregulated input voltage.

6.3. A stabilized DC voltage adjustable in the range 12–24 V is to be made. A 12 V three-terminal fixed voltage regulator chip is available. Show how the three-pin fixed voltage regulator can be used with external programming resistors to implement the required adjustable voltage regulator. Determine suitable values of the programming resistors, given that the ground pin current of the three-pin regulator is 50 μA.

6.4. Figure P6.4 shows a 5 V three-pin linear regulator chip being used to provide a regulated output of 15 V. The current in the ground pin of the three pin regulator is 0.1 mA. Determine the value of the resistor labeled R.

6.5. The front end rectifier of a power supply unit is supplied from 120 V, 60 Hz mains. The current to be supplied by it is 1.5 A DC. Choose an approximate value for the filter capacitance so that the peak-to-peak voltage ripple across it is limited to 3 V.

6.6. The front end recifier of a SMPS nominally feeds 162 V DC to the converter section. The latter will function satisfactorily as long as the DC voltage does not fall below 120 V. The filter capacitor of the front end rectifier is 5000 μF. Determine the maximum time duration of a power supply interruption that the SMPS can ride over if it is drawing a steady current of 1 A DC from the rectifier–filter section. Express the answer in number of cycles of the 60 Hz input to the unit.

6.7 A switch mode power supply has a high power factor preregulator, which may be treated as ideal. The voltage at the output of the preregulator is maintained constant at 400 V. If the AC input to the unit is from the 120 V, 60 Hz mains, determine the range of values of the duty cycle of the voltage step up chopper

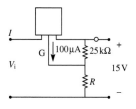

FIGURE P6.4

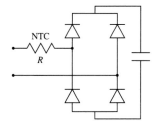

FIGURE P6.8

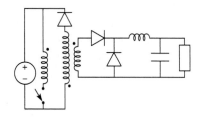

FIGURE P6.9

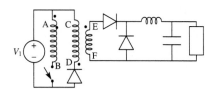

FIGURE P6.10

that does the preregulation. How will this range change if the input is from a 240 V AC supply?

6.8. An OFF LINE switch mode power supply is working from a 120 V, 60 Hz supply. A NTC resistor is used as shown in Fig. P6.8, to limit the peak magnitude of the inrush current to 20 A when the equipment is turned ON. What should be the cold resistance of the NTC resistor? The AC current drawn from the mains by the SMPs during steady operation is 2 A r.m.s. What should be the hot resistance if the power dissipation in the resistor is to be limited to 10 W.

6.9. The input DC voltage to the forward converter of Fig. P6.9 is 170 V. The switching frequency is 100 kHz and the duty cycle is 40%. The converter transformer has 100 turns each on the primary and feedback windings.

(a) Determine the time duration in one switching period when the feedback diode will be ON.

(b) Sketch the waveform of the voltage across the primary, indicating related magnitudes and time durations.

(c) Determine the maximum voltages that have to be withstood by (i) the static switching device and (ii) the feedback diode.

6.10. For the isolated buck converter shown in Fig. P6.10, the data are as follows: $V_1 = 160$ V, switching frequency = 50 kHz, $N_{AB} = 80$ turns, $N_{CD} = 60$ turns, $N_{EF} = 20$ turns and duty cycle = 20%. Sketch the waveforms of V_{AB}, V_{CD} and V_{EF}, indicating relevant intervals and magnitudes. Assume ideal elements and repetitive conditions.

6.11. For the SMPS of Problem 6.10, determine

(a) the maximum permissible duty cycle;

(b) the output voltage V_o at this duty cycle.

6.12. The data for the forward converter of Fig. P6.12 are as follows: $V_1 = 170$ V, switching frequency = 20 kHz, duty cycle = 30%, number of turns on the primary winding = 200, number of turns on feedback winding = 100. Determine

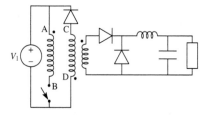

FIGURE P6.12

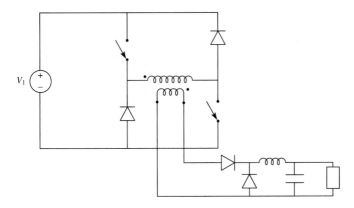

FIGURE P6.14

    (*a*) the duration of the energy feedback interval in one switching cycle;
    (*b*) the peak voltage to be withstood by the switching element;
    (*c*) the maximum permissible duty cycle.

6.13. In Problem 6.12, the number of turns on the feedback winding is increased to 450. Determine the new values of
    (*a*) the energy feedback interval,
    (*b*) the peak voltage that the switching element and the feedback diode will have to withstand,
    (*c*) the permissible maximum duty cycle.

6.14. The two-switch forward converter of Fig. P6.14 has the following data: $V_1 = 340$ V, frequency $= 20$ kHz, duty cycle $= 40\%$, number of turns on the converter transformer primary $= 170$, and number of turns on the secondary $= 37$. Determine
    (*a*) the energy recovery time in one switching period;
    (*b*) the peak voltage that has to be withstood by the switching element;
    (*c*) the DC output voltage.

6.15. For the isolated buck–boost converter of Fig. P6.15, assume ideal elements and repetitive conditions. The data are as follows: $V_1 = 160$ V, frequency of switching $= 25$ kHz, duty cycle $= 80\%$, $N_{AB} = 80$ turns and $N_{EF} = 20$ turns. Assume continuous current through D during the OFF period of the switch.

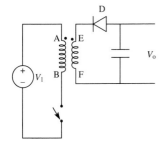

FIGURE P6.15

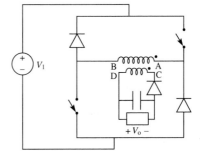

FIGURE P6.17

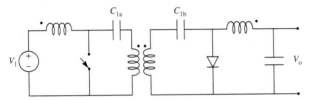

FIGURE P6.18

(a) Sketch the waveforms of the voltages V_{AB} and V_{EF}, indicating relevant magnitudes and time intervals.

(b) Determine V_o.

6.16. In the isolated buck–boost converter of Problem 6.15, the data are as follows: $V_1 = 170$ V, switching frequency = 50 kHz, duty cycle = 0.25, $N_{AB} = 85$ turns and $N_{EF} = 10$ turns. Determine the output voltage V_o. State any assumptions made.

6.17. For the buck–boost converter of Fig. P6.17, the switching frequency is 100 kHz and the duty cycle is 30%. The transformer turns are AB = 300 and CD = 15. $V_1 = 340$ V. Determine the output voltage V_o.

6.18. In the isolated Ćuk converter shown in Fig. P6.18 assume a turns ratio of unity 100% magnetic coupling for the isolation transformer and the coupled reactor. For an input voltage of 170 V and a duty cycle of 0.35, determine the output voltage V_o and indicate the polarity of the output terminals. Assume ideal elements and repetitive operating conditions.

6.19. For the Ćuk converter of Problem 6.18, take $V_1 = 160$ V, with a switching frequency of 50 kHz at a duty cycle of 30%. Determine the output voltage and the voltages across the capacitors C_{1a} and C_{1b}. The coupled reactor and the isolating transformer may be treated as ideal with turns ratio of unity. Assume repetitive conditions and ideal elements.

6.20. For the half-bridge resonant converter shown in Fig. P6.20, $L = 40 \mu H$ and $C = 0.25 \mu F$. Determine the maximum value of R below which the current will be oscillatory.

6.21. In the resonant full-bridge circuit of Fig. P6.21, $L = 100 \mu H$, $C = 0.1 \mu F$ and $R = 20 \Omega$. Determine

(a) the undamped resonant frequency;

(b) the actual oscillatory frequency;

(c) the maximum switching frequency possible for the discontinuous mode of operation.

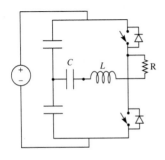

FIGURE P6.20

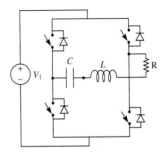

FIGURE P6.21

CHAPTER
7

ADJUSTABLE SPEED DC MOTOR DRIVES

7.1 INTRODUCTION

Historically, the DC motor was almost universally used for most adjustable speed drives from very early times, till the recent developments in static inverters initiated the use of AC motors for some drives. A DC motor control scheme known as the Ward Leonard system was the well-established type of control for all high performance drives. In this introductory section, we shall briefly outline the Ward Leonard system, because modern solid state power electronic controllers for DC motors are, by and large, solid state adaptations of the well proven Ward Leonard control scheme.

Figure 7.1 shows the typical arrangement in a Ward Leonard control scheme. The rotating machines shown inside the broken lines constitute the Ward Leonard set. This is a motor generator set consisting of a three-phase AC motor labeled 1, driving two DC generators, which are the main DC generator labeled 2 and a smaller DC generator called the exciter and labeled 3. The main generator 2 provides the adjustable DC for the motor to be controlled. The exciter is for the purpose of providing the DC for control purposes. An alternative DC source may also be used for this purpose, instead of an exciter coupled to the Ward Leonard set. The motor to be controlled is

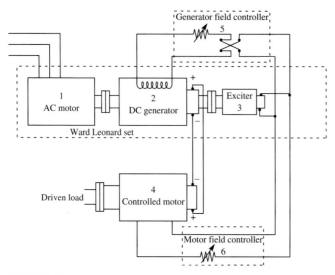

FIGURE 7.1
The Ward Leonard control scheme for a DC motor.

labeled 4. It is coupled to the load to be driven. The AC motor 1 of the motor generator set draws power from the three-phase system bus, and converts it into DC by means of the DC generator coupled to it. This motor generator set must have a sufficient power rating to supply the full load power for the DC motor being controlled, and must be running continuously during the period of use of the DC motor. The output of the DC generator is directly connected to the armature terminals of the DC motor. The voltage applied to the DC motor is adjustable by means of the field current of the generator. The field current of the generator can be adjusted or reversed by means of the reversing switch and the adjustable rheostat shown in series with the field, inside the field control block of the generator, labeled 5. In this manner, the magnitude and polarity of the DC voltage applied to the controlled motor can be adjusted.

The field of the DC motor is also supplied from the same DC excitation bus, and can be independently adjusted by means of the series rheostat of the motor field control block labeled 6. The manner of operation of the controller may be described as follows.

To start the motor from rest, the generator field is initially set at close to zero, so that its voltage output, which is the DC voltage provided to the motor, is close to zero. The field current of the motor is adjusted to its maximum rated value by making the resistance of 6 equal to zero. The field current of the generator is now progressively increased (using the generator field control block labeled 5) in the required direction, to give the proper polarity of the DC motor voltage, for the wanted direction of its rotation. During this period, the motor accelerates and attains a speed that will be determined by the voltage applied to its armature, which is the voltage output of the generator. The manner in which the field is adjusted should be such that the motor draws the full rated current to enable it to develop the full rated torque and accelerate quickly to reach the required speed. As the motor spins, it generates a "back e.m.f." at its armature terminals, which opposes the applied e.m.f. It is the difference between this back e.m.f. and the applied e.m.f. that drives the current through the motor armature. Therefore, to maintain the constant maximum armature current, the generator voltage is continuously adjusted so that its value is higher than the back e.m.f. of the motor by the correct amount needed to maintain the armature current at the maximum rated value. Notice that we have maintained the field flux of the motor at its full maximum value. This will enable the motor to develop the maximum torque, because the torque is directly proportional to its field flux and its armature current. The speed will be dependent on the armature voltage, and this can be finally set at the value needed for the required speed.

To lower the speed or to bring the motor to rest, this scheme allows electric braking. For this, the field of the generator is adjusted to make the generator voltage less than the motor induced voltage. This will cause the motor current to reverse. The mode of operation of the motor will now change into the generating mode, and the mode of operation of the Ward Leonard generator will change into the motoring mode. In this way, the controlled motor will operate as a generator during braking, being driven by the stored kinetic energy of the rotating system consisting of the motor and all inertia coupled to it. In this way, energy will flow from the motor to the Ward Leonard generator machine 2, which is now in the motoring mode. This energy will be fed back to the AC system bus through the AC motor, which will now go into the generating mode. As the motor 4 slows down because of this, its induced voltage keeps dropping, and the terminal voltage must be continuously lowered by the adjustment of block 5 to keep the braking current at its maximum permissible value, to achieve fast deceleration to the new wanted speed. If the motor is to be stopped, this means that the new speed is zero, and therefore the field current of the Ward Leonard generator 2 should be stopped at zero. If a reversal of the direction of rotation is needed, the field of the generator should be increased again in the reverse direction, so that the terminal voltage is reversed and the motor will go into the motoring mode and accelerate in the reverse direction.

Till now, we have kept the field current of the motor at its maximum value, and have not made any adjustment to it. The reason for this was to enable the maximum torque to be developed so that the motor can accelerate or decelerate to reach the required speed in the shortest possible time. However, if we stick to this policy of keeping the field current of the motor at its maximum value, and increase the speed only by increasing the generator voltage, we shall reach the highest speed possible when the generator field reaches the maximum value. This maximum speed is called the base speed of the motor. Base speed is the highest speed that can be achieved by keeping the field flux at the rated maximum and the armature voltage also at the rated maximum. But DC motors are invariably designed to run above the base speed, typically about 2–3 times the latter. To increase the speed beyond the base speed, we shift the control to the field controller of the motor, labeled 6. The speed of a DC motor varies inversely as the field flux. Therefore, by weakening the field current, and therefore the field flux, we can make the speed go up. Speeds above the base speed are achieved by field-weakening. However, the torque is proportional to the product of the field flux and the armature current. Therefore, in the field-weakening range, the maximum torque that the motor can develop decreases, because we have to limit the armature current at the rated maximum. Below the base speed, the motor can develop the same maximum torque at all speeds, because the field flux is maintained at its maximum value. Therefore the speed control range below the base speed is called the constant-torque range. Above the base speed, the torque drops off as the speed is increased by field-weakening. Therefore, in this range, the maximum power output is approximately constant. The speed control range above the base speed is often described as the constant-horsepower region of control for this reason. In the Ward Leonard control scheme, the DC motor field and armature are supplied from separate sources. In such a separately excited motor, there are two distinct regions of control: the constant-torque region, where the control is by adjustment of the armature voltage, and the constant-horsepower region, where the control is by field-weakening. The base speed is the boundary that separates the two control ranges. This is exactly the same as the control policy followed in the case of a separately excited DC motor controlled by means of static power electronic converters, to be described in this chapter. The static converters basically serve the same purpose as the rotating motor generator set of the Ward Leonard scheme. For this reason, such a control system may be described as the static Ward Leonard system.

In recent years, static Ward Leonard systems have totally replaced the rotary Ward Leonard system, which was the established means of control in the past, for high performance DC motor drives. The rotary Ward Leonard scheme gives very good performance, but has many disadvantages as regards cost and maintenance. The generator and the motor of the Ward Leonard set have to be each rated at a higher power than the motor itself, as we show in the illustrative example to follow. Also, there are the requirements of foundations and the space needed for the large rotating Ward Leonard set, and the large maintenance requirements, especially for the DC generator, involving commutator maintenance and brush replacement. Noise and vibration are other disadvantages that are eliminated when the rotating set is replaced by a static solid state power electronic converter.

Illustrative example 7.1. A DC motor is rated at 600 kW for a rolling mill drive in a steel mill. It uses a rotary Ward Leonard controller as in Fig. 7.1. The total DC power needed for the field control is approximately 10% of the motor rating.

Determine the total power rating of the machines of the Ward Leonard set. Assume an efficiency of 85% for each machine.

Solution. We assume that the stated efficiency of 85% for the motor is for the conversion efficiency through the armature. Therefore the output power needed from the Ward Leonard generator will be

$$P = 600/0.85 = 706 \text{ kW}.$$

The rating of the exciter machine is $600 \times 0.10 = 60 \text{ kW}$.

The driving power needed to drive the exciter is $60/0.85 = 71 \text{ kW}$. The driving power needed for the generator of the Ward Leonard set is $706/0.85 = 831 \text{ kW}$. The total output power needed for the AC motor is $831 + 71 = 902 \text{ kW}$.

The input power from the AC source is $902/0.85 = 1061 \text{ kW}$.

The total output rating of all the machines of the Ward Leonard set is $706 + 902 + 60 = 1668 \text{ kW}$.

The above example shows that machines whose output ratings add up to a total over 1600 kW are needed to control the 600 kW motor. This is because the main power has to pass through two machines before it reaches the motor. This fact also can be seen to grossly lower the overall power efficiency.

7.2 DC MOTOR BASICS

7.2.1 Structure and Circuits

The DC motor basically has two electric circuits: the field circuit and the armature circuit. The field circuit is located on the stationary part or stator of the machine. It consists of coils surrounding the magnetic poles of the stator. These poles project inwards in the stator, as shown in the sectional diagram of Fig. 7.2(*a*). There will be an even number of poles, of alternately north and south polarity. Figure 7.2(*a*) shows a two-pole machine. All the poles will be of the same strength; that is, the magnetic flux of every pole will be the same. The field windings will be identical for all poles, and the current will be the same in each. The field coils of all the poles may be connected in series, and the field circuit ends up in two terminals labeled F_1 and F_2, which are the field terminals of the machine, shown so labeled in Fig. 7.2(*c*). The purpose of the field circuit is to magnetize the poles of the motor and thereby create the magnetic flux in the airgap of the motor between the stator and the rotor. The current supplied to the field circuit serves only to create and maintain the magnetic flux. No conversion of power to mechanical power takes place from the field circuit, although the field flux is necessary as the medium through which the actual conversion of power takes place in the motor. The field circuit will not be

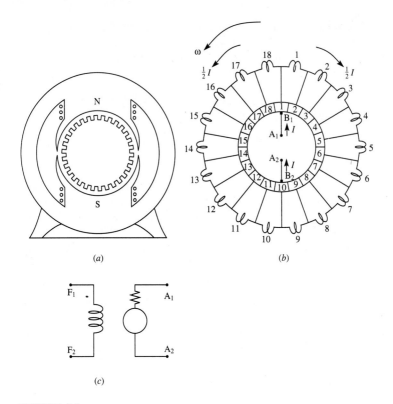

(a)   (b)

(c)

FIGURE 7.2
Structure and circuits of the DC motor.

necessary if permanent magnets are used for the magnetic poles. There are such permanent magnet DC motors also, without any field windings. But in this chapter, we shall consider only machines with field windings.

The armature circuit. The main power circuit of the DC motor is the armature circuit, and this is on the rotor. The armature circuit consists of coils housed in the slots that are shown around the armature periphery in the sectional representation of Fig. 7.2(a). The width of a coil, known as the "coil pitch," is equal to the width of the zone of one pole, called the "pole pitch." Therefore, if one side of a coil is under a north pole, the other side will be at the corresponding location under the adjacent south pole. When an armature coil carries a current, the forces due to the interaction of the current and the field flux will be equal but opposite on the two sides of the coil, and together they produce the torque that the coil contributes to the total motor torque. The coils on the armature are interconnected in the manner shown symbolically in Fig. 7.2(b) for a machine with 18 armature coils. These coils are labeled 1–18.

The end of a coil is connected to the beginning of another coil. Proceeding in this manner, the end of the last coil is connected to the beginning of the first, so that the armature circuit becomes a closed circuit, with no free beginning or end terminals. The armature has another structure mounted on it called the "commutator," This has a cylindrical surface, consisting of narrow copper "segments" fixed lengthwise on the outside. Each segment is insulated from the others by mica strips located between the adjacent segments. The commutator is shown in a sectional view in Fig. 7.2(b) in the center of the figure. There are 18 segments on this commutator—the same as the number of coils. The commutator segments are labeled 1–18 in the figure. In an actual motor, the commutator segments serve as binding posts to which the armature coils are connected. The end terminal of one coil and the beginning terminal of another are both connected to the same commutator segment. In this manner, each commutator segment serves to series-connect the coils and achieve the final connection pattern shown in Fig. 7.2(b). The connections of the individual coil ends to the same commutator segment are not explicitly shown in the figure. The commutator is fixed to the rotor, and rotates with the armature. Electric current is fed to the rotating armature circuit through the commutator, by means of two stationary graphite brushes, each of which makes rubbing contact with the commutator surface. The commutator and brush assembly constitute an ingenious and unique means of switching, which makes the pattern of current flow in the zone under each stator pole unaffected by the angular movement or rotation of the armature. Figure 7.2(b) has been drawn in a manner that will bring out this aspect of the DC motor. In the position of the rotor shown in Fig. 7.2(b), the brush labeled B_1 is making contact with the commutator segment labeled 1, and the other brush is making contact with the commutator segment 10. Current enters the armature circuit from brush B_1 through the commutator segment 1 and returns through the segment 10 to the brush B_2. Between these two points, the armature in effect has two parallel paths, and each path carries half the total armature current. One path consists of the nine coils labeled 1–9. Each coil in this group of nine coils starts under, say the north pole, and ends under a south pole. Therefore all of them contribute to torque in the same direction. The other parallel path consists of the remaining nine coils, which are labeled 10–18. These coils, which constitute the second parallel path, start under a south pole and end under a north pole. Since the direction of current is in reverse in these coils, they also contribute to torque in the same direction. As the armature rotates through an angle equal to the zone of one commutator segment, the commutator segment that makes Contact with the brush B_1 will change from 1 to 2, and the commutator segment that makes contact with brush B_2 will change from 10 to 11. But the pattern of current flow under a pole will not change. The role of coil 1 will be taken over by 2 and the role of coil 10 will be taken over by 11. We shall still have nine coils in one parallel path, and nine coils in the other, all contributing to torque production in exactly the same manner as before. The commutator functions in this way to ensure that all the coils under one pole will be carrying current in the same direction, irrespective of rotation. As a coil moves from the zone of one pole and enters the zone of the opposite pole, it automatically moves to the other parallel path and continues to contribute torque in the same direction as before. The current flow pattern in the conductors under a pole does not change because of rotation, although the individual conductors change. If we only change the labeling of the coils and the commutator segments in Fig. 7.2(*b*) by incrementing each number by one every time the motor rotates through the angular zone of one commutator segment, then the condition represented by Fig. 7.2(*b*) will be valid for all positions of the motor, unaffected by its speed of rotation. This is the unique feature of the switching achieved by the commutator and brush assembly in a DC machine.

The commutator is the device that makes the machine capable of operating from a DC source. Actually, the current flowing in the individual coils of the armature are alternating in nature, because every time a coil shifts from one parallel path to the other; that is, every time a coil moves from the zone of one pole to the zone of the opposite pole, there is a reversal of current through it. Therefore the commutator and brush assembly constitute the means of converting the DC current input into the brushes to AC current in the armature coils. The necessity to have the commutator and brush assembly may be considered also as a major disadvantage of the DC motor. They add to the weight and cost of the machine, and above all to its maintenance requirements. The brushes wear out after a period of use, and may have to be replaced. In an industrial drive where the machines operate continuously for long periods of time, there will be some wear of the commutator surface also. It may have to be resurfaced, and then it becomes an expensive maintenance operation. Also, the switching taking place between the commutator and the brushes as the machine rotates is not always perfect, and there is invariably a certain amount of sparking at the edges of the brushes when the current flowing between a brush and a commutator segment is interrupted as the segment moves past the commutator. This makes the use of an exposed commutator and brush structure unsafe in hazardous environments. While the DC motor has excellent controllability compared with other types of motors, these limitations are contributing to an increasing trend towards the use of AC motors in adjustable speed drive systems.

7.2.2 Magnetic Decoupling between the Field and Armature Circuits

An inherent feature of the DC motor that is responsible for its high performance in adjustable speed applications is the fact that the magnetic circuits of the field and armature are mutually decoupled. By this, we mean that the magnetic flux created by the field current does not link with the armature coils. Similarly, the magnetic flux resulting from the armature current does not link with the field winding. This can be explained by reference to Fig.

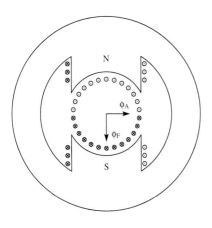

FIGURE 7.3
Spatial orientations of the field flux and the armature
m.m.f. in a DC motor.

7.3, which shows a sectional representation of the field and armature in a DC
motor. The north and south poles are shown vertically. The magnetic flux of
these poles are created by the field winding shown around the poles which
carry currents in the directions shown. A ⊗ symbol indicates that the direction
of current is downwards into the surface of the drawing on the paper, and the
dot ⊙ indicates that the direction is upwards out of the surface of the drawing.
The current directions in the armature conductors under the zone of each pole
are also shown in this manner. We notice that all the conductors carry currents
upwards under the north pole and downwards under the south pole. As the
armature rotates, the individual conductors keep moving from the zone of one
pole to the zone of the other pole. But when a conductor crosses the zone in
this manner, the direction of its current also reverses, by the switching action
of the commutator. Therefore the current flow pattern depicted in Fig. 7.3 does
not change because of the rotation of the armature. The spatial field is the
same whether the armature is stationary or rotating. Inspection of Fig. 7.3
shows that the m.m.f. created by the armature currents is in the direction
labeled ϕ_A. The armature is like a coil oriented with its axis in the direction
ϕ_A. The direction of the field flux is vertically downwards, as indicated by ϕ_F
in the figure. Notice that these two spatial directions are fixed and perpendicular to each other. This means that the flux created by the armature current
does not link with the field circuit, and vice versa. In other words, the field
circuit and the armature circuit are magnetically decoupled.

This mutual magnetic decoupling has considerable significance on the
"dynamic" performance of a DC motor in an adjustable speed drive system,
which may be explained as follows. When a speed change is to be implemented
by changing the voltage applied to the armature circuit, the armature current
has to build up fast for the machine to develop a high torque to accelerate or
decelerate quickly to reach the wanted speed. Inductive time constants
introduce some delay in the rate at which currents can build up in the circuits
of the motor. The field circuit of a motor is highly inductive, whereas the

inductive time constant of the armature circuit is small. Because the field circuit is magnetically decoupled from the armature circuit, it becomes possible for the armature current to build up very quickly without being hampered in this by the need to create a change in the field flux. The decoupling between the field and armature circuit exists naturally in the DC motor. Therefore, by supplying the field and armature from separate converters, it becomes possible to make fast changes in the armature current. The "transient" or "dynamic" performance of drive systems that use DC motors with separately controlled static converters for the field circuit and the armature circuit is exceptionally good. By this, we mean that the motor can respond in an extremely fast manner to changes in demanded speed or demanded torque.

7.2.3 Equations for the torque and the Induced e.m.f. in a DC Motor

From Fig. 7.3, it can be seen that the direction of current flow in an armature conductor is perpendicular to the direction of the field flux. Therefore each conductor has an electromagnetic force induced on it that is proportional to the armature current and the flux density in the air gap, which is proportional to the total flux of a pole. The resulting torque of the individual conductors add up to give the total torque of the motor, which may be expressed by the following equation:

$$T = K_1 \Psi I \quad \text{N M} \tag{7.1}$$

where Ψ is the flux per pole in W, I is the armature current in A and K_1 is a machine constant that depends on the dimensions, number of poles and the armature winding details such as number of conductors and the number of parallel paths.

When the armature spins, each armature conductor will have a tangential velocity that is perpendicular to the radial magnetic flux. Therefore there will be an electomagnetically induced e.m.f. in each conductor, and such e.m.f.s will add up to give the total "induced e.m.f." or "back e.m.f." of the motor armature. This e.m.f. will be proportional to the velocity and the flux per pole, and may be written as

$$E = K_2 \Psi \omega \quad \text{V} \tag{7.2}$$

where, Ψ is the flux per pole in W, ω is angular velocity in rad/s and K_2 is a proportionality constant determined by the dimensions and winding details of the machine.

Equations (7.1) and (7.2) are respectively the torque and voltage equations of the motor. When the terms are expressed in standard units, the constants K_1 and K_2 will be found to be equal. This can be verified by considering the conservation of power, as shown below.

The power converted to mechanical power is the electrical power fed into the machine in opposition to the induced e.m.f. This power will be given by

$$P = EI = K_2 \Psi \omega I \quad W$$

The converted mechanical power will be given by the torque in N m multiplied by ω in rad/s:

$$P = T\omega = K_1 \Psi I \omega \quad W$$

Equating the two powers, we get

$$K_2 \Psi \omega I = K_1 \Psi I \omega$$

From this, we find that the constants in the e.m.f. equation and the torque equation are the same, and we may write these equations as

$$E = K\Psi\omega \tag{7.4}$$

$$T = K\Psi I \tag{7.5}$$

where K is a constant for a given motor.

The electrical circuit equations for the armature and the field circuits can be written by reference to Fig. 7.4. The terminals of the armature circuit are labeled A_1 and A_2 and those of the field F_1 and F_2. The voltage, labeled V, that we measure at the terminals of the armature is the voltage applied by the external DC source that supplies power to the armature circuit. It is different from the induced voltage or the back e.m.f. of the armature, which is labeled E in the figure. The total resistance of the armature circuit, which is actually distributed along the armature conductors, brushes and the contact zones between the brushes and the commutator surface, is shown as a lumped resistance labeled R_a in the figure.

The DC voltage equation for the armature circuit may be written as

$$V = E + R_a I \tag{7.6}$$

where I is the armature current. Similarly, the DC voltage equation for the field may be written as

$$V_f = R_f I_f \tag{7.7}$$

where the suffix f denotes the field quantities.

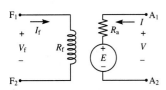

FIGURE 7.4
Armature and field circuits.

7.2.4 The Saturation Curve

The flux per pole, Ψ, which is present in both the voltage and torque equations, is actually determined by the field current I_f. The relationship between the field current I_f and the flux depends on the reluctance of the magnetic circuit of the field. For low values of field current, and therefore low values of the flux, the magnetic circuit will be linear. But magnetic saturation will commence when I_f is progressively increased, and begins to exceed the linear region. The working range of a motor usually extends into the saturation region. Therefore the relationship between I_f and the flux is needed for calculations. This relationship is available usually in an indirect form, or can be determined by experiment. Equation (7.4) shows that the induced e.m.f. E is proportional to the flux Ψ if the machine is rotating at a constant speed ω. Therefore the test performed to obtain a relationship equivalent to that between the flux and the field current is to spin the machine at constant speed by means of another motor coupled to it, and note the voltage induced by means of a voltmeter connected to the terminals of the armature for different values of the field current. There should be no current flowing in the armature, and, under this condition, the voltage at the armature terminals will be equal to the induced e.m.f. The graph showing the induced e.m.f. against the field current, at a known fixed speed, is called the "magnetization curve" or the "saturation curve." A typical saturation curve for a DC machine is shown in Fig. 7.5. The magnetization curve shows the nature of the relationship between the magnetic flux and the field current, although it is actually plotted as the induced voltage versus the field current. The curve is always for a fixed speed, the value of which should be stated. The magnetization curve obtained for one fixed speed is sufficient. Equation (7.4) shows that the induced e.m.f. is directly proportional to the speed. Therefore, if the magnetization curve is available for one speed, the curve for any other speed can be obtained by simply scaling the voltages, proportional to the speed.

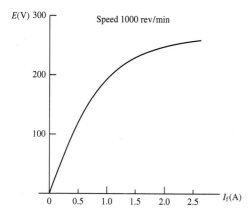

FIGURE 7.5
The magnetization curve for a DC machine.

7.2.5 Method of Exciting the Field of a DC Motor

We have seen that the DC motor has two circuits: the armature circuit, through which the power converted to mechanical power is transformed, and the field circuit, which provides the magnetic flux in the air-gap of the machine. These two circuits may be connected in parallel and supplied from a single DC source. When a motor is operated in this manner, it is called a "shunt motor." Alternatively, the two circuits may be fed from two separate DC sources. When the motor is operated in this manner, it is called a "separately excited motor." The field circuit can also be connected in series with the armature, and the motor may be operated in this manner with only a single series circuit consisting of the armature and the field. Such a motor is called a "series motor". The field flux of the field windings is determined by the total ampere-turns (m.m.f.) created by the field windings. The required ampere-turns NI can be produced by a large number of turns carrying a small current or a small number of turns carrying a large current. If the motor is intended for series operation, the field is wound with a small number of turns designed to carry the large armature current that is in series with it. In this way, the resistance and resulting voltage drop as well as the power dissipated in the field windings are all kept low. For separately excited and shunt excited operations, the source of the motor field functions effectively as a voltage source, and the field should be designed accordingly. The field coils of such a motor are designed with a large number of turns rated for a small current. In this way, the power needed for the field circuit is kept low. In another type of DC motor, called the "compound motor," there are two separate field coils for each pole; one intended for series connection and the other for shunt or separate connection. The type of motor that gives the maximum flexibility for speed control is the separately excited motor. Such motors are the most widely used for static adjustable speed systems.

7.3 SPEED CONTROL OF A SEPARATELY EXCITED MOTOR

7.3.1 Converter Connections

Figure 7.6 shows the converter arrangement for supplying power separately to the field and armature circuits. The type of converter to be used in a static

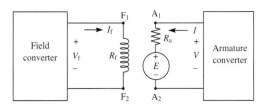

FIGURE 7.6
Converter arrangement for a separately excited DC motor.

speed control scheme for a separately excited motor depends on the type of power source available. If the available power source is DC then DC-to-DC converters, that is, choppers, will be used. This will be the case if DC motors are to be used in a battery powered automobile or in an electric train in a DC railway network. If, on the other hand, the power is from a three-phase AC utility supply, as is the case in the large majority of industrial DC motor drives, the converters will be line commutated phase controlled converters. Two separate converters will be used: one of small capacity for the field and another of larger capacity for the armature circuit, as shown in Fig. 7.6. Both converters must be capable of providing variable voltage output if both field and armature currents need to be varied. For adjustable speed applications, the main control is in the armature circuit, and therefore the armature converter is always designed for variable output. The field converter has to provide variable output only if the control scheme involves field adjustment. The type of converter is not explicitly shown in Fig. 7.6. They may be choppers or line commutated converters, depending on the nature of the available power source.

7.3.2   Speed Control Ranges with Separate Excitation

Equation (7.6) may be rewritten to give the induced e.m.f. as

$$E = V - R_a I$$

This value of E may be substituted in the induced e.m.f. equation (7.4) to give

$$V - R_a I = K\Psi\omega \tag{7.8}$$

From (7.8), we can write the following expression for the speed

$$\omega = \frac{V - R_a I}{K\Psi} \quad \text{rad/s} \tag{7.9}$$

We can use (7.9) to explain the two distinct ranges of speed that exist in the practical speed control of a separately excited DC motor.

The torque developed by the motor is given by (7.5) as proportional to the product of the flux Ψ and the armature current I. Therefore the armature current I in the numerator on the right-hand side in (7.9) is determined by the torque required to drive the load. The maximum value of I will be determined by the maximum torque required to be developed. $R_a I$ is the ohmic drop in the resistance of the armature circuit. This voltage drop will be limited by the maximum value of the armature current I. Therefore the numerator on the right-hand side increases when we increase V. Equation (7.9) shows that we can increase or decrease the speed by adjustment of the impressed voltage V across the armature if we maintain the field flux Ψ at a fixed value. In practice, when we adjust the speed in this manner, we keep the flux Ψ at the highest possible value. This is because the torque is determined by the product of the flux and the armature current. Therefore, by keeping the flux at the maximum value, the current that needs to be drawn from the armature converter will be the minimum for developing the required torque. To maintain the flux at the maximum value, we keep the field current at the maximum value by adjusting the voltage output of the field converter at the upper ceiling limit. The lower range of speed control of a separately excited motor consists of all the speeds from zero up to the highest possible, with the field flux at the maximum value. In this range, all speed adjustments are made by varying the armature voltage, that is, by varying the output of the armature converter. The highest speed in this range is obtained when the armature converter is adjusted to give the maximum permissible output, which should correspond to the rated voltage of the motor. This speed is called the "base speed" of the motor. At the base speed, and any other speed below it, the motor can develop the same maximum torque, which is determined by the maximum flux and the rated maximum armature current, in accordance with the torque equation (7.5).

This range of speed control is described as the constant-torque region. In this region, the power output capability of the motor increases proportionally to the speed, because the power is given by the product of the speed and the torque.

Above the base speed limit, the impressed armature voltage is maintained at the upper ceiling limit, and all speed adjustments are made by decreasing the field flux, that is, by decreasing the output of the field converter. The speed increases inversely as the flux is decreased, according to (7.9). Because the torque is proportional to the product of the flux and the armature current, the reduction in the flux for speeds above the base speed causes a proportionate reduction in the maximum torque that the motor can develop. An increase in speed with a proportionate reduction in the torque capability makes the power output capability constant. Therefore the speed control range, achieved by field weakening, above the base speed is described as the constant-horsepower range. The maximum speed attainable in the field-weakening range will be determined by the minimum permissible field for the satisfactory running of the machine. This may be typically extended up to about two to three times the base speed.

Illustrative example 7.2. The speed of a DC motor is controlled by means of separate converters for the field and armature circuits. The armature resistance is $0.5\,\Omega$ and the field circuit is $125\,\Omega$. The magnetization curve for this machine at 1000 rev/min is as given in Fig. 7.5. The field voltage is kept unchanged at 250 V. Assume that the load torque is the same at all the speeds and that this corresponds to an armature current of 20 A. Determine the speed of the motor for the following values of the output voltage of the armature converter: (a) 50 V; (b) 100 V; (c) 150 V; (d) 200 V; (e) 250 V.

Solution. The field current is $250/125 = 2\,A$. Reference to the magnetization curve of Fig. 7.5 shows that for this field current the flux is such as to induce a voltage E = 250 V at a speed of 1000 rev/min.

(*a*) Here

$$E = V - R_a I = 50 - 0.5 \times 20 = 40 \text{ V}$$

Since the induced e.m.f. is 250 at 1000 rev/min, to induce 40 V, the speed will be given by

$$N = \frac{40}{250} \times 1000 = 160 \text{ rev/min}$$

(*b*) Here

$$E = 100 - 0.5 \times 20 + 90 \text{ V}$$

and

$$N = \frac{90}{250} \times 1000 = 360 \text{ rev/min}$$

(*c*) Here

$$E = 150 - 0.5 \times 20 = 140 \text{ V}$$

and

$$N = \frac{140}{250} \times 1000 = 560 \text{ rev/min}$$

(*d*) Here

$$E = 200 - 0.5 \times 20 = 190 \text{ V}$$

and

$$N = \frac{190}{250} \times 1000 = 760 \text{ rev/min}$$

(*e*) Here

$$E = 250 - 0.5 \times 20 = 240 \text{ V}$$

and

$$N = \frac{240}{250} \times 1000 = 960 \text{ rev/min}$$

We may tabulate the speed versus armature voltage as follows:

Armature converter voltage (V)	50	100	150	200	250
Speed (rev/min)	160	360	560	760	960

The above example illustrates the speed control below base speed by armature voltage control. The following example illustrates control above base speeds by field weakening.

Illustrative example 7.3. The motor of example 7.2 is rated for a maximum terminal voltage of 250 V for both field and armature. The minimum field voltage for satisfactory operation of the motor is 50 V. In the field-weakening mode of control, the armature voltage is kept constant at 250 V and the field voltage is changed. Determine the speed for the following values of the field input voltage: 200 V 150 V 100 V and 50 V. Specify the base speed and the speed control ranges for armature voltage control and field weakening. In all cases, assume that the motor is drawing its full rated current of 20 A. What is the ratio of the maximum speed to the base speed?

Solution. Since the armature voltage is fixed at 250 V and the armature is drawing 20 A in all cases, the induced voltage of the armature will be the same for all the cases, and will be

$$E = 250 - 20 \times 0.5 = 240 \text{ V}$$

For $V_f = 200$ V, $I_f = 200/125 = 1.6$ A. At the flux corresponding to this field current, reference to the magnetization curve shows that the induced e.m.f. is 238 V at 1000 rev/min. Therefore, to induce 240 V, the speed will be

$$N = \frac{240}{238} \times 1000 = 1008 \text{ rev/min}$$

For $V_f = 150$ V, $I_f = 150/125 = 1.2$ A, and the induced e.m.f. at 1000 rev/min from the graph is 215 V. Therefore the speed for $E = 240$ V will be

$$N = \frac{240}{215} \times 1000 = 1116 \text{ rev/min}$$

For $V_f = 100$ V, $I_f = 100/125 = 0.8$ A and the induced e.m.f. at 1000 rev/min for this value of I_f from the graph is 173 V. Therefore the speed for $E = 240$ V is

$$N = \frac{240}{173} \times 1000 = 1387 \text{ rev/min}$$

For $V_f = 50$ V, $I_f = 50/125 = 0.4$ A, and the induced e.m.f. at 1000 rev/min for this value of I_f from the graph is 100 V. Therefore

$$N = \frac{240}{100} \times 1000 = 2400 \text{ rev/min}$$

The base speed is the speed for $V_f = 250$ V. This was found in the previous example as 960 rev/min.

The speeds in the field weakening range are tabulated below

Field voltage (V)	250	200	150	100	50
Speed (rev/min)	960	1008	1116	1387	2400

The speed control ranges are as follows:

armature voltage control: 0–960 rev/min
field weakening: 960–2400 rev/min

The maximum-to-base speed ratio is 2400/960 = 2.5.

7.3.3 The Closed-Loop Control Strategy for a Separately Excited Motor

Figure 7.7 shows a closed-loop control scheme for a separately excited DC motor that is based on armature voltage control below base speed and field weakening above base speed. The operation of the controller may be described as follows.

We shall first describe the armature control part. There are two control loops for control of the armature current. One is an outer "speed loop" and the other is an inner "current loop." The wanted speed is given as reference input to the outer speed loop. This speed reference signal is labeled ω_{ref} in Fig. 7.7. This will be an adjustable voltage to be set on the basis of certain number of volts per rev/min. The actual prevailing speed is sensed by a tachometer or some kind of speed sensing device, and fed back to the controller. This sensed speed is labeled ω_f in the figure, and will also be scaled at the same number of volts per rev/min. The difference $\omega_{\text{ref}} - \omega_f$ is the speed error. This is amplified by the speed control amplifier, labeled S_a in the figure. The output of the speed control amplifier does not directly control the armature converter. Instead, its output is given as the reference input to a second control loop, called the current control loop. This current reference to this control loop is labeled

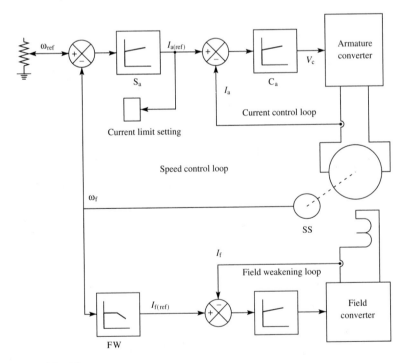

FIGURE 7.7
Closed-loop control scheme for a separately excited DC motor.

$I_{a(ref)}$ in the figure. The actual armature current is sensed by a fast current sensing device and fed back to the current controller as the signal labeled I_a in the figure. The current error is $I_{ref}-I_a$. This error is amplified by the current control amplifier, C_a in the figure. The output of the current control amplifier serves as the control input to the armature power converter. This signal is labeled V_c in the figure. It controls the armature converter in such a way as to make the current error zero. If the armature converter is a phase controlled line commutated converter, such as what we have studied in Chapter 2, the output V_c of the current amplifier varies the firing angle of the converter suitably. If, on the other hand, the armature converter is a chopper then V_c serves to vary the duty cycle of the chopper suitably.

In this scheme, the outer speed loop gives the reference input to an inner current loop. This type of nested closed-loop control technique can also be extended for position control in a servo drive. For a position control system, there will be three control loops. The outermost will be a position loop, for which the reference signal given will be the required angular position of the motor. The actual position should be sensed and given as the position feedback signal to the position control amplifier. The output of the position control amplifier serves as the reference input for the speed control loop. The speed control loop and the current control loop will be organized in the same manner as shown in Fig. 7.7.

Field-weakening. In Fig. 7.7, a separate converter is shown, labeled the "field converter," which supplies the field current. A closed-loop controller is shown for the field current also. The reference input to this controller is from the circuit board labeled FW to signify field weakening. The field-weakening module also has the speed signal ω_f input from the speed sensor. The FW module is designed to give a fixed output, corresponding to the maximum field current of the motor, as long as the actual speed as determined by the speed sensor, is below the base speed of the machine. When the speed exceeds the base speed limit, the output of FW is designed to start decreasing, as shown symbolically, inside this block in Fig. 7.7. Since the output of FW is the reference for the field current, the field-weakening circuit module implements its objective by progressively giving a lower field current reference for speed adjustments above the base speed. For closed-loop control of the field current, there is shown in Fig. 7.7 a current sensing device in the field circuit, which provides the feedback signal I_f of the field current. The field error amplifier amplifies the error between $I_{f(ref)}$ and I_f, and controls the field converter in such a way as to cancel the error. For this, the output of the field amplifier will vary the firing angle of the field converter if it is phase controlled converter, or the duty cycle if it is a chopper.

Current limit for armature current. A current limiting feature is invariably included for the armature current controller section. This is to avoid giving an excessively high current reference at starting or whenever the speed error is very large. The current limit is easily set by clamping the maximum value of the output of the speed control amplifier to the value corresponding to the desired current limit. In this way, the reference current for the current amplifier will not exceed the maximum safe current permissible for the motor. When the speed error is sufficiently large, such as at starting, or when a large change of speed is called for, the speed amplifier output will stay at the clamped current limit value. The machine will develop the maximum torque corresponding to the maximum current, to achieve the demanded speed quickly. As the speed of the machine comes close to the demanded speed, the speed error becomes small, and the output of the speed error amplifier comes out from the current limit causing the motor to smoothly reach the desired speed. The current limit feature is also symbolically shown in Fig. 7.7 as an adjustable limit at the reference input to the current control amplifier.

7.3.4 Regenerative Braking and Reversing

When a motor spinning in one direction has to be regeneratively braked, it is made to function as a generator and supply power from the armature back into the source through the converter. For reversal of power flow, either the armature current or the armature voltage has to be reversed. We shall consider two situations that occur in practice:

1. When the armature converter is a chopper, such as is the case in a battery powered vehicle;
2. when the armature converter is a phase controlled line commutated converter, such as will be in a typical industrial rolling mill drive fed from an AC utility bus.

7.4 CHOPPER CONTROLLED DRIVES

7.4.1 Driving and Braking Configurations

Figure 7.8(*a*) shows the circuit in the driving mode. The chopper configuration is for the voltage step down mode of the operation. To move into the braking

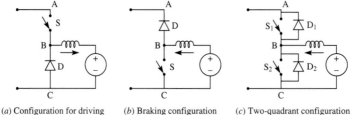

(*a*) Configuration for driving   (*b*) Braking configuration   (*c*) Two-quadrant configuration

FIGURE 7.8
Chopper circuits for regenerative braking.

mode, it is necessary to change the chopper configuration into the voltage step up mode. This configuration is shown in Fig. 7.8(*b*). This involves a change of connections, which has to be implemented by means of contactors or switches. For fast operation, it is more convenient to use a two-quadrant chopper as shown in Fig. 7.8(*c*). With the two-quadrant chopper, as soon as braking is to be implemented, the controlled switch S_1 is turned OFF by blocking its gate pulses. The motor current falls to zero. The gating pulses are then channeled to the static switch S_2, which is pulse width modulated in such a way as to provide the required braking current. The braking torque will be proportional to the armature current, which can be programmed by a suitable reference input to a closed-loop current controller. By continuous adjustment of the duty cycle it will be possible to achieve regenerative braking down to a very low speed.

Illustrative example 7.4. A battery powered car has a DC motor drive. The armature is supplied from a two-quadrant chopper. The field is directly supplied from the battery through a field weakening rheostat. The battery voltage is 120 V. The armature resistance is 0.4 Ω. After the vehicle has been running at the maximum armature voltage and field current for some time, it is required to implement regenerative braking. The armature current prior to the commencement of braking was 80 A. Assume that the chopper is ideal. Determine the initial duty cycle of the chopper to give a braking current of 50 A.

Solution. The motor induced e.m.f. at the commencement of braking is $120 - 80 \times 0.4 = 88$ V. At the commencement of braking, the motor e.m.f. is therefore 88 V. For a braking current of 50 A, the initial voltage at the terminals B and C of the chopper (see Fig. 7.8) will be $V_1 = 88 - 50 \times 0.4 = 68$ V. Therefore the initial voltage step up ratio is $120/68 = 1.765$.

The voltage step up ratio is $1/(1 - D)$, where D is the duty cycle. Therefore $1/(1 - D) = 1.765$. This gives the initial duty cycle as $D = 0.433$, or 43.3%.

7.4.2 Reversal of Direction of Rotation

To reverse the direction of rotation of the motor, it is necessary to reverse either the field or the armature terminals. The field circuit is highly inductive, and for fast reversals it is usually more convenient to reverse the armature terminals. The reversal of the armature terminals can be achieved through the use of an electromagnetically operated switch (contactor) as shown in Fig. 7.9.

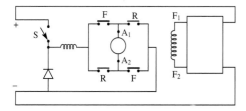

FIGURE 7.9
Reversal of armature terminals using electromagnetic contactors.

The forward contactor is an electromagnetically operated switch with two poles labeled F and F in the figure. Similarly the reverse contactor also has two poles labeled R and R. There should be suitable interlocking to ensure that only one contactor can be ON at any time, and that both forward and reverse contactors cannot be simultaneously ON at anytime. In Fig. 7.9, the field is separately excited, and the field source is shown as a separate converter fed from the same bus as the armature converter. When the forward contactor is ON, the contacts labeled F and F are ON, and therefore the positive terminal of the chopper output is connected to armature terminal A_1 and the negative to terminal A_2 of the motor. For reverse running, the contacts F and F are turned OFF, and the contacts labeled R and R of the reverse contactor are kept ON.

7.4.3 Static Four-Quadrant Chopper

For very fast reversals, in a fully static manner, without the use of mechanical switching, it is best to use a four-quadrant chopper, the circuit configuration of which is shown in Fig. 7.10. The static four quadrant chopper has four switching modules, labeled by subscripts $1, \ldots, 4$ in Fig. 7.10. Each switching module consists of a controlled static switch, together with an antiparallel diode. The manner of switching control by pulse with modulation may be described as follows.

For one direction of rotation, which we shall call the forward direction, the switches S_2 and S_3 will be kept OFF. S_4 will be kept ON. The motor speed will be controlled through pulse width modulation of the switch S_1. If now a speed change to a target speed in the reverse direction is to be implemented, the motor is first braked regeneratively until the speed falls to zero, and then accelerated in the reverse direction to achieve the target speed. The sequence of operation will be as follows.

All the switches will be turned OFF at first. This will cause the motor current to flow through the diodes D_2 and D_3, causing the current in the motor

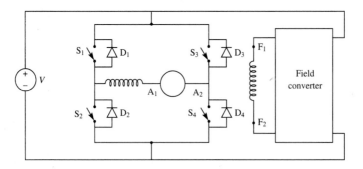

FIGURE 7.10
Static four-quadrant chopper.

circuit, consisting of the motor armature and the smoothing inductor, to fall to zero. After this, braking is initiated by the pulse width modulated switching of the static switch S_3. During the ON period of S_3, current will build up in reverse through D_1 and S_3 because of the induced e.m.f. of the armature. During the OFF period of S_3, energy will be fed back to the source by current flow through D_1 and D_4. The braking current will be controlled to give the desired braking torque characteristic by appropriate control of the duty cycle of the static switch S_3 during the braking period. At the end of the braking period, when the motor speed has fallen to zero, the static switch S_2 will be turned ON, and the operation will shift to the driving mode in the reverse direction by continuing the pulse width modulated switching of S_3. The motor will accelerate and come to the desired speed in the reverse direction.

7.5 DC MOTOR DRIVES USING PHASE CONTROLLED THYRISTOR CONVERTERS

The large majority of industrial DC motor drives employ line commutated thyristor converters, because these can work directly from the AC utility bus. The armature converter will be typically a three-phase thyristor bridge. The field converter has to handle only the small power needed for excitation. For a small drive, this may be a single-phase bridge. In many cases, for the sake of economy, a semicontrolled bridge may be chosen, whether single- or three-phase, since regeneration is not needed for the field circuit. A semicontrolled bridge consisting of a combination of thyristors and diodes will provide the necessary control range for field-weakening. Speed control is achieved by armature voltage control below the base speed and field-weakening at higher speeds, as we described earlier. However, implementation of regenerative braking and reversal of the direction of rotation may not be possible by static means if only a single converter bridge is used for each of the armature and field circuits. The thyristor bridges can carry current only in one direction. Reversal can be achieved either by reversal of the field terminals or by reversal of the armature terminals. We shall describe both techniques, but reversal of the field terminals, which will mean reversal of field current, will take a longer time because of the large inductance of the field circuit. The armature reversal is faster, and the manner of doing this by contactors is similar to what we described for the chopper controlled drive. We shall first describe armature reversal by means of mechanical switches. Later, we shall describe fully static methods based on field current reversal and also armature current reversal.

7.5.1 Armature Reversal by Means of Contactors

There are two contactors shown in Fig. 7.11, which may be electromagnetically operated. They are interlocked in such a way that only one of them can be ON

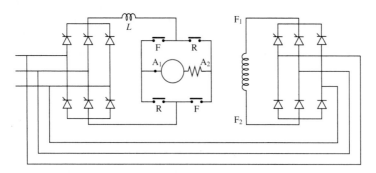

FIGURE 7.11
Armature reversal by means of contactors.

at any time. Both are two-pole switches, and the poles of the forward contactor are labeled F and F. Similarly, the poles of the reverse contactor are labeled R and R.

Let us assume that the forward contactor is ON and that the motor has been driving a load in the forward direction. It is now desired to reverse the motor and to achieve a target speed in the reverse direction. The following will be the sequence of events which typically will be executed by means of a closed-loop controller.

The armature converter is first phased back. This means that the firing angle of the converter is made large, so that its output voltage falls to a low value. This causes the motor current to decrease and fall to zero. A zero-current detector for the armature current initiates the opening of the forward contactor, which means that the contacts F and F will be opened. The firing angle of the converter is now adjusted to a large value, corresponding to inversion mode, so that the "back voltage" of the converter is larger than the motor induced voltage at the armature terminals. The reverse contactor is now closed, which implies the closing of the contacts R and R in Fig. 7.11. Initially, no armature current can flow, because the back voltage of the inverter is larger than the motor voltage. The firing angle of the inverter is now progressively decreased. This causes the braking current to flow and energy to be fed back to the AC bus through the converter, which is now operating in the inversion mode. The motor will slow down because of regenerative braking. During this period, the firing angle is continuously decreased in such a way that the desired braking torque characteristic is obtained. A firing angle of 90° will correspond to zero output voltage of the inverter and practically zero speed. Further decrease of the firing angle will cause the converter to go into the rectification mode, and the motor will now accelerate in the reverse direction, because the terminal polarities of the armature correspond to reverse rotation. With closed-loop control, at zero speed and at speeds well below the target speed, the speed error will be large, and in all probability the motor will be operating under the current limit set on the controller. As the speed comes very close to the reference setting, the speed error will become small, and the output of the speed amplifier, which is the reference input to the current controller and which remained clamped at the current limit setting, will now come out of the current limit. The current reference will become small, slowing down the acceleration and enabling the machine to smoothly reach and settle down at the target speed.

In Fig. 7.11, the inductance L shown in the armature converter output is the smoothing inductance. An adequate value of this inductance is necessary for the satisfactory operation of the converter in the inversion mode. The resistance of the armature also is shown, in series with the armature terminals. The field converter shown is a semicontrolled bridge, which will be more economical than an all-thyristor fully controlled bridge, and will be sufficient for the purpose of field-weakening.

7.5.2 Static Field Current Reversal

Figure 7.12 illustrates the field reversal technique by static means. The field has a large time constant, and it may take several seconds to implement field current reversal by direct switching. The reversal can be speeded up by having two converters, as shown in Fig. 7.12, for the field circuit. Both are fully controlled thyristor bridges. If the motor is spinning in the forward direction, and a target speed in the reverse direction is to be implemented with regenerative braking of the motor, the following will be the sequence of events.

First, the firing angle of the armature converter is made large, to bring down the voltage applied to the armature, and make the current drop to zero in the armature circuit. It is assumed that initially the field is being supplied

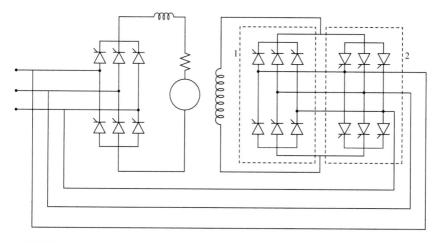

FIGURE 7.12
Static reversal of field current.

positive current from the positive current converter labeled 1. The firing angle of this converter is now delayed by a large angle so that it goes into the inversion mode. This causes the stored energy in the field to be fed back into the AC mains, through the converter, which is now in the inversion mode, and the field current rapidly falls to zero. Because the field current has fallen to zero, the induced voltage of the armature also falls to zero, even though the machine is still spinning in the forward direction.

After the field current has fallen to zero, and a few milliseconds of "dead time" has been provided, the negative field current converter labeled 2 in the figure is gated, after disabling the positive current converter by stopping the gate pulses to its thyristors. This causes the field current to build up and increase in the reverse direction, at the same time causing the armature voltage to build up with reversed polarity, because of the reversed field flux. Simultaneously, the armature converter is gated in the inversion mode. Regenerative braking of the motor commences, and the machine slows down, as the armature delivers braking power into the AC mains, through the converter, which is now operating in the inversion mode. As the braking continues and the machine is slowing down, the firing angle of the inverter is progressively decreased towards 90°, at which angle the back voltage of the inverter will be zero. When the speed has come down to zero, the firing angle is further reduced, and the converter operation changes into the rectification mode. The machine now goes into the motoring mode, and accelerates in the reverse direction to reach the target speed as the firing angle is decreased. In a closed-loop control scheme, this will be happening under current limit, as long as the speed error is large. It comes off from the current limit as the speed comes close to the target speed, and the machine finally reaches the target speed smoothly and settles down at this speed.

The field reversal scheme provides fully static four-quadrant operation without the use of electromechanical contactors. That is to say, driving and braking is possible for both directions of rotation. The provision of the two fully controlled converters for the field speeds up the field current reversal by making it possible to quickly bring the field current to zero through feedback of the energy stored in the magnetic field of the field circuit, by inverted operation of the appropriate converter. The two converters together constitute a "dual converter." The dual converter is a very convenient means of supplying current in either direction in a load with the facility for regeneration. A dual converter in the field circuit is more economical than having such a converter for the armature, because the power requirement of the field is much smaller than that of the armature. However, the closed-loop control circuitry is somewhat more complex in a scheme using the field reversal technique. Even with a dual converter, the field reversal scheme is still not very fast, because of the large inductance of the field. To achieve fully static four-quadrant operation, with the fastest possible response to commands for speed changes and reversals, it is necessary to use a dual converter in the armature circuit. This is the most favored scheme for industrial DC motor drives. Such a scheme is really the static equivalent of the Ward Leonard control system. We shall now proceed to describe the dual converter.

7.6 THE PHASE CONTROLLED DUAL CONVERTER

A dual converter is a combination of two identical fully controlled line commutated converters. Both the midpoint and the bridge configurations are possible for the converters of the dual converter. The three-phase bridge is the most commonly used configuration for the individual converters. However, we have shown in Chapter 2 that the phase controlled bridge configuration is a combination of two phase controlled midpoint converters. Similarly, a dual converter consisting of two bridges is essentially a combination of two dual converters each consisting of two midpoint dual converters. Therefore, for the sake of simplicity, we shall first consider a dual converter consisting of two midpoint converters before considering the bridge itself.

7.6.1 Dual Midpoint Converter

Figure 7.13 shows two midpoint converters with the same three-phase AC input voltages V_a, V_b and V_c. The converter on the left gives an output DC voltage across its common cathode terminal K and the AC source neutral N. This converter will give a positive output voltage if the firing angle is less than 90°, that is, if it is in the rectification mode. Its DC terminal voltage will be negative if the firing angles are greater than 90°, corresponding to the inversion mode. The second midpoint converter has the common-anode configuration. Its DC terminals are the common anode terminal A and the AC source neutral N. This converter is identical to the first, except for the fact that it has the common-anode configuration instead of the common-cathode configuration. The second converter will therefore give a negative output voltage if it is operated in the rectification mode with firing angles less than 90°, and a positive DC voltage if it is operated in the inversion mode with firing angles greater than 90°. The direction of the DC current for the first converter, that

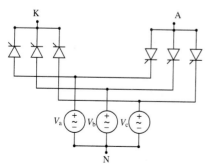

FIGURE 7.13
Two three-phase midpoint converters supplied from the same AC bus.

is, the common-cathode converter, will be away from the terminal K into the DC load. We shall consider this as the positive direction of the load current. Therefore we shall call the common-cathode converter the positive-current converter. Similarly, the second converter, which is the common-anode converter, will be called the negative-current converter, because the direction of the load current is in reverse for this converter. Let α_p be the firing angle of the positive-current converter and α_n that of the negative-current converter. Let us also assume that the positive converter is in the rectification mode, so that α_p is less than 90°.

The DC output voltage of the positive-current converter is

$$V_{KN} = V_{d0} \cos \alpha_p$$

where $V_{d0} = 1.17V_p$ is the voltage for zero firing delay, with V_p the r.m.s. phase voltage (see Table 2.1 for the case of the three-phase midpoint converter).

The DC output voltage of the negative current converter is

$$V_{AN} = -V_{d0} \cos \alpha_n$$

The negative sign is because this converter has a common-anode configuration.

If now we adjust the firing angle α_n of the negative-current converter in such a way that

$$\alpha_p + \alpha_n = 180° \tag{7.10}$$

we have

$$\cos \alpha_n = \cos (180 - \alpha_p) = -\cos \alpha_p$$

Therefore the DC terminal voltage of the negative current converter becomes

$$V_{AN} = -V_{d0}(-\cos \alpha_p)$$
$$= V_{d0} \cos \alpha_p$$

Therefore we find that both the positive- and negative-current converters have exactly the same voltage, with the same polarity, if the condition (7.10) is satisfied. This condition implies that if one of the converters is in the rectification mode then the other is in the inversion mode.

The question that we now wish to pose is this. If the two converters are controlled in such a way that $\alpha_p + \alpha_n = 180°$, the DC voltages V_{KN} and V_{AN} of the two converters will be the same in both magnitude and sign. Therefore, are we justified in connecting the two converters in parallel by solidly connecting the terminals K and A and operating both converters together, satisfying (7.10) regarding the firing angles?

The answer to this question is negative. Although the DC voltages are the same, we are still unable to connect them solidly in parallel, because, besides the DC voltage, each converter has an AC ripple voltage also. The ripple voltages may not be the same. Therefore the voltages will not match at every instant. The unequal instantaneous voltage difference can cause an excessive current to flow between the converters if both converters are in operation simultaneously, unless steps are taken to limit this current. The instantaneous voltage difference between the two converters is illustrated by the following numerical example.

Illustrative example 7.5. The two midpoint converters of Fig. 7.13 are both fed from the secondary of a three-phase transformer with a phase-to-neutral voltage of 120 V. The firing angle of the positive current converter is 30° and that of the negative current converter 150°.
(*a*) Determine the DC voltage of each converter and sketch the waveforms at the DC terminals KN of the positive-current converter and also that at AN of the negative-current converter.
(*b*) Sketch the waveform of the ripple voltage $v_{KN} - v_{AN}$ and determine the peak value of this ripple voltage.

Solution. The waveform of v_{KN} is plotted in Fig. 7.14(*a*). For this, the three-phase voltage waveforms v_{aN}, v_{bN} and v_{cN} are first sketched as three sine waves displacled by 120°. The reference instants for the firing delay angles are the intersection points of these sine waves on the positive side. The waveform of v_{KN} is shown by thickening the appropriate segments of the sine waves, starting with a 30° delay from the reference instants. The DC voltage is given by (see Table 2.1):

$$v_{KN} = 1.17 \times 120 \times \cos 30° = 121.6 \text{ V}$$

The waveform of v_{AN} is plotted in Fig. 7.14(*b*). For this also, we first plot the three sine waves representing the three phase voltages. Since this is a common-anode circuit, the reference instants for the firing delay are the intersection instants on the negative side of the sine waves. The waveform of v_{AN} is shown by thickening 120° segments of the sine waves, starting with 150° delay from the reference instants. The DC component of the voltage is given by

$$V_{AN} = -1.17 \times 120 \cos 150° = 121.6 \text{ V}$$

The waveform of the ripple voltage v_{KA} is plotted in Fig. 7.14(*c*). This is obtained by taking the difference $v_{KN} - v_{AN}$. This waveform consists of 60° segments of the line voltage waveforms, 30° on either side of the negative-going zero-crossing. Therefore the peak value of the ripple voltage is $\sqrt{2} \times \sqrt{3} \times 120 \sin 30° = 147$ V.

7.6.2 Operating Schemes for Dual Converters

The purpose of using a dual converter is to make it possible for the load current to flow in either direction without having to use electromechanical contactors for reversing the load terminal connections. This objective is best served if we can connect the two individual converters in parallel, so that the load can be connected across this parallel combination. But we have seen that although we can make the two voltages equal, parallel connection is not possible, because there will be a large ripple voltage in the parallel loop. Because of this, if we connect the two converters in parallel, it is necessary to

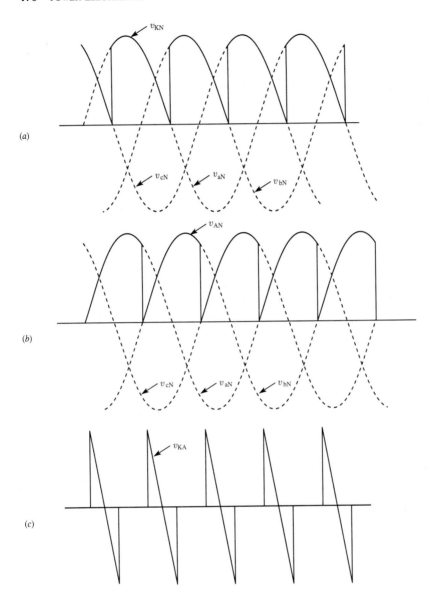

(a)

(b)

(c)

FIGURE 7.14
Individual converter voltage waveforms and the ripple voltage waveform in a midpoint dual converter.

introduce an inductance to limit the current that will circulate because of the instantaneous difference in the voltages. Alternatively, we can connect them solidly in parallel, without any current limiting reactor, if we decide to energize only one converter at a time. Thus we arrive at the two basic types of dual

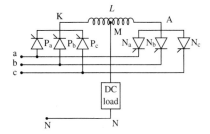

FIGURE 7.15
Three-phase midpoint dual converter with circulating current.

converters, based on how they are operated, which may be described as

1. the "circulating current" type dual converter;
2. the "circulating current-free" type of dual converter.

Figure 7.15 shows the circulating current type dual converter. Each of the two converters that constitute the dual converter is a midpoint converter. The thyristors labeled P_a, P_b and P_c constitute the positive-current converter, which has the common-cathode configuration, the terminal K being the common cathode. The thyristors labeled N_a, N_b and N_c constitute the negative-current converter. This converter has the common-anode configuration, the terminal A being the common anode. The two converters are operated so that the DC voltages of both are the same. By making $\alpha_p + \alpha_n = 180°$, the DC voltages of the two converters will always have the same magnitude and polarity. However, we have seen that there will be a ripple voltage appearing between the terminals K and A. To limit the ripple current to a small value, we insert an inductor L between K and A and take the output at M, which is the midpoint of the inductor, as shown in Fig. 7.15.

To illustrate the operation of the dual converter, and to highlight the purpose of the reactor L, we shall take a specific operating condition. We shall consider the operating condition when the firing angle α_p of the positive current converter is 30° and α_n of the negative converter is 150°.

Figure 7.16(a) shows the output voltage waveform of the positive current converter at the terminals K–N for the firing angle 30° stated above. Similarly, Fig. 7.16(b) shows the output voltage waveform of the negative current converter at the terminals A–N for the stated firing angle of 150°. From the waveform, we can see that the DC voltages are the same for both converters. The conducting thyristors during different intervals of time are also indicated on the top of Fig. 7.16. Inspection of this shows that during the interval t_1 to t_2, the c-phase thyristor is conducting in the positive-current converter and the a-phase thyristor is conducting in the negative-current converter. During the first half of this, the c-phase voltage is more positive than the a-phase voltage, and therefore a circulating current will build up in the direction K to A. During the second half of this interval, the a-phase voltage becomes higher than the c-phase voltage. Therefore the circulating current that was built up will fall to zero. But the current cannot reverse, because current can only flow in one direction owing to the unidirectional nature of the thyristors. We notice that

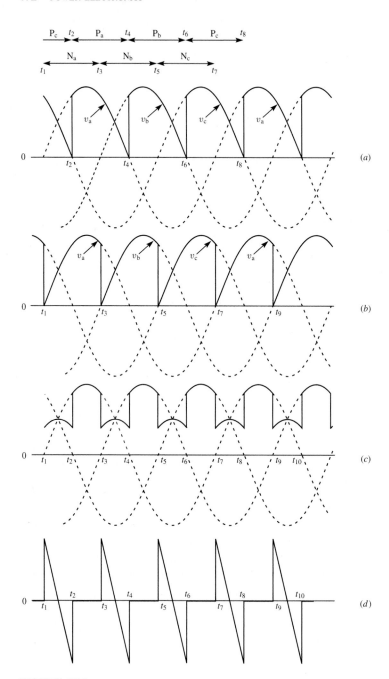

FIGURE 7.16
Voltage waveforms in three-phase midpoint dual converter with circulating current.

had it not been for the inductance L, there would be a direct short-circuit between the c-phase and the a-phase during the first half of the interval t_1 to t_2. The circulating current during this interval is limited to a low value by the presence of the inductor L.

In the interval t_2 to t_3, the a-phase thyristors are in conduction in both the positive-current and negative-current converters. Therefore both K and A are at the same potential, and the voltage across L will be zero. There will be no voltage to support circulating current during this interval.

During the interval t_3 to t_4, the situation is similar to that in the interval t_1 or t_2. The phases that will be connected will be the a-phase in the positive-current converter and the b-phase in the negative-current converter. During the first half of this interval, the a-phase voltage is higher, and a circulating current will build up. This current will decay to zero during the second half of this interval, because the b-phase voltage will be higher during this half. As stated earlier, the current cannot reverse, because the circulating current can flow only in one direction.

The waveform of the voltage appearing across the circulating current reactor L is shown in Fig. 7.16(d). This waveform is obtained by taking the instantaneous voltage difference between the waveforms of the positive-current converter given in (a) and the negative-current converter given in (b) in the same manner as we did in Example 7.5. During an interval when the same phase is conducting in both converters, the voltage across the circulating current reactor will be zero. During an interval when the conducting phases are different in the two converters, the voltage will be the line voltage between these two phases. Therefore the waveform of v_{KA} will actually consist of segments of the line voltage waveforms of the three-phase supply and zero-voltage segments. The v_{KA} waveform is seen to be an AC voltage. However, an AC circulating current cannot normally flow, because the converters can conduct current only in one direction. If an AC circulating current flow is to be made possible, it is necessary to make a DC current to circulate between the converters. The minimum magnitude of this DC circulating current must be equal to the peak value of the AC circulating current, to ensure that the resultant current is unidirectional. By maintaining a minimum DC circulating current of this magnitude, this together with the superimposed AC ripple current will be continuous and unidirectional in both converters. To make a DC current circulate, it is necessary to make $\alpha_p + \alpha_n < 180°$. A closed-loop current control scheme can be used to maintain the minimum DC circulating current. The use of a large value for the circulating current reactor L will make the AC circulating current less. Therefore the minimum value of the DC circulating current necessary to ensure continuous conduction of the two converters will also be correspondingly less. This is illustrated by Example 7.6 below.

The waveform at the DC terminals of the dual converter, which are the AC source neutral N and the midpoint M of the circulating current reactor, is sketched in Fig. 7.16(c). This will be equal to the conducting phase voltage during intervals when the same phase is conducting in both converters. During intervals when the conducting phases are different, the voltage at the midpoint M will be the mean of the voltages of the two conducting phases. The DC load voltage waveform in Fig. 7.16(c) is sketched on this basis. The DC component of this will be equal to the DC component of the voltage of the positive-current converter, which will also be equal to the DC component of the negative-current converter, because the waveform at the midpoint M is the average of the individual converter waveforms and the DC voltage across L is zero.

We shall assign the positive reference direction for the DC load current as being from M to N. On this basis, a positive load current will be supplied by the positive-current converter, which is the converter shown on the left in Fig. 7.15. When the load current is in the negative direction, this load current will be supplied through the negative-current converter. In the dual converter with circulating current, both converters are always available for carrying the load current, and it becomes possible for the load current to change direction in a natural manner. This is the main advantage of the circulating current type of dual converter. In the non-circulating current type of dual converter, which we shall consider later, we have to have special logic circuitry to implement the interchange of the converters, when a current reversal is to take place. There will be a longer delay for implementing the changeover between the converters that will be needed for a current reversal. The circulating current type of dual converter is the one best suited for fast and natural changes of the direction of the load current.

Illustrative example 7.6. A dual midpoint converter is supplied from a three-phase 60 Hz bus with a line voltage of 400 V. The firing angles are $\alpha_p = 30°$ and $\alpha_n = 150°$. The circulating current reactor has an inductance $L = 25$ mH. Assume that the circulating current builds up from zero at the commencement of the negative-going zero-crossing of the ripple voltage across L. Determine the minimum value of the DC circulating current necessary to permit the unrestricted build up of such a negative segment of ripple current.

Solution. In this problem, the given firing angles are the same, corresponding to Fig. 7.16. Therefore the voltage across the circulating current reactor will have the same waveform as plotted in Fig. 7.16(d). This will consist of segments of the line voltage waveform in the interval $-30°$ with respect to the negative-going zero-crossing to $+30°$ with respect to the same zero-crossing. Therefore a segment of the inductor voltage waveform can be expressed in this interval as

$$v_1 = 400 \times \sqrt{2} \times (-\sin \omega t)$$

The negative circulating current is to be assumed to build up from $t = 0$, starting from an initial zero value. During this period,

$$L\frac{di}{dt} = 400 \times \sqrt{2} \times (-\sin \omega t)$$

The negative current build-up will be from $\omega t = 0$ to $\omega t = \frac{1}{6}\pi$.

Therefore, by integration,

$$I = \frac{400 \times \sqrt{2}}{\omega L} \int_0^{\pi/6} -\sin \omega t \, d(\omega t) = \frac{400 \times \sqrt{2}}{\omega L} [\cos \omega t]_{\omega t=0}^{\omega t=\pi/6}$$

$$= \frac{400 \times \sqrt{2}}{2\pi \times 60 \times 0.025} (0.866 - 1)$$

$$I = -8.04 \text{ A}$$

Since the converters can carry only unidirectional current, the positive converter must have a minimum DC circulating current flowing in it of magnitude not less than 8.04 A, to permit the unrestricted growth of negative current up to this magnitude without the need for reversal of the direction of the resultant current. Therefore the minimum level of the DC circulating current needed to meet the stated condition is 8.04 A.

We may state here that, under steady-state operation, the circulating current flowing between the converters will settle down to a steady AC current. The DC circulating current between the converters is needed only to permit the negative segment of this AC current to flow. Under steady-state operation, the required value of the minimum DC circulating curent will be less, because the value calculated above is indicative of the total decrease of current from the positive peak to the negative peak.

We may also state here that, to achieve this minimum DC circulating current, the firing angles of the converters will need readjustment. For example α_p may be changed from 30° to 30° − δ and α_n from 150° to 150° − δ. The angle δ will be negligible for the ideal situation of negligible circuit resistance and voltage drops across the thyristors.

7.6.3 The Dual Bridge Configuration

We chose the midpoint configuration to explain the basic features of the dual converter of the circulating current type, because of its simplicity. A transformer is needed for the midpoint configuration to provide the neutral terminal, which is one of the output terminals of the dual converter. The bridge configuration does not need a transformer, and, for this reason, the bridge configuration is often a more convenient choice for the dual converter than the midpoint circuit. A dual converter that has the bridge configuration for the individual converters and that uses the circulating current principle is shown in Fig. 7.17.

The bridge version of the dual converter is actually a combination of two midpoint dual converters. (To emphasize this we have enclosed each of the two midpoint dual converters inside boxes of broken lines.) Because of this, it is necessary to have two circulating current reactors—one for each midpoint dual converter. If only one circulating current reactor is used, a short-circuit can occur between the AC lines when two different phases are conducting in the other dual converter. Therefore two separate coils are shown in Fig. 7.17 for

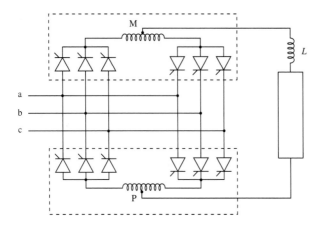

FIGURE 7.17
Three-phase dual bridge converter with circulating current.

the circulating current. A separate smoothing inductance for the load, labeled *L,* is also shown in Fig. 7.17.

7.6.4 Speed Control Scheme for a Reversible DC Motor Drive, Using a Dual Converter with Circulating Current

Figure 7.18 shows a fully static speed control scheme for an adjustable speed four-quadrant drive, using a separately excited DC motor and a phase

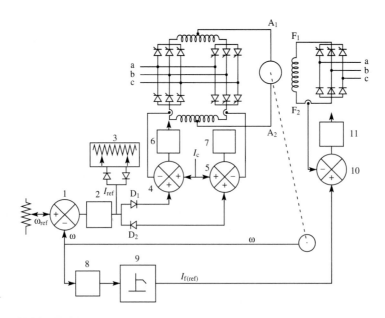

FIGURE 7.18
Four-quadrant DC motor drive using a dual bridge converter with circulating current.

controlled dual armature converter bridge. In Fig. 7.18, the armature of the motor is supplied from a three-phase dual bridge with circulating current. There are two circulating current reactor coils—one on each side of the bridges. The output terminals of the dual bridge, from which the motor armature is fed, are the midpoints of the two circulating current reactors. The motor field is fed from a three-phase semicontrolled bridge. The operation of the circuit may be described as follows. The demanded speed is set on a potentiometer and given as the speed reference ω_{ref}. The actual speed is labeled ω and is sensed by means of a tachogenerator. The module labeled 1 is the speed error detector. The output of 1 is the speed error $\omega_{\text{ref}} - \omega$. The circuit block labeled 2 is the speed error amplifier. The output of the speed error amplifier serves as the current reference I_{ref} for the dual converter. A current limit feature is introduced by the clamping circuit labeled 3. The clamping circuit labeled 3 serves to clamp the maximum positive and negative values of the current reference at adjustable limiting values, and thereby to set the positive and negative current limits. If I_{ref} is positive then it is channeled through the steering diode D_1 to the block labeled 4, which is the current error detector. In addition to the current reference input I_{ref}, there is also an additional input to this block, which is the circulating current reference I_c. In the scheme of Fig. 7.18, a small DC circulating current is maintained in the converters, to enable the AC circulating current due to the ripple voltage between the converters to be superposed without the need for reversal of the resultant current in a converter. For this to happen, the DC circulating current has to be greater than or equal to the peak value of the AC ripple current. In this way, the AC current will be able to flow in both converters continuously. Thereby, both converters will always be maintained in continuous conduction. For this, the circulating current reference I_c should be adjusted to give a minimum DC circulating current, equal to the calculated peak value of the AC ripple current. If the demanded load current is a positive current (as will be the case when the speed error is positive), the total current to be supplied by the positive-current converter is the load current, plus the circulating current. The current of the negative-current converter will be only the circulating current. This is in conformity with what we see from Fig. 7.18—that the total reference current input for the positive current converter, which is given to the module 4, will be the output of the speed amplifier plus the circulating current reference. At the same time, the total reference current input to the negative-current converter (to the block labeled 5) is only the circulating current reference. The feedback current I_f to the block 4 is the sensed current of the positive-current converter. The output of the current error detector block is the current error $I_{\text{ref}} + I_c - I_f$. This is the input to the current error amplifier block labeled 6. The output of the current error amplifier is the control voltage that determines the firing angle of the positive-current converter. It changes the firing angle α_p of the positive-current converter as to eliminate the current error.

If the speed error is negative, that is, if the actual speed is higher than the wanted speed, then the output of the speed amplifier, that is, block 2, will be negative. This means that I_{ref} becomes negative. The negative I_{ref} is channeled through the steering diode D_2 to the block labeled 5, which is the current error detector of the negative current converter. Now the total current reference for the negative current converter will be I_{ref} plus the circulating current reference. The total current reference for the positive-current converter will be only the circulating current reference. For negative values of I_{ref}, the current control loop of the negative-current converter functions exactly like the functioning of the current control circuit of the positive-current converter for positive values of I_{ref}. The negative-current converter also has a current sensing unit that sends the feedback current signal to the current error detector. The current error amplifier for the negative-current converter is labeled 7 in Fig. 7.18. It functions in a similar manner to the block 6.

The scheme shown in Fig. 7.18 also incorporates field-weakening. The circuit block labeled 8 is an absolute-value module, which converts the output of the speed sensor to its absolute value, eliminating the negative sign if any. The output of the block 8, which is the absolute value of the speed, is fed to the field-weakening block, which is labeled 9. The output of the field-weakening module is the field current reference. The field-weakening module functions in the following manner. As long as the absolute value of the speed is below the base speed limit, its output is constant at the value corresponding to the maximum field current of the motor. As soon as the absolute value of the speed begins to exceed the base speed, the output of the module begins to drop, thereby initiating field-weakening. This is shown symbolically by the functional representation shown inside the block 9. The field current is maintained at the value set by the field current reference $I_{f(ref)}$ output by the field weakening module. The field current error detector is labeled 10 and the field current error amplifier 11.

7.6.5 Sequence of Operations during a Speed Reversal

Assume that the motor is spinning at a steady speed in the forward direction, driving a load. The motor current will be supplied by the positive-current converter, which will be working in the rectification mode. The negative-current converter will only be carrying the circulating current, and it will be in the inversion mode. The terminal voltages of both converters will be the same. Let us now assume that a particular speed is wanted in the reverse direction. For this, we change the speed reference to the value corresponding to the new speed. This will instantly cause a large negative speed error. The output of the speed amplifier will instantly go into the negative-current limit value. The steering dioide D_1 will block this negative voltage from the current control loop of the positive-current converter and channel it through the diode D_2 to the current control loop of the negative-current converter. Therefore the total reference current for the positive-current controller will fall to the value corresponding to the circulating current, whereas the total reference current for the negative-current converter will have the maximum magnitude. Therefore the positive current will fall to zero, and the negative current will build up in the motor. Since the motor will initially be spinning in the same direction as before, the reversal of current direction will initiate regenerative braking. The negative-current converter will be operating in the inversion mode as the machine slows down through regenerative braking. The firing angle will progressively decrease from the initial value in the inversion region to about 90°, when the machine speed falls to zero. After this, the firing angle will continue to decrease as the machine accelerates in the reverse direction. The machine will accelerate under current limit, and as the speed becomes very close to the demanded speed, the speed error becomes very small, and therefore the output of the speed amplifier falls below the clamping level, and the machine comes out of current limit and attains the steady commanded speed.

7.6.6 Dual Converter without Circulating Current

In a dual converter, only one of the two converters need be in operation at any given time to handle the load current. Therefore the purpose of a dual converter can be served by operating only one converter at a time and keeping the other in readiness to take over when the current reversal is needed. Such is the arrangement in the dual converter shown in Fig. 7.19. This is a "circulating current-free" dual converter. The positive and negative converters are shown separately boxed inside broken lines. Both converters are solidly connected in "antiparallel" without any circulating current reactor in between. Only one

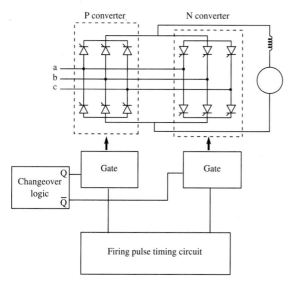

FIGURE 7.19
The circulating current-free dual converter.

converter carries the load current, depending on the direction. If the load current is positive, the positive converter is operated with the appropriate firing angle. The firing pulses to the negative converter can be either blocked or phased back so the negative converter cannot carry current. For example if the positive converter is in the rectification mode with a firing angle less than 90°, we can prevent conduction in the negative converter by either blocking the gate pulses to the thyristors or by delaying the gate pulses so that the e.m.f. presented by it is in opposition to the flow of circulating current and is large enough to prevent any circulating current. In the scheme shown in Fig. 7.19, the gate timing pulses of both converters are generated by the circuit block labeled "Firing pulse timing circuit." The gate timing pulses for each converter bridge are channeled through a separate "gate" circuit block. Only one of the gates will be open at any time, so that the firing pulses can reach only one converter at any time. The circuit block labeled "Changeover logic" determines which gate is "open." For example, if the required load current direction is positive then the Q output of the changeover logic block will be high, and so the gate for the positive converter will be open, causing the firing pulses to be applied to the thyristors of the positive converter. At the same time, the $\bar{Q}$ output of the changeover logic will be low, so that the gate for the negative converter will be closed, preventing the firing pulses of this converter being applied to the thyristors.

How well this system will work depends on how accurately and quickly the changeover logic circuit block can ensure that the gate pulses are channeled to the proper converter. For example, let us assume that the positive converter is functioning in the rectification mode and the motor is driving a load in the forward direction. It is now required to implement a particular speed in the reverse direction. For this, we change the speed reference input to the speed error amplifier. Its output will immediately cause the firing angle of the positive-current converter to increase and bring the voltage to zero, causing the motor armature current to fall to zero. A zero-current detector in the motor circuit will cause the change-over logic block to block the firing pulses to the positive-current converter, and cause the firing pulses to be channeled to the negative-current converter. As a safety precaution, a finite time delay of several milliseconds will be provided after the current zero is detected before the firing pulses are released to the incoming negative current converter. To avoid a large current inrush, it is necessary for the firing angle to be adjusted so that the incoming converter voltage will closely match the motor voltage. The firing circuit will now progressively advance the firing angle of the negative current converter so that the motor will regeneratively brake down to zero speed and then accelerate in the reverse direction. The motor will be operating under current limit till the speed becomes close to the set speed. After this, the output of the speed amplifier will come out of current limit, and the motor will settle down at the demanded speed.

In our description, we have implied that the changeover logic was actuated by the zero-current detector. The system will work satisfactorily on this basis if the motor current is continuous at all times. But there will be difficulties if the discontinuous mode of current flow happens to set in. If the motor is working with discontinuous current flow, a current zero does not necessarily call for a changeover of the converters. Changeover is needed only if the current zero is to be followed by a current reversal. If the changeover logic is solely dependent on zero-current detection then an erratic changeover may get implemented under circumstances when no changeover between converters is called for. To ensure that this does not happen, the changeover logic block must have other inputs, additional to the zero-current detector, to ensure that a changeover is really needed after the zero-current detection. A more elaborate scheme will be needed for the changeover logic, which we shall not further discuss here. A motor voltage feedback is also necessary for the firing pulse timing circuit to ensure that the voltage of the incoming bridge always matches the motor voltage.

Compared with the circulating current type of dual converter, the circulating current-free type is more efficient, because the converters need handle only the load current. The absence of the circulating current reactors makes it more economical in cost and weight. But the changeover between converters takes longer, because of the dead time provided after current zero. The changeover does not take place "naturally" as in the circulating current type, and, for satisfactory working, requires more complex logic circuitry.

7.7 CONTROL OF SERIES MOTORS

Our treatment so far has dealt with the separately excited motor. In a DC series motor, the field circuit and the armature circuit are connected in series. For such operation, the field coils are designed differently.

Separate and shunt excitation. DC motors intended for use in separately excited drives are usually designed for field currents that are small compared with the rated armature currents. This means that to create the required ampere-turns corresponding to the full field flux, the number of turns of the field coils will be large. The cross-section of the wire used for the field coils will be small, because the current is small. The resistance of the field circuit will therefore be large. The voltage rating of the field circuit will be high—often comparable to or the same as the voltage rating of the armature. When the voltage rating of the field circuit and the armature circuit are the same, the motor is also capable of "shunt" operation. A shunt motor is one that is operated with the field circuit and the armature circuit connected in parallel. But a shunt motor is not very convenient for adjustable speed application. If the voltage applied to a shunt motor is varied, the field circuit and the armature circuits react in an opposite manner as regards speed changes, If, for instance, the voltage is reduced, the decrease in armature voltage tends to lower the speed. At the same time, the decrease in the field current tends to increase the speed. Therefore, if a machine designed for shunt operation at the rated voltage is to be used for an adjustable speed application, it is more convenient to disconnect the field terminals from the armature circuit, and use the machine as a separately excited motor, with separate controls for the armature and field inputs.

Series excitation. The second major category of DC motors is the "series motor." A DC series motor is one that is operated with the field and armature circuits in series. Therefore there is only one circuit and one current, which is the same for the field and the armature. Since the field current is the same as the armature current, its rated value will be large. Therefore the number of turns on the field coils required to create the necessary ampere-turns will be small. The field coils will therefore be designed with a relatively small number of turns, using wires of large cross-section, sufficient to carry the full field current. Therefore the resistance of the field circuit will be small. This is appropriate for series operation, because a low value of the field circuit resistance makes the voltage drop across it small, and enables the major part of the DC line voltage to be available across the armature.

The voltage and torque equations (7.4) and (7.5) are general, and independent of the manner in which the field circuit is excited. The torque equation shows that the torque is proportional to the product of the flux and the armature current. Therefore, if the field is not saturated, the field flux will be proprtional to the field current, which in the series machine is the same as the armature current. Therefore the torque developed will be proportional to the square of the armature current. Therefore, if the armature current is made sufficiently large, it will be possible to develop large torques. At starting and at low speeds, when the motor e.m.f. induced is small, it will be possible to make the armature current large, and produce a large torque. When the motor speed rises to a large value, the induced voltage will be large, and therefore the current becomes small, as does the torque. This type of torque characteristic is what is wanted for a traction drive. A large number of electric suburban rapid transit railway systems are based on DC power. Traditionally, such systems have used the DC series motor. At starting and during the acceleration phase, a large torque is needed, because of the large inertia of the train. When the acceleration phase is over, the torque requirement for free running on level tracks will be small, equal only to what is needed to overcome the train resistance due to friction and air resistance. This torque requirement matches the torque characteristic of the series motor. Because of this, historically, the DC series motor was invaribly chosen for electric traction drives.

Before the era of power semiconductor switches and static converters, the method employed for the control of series traction motors, was by inserting series resistances. This was done in steps, using switches that were operated electromagnetically or electropneumatically to switch the large currents involved. When there are two or more motors, they are initially connected in series, and then in parallel. For example, if there are two identical motors driving a motor coach of the train, the sequence of switching will be as follows. Both motors will initially be connected in series along with an appropriate number of resistance sections in series. The total resistance in the circuit will be such as to limit the starting current to the desired value. As the motors accelerate and develop back e.m.f., the resistances will be sequentially cut OFF, section by section, so that the average current during acceleration will be the value needed for the accelerating torque needed. Each resistance is cut OFF from the circuit by means of an electromagnetically operated switch, which short-circuits its terminals. When the full series configuration without resistances is achieved, the voltage across each motor will have risen to half the total system voltage. To continue with the acceleration phase, the motors are again connected in parallel with an appropriate value of resistance steps to limit the current per motor to the previous value. The acceleration process is continued, by cutting off resistance sections in such a way that the average motor current continues to be the same as before. When the full parallel configuration is finally achieved, each motor will have the full line voltage across it. Further speeding up will cause the motor current to fall, and the acceleration will decrease. The torque also will decrease, and if the motors are continued to operate in this manner, the speed will stabilize at the value at which the torque developed will balance the train resistance due to friction, wind and gravitational force due to slope, if any. The latter can be positive or negative, depending on whether the slope is an ascent or a descent. This type of control, which came to be known as "series–parallel control," mitigates the wastage of power in the resistances, as compared with total resistance control. Even so, this form of control is very inefficient, because there is still a large wastage of energy in resistances. With the advent of power electronics and the development of DC-to-DC chopper converters, it became possible to change over to the more efficient method of chopper control. Thyristors with large current ratings, that are capable of handling the full current of railway traction motors are available. Figure 7.20 shows the power circuit for the control of a DC series motor using a chopper converter. The controlled switch labeled S will typically be a force commutated thyristor switch.

Chopper control eliminates the need for inserting series resistances, and

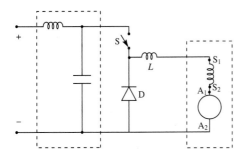

FIGURE 7.20
Chopper control of a DC series motor.

is therefore much more efficient, although there will be some loss of power in the switching devices. It also enables smooth and continuous control by variation of the duty cycle of the chopper, whereas with resistance control the adjustment has to be in discrete steps.

To achieve reversal of the direction of rotation, it is necessary to reverse the terminal connections of either the field or the armature alone, but not both. If the terminal connection of the motor itself is reversed, both the armature current and the field current will reverse. Since the torque depends on the product of the armature current and the field flux, reversal of the motor terminal connections will not reverse the direction of the motor, because the product of the two negative quantities will still be positive. The motor direction will remain the same with either polarity of the supply voltage. It is necessary to have a reversing switch, in either the field or the armature circuit, to implement the reversal of rotation.

The implementation of regenerative braking is very difficult with resistance control. Before the advent of chopper control, some techniques for regenerative braking did exist. To feed power back from the motor to the DC lines, it is necessary for the motor to develop a larger voltage than the DC line voltage. One scheme that was employed in the case of several motors driving a train was to connect the armatures of the motors in series and then control the fields separately, so that the motors together build up more than the line voltage, and drive reverse current into the lines. Such schemes are rather complicated, and provide regenerative braking only in a limited speed range. With chopper control, it is possible to implement regenerative braking down to very low speeds, with the chopper in the voltage step up configuration. Even so, the series connection of the field is not convenient from the point of view of regenerative braking, because the machine has to function as a DC series generator. It has been found more convenient to use separate excitation of the field during regenerative braking, in contrast to the series operation during driving. If this is done, and the field current is adjusted to a fixed value, the motor induced voltage becomes dependent only on the speed, and its dependence on the braking current vanishes because the field is separately excited.

With the use of separate static converters for the field and armature, it is possible to make rapid adjustments of the field and armature current to give the optimum torque characteristics. Therefore the torque characteristics of the series motor no longer have the significance they had before the development of static converters. The trend towards the use of series motors for traction drives has become less and less. In fact, the present trend is to totally eliminate the use of DC motors, and go in for three-phase AC motors fed from inverters. When an AC synchronous or induction motor is used as the traction motor, in a system fed by DC power, the chopper converter has to be replaced by an inverter. With AC synchronous or induction motors, the implementations of regenerative braking and reversals of direction can be done in a simpler manner, without the need for mechanical switching or change of the converter configuration. These aspects will become evident when we take up the study of AC motor drives in the next chapter.

7.8 CHAPTER 7 SUMMARY

Historically, the DC motor was extensively used for adjustable speed drives, before the development of static variable frequency inverters made the use of AC motors possible. We began with a description of the Ward Leonard system, which was the established method of control for adjustable speed DC motor drive systems, and which demanded four–quadrant operation and high performance. In the Ward Leonard system, the power is fed to the motor in a controlled manner, using a separate DC generator. The system is very expensive in initial and maintenance costs, because of the need to use a large rotating motor–generator set. But it gave very good controllability for reversible DC motor drives using regenerative braking. Static speed control systems, which have replaced the Ward Leonard scheme, are basically solid state adaptations of the same basic control strategy, with static converters in place of the rotary motor–generator set.

To bring out the theoretical aspects of the speed control of the DC motor, we presented the voltage and torque equations of the machine and the electrical loop equations for the field and the armature circuits. We highlighted the inherent magnetic decoupling that exists between the field circuit and the armature circuit in DC machines. This decoupling is the consequence of the fact that the spatial orientations of the field flux and the armature m.m.f. are in quadrature. These orientations do not change with the rotation of the machine, because the field coils are fixed, and the switching action of the commutator and brush assembly ensures that the armature m.m.f. also has a fixed orientation in space. For a correctly positioned brush assembly, this is in quadrature with the field flux. This magnetic decoupling makes it possible to implement very fast changes in the armature current without being burdened by the large time constant of the field circuit.

The speed control scheme that utilizes the features of a DC motor for the maximum advantage as regards controllability and quickness of response is one in which the motor is separately excited by having independent converters for the field and armature circuits. Such a scheme has two distinct speed control ranges, which can be derived using the motor equations. The lower speed range extends from zero to the "base speed." In this range, the field current is kept fixed at the maximum value, and all speed adjustments are made by means of the armature voltage. In this range, the motor can give the fastest response and develop the maximum torque at all speeds. This range is called the constant-torque range. The second range of adjustment extends from the base speed to the speed corresponding to the lowest permissible field current. In this "field-weakening" range, the speed adjustments are made by adjustment of the field current below its maximum value. In the field-weakening range, the maximum possible torque drops off as the speed is increased, and the maximum possible power output is approximately constant. We described an automatic closed-loop control scheme based on armature voltage control below base speed and field-weakening above this speed. The control strategy is such that the control automatically shifts into field-weakening above the base speed.

The converters for the field and armature in a separately excited control scheme may be choppers or phase controlled line commutated converters. We first described schemes employing choppers. For a reversible drive with regenerative braking, we may use mechanical switches to implement the necessary reversal of the armature current. But for a fully static four-quadrant operation, without any mechanical switching, it is necessary to use a four-quadrant chopper. We described a reversing operation with regenerative braking using such a chopper.

Industrial drive systems that operate from the AC utility bus employ phase controlled line commutated converters for the field and armature circuits. We described systems that use a two-quadrant converter for the armature, together with mechanical switching, to implement operation in all four quadrants. Fully static four-quadrant operation can be implemented economically using static field reversal by having a dual converter for the field circuit. However, even such systems may not be fast enough in some applications, because of the time delays involved in the field reversal. To achieve the fastest response and fully static operation, the choice falls on a dual converter for the armature circuit. This is the favored choice in high performance systems.

We described the functioning of the dual phase controlled converter, in view of its importance for industrial DC motor drives. We first described the simpler version of the dual converter that employs the midpoint configuration, and explained the two categories of dual converters, namely those that use circulating current and those that do not use circulating current. Dual converters that use the circulating current principle keep both converters always in readiness for current reversals, and therefore enable very fast reversal of current direction. The circulating current-free type of converter employs a changeover logic circuitry to change over between converters when a current reversal is needed. This circuitry must be able to identify without error the situations when a changeover is actually required. A fall of the current to zero may not always be a prelude to a changeover, if the converter is operating with discontinuous current flow caused by the motor back e.m.f. In spite of the more complex logic circuitry needed to ensure reliable changeover, the circulating current-free type of dual converter is more often preferred, because it is more efficient and economical. It is efficient because of the elimination of the power due to the circulating current. It is economical because of the elimination of the cost and weight of the circulating current reactors. The bridge configuration eliminates the need for a converter transformer. Therefore the dual bridge converter without circulating current is the preferred configuration for many industrial drive systems.

PROBLEMS

7.1. The full load power drawn by the armature circuit of a DC motor is 500 kW. It is controlled by a rotating Ward Leonard machine set consisting of an AC motor, a DC generator and a DC exciter. The total excitation power needed for control of the generator and the motor is 75 kW. Assume an efficiency of 88% for each rotating machine, and determine the output power ratings required for each machine of the Ward Leonard set. Sketch how these machines are connected in a typical rotating Ward Leonard control system.

7.2. The voltage rating of a DC machine when used as a shunt motor is 240 V. The machine is used as a separately excited motor in an adjustable speed drive system with separate converters for the armature and the field. The armature resistance of the machine is 0.4 Ω and the field circuit resistance is 120 Ω. The magnetization curve of the machine at 800 rev/min is given in Fig. P7.2. Determine
  (*a*) the base speed of the machine at no load;
  (*b*) the base speed on load when the motor drives a load and the armature draws 60 A from the armature converter.

7.3. The field converter for the motor in Problem 7.2 is adjusted to give the maximum output of 240 V. Determine the output voltage of the armature converter needed for a speed of 600 rev/min if
  (*a*) the motor is unloaded and draws negligible armature current;
  (*b*) the motor is loaded and the load torque requires an armature current of 60 A.

7.4. A DC machine has an armature resistance of 0.2 Ω and a field circuit resistance of 100 Ω. As a shunt motor, it is rated at 200 V. Its magnetization curve at 800 rev/min may be assumed to be the same as given by Fig. P7.2. It is used as a separately excited motor with a voltage step down chopper to supply power to the armature circuit and an adjustable resistance in the field circuit. The DC power supply voltage is 200 V. The arrangement is shown in Fig. P7.4. The external adjustable resistor in the field circuit is set at zero. The chopper duty cycle is 60%.

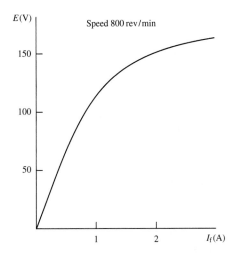

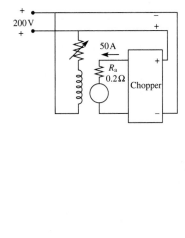

FIGURE P7.2   **FIGURE P7.4**

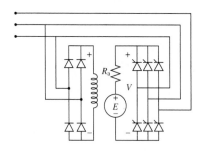

FIGURE P7.5

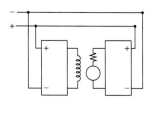

FIGURE P7.6

(a) If the motor is drawing an armature current of 50 A under these conditions, determine the speed of the motor.

(b) If the motor speed is to be adjusted to 700 rev/min for the same armature current, what should be the duty cycle of the chopper.

7.5. A DC machine, whose voltage rating as a shunt motor is 400 V, is used in an adjustable speed DC drive system. The armature is supplied from a three-phase phase controlled bridge rectifier. The field is fed from a single-phase diode bridge rectifier. The field rectifier is supplied from two AC lines of the same three-phase AC bus that supplies the armature bridge. The power circuit arrangement is shown in Fig. P7.5. Determine the AC line voltage required. Also determine the minimum firing delay angle necessary for the armature bridge so that the armature voltage does not exceed the rated value.

7.6. As a shunt motor, a DC machine is rated at 240 V. Its armature resistance is 0.3 Ω and the field circuit resistance is 120 Ω. Its magnetization curve is the same as given in Fig. P7.2.

It is used as a separately excited motor with two separate chopper converters for the armature and field circuits. Both choppers are fed from a 300 V DC bus. The power circuit arrangement is shown in Fig. P7.6. The control system employs armature voltage control below the base speed limit and field-weakening for higher speeds. Determine

(a) the base speed limit;

(b) the duty cycles of the two choppers at the base speed;

(c) the duty cycles of the choppers if the speed is to be increased to 50% above the base speed limit.

Assume an armature current of 30 A in all cases.

7.7. For the system in Problem 7.6, the ratio of the maximum speed to the base speed in the field weakening range is equal to two, with a constant armature current of 60 A at both speeds. Determine the duty cycles of the two choppers at the base speed and the range of duty cycle for the field chopper for the total control range.

7.8. The magnetization curve of a DC machine at 1000 rev/min is as given in Fig. 7.5. The rated armature and field voltage of this machine is 250 V. Its armature circuit resistance is 0.6 Ω and its field circuit resistance is 125 Ω. It is used as a separately excited motor in a static speed control system with independent converters for the field and the armature. The controller is based on armature voltage control below base speed and field-weakening for higher than base speeds. Determine the voltage output required from the field converter to give a no-load speed of

2000 rev/min in the field-weakening range. By how much will this speed fall if the motor is loaded under these conditions so that it draws an armature current of 50 A?

7.9. The base speed at full load current for a separately excited DC motor in an adjustable speed drive system is 1000 rev/min. The torque developed by the motor at the base speed at the rated armature current is 955 N m. Determine the torque that the motor can develop for the rated armature current in the field-weakening range if the speed is increased to 1600 rev/min by field-weakening. Calculate also the torque that the motor will develop at 2000 rev/min at the rated armature current.

7.10. Determine the mechanical power output of the motor in Problem 7.9 in kilowatts at the rated armature current at the base speed of 1000 rev/min and at the speeds of 1600 rev/min and 2000 rev/min achieved by field-weakening.

7.11. A battery powered vehicle has a DC motor drive. The field of the motor is fed from the fixed battery voltage of 120 V. The armature is supplied through a chopper converter from the same battery. The resistance of the armature is 0·4 Ω. The vehicle was being driven on level road at a constant speed, with the motor armature drawing 50 A from the chopper. The chopper duty cycle is 80% under these conditions. The vehicle then goes down a gradient, but the speed is maintained unchanged by regenerative braking. The armature current during this period is 2.5 A in reverse. Determine the duty cycle of the chopper, which is now in the operating mode for braking.

7.12. The armature circuit of a DC motor is controlled by a two-quadrant chopper working from a 250 V DC supply. The field of the motor is supplied at a fixed voltage at the rated value. The armature resistance is 0.8 Ω. The motor has been brought up to a speed of 1000 rev/min and the armature is drawing 40 A. Under these conditions, the duty cycle of the chopper is 90%. The motor is now regeneratively braked by changing the chopper operating mode appropriately. The target speed at the end of the braking operation is 400 rev/min in the same direction. During the entire period of braking, the braking torque is kept constant by maintaining a constant reverse armature current of 20 A. For this, the duty cycle of the chopper is automatically adjusted by closed-loop control. Determine the range of variation of the chopper duty cycle in the braking mode from the commencement of braking till the target speed is achieved.

7.13. The voltage rating of a DC shunt motor is 500 V. It is used as a separately excited motor in an industrial adjustable speed drive system. The armature converter is a fully controlled three-phase thyristor bridge. The field converter is a semi-controlled three-phase bridge. Both bridges are supplied with AC power from a 440 V utility bus. The armature resistance of the motor is 0.15 Ω. The no-load speed at the rated armature voltage and full field is 1800 rev/min. Determine
(a) the minimum permissible firing angle of the armature converter;
(b) the firing angle of the armature converter to achieve a speed of 800 rev/min under (i) no load conditions and (ii) under full load conditions when the armature is drawing 100 A.

7.14. In Problem 7.13, determine the firing angle of the field converter to give the full field current. Also determine the firing angles of the field converter in the field-weakening range that will give the following speeds, assuming that the magnetic circuit of the field is linear:

(*a*) 2000 rev/min;

(*b*) 2400 rev/min.

In each case, give the answers for (i) the no-load condition when the armature current is negligible and (ii) under full load conditions when the armature is drawing the full load current of 100 A.

7.15. The magnetization curve of a DC machine may be taken to be as given by Fig. 7.5 at 1000 rev/min. The rated armature and field voltage of the machine is 250 V. It is used as a separately excited motor for an adjustable speed application. The closed-loop speed control strategy is based on armature voltage control below base speed and field-weakening above base speed. The armature converter is a fully controlled thyristor three-phase bridge, and the field converter is also a three-phase bridge, but of the semicontrolled category. The AC line voltage input to the bridges are at 208 V. The rated full load current of the armature is 120 A and the armature resistance is 0.2 Ω. The field circuit resistance is 100 Ω. Determine

(*a*) the firing angles of the two bridges at base speed under no-load conditions;

(*b*) the firing angles needed for the field converter bridge to give load speeds of
(i) 1500 rev/min, (ii) 2000 rev/min and (iii) 2500 rev/min.

Assume that the converters are ideal, with negligible voltage drops and negligible overlap.

7.16. For the motor of Problem 7.15, consider the speed range below base speed, in which the field voltage is maintained at the rated value. Determine

(*a*) the firing angles of the armature bridge to achieve no-load speeds of (i) 400 rev/min, (ii) 600 rev/min and (iii) 800 rev/min;

(*b*) the firing angles needed to achieve the above speeds when the armature carries the full load current of 120 A.

7.17. The voltage rating of a DC motor is 150 V for armature as well as field. Its armature resistance is 0.4 Ω and its field circuit resistance is 75 Ω. Its magnetization curve may be taken to be as in Fig. P7.2. It is used in an adjustable speed drive with armature voltage control below base speed and field-weakening at higher speeds. The armature converter is a fully controlled three-phase thyristor bridge and the field converter is a semicontrolled three-phase bridge. Both bridges are fed from a three-phase AC bus of line voltage 120 V. The rated full load armature current is 50 A. Determine

(*a*) the firing angles of the two bridges at the base speed with the armature drawing the full rated current;

(*b*) the speeds of the machine at the following values of the firing angle of the armature bridge: (i) 30°, (ii) 45° and (iii) 60°.

7.18. In Problem 7.17, assume that the armature terminal voltage is at the rated value and that the armature is carrying the full rated current. Determine the speed of the motor for the following values of the firing angle of the field converter: (i) 45° and (ii) 60°. Assume that the converter is ideal, and neglect overlap.

7.19. A DC motor with a voltage rating of 400 V for both armature and field is used as a separately excited motor. The field is supplied at a fixed DC voltage at the rated value. The armature is supplied from a separate thyristor bridge fed from a 460 V three-phase 60 Hz AC bus. The inductance per phase of the input AC lines is 400 μH. The rated armature current is 120 A and the armature resistance is

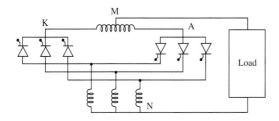

FIGURE P7.21

$0.12\,\Omega$. The no-load base speed is 1200 rev/min. Take the commutation overlap of the armature bridge into consideration, but otherwise assume the converter to be ideal. Determine the firing angle of the converter needed to achieve a speed of 800 rev/min when the armature is carrying the full load current of 120 A.

7.20. The motor of Problem 7.19 is being regeneratively braked by reversing the armature terminals and operating the armature converter in the inversion mode. Prior to the commencement of braking, the motor is spinning in the forward direction at 1200 rev/min on no load. Determine the firing angle of the converter at the commencement of braking if the braking current corresponds to the full load armature current of 120 A.

7.21. A phase controlled dual converter consists of two three-phase midpoint converters—a common-cathode circuit and a common-anode circuit—as shown in Fig. P7.21. Both converters are supplied from the same secondary terminals of a three-phase transformer. The phase-to-neutral secondary voltage of the transformer is 120 V and the frequency 60 Hz. The firing angles of the converters are 45° for the common-cathode converter and 135° for the common-anode converter. On the same lines as in Fig. 7.16, draw the following waveforms:

(*a*) the individual converter output voltage waveforms v_{KN} and v_{AN};

(*b*) the output voltage v_{MN} at the midpoing of the circulating current reactor;

(*c*) the voltage v_{KA} across the circulating current reactor.

7.22. In Problem 7.21 obtain a mathematical expression for the segments that constitute the waveform of the voltage v_{KA} across the circulating current reactor. Specify the phase angle range within which this expression is applicable. The circulating current reactor has a total inductance of 50 mH.

(*a*) Determine the minimum value of the DC circulating current to be maintained to facilitate the continuous flow of the AC circulating current due to the ripple voltage across the reactor.

(*b*) If the DC circulating current is to be limited to 5 A under these conditions, what should be the inductance value of the reactor?

CHAPTER
8

ADJUSTABLE SPEED AC MOTOR DRIVES

8.1 INTRODUCTION

It is in the application area of adjustable speed AC motor drives that modern progress in Power Electronics is having the most significant impact. Historically, the two most important categories of AC motors, namely the induction motor and the synchronous motor, were considered unsuitable for adjustable speed applications, because the AC power system frequency is fixed at 60 Hz or 50 Hz. For a fixed frequency input, the synchronous motor has a fixed speed and the induction motor also has a nearly fixed speed. However, AC motors have definite advantages in cost, size and weight, and require much less maintenance compared with DC motors. Even so, the DC motor was the favored choice for high performance adjustable speed drives, because of its easy controllability and fast response to speed control commands. Progress in Power Electronics has made it possible to build inverters that could provide AC power with adjustable frequency and adjustable voltage. This gave the impetus to the development of adjustable speed AC motor drives using induction and synchronous motors. Great strides have been made in recent years in the area of AC motor drives using static power electronic converters. In this chapter, we intend to present a study of such drives, which have gained popularity for practical applications. These drives are providing very dependable performance and very accurate speed control. We intend to make a broad coverage of conventional AC motor drives using the two most common types of AC motors, namely the induction and synchronous types. In practice, the three-phase versions of these motors are the ones that are invariably used. Therefore we shall confine our treatment to three-phase motors only. There are three aspects that are relevant to our study of adjustable speed AC motor drives:

1. the static converters employed to control such drives;
2. the motors employed;
3. the control strategies used, which depend on the motor type and the converter type.

In this chapter, our plan of presentation will be to first describe the types of converters that are used in AC motor drive systems. We shall then take up the study of different drive systems using each type of motor and the appropriate converter. Specific examples of the control techniques used will be described for each type of drive system. The types of converters are the following:

1. The voltage source inverters
   a. the three-phase six-step type
   b. the pulse width modulated type
2. The current source and current regulated inverters
   a. the auto sequentially commutated type
   b. the current controlled hysteresis type
   c. the current controlled PWM type
3. The cycloconverter
4. The load commutated type of inverter

After a presentation of the different types of inverters in the first part of this chapter, we shall take up the study of the drive systems. The types of motors to be considered are the induction motor, both squirrel cage and slip ring types, and the synchronous motor, which may have one of the three types of rotors, namely the permanent magnet type, the electrically excited type and the reluctance type.

8.2 VOLTAGE SOURCE INVERTERS

8.2.1 The Three-Phase Six-Step Voltage Source Inverter

Voltage source inverters were introduced in Chapter 5. Three-phase bridge inverters feeding Y-connected three-phase loads were dealt with in Section

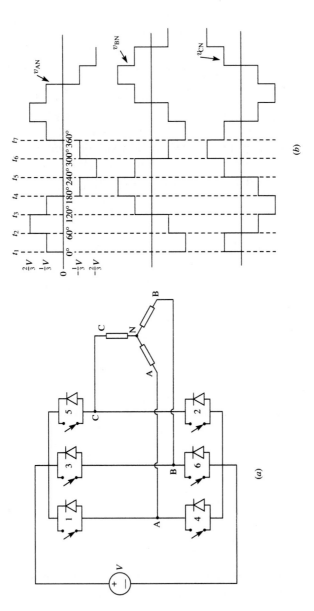

FIGURE 8.1
(a) Circuit configuration and (b) phase voltage waveforms of a three-phase six-step inverter feeding a Y-connected load.

5.10.3. The circuit configuration given in Fig. 5.36 and the phase voltage waveforms across the load, which were shown in Fig. 5.39, are reproduced in Fig. 8.1 for ready reference.

Because of the shape of the output voltage waveform, this inverter is often called a "six-step" voltage source inverter. Voltage adjustment is not possible within the inverter. To adjust the output voltage, it is necessary to adjust the DC input voltage to the inverter. The manner in which this is done depends on how the DC is provided. In the case of a DC link inverter in which the DC link is provided by a phase controlled thyristor bridge, the DC link voltage, and therefore the AC output voltage of the inverter, can be conveniently made adjustable by phase control of the thyristor rectifier bridge. If the DC link is provided by a diode rectifier bridge then a voltage step down chopper may be inserted between the rectifier and the inverter. Voltage adjustment may then be made by adjustment of the duty cycle of the chopper. The same arrangement, that is, a chopper of variable duty cycle, may be used in all cases where the DC is a fixed voltage source such as a battery. These arrangements are shown in Figs. 8.2(a) and (b). The line filter on the input side is also shown.

8.2.1.1 ELECTRIC BRAKING IN CONVERTER-FED DRIVES. It is relatively easy to incorporate the facility for electric braking in inverter-fed drives using induction and synchronous motors. During braking, the operation of the motor changes into the generating mode. The generator is being driven by the kinetic energy of the moving system or by gravitational energy if the braking is being done on an electrically driven vehicle that is going down a gradient. For the braking to be successful, it is necessary to have some means to channel the energy that is being recovered. We have shown in Chapter 5 that the inverter

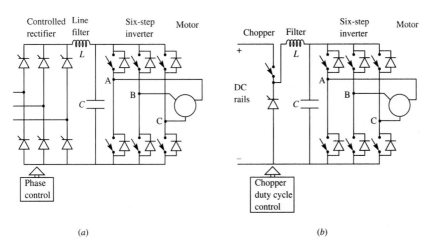

FIGURE 8.2
Voltage adjustment methods for the three-phase six-step voltage source inverter.

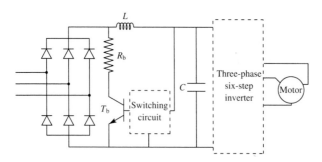

FIGURE 8.3
Automatic switching ON of the braking resistor.

can function with reverse power flow, that is, with power flow from the AC side to the DC side, when there is a source of power on the AC side. When this happens in a voltage source inverter, there is a reversal of the direction of the DC current on the input side of the inverter. If there are several inverter-fed drives from the same DC rails, the power returned by an inverter whose motor is braking may be absorbed by the other inverters whose motors are in the driving mode, and therefore there may not be any net reversal of DC current and no need to make special provision for absorbing the power that is being returned by the braking. But in the more common situation where there is only a single inverter drive connected to the DC rails, means should be provided to absorb the power being returned through the inverter during braking. In the case of a voltage source inverter, the DC link voltage cannot reverse. It is the direction of the DC current that reverses for reversal of power flow. If the DC link voltage is provided by a thyristor or diode rectifier, reversal of current direction is not possible on the DC side, because of the unidirectional nature of the thyristors and diodes. If the DC is provided by a battery, the battery can be made to recharge by the regenerated current. However, if there is a chopper between the battery and the inverter, the chopper will have to be a two-quadrant one, so that the direction of power flow can be reversed. Thereby power can be made to flow from the lower voltage motor side of the chopper to its higher voltage battery side. The use of a braking resistor is a simple means that is often employed to dissipate the energy that is returned through braking, when a simple means of channeling this energy back into the power source is not convenient. This is shown in Fig. 8.3. In this figure, the braking resistor is labeled R_b. This resistor is automatically switched ON during braking by turning ON the transistor switch labeled T_b. During braking, the reversal of DC current through the inverter causes the voltage across the filter capacitor to rise. The rise in voltage is sensed by means of a voltage sensing circuit, which automatically turns ON the transistor switch T_b. This is illustrated by the following numerical example.

Illustrative example 8.1. The DC input voltage to a three-phase six-step inverter feeding an AC motor is 240 V. The capacitor of the input filter on the DC side has a value of 1000 μF. During a braking operation of the motor, the reversed value of the DC current of the inverter is 10 A. The primary DC source is from a diode bridge rectifier, and no reversal of DC current is possible through it. Determine the rate at which the DC voltage may be expected to rise on the DC side of the inverter when braking commences. If the rise in voltage is not to exceed 25 V, what is the maximum permissible delay for switching ON the braking resistor. Also determine the approximate value of the braking resistor. What will be the disadvantage if the value of the braking resistor is lower?

Solution. Since reversal of the DC current is not possible, the braking current will charge the filter capacitor, and the voltage of the DC rails will rise. The rate of rise of voltage may be determined as follows by determining the rate of charging of the filter capacitor.

$$\Delta v = \frac{I\,\Delta t}{C} = \frac{10\,\Delta t}{1000 \times 10^{-6}}$$

This gives the rate of rise of voltage as:

$$\frac{\Delta v}{\Delta t} = 10\,\text{V/ms}$$

At this rate, it will take 2.5 ms for the voltage to rise by 25 V. Therefore the static switch should turn ON the braking resistor within 2.5 ms from the instant of commencement of braking, to limit the voltage rise to within 25 V.

Under steady braking conditions, the DC voltage will be 240 V. For the braking resistor to draw 10 A, its value should be

$$R_b = 240/10 = 24\,\Omega$$

Use of a lower value of R_b will cause it to draw more than 10 A. Since the current fed through the inverter is only 10 A, the additional current will have to come from the rectifier bridge. This will therefore mean unnecessary wastage of power from the diode bridge rectifier.

Example 8.1 shows that the correct value of the braking resistor to be used depends on the value of the DC current fed back by braking. The current I_R drawn by the braking resistor, from the DC rails of voltage V, is equal to V/R. If this current is less than the DC current I fed back by braking, then the excess current will charge the filter capacitor and cause the voltage of the DC rails to rise. On the other hand if the current drawn by the braking resistor is larger than I, the difference current will be drawn from the input side, resulting in wastage of power. During normal operation, the braking current I may be expected to have different values. It is not practical to insert different values of R to suit this current. The practical solution to this problem is to use a fixed value of R, which is lower than the lowest expected value. The static switch labeled T_b in Fig. 8.3 is operated as a chopper at a high repetitive switching frequency. The effective value presented by R can be varied by adjustment of the duty cycle of this chopper.

If the switching period, the ON time, and the duty cycle of the switch are respectively labeled as T, T_{ON} and D, the resistance R will draw current only during the ON time T_{ON} the switch. Therefore the DC current drawn by R, from the DC rails will be given by:

$$I_R = (V/R)(T_{ON}/T)$$
$$= (V/R)D.$$

The effective value of R presented by the chopper will therefore be equal to:

$$R_{eff} = (V/I_R) = R/D$$

This shows that the effective value of the braking resistor is continuously adjustable, in the range R to infinity, by adjustment of the duty cycle D, from unity to zero. This adjustment can be automatically implemented by a controller which will monitor the voltage of the DC rails and automatically maintain this voltage at the correct value, by adjustment of the duty cycle of the chopper.

8.2.1.2 DATA RELATING TO THE SIX-STEP VOLTAGE WAVEFORM

The r.m.s. value of the voltage. The r.m.s. magnitude of the six-step phase voltage waveform of the inverter may be determined from Fig. 8.1 as follows

$$\text{mean square value} = \frac{1}{\pi}[(\tfrac{1}{3}V)^2 \times \tfrac{1}{3}\pi + (\tfrac{2}{3}V)^2 \times \tfrac{1}{3}\pi + (\tfrac{1}{3}V)^2 \times \tfrac{1}{3}\pi]$$

$$= \tfrac{2}{9}V^2$$

This gives the r.m.s. voltage as

$$V_{rms} = \frac{\sqrt{2}}{3}V = 0.4714V \tag{8.1}$$

where V is the DC side voltage of the inverter.

Fundamental sinusoidal component. The fundamental sinusoidal component of the voltage may be obtained by Fourier analysis of the six-step voltage waveform as follows:

$$V_1 = \frac{2}{\pi}\left(\int_0^{\pi/3} \tfrac{1}{3}V \sin\theta \, d\theta + \int_{\pi/3}^{2\pi/3} \tfrac{2}{3}V \sin\theta \, d\theta + \int_{2\pi/3}^{\pi} \tfrac{1}{3}V \sin\theta \, d\theta\right)$$

$$= \frac{2}{\pi}V$$

the r.m.s. value of the fundamental component will therefore be

$$V_{1rms} = \frac{\sqrt{2}}{\pi}V = 0.4502V \tag{8.2}$$

Total harmonic voltage. The r.m.s. value of the total harmonic voltage may be found using the values of the r.m.s. total voltage and the r.m.s. fundamental voltage as follows:

$$V_{\text{Hrms}} = (V_{\text{rms}}^2 - V_1^2)^{1/2} = [(0.4714V)^2 - (0.4502V)^2]^{1/2} = 0.1398V \qquad (8.3)$$

Total harmonic distortion (THD). We may define the total harmonic distortion as

$$\text{THD} = \frac{\text{r.m.s. total harmonic voltage}}{\text{r.m.s. total voltage}}$$

For the six-step inverter phase voltage, this becomes

$$\text{THD} = 0.1398/0.4714 = 0.2966 = 29.66\% \qquad (8.4)$$

Illustrative example 8.2. The DC link voltages of a six-step three-phase inverter is provided by a three-phase diode bridge from a 208 V three-phase supply with large inductive smoothing. Determine the r.m.s. value of the AC output voltage of the inverter and the r.m.s. value of the fundamental AC phase voltage component. Also determine the total harmonic distortion in the phase voltage.

Solution. For a three-phase diode bridge rectifier, the DC voltage is $1.35V_L = 1.35 \times 208 = 280.8$ V. For this value of the DC voltage, the r.m.s. phase voltage of the inverter will be, from (8.1),

$$V_{\text{rms}} = \frac{\sqrt{2}}{3} \times 280.8 = 132.37 \text{ V}$$

The fundamental sinusoidal component will be, from (8.2),

$$V_{\text{1rms}} = \frac{\sqrt{2}}{\pi} \times 280.8 = 126.40 \text{ V}$$

The total harmonic voltage is

$$[(132.37)^2 - (126.40)^2]^{1/2} = 39.30 \text{ V}$$

so

$$\text{THD} = 39.30/132.37 = 29.69\%$$

8.2.2 The Pulse Width Modulated Three-Phase Voltage Source Inverter

The favored type of pulse width modulation is sinusoidal pulse width modulation (SPWM), which was described in Chapter 5. This is usually implemented by the sine triangle comparison method, either by an analog method such as that described in Section 5.9.5 or by digitally computing the individual pulse widths by means of a microprocessor and implementing the switching accordingly. In our treatment of SPWM in Chapter 5, we treated the single-phase full bridge as a combination of two half-bridges. We used the same triangular carrier wave for each of the two half-bridges. But we used two separate reference sine waves with a phase difference of 180° for the two half-bridges. We extended this approach to the three-phase bridge and viewed the latter as a combination of the three half-bridges. We used the same triangular carrier for the three half-bridges. But the reference sine waves used had a mutual phase difference of 120°. To determine the voltage magnitude of the three-phase bridge, we shall first determine the voltage waveform for a half-bridge. The line voltage of the single-phase full bridge may be viewed as the difference between the individual half-bridge voltages. It is actually two times the half-bridge voltage because of the 180° phase difference between the half-bridge voltages. When we take the difference in this manner, the line voltage for a three-phase inverter will be $\sqrt{3}$ times the individual half-bridge voltage, because of the 120° difference between the half-bridge voltages. We shall first consider the single-phase full bridge. A typical SPWM waveform is shown in Fig. 8.4. This consists of a train of pulses of varying width. The widest pulse occurs at $\theta = 90°$. The widths of the pulses decrease at a rate proportional to $\sin \theta$, where θ is the angle measured from the zero-crossing of the reference waveform. The maximum pulse width, which occurs at $\theta = 90°$, will depend on the modulation index, which we defined in Chapter 5 as

$$M = \frac{\text{amplitude of reference sine wave}}{\text{amplitude of triangular carrier wave}}$$

We shall consider an ideal SPWM waveform for which the number of pulses per half period is arbitrarily large, which means that the frequency of

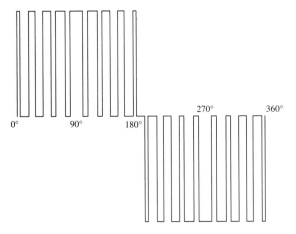

FIGURE 8.4
SPWM waveform.

the triangular carrier wave is arbitrarily large. We shall assume that one half-period of the AC, π rad in angular measure, is divided into a large number of equal intervals each of width $d\theta$. The width of a pulse period is arbitrarily small, because the triangular carrier frequency is arbitrarily high. The pulse width will be widest at $\theta = \frac{1}{2}\pi$. It will decrease as $\sin \theta$ in either direction from the middle. For 100% modulation index, the middle pulse will have full width equal to $d\theta$. For any other pulse, the width will be $\sin \theta \, d\theta$. The r.m.s. value of the total voltage may be calculated as follows.

8.2.2.1 DATA RELATING TO THE SPWM VOLTAGE WAVEFORM

The r.m.s. value of SPWM voltage. We first determine the mean value of the square of the voltage. If the squared voltage is plotted as a function of θ, this will have the same waveform except that the pulse height will now be V^2. The width of the squared-voltage pulses will still be the same. Therefore the area under each pulse will be

$$V^2 \sin \theta \, d\theta, \quad \text{where} \quad \sin \theta \, d\theta \text{ is the width of the pulse}$$

The total area under the V^2 graph in one half-period will be

$$\int_0^\pi V^2 \sin \theta \, d\theta = 2V^2$$

Therefore the mean-square value of the total voltage will be $(2/\pi)V^2$, and the r.m.s. value of the total voltage becomes

$$V_{\text{rms}} = \sqrt{\frac{2}{\pi}} V = 0.7979V$$

For the half-bridge, the value will be half of this. Therefore the r.m.s. value of the SPWM voltage of the equivalent half bridge will be:

$$V_{\text{rms}} = 0.7979V / 2 = 0.3990V \tag{8.5}$$

Fundamental sinusoidal component. The fundamental sinusoidal component may be determined by Fourier analysis as follows. The amplitude V_1 of the fundamental component will be given by Fourier's theorem as

$$V_1 = \frac{2}{\pi} \int_0^\pi V \sin \theta \sin \theta \, d\theta$$

$$= V$$

So the r.m.s. value

$$V_{\text{1rms}} = \frac{V}{\sqrt{2}} = 0.7071V$$

The voltages corresponding to a half-bridge will be half of the above values. Thus the equivalent half-bridge value will be

$$\text{r.m.s. fundamental voltage} = \frac{V}{2\sqrt{2}}$$

The half-bridge voltages will be the equivalent phase voltages for a Y-connected load:

$$\text{r.m.s. fundamental voltage} = \frac{0.7071}{2} V = 0.3535V \tag{8.6}$$

the fundamental line voltage of the three phase bridge will be

$$\sqrt{3} \times 0.3535V = 0.6123V \tag{8.7}$$

The r.m.s. fundamental line voltage given by (8.7), namely 61.23% of the DC link voltage, is the ideal theoretical value based on infinite frequency of the triangular carrier wave. In an actual inverter, the voltage will be less, because it is impossible to implement the very narrow pulses called for by the SPWM strategy, owing to limitations of the switching elements.

Total harmonic voltage and total harmonic distortion factor (THD). The total harmonic voltage (r.m.s. value) can be determined from the values of the r.m.s. total voltage and the r.m.s. fundamental voltage. For the half-bridge, these are

$$V_{\text{rms}} = 0.3990V, \quad \text{from (8.5)}$$
$$V_{\text{1rms}} = 0.3535V, \quad \text{from (8.6)}$$

This gives the total harmonic voltage (r.m.s.) as

$$V_{\text{Hrms}} = [(0.3990V)^2 - (0.3535V)^2]^{1/2} = 0.1850V$$

This gives the total harmonic distortion factor as

$$\text{THD} = 0.1850/0.3990 = 0.4637 = 46.37\%$$

Illustrative example 8.3. The DC link voltage of a three-phase SPWM inverter is obtained from a three-phase diode bridge rectifier fed from a three-phase 208 V bus. This bridge supplies power to a three-phase Y-connected motor.
(*a*) Determine the following for a modulation index of 100%:
    (i) the r.m.s. phase voltage;
    (ii) the r.m.s. value of the fundamental frequency sinusoidal component of the motor phase voltage;
    (iii) the total harmonic content in the phase voltage and the total harmonic distortion factor.
(*b*) Recalculate the above for a modulation index of 60%.

Solution. The DC link voltage is $1.35 \times 208 = 280.8$ V.

(a) The required answers may be calculated by substitution into the formulae derived above.

(i) The r.m.s. phase voltage is $0.3990 \times 280.8 = 112.04$ V.

(ii) The r.m.s. value of the fundamental sinusoidal component is $0.3535 \times 280.8 = 99.26$ V.

(iii) The total harmonic voltage is

$$\sqrt{(112.04^2 - 99.26^2)} = 51.97 \text{ V}$$

and the total harmonic distortion factor is

$$\text{THD} = \frac{51.97}{112.04} = 0.4639 = 46.39\%$$

(b) For a modulation index of 60%, the voltage magnitudes calculated above, which are for 100% modulation index, will have to be multipled by 0.6. Thus

$$V_{\text{rms}} = 112.04 \times 0.6 = 67.22 \text{ V}$$

$$V_{1\text{rms}} = 99.26 \times 0.6 = 59.56 \text{ V}$$

$$V_{\text{Hrms}} = 51.97 \times 0.6 = 31.18 \text{ V}$$

$$\text{THD} = 46.39\%$$

Illustrative example 8.4. In the above example, if the motor is Δ-connected, what will be the value of the funamental sine wave component of the line voltage for a modulation index of 60%?

Solution. The fundamental sinusoidal component of the line voltage will be $\sqrt{3}$ times the phase-to-neutral voltage:

$$\text{r.m.s. fundamental line voltage} = \sqrt{3} \times 59.56 = 103.16 \text{ V}$$

8.2.2.2 COMPARISON OF THE SIX-STEP AND SPWM WAVEFORMS. A

comparison of the voltage magnitudes of the six-step and SPWM waveforms show that the AC output voltage is higher from the six-step inverter. In Examples 8.2 and 8.3, both inverters are supplied from the same DC link voltage of 280.8 V. The six-step inverter gives a phase voltage of 132.4 V r.m.s., whereas the SPWM inverter gives a phase voltage of 112.04 V. Therefore the six-step inverter voltage is nearly 18% higher than the SPWM waveform. Comparing the fundamental sine wave components, we find that, whereas the SPWM inverter gives 99.26 V, the six-step inverter gives 126.6 V, which is about 27% higher. The total harmonic distortion factor is also actually less for the six-step inverter: while it is 46.4% for the SPWM waveform, it is only 29.8% for the six-step waveform.

The values that we have stated are for an ideal SPWM waveform, where the pulse number is infinite, corresponding to a carrier frequency of infinity. The actual value of the carrier frequency used depends on the frequency capability of the switching device. The upper values of the carrier frequency used is usually in the neighborhood of 20 kHz when the switching elements are power MOSFETS or IGBTs. Thyristors with force commutation circuits and GTOs can handle much larger powers, but their switching frequency capabilities are much smaller, and so the carrier frequency may have to be limited within a few kilohertz. For a finite number of pulses per half-cycle, the voltage values will have to be recalculated. See Problems 8.4 and 8.5.

The equivalent circuits of AC motors are based on a sinusoidal waveform for the voltage. Therefore, for current calculations using the motor equivalent circuits, the fundamental sinusoidal component of the motor voltage is the one that is most significant. We notice that, from this point of view, the SPWM waveform has a smaller fundamental frequency component.

The main advantage of the SPWM waveform is the fact that, with a large switching frequency, the harmonic frequencies of the voltage are very much higher. Although there is a signficant component of harmonic voltages, all these harmonics are at very high frequencies. Motor equivalent circuits are typically inductive circuits. Therefore the current contributions due to the harmonic voltages are very small, because the impedance offered by the motor inductance is very high for the high frequency voltage harmonics. Therefore the resultant motor currents are pretty close to the ideal sine wave. With the six-step waveform, the lower order harmonics have significant amplitudes, and therefore the current waveforms may not be as good sine waves as compared with the SPWM inverter.

Voltage control. The SPWM inverter provides the facility for easy adjustment of the AC output voltage by variation of the modulation index. Therefore no separate converter to adjust the DC link voltage, is needed on the input side for the AC output voltage adjustment, as will generally be necessary in the case of the six-step inverter. We shall see that voltage adjustment is necessary in AC motor speed control, besides frequency adjustment, for maintaining the proper flux level in the motor when the speed is varied by frequency variation. In the SPWM inverter, the voltage adjustment is done by variation of the modulation index. This is achieved by varying the amplitude of the reference sine wave. The frequency adjustment is done by variation of the frequency of the reference sine wave. These adjustments can be arranged to be made independently.

In a typical motor control application, the motor voltage is increased as the motor frequency is increased for increasing the speed. The voltage increase is for the purpose of maintaining the magnetic flux level constant within the motor. In the SPWM inverter, the maximum output voltage is achieved when the modulation index is increased to unity (100%). At 100% modulation index, the reference sine wave will have an amplitude equal to the amplitude of the triangular carrier. Usually, there will be a need to increase the voltage further for further increase in frequency and speed. If the reference waveform is increased further, the middle pulses will begin to merge and the number of pulses will begin to drop. After the maximum voltage with SPWM operation is reached, it is a typical practice to change over to the six-step mode of operation of the inverter. In this way, it will be possible to utilize the maximum voltage capability of the inverter bridge.

Braking operation. When the motor fed from the inverter is being braked electrically, the direction of power flow reverses. Power will be flowing from the AC side to the DC side of the inverter. This is possible in the SPWM inverter as in the case of the six-step inverter. When this happens, the DC current reverses on the DC side of the inverter. If the power source of the DC rails is not able to absorb the reverse power, a braking resistor may be turned ON automatically by means of a voltage sensing circuit in the same way as we described for the six-step inverter. In this respect, the statements made earlier relating to the six-step inverter are also applicable to the SPWM inverter.

8.3 THE CURRENT SOURCE AND CURRENT REGULATED TYPES OF INVERTERS

Inverters belonging to this category generally behave as adjustable current sources on the AC output side. The term "current source inverter" is generally used to describe a force commutated thyristor inverter that has been widely used for the speed control of three-phase AC motors. The DC input to this type of inverter behaves as a current source. The current regulated category of inverters may have a DC voltage source on the input side. But their switching is controlled in such a way that they behave as a constant current source rather than as a constant voltage source on the output (AC) side. We shall first describe the current source inverter and then proceed to describe the current regulated types.

8.3.1 The Current Source Inverter

The switching elements of this inverter are thyristors. The turn ON switching of each thyristor is achieved by a firing pulse on its gate. For turn OFF switching, there is a force commutation circuit consisting of capacitors and diodes, which are part of the circuit topology of the inverter. But there are no auxiliary thyristors that have to be fired to actuate the force commutation circuit. The force commutation circuit functions in such a way that the firing of an incoming thyristor automatically commutates the outgoing thyristor. This type of force commutation, in which each outgoing thyristor is sequentially turned OFF, automatically, by the firing of an incoming thyristor, has been described as "auto sequential commutation". The inverter itself is often described as an "auto sequentially commutated" one.

What makes this inverter function like an AC current source on its output side is not the switching strategy within the inverter but rather the fact that the input DC to the inverter itself functions as a constant current source.

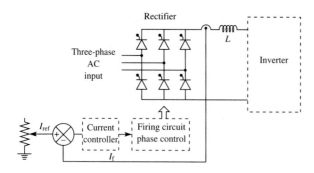

FIGURE 8.5
The adjustable current source section.

In the voltage source inverters described in the previous section, the DC rails functioned as a voltage source. The switching process within the inverter only serves to make this DC voltage source appear as an alternating three-phase AC source on the output side. Therefore the inverter functions like a voltage source on the output side. In a similar manner, the input to the current source inverter functions as a DC current source. The switching operation within the inverter only serves to channel this DC current source in such a way that it appears as a three phase AC on the output side. Therefore this inverter functions like a current source on the output side.

The auto sequentially commutated current source inverter basically consists of two converters. There is a front end rectifier, which is a phase controlled thyristor rectifier. This is phase controlled in such a way that it functions as an adjustable DC current source. Additionally, a large inductance is inserted in series on the DC side, which further improves its characteristic as a current source. The operation of this front end rectifier current source section may be explained by reference to Fig. 8.5. The front end thyristor rectifier is made to function as a current source by means of automatic closed-loop control of the firing angle of the thyristors. The desired value of the current is set by means of an adjustable signal voltage I_{ref}. The actual instantaneous value of the current is sensed by means of a fast current sensing circuit and the sensed value of the DC current is I_f. In the figure, the current sensing device is shown on the DC side, because it is the DC current that needs to be controlled. But in many practical pieces of equipment, the current is sensed on the three-phase AC input lines of the rectifier by means of current transformers. By rectification of the output of these current transformers, we can get a signal that is identical to the rectified DC current of the main converter. The current transformers provide a simple and inexpensive means of electrical isolation between the high voltage power circuit and the gating control circuit. Isolated current sensing is a requirement in most motor control applications, and at present devices are available specifically for this purpose. One such scheme uses Hall effect devices. The Hall device senses the magnetic field due to the current and provides an electrical output proportional to this field. In an alternative scheme, the DC current is sensed by means of a resistance shunt, and this information is transmitted across an optocoupler—digitally, to avoid the nonlinearities and temperature effects of the optocoupler device. For the present application, current transformers are most convenient, and provide the required signal, although the actual sensing is done on the AC side of the rectifier.

The difference $I_e = I_{ref} - I_f$ is the current error. This is detected by the error detector, and serves as the input to the current controller, which is an error amplifier. The output of the error amplifier is the control voltage, for the firing angle in the gate firing circuit of the converter. It changes the firing angle in such a way as to nullify the current error and make the DC current conform to the value set by I_{ref}. In this way, the rectifier is made to provide a constant DC current, the value of which is adjustable by adjustment of I_{ref}. The switchings of the phase controlled bridge occur only at 60° intervals. Therefore the closed-loop controller can correct the current error only at 60° intervals. To minimize the current variations within the intervals of switching, a large DC inductor is always placed in series with the DC output side of the rectifier. This inductor is labeled L in Fig. 8.5. It is invariably a large iron-cored reactor. This inductor makes the circuit function as a better current source, but also introduces the disadvantage that it makes the circuit slow in responding to changes in the demanded current, because of the large time constant introduced by the inductor, thereby making the overall response of the control system slower.

8.3.1.1 INVERTER SECTION. The main switching elements of the inverter section are six thyristors, in the conventional three-phase bridge topology. Additionally there are six diodes and six capacitors, for the purpose of implementing force commutation. In Fig. 8.6, we have omitted these circuit elements, which serve to implement force commutation. Each bridge thyristor is symbolically indicated as a static switch. The complete power circuit will be shown later, when we describe the manner in which force commutation takes place. To determine the theoretical waveform of the AC output current, it is only necessary to know the conduction periods of the bridge thyristors. Each thyristor is made to conduct for an interval equal to 120° of the output AC frequency of the inverter. The conduction sequence is indicated in Fig. 8.6. Each thyristor switch is labeled in the order in which it conducts. Based on the conduction sequence and the switching instants, as indicated in Fig. 8.6, the waveforms of the line currents on the AC output side of the inverter are sketched in the figure. The six thyristor switches are shown symbolically as unidirectional controlled switches. The force commutation circuit is not shown in this figure. Two separate circuits are shown for Y-connected and Δ-connected loads in (*a*) and (*b*) respectively.

One complete period of the output AC is divided into six equal intervals, and the boundary instants of these are labeled $t_1, \ldots, t_6$. The conduction intervals of the thyristor switches are also indicated in the figure. Each thyristor switch conducts for one third of the AC period. For example, the

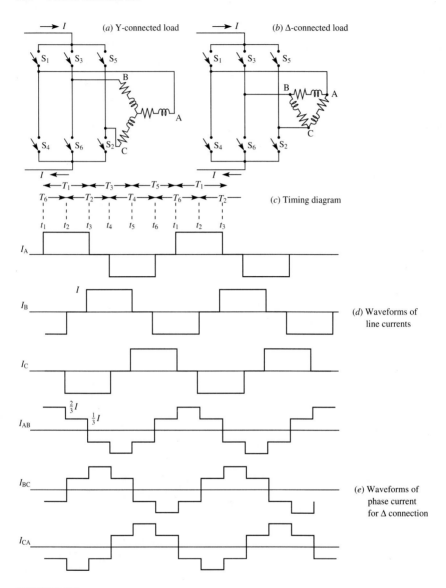

FIGURE 8.6
Circuit topology, conduction sequence and the current waveforms of the auto sequentially commutated current source inverter.

upper leg thyristors S_1, S_3 and S_5 conduct respectively from t_1 to t_3, t_3 to t_5 and t_5 to the next t_1. Similarly, the lower leg thyristors S_2, S_4 and S_6 conduct respectively in the intervals t_2 to t_4, t_4 to t_6 and t_6 to the next t_2. These intervals are indicated by the timing diagram (c).

The input to the bridge is a current of magnitude I. The waveform of the output current in each load impedance can be plotted by tracing the current path in each $60°$ interval through the closed switches according to the timing diagram (c). This is done in (d) for the Y-connected load. For this load, the line current waveform is the same as the phase current waveform, since both currents are the same. The current for the Y-connected load is seen to consist of rectangular pulses of $120°$ duration and amplitude equal to the magnitude of the input current. The three-phase current waveforms are displaced by $120°$.

The phase current waveforms for a Δ-connected load are plotted in Fig. 8.6(c). The phase current waveforms for the Δ-connected load are derived in the following manner. The line current waveforms are the same for both Y- and Δ-connected loads. However, for the Δ connection, the line current divides into two parallel paths when flowing through the load. One path consists of two phase impedances in series, whereas the other path has only one phase impedance. For a balanced three-phase load, the three impedances are equal to one another. Therefore the current division will be one-third of the current through the path consisting of two phase impedances in series and two-thirds of the current through the single impedance. The phase current waveforms are drawn in (e) on the basis of such a current division. The phase current waveform for the Δ-connected load is seen to be a six-step waveform, similar to the six-step voltage waveform for the Y-connected load in the case of a voltage source inverter.

The waveforms derived above are on the basis of ideal elements and on the basis of the conduction sequence and timing indicated in Fig. 8.6(c). The actual waveform may have spike voltages created as a consequence of the commutation process. We shall now proceed to describe the auto sequential force commutation technique employed in this type of current source inverter.

8.3.1.2 AUTO-SEQUENTIAL COMMUTATION.

The three-phase bridge inverter circuit together with the six capacitors and the six diodes used for force commutation is shown in Fig. 8.7(a). In this circuit, the firing of an incoming thyristor automatically commutates the outgoing thyristor. This takes place sequentially in the order in which the thyristors are fired. For this reason, the commutation process is referred to as auto sequential commutation. To explain the commutation process, it will be simpler if we first take a look at the single-phase version of the inverter. The single-phase bridge version is shown in Fig. 8.7(b). The sequence of circuit changes during a commutation of the thyristor T_1 in the single-phase version are shown in (c), (d), (e) and (f). Figure 8.7(c) shows the circuit condition just prior to the turning ON of the thyristors T_3 and T_2. The path of current flow is highlighted. It is from the positive pole of the DC source through T_1, the diode D_1, the load, the diode D_4 and the thyristor T_4, back to the negative pole of the DC source. The polarities of the two commutating capacitors C_{13} and C_{24} will be as shown. The validity of this statement regarding the capacitor voltage polarity will be evident when we go through the sequence of changes during the commutation.

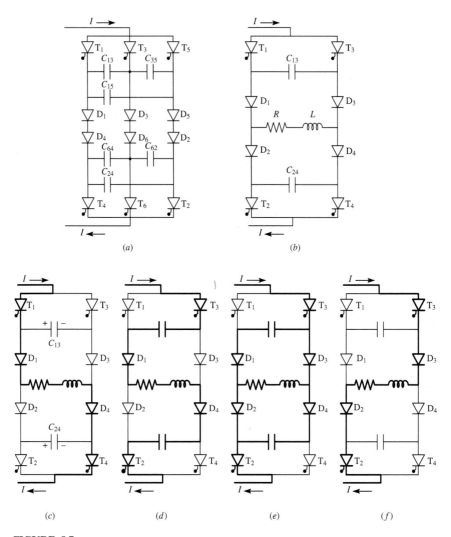

FIGURE 8.7
Current source inverter circuit topology and the circuit changes during commutation in the single phase version of the inverter.

If we look at the loop consisting of the commutating capacitor C_{13} and the thyristors T_1 and T_3, we notice that the thyristor T_3 is forward-biased because of the capacitor voltage. For similar reasons, the thyristor T_2 is also forward-biased by the voltage of the capacitor C_{24}. The commutation process commences when the thyristors T_3 and T_2 are simultaneously fired. When T_3 turns ON, the voltage of the capacitor C_{13} automatically reverse-biases the thyristor T_1 and turns it OFF. Similarly, the voltage of the capacitor C_{24} reverse-biases the thyristor T_4 and turns it OFF. The path of current flow immediately after the turning ON of T_3 and T_2 will be as shown in (d). The current will now charge the two capacitors with the polarity opposite to their previous polarity. When the voltage of the capacitor C_{13} reaches the necessary magnitude to forward-bias the diode D_3, this diode turns ON. A similar process will take place on the negative side, and will turn ON the diode D_2. Assuming the load to be inductive, which is typical, the capacitor C_{13} and the load inductance will constitute an oscillatory circuit with a certain initial capacitor voltage and a certain initial inductive current, when the diode D_3 turns ON. The circuit will be as highlighted in (*e*). The oscillatory transient current will turn OFF the diode D_1. A similar process will turn OFF the diode D_4 on the negative side of the bridge. The current path will now be as shown highlighted in (*f*). The commutation sequence is now complete, and the capacitors C_{13} and C_{24} are now charged with the right polarity for the next commutations of T_3 and T_2 the next time when T_1 and T_4 are fired. The load current will now be flowing in the reverse direction.

The commutation process in the three-phase bridge is illustrated by Fig. 8.8. In the three-phase bridge, the commutations on the top and bottom sections of the bridge do not take place at the same time. There is a 60° phase displacement between the commutation on the top and bottom sections of the inverter. Figure 8.8 shows the sequential circuit changes during a commutation in a top leg of the inverter. The commutation process shown in the figure is for the commutation of the thyristor T_1. This is initiated by the firing of thyristor T_3 in the top section of the inverter. Prior to the turning ON of T_3, the path of current flow is as shown highlighted in (*a*). The polarities of the capacitor

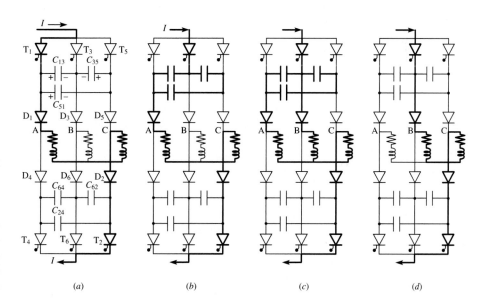

(*a*)   (*b*)   (*c*)   (*d*)

FIGURE 8.8
Circuit changes during a commutation in the three-phase bridge.

voltages that are involved in this commutation are also indicated in (*a*). In the three-phase bridge, there are three capacitors that are involved in a commutation. Actually, during any one commutation, one capacitor is effectively in parallel with the series combination of the other two capacitors. For example, in the case of the commutation of T_1, when T_3 is fired, the parallel combination is the capacitor C_{13} in parallel with the series combination of C_{35} and C_{51}. The effective value of the equivalent capacitor block is 1.5C, where C is the value of a single capacitor, all capacitors being assumed to be equal. It is the charge on this equivalent series parallel capacitance that serves to commutate T_1 when T_3 is fired. This turn OFF switching takes place in the same way as we described for the single-phase bridge. Immediately after T_3 is fired, the thyristor T_1 is commutated by the equivalent capacitor voltage, and the circuit changes to that shown in (*b*). The current flowing through T_3 charges the capacitor block in reverse, and, as soon as the diode D_3 gets forward-biased, it turns ON, and the circuit changes to that shown in (*c*). This causes the capacitor block to come in parallel with the series combination of the A and B phase loads, which, being inductive, will create an underdamped oscillatory circuit. The oscillatory transient will turn OFF the diode D_1, and the circuit will change to that shown in (*d*). The commutation process is now complete, and this circuit configuration continues until the next commutation, which is from T_2 to T_4, when T_4 is fired in the lower section of the bridge.

8.3.1.3 CURRENT WAVEFORMS. The phase current waveforms for the Y-connected and Δ-connected loads were derived in Figs. 8.6(*d*) and (*e*) respectively. For the Y-connected load, the phase current properties are calculated below on the basis of the waveforms in Fig. 8.6.

The r.m.s. current. This is

$$I_{rms} = \sqrt{\tfrac{2}{3}}\, I = 0.8165I$$

Fundamental sinusoidal component. The amplitude of the fundamental sinusoidal component may be determined by harmonic analysis as follows:

$$I_1 = \frac{2}{\pi} I \int_{\pi/6}^{5\pi/6} \sin\theta\, d\theta = \frac{2\sqrt{3}}{\pi} I$$

The r.m.s. value of the fundamental sinusoidal component will therefore be

$$\frac{\sqrt{2}\sqrt{3}}{\pi} I = 0.7797I$$

The r.m.s. total harmonic current may be determined using the values of the r.m.s. total and the r.m.s. fundamental:

$$I_{Hrms} = 0.2424I$$

The total harmonic distortion factor will be

$$\text{THD} = \frac{0.2424}{0.8165} = 0.2968 = 29.68\%$$

Δ-connected load. The phase current waveform for a Δ-connected load has been derived in Fig. 8.6(e). This is seen to be a six-step current waveform similar to the six-step voltage waveform for a Y-connected load in a voltage source inverter. This was a voltage waveform whose step amplitudes were $\frac{1}{3}V$ and $\frac{2}{3}V$. Here we have a current waveform with step amplitudes $\frac{1}{3}I$ and $\frac{2}{3}I$. The values for the total r.m.s., the fundamental r.m.s. and the total harmonic currents, and the distortion factor may be determined in the same way as we did for the six-step voltage waveform. These values are as follows:

total current (r.m.s. value) $= 0.4714I$

fundamental sinusoidal current (r.m.s. value) $= 0.4502I$

total harmonic current (r.m.s. value) $= 0.1398I$

total harmonic distortion factor $= 29.66\%$

Illustrative example 8.5. The current input of a three-phase current source inverter is 20 A DC. The inverter frequency is 40 Hz. The load is Y-connected, and each phase consists of a resistance of 2 Ω in series with an inductance of 10 mH. Determine the r.m.s. magnitude of the fundamental sinusoidal component of the phase voltage.

Solution. The complex phase impedance at 40 Hz is $2 + j2\pi 40 \times 10 \times 10^{-3} = (2 + j2.516) = 3.212\angle 51.49°$.

The sinusoidal 40 Hz component of the phase current is $0.7797 \times 20 = 15.594$ A.

The sinusoidal 40 Hz component of the phase voltage is $15.594 \times 3.212 = 50.09$ V.

8.3.1.4 REGENERATIVE BRAKING IN A CURRENT SOURCE INVERTER DRIVE. When a three-phase induction or synchronous motor fed from a current source inverter is regeneratively braked, the direction of the current does not reverse. What happens is a reversal of the DC voltage at the inverter input terminals. Since there is no current reversal, the controlled current section, namely the phase controlled thyristor front end converter, automatically goes into the inversion mode during regenerative braking, and power is fed back to the AC bus. There is no need for a braking resistor on the DC side of the inverter. This is a definite advantage of this type of current source inverter. Another advantage is that if a motor fed from a current source inverter stalls owing to excessive load torque, there will be no overcurrent, because the current is limited by the current source. Therefore the motor will not be overloaded. There may be an increase in temperature, because the cooling fan, if any, on the shaft of the motor, will not be operative while the motor is stalled. But the possibility of damage is very much less. If the drive is jammed, it may be possible to remove the jam by reversing the motor through reversing the phase sequence of the inverter using the switching control circuit. These advantages have contributed to the popularity of the auto sequentially commutated current source inverter in many industrial AC drives.

8.3.2   The Current Regulated Inverter with Hysteresis-Type Current Control

This type of inverter was briefly introduced in Section 5.3.3. The circuit topology is the same as that of a conventional voltage source inverter. As in the latter, the DC rails also function as a voltage source. The inverter is made to behave as a constant current AC source by the closed-loop switching control strategy, to be described below with reference to Fig. 8.9(a), which shows one phase leg of the inverter.

The switching blocks of the inverter are the conventional types, such as an IGBT with an antiparallel power diode or a power MOSFET. The required instantaneous value of the current is provided to the control circuit as a reference waveform. This is labeled I_{ref} in the figure. The actual value of the

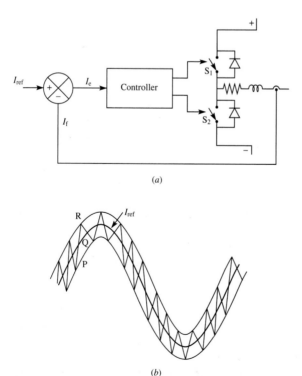

(a)

(b)

FIGURE 8.9
Controlled current inverter with hysteresis-type controller.

instantaneous phase current is sensed by means of a fast current sensor and fed back to the control circuit. This sensed value of the current is labeled I_f in the figure. The error detector compares the reference value with the actual value, and determines the current error. This current error is labeled I_e, where $I_e = I_{ref} - I_f$, and is fed to the current controller. The current controller is of the "hysteresis" type, and works in the following manner.

If the current at a given instant is given by a point such as P in Fig. 8.9(b), where the error is positive; that is, if the actual current I_f is less than the required value I_{ref} then the current controller will turn OFF the switch labeled S_2 and turn ON the switch labeled S_1. This will cause the load to be connected to the positive DC rail, and will cause the current to increase. As the current increases, the positive current error becomes less and less, and becomes zero when the instantaneous current value falls on the reference waveform at the point labeled Q in (b). The current error then begins to be negative as the current continues to rise. At the instant when the magnitude of the negative error reaches a preset limit, corresponding to the point R in (b), the controller causes the switch S_1 to turn OFF and S_2 to turn ON, thereby connecting the output terminal to the negative DC rail. Therefore the current will now start to decrease. The current error will decrease in magnitude, and becomes zero again when the instantaneous current value falls again on the reference waveform. The error will then begin to be more and more positive. As soon as the error magnitude reaches the preset limit, the controller will again operate the switches to increase the current. In this manner, the actual current waveform will be made to follow the serrated shape shown in (b) within an upper and a lower boundary, as sketched in this figure. The maximum positive and negative current error, as compared with the reference waveform, will be the two preset error limits, which are called hysteresis limits. Figure 8.9(b) shows the reference waveform in the middle, between the upper and lower waveforms defining the hysteresis limits. If the preset hysteresis limits are made small, the actual current waveform will conform more closely to the reference wave. But this will mean a higher switching frequency for the inverter switching elements. The switching frequency is not fixed, but will vary depending on the rate at which the current grows and decays. This in turn will depend on the bridge voltage, the circuit constants of the load and the instantaneous value of the reference waveform. The hysteresis limits will have to be set in such a way that the switching devices used on the inverter legs are capable of switching at the maximum rate of switching that will be required. The power MOSFET has the highest switching frequency capability among the power devices presently available. Therefore the use of this device as the switching element will permit the use of low hysteresis limits. If the reference waveform provided to the hysteresis-type controller is a sine wave, the inverter output will conform to this waveshape, within the hysteresis limits. It is to be noted that in this type of inverter, there is a reference value for the current at every instant, which we may call the commanded value. The inverter attempts to make the current conform to the commanded value at every instant. This

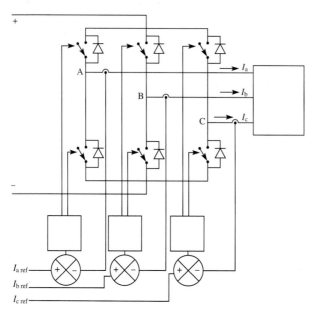

FIGURE 8.10

Independent control of phase currents in a controlled current inverter.

feature makes it possible to use such an inverter in a closed-loop speed control strategy in a motor control application, in which the controller is designed to determine the required current at every instant, and provide this as the commanded value of the current. For this reason, this type of controlled current inverter is the one that is most suitable for the speed control strategy called "vector control," which is to be described in Chapter 9. In vector control, the controller determines the instantaneous current needed in each phase of the motor, and commands the inverter to implement this current. Each phase current in a three-phase motor can be independently controlled by having a separate hysteresis controller for each phase, as shown in Fig. 8.10. There is a separate current reference for each phase. Each phase current is separately sensed and compared with the corresponding reference. There is a separate controller for each phase leg of the inverter, which serves to independently force the output current of that phase to conform to the reference value for that phase. In vector control of a motor, the individual reference values of the current will be computed by the controller, on the basis of the vector control strategy, and provided to the switching control circuit, for the purpose of making the actual current conform to the reference.

8.3.3   The Current Regulated Inverter with PWM Control

This type of inverter is also designed to make the output current conform to a reference value at every instant. But the difference between this and the hysteresis type is in the manner of switching employed to make the output

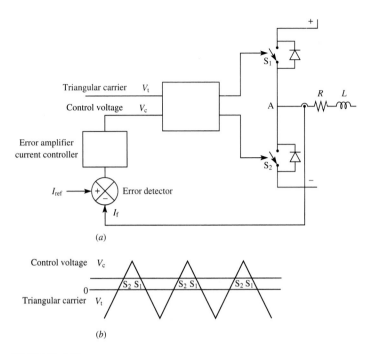

(a)

(b)

FIGURE 8.11
Controlled current inverter using the PWM type of control.

current conform to the reference. In the PWM type of control, the switching frequency is fixed. The pulse width is determined by comparing the triangular waveform with a control voltage. This may be explained by reference to Fig. 8.11. Figure 8.11(a) shows one leg of the inverter bridge and one load phase (phase A). The load phase is shown as a series R–L circuit. The two switches of the inverter leg are labeled S_1 and S_2. Pulse width modulation is implemented by the comparison of a triangular carrier waveform V_t and an adjustable control voltage V_c. The comparator compares these two voltages, and whenever V_c is higher than V_t, the static switch S_1 is kept ON and S_2 OFF. The switching instants correspond to the intersections of the triangular wave and the control voltage, as illustrated in Fig. 8.11(b). At the instants labeled S_1 in this figure, S_1 is turned ON and S_2 turned OFF. Similarly, at the instants labeled S_2, S_2 is turned ON and S_1 is turned OFF. When S_1 is ON and S_2 OFF, a positive pulse will appear at the load terminal A. A negative pulse will appear at the load terminal during the intervals when S_2 is ON. As may be seen from (b), for a zero value of the control voltage, the durations of the positive and negative voltage pulses will be equal. For a positive value of the control voltage, the positive pulses will be of longer duration than the negative pulses. Similarly, for a negative value of the control voltage, the negative voltage pulses will be of longer duration than the positive pulses.

To implement current controlled operation of the inverter, a reference current value is provided to the control circuit. This is the signal voltage labeled I_{ref} in Fig. 8.11. A fast current sensing device senses the actual instantaneous current and feeds this signal back to the controller. This signal voltage is labeled I_f in the figure. The error detector compares the reference I_{ref} and the feedback signal I_f. Its output is the current error $I_{ref} - I_f$. This error signal is fed into the circuit block labeled "Error amplifier current controller." The output of this error amplifier serves as the control voltage V_c for pulse width modulation. The comparator compares the control voltage and the triangular carrier voltages, and implements the switching in the manner indicated in (b). If the current error is positive, that is, if I_{ref} is more than I_f, then S_1 will be ON for a larger portion of the switching period, and the positive pulse will be wider than the negative pulse, resulting in a net increase of load current. If the current error is negative, the negative pulse at the load terminal will be of longer duration, resulting in a net decrease of the load current. In this manner, the current controller forces the load current to conform to the reference value.

It will be noticed that both the hysteresis type of controller and the PWM type are designed for achieving the same purpose of making the actual current conform to the reference value at each instant. But the PWM type has the advantage that the switching frequency has a fixed value as determined by the frequency of the triangular carrier wave. The PWM-type controlled current inverter is also suitable for applications such as vector control, where the controller computes the value of the motor current required at each instant, and calls upon the inverter to provide this current. In a three-phase inverter, the individual phase currents can be controlled independently, in the same way as we described for the hysteresis-type controller.

Illustrative example 8.6. A controlled current inverter employs the PWM current control strategy shown in Fig. 8.11. The frequency of the triangular carrier is 10 kHz. The inverter AC frequency is 40 Hz. The triangular carrier waveform used has a peak-to-peak amplitude of 12 V. What will be the number of positive and negative voltage pulses for one period of the inverter AC.

Determine the width of widest positive pulse width and the value of the control voltage V_c at which it occurs. Determine also the widths of the positive and negative pulses for a control voltage of +5 V.

Solution. The period of the triangular carrier waveform $= 1/(10 \times 10^3) = 100\ \mu s$. there will be one positive pulse and one negative pulse in one period of the triangle. The Period of the inverter AC $= 1/40 = 25\,000\ \mu s$. Therefore the number of pulse repetition periods in one AC period $= 25\,000/100 = 250$. Therefore there will be 250 positive pulses and 250 negative pulses in one period of the inverter AC.

Reference to Fig. 8.11 shows that the widest positive pulse will ideally have a width of $100\,\mu s$, and this will occur when the value of the control voltage is equal to the amplitude of the triangle, that is, for

$$V_c = 6\text{ V}$$

For $V_c = 4$ V, the positive pulse width, determined by the use of similar triangles, will be

$$\frac{6+4}{12} \times 100 = 83.3\;\mu s$$

The width of the negative pulse is $100 - 83.3 = 16.7\;\mu s$.

8.4 THE PHASE CONTROLLED CYCLOCONVERTER

The cycloconverter is a direct frequency converter that converts from an AC input, which is typically the three-phase fixed frequency utility bus, to an adjustable AC output, whose voltage and frequency are independently adjustable. The phase controlled cycloconverter has the same exact circuit topology as the phase controlled dual converter. The difference is only in the manner in which the switching is controlled.

The phase controlled dual converter, which was initially introduced in Chapter 2, was described in more detail in Chapter 7 in connection with its use for DC motor drives. The reader is advised to review this description. We shall, however, here recall the main features, in so far as they are relevant to our description of the cycloconverter, that is, the use of the dual converter topology to provide an AC output of adjustable frequency and adjustable voltage.

The circuit configuration consists of two identical phase controlled thyristor converters, which are labeled P and N in Fig. 8.12. They may have

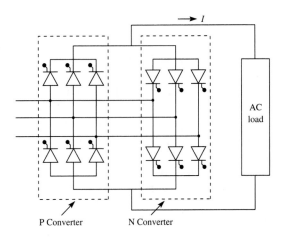

P Converter   N Converter

FIGURE 8.12
Cycloconverter circuit topology.

either the midpoint or the bridge configuration. The midpoint configuration needs a transformer, which makes the equipment heavy and bulky. Therefore the favored configuration is the bridge. The input to the converter may be single-phase or polyphase. The three-phase bridge is the most popular configuration, and is shown in Fig. 8.12.

From the theory of the three-phase line commutated bridge converter presented in Chapter 2, we recall the relationship between V_d, the DC component of the output, and V_L, the AC r.m.s. line voltage of the three-phase bridge, as

$$V_d = 1.35 V_L \cos \alpha, \quad \text{where } \alpha \text{ is the firing delay angle of the bridge}$$

If we operate the P bridge in Fig. 8.12 with a firing delay angle α_p and the N bridge with a firing delay angle $\pi - \alpha_p$ then the DC output of the two bridges will be as follows:

$$\text{output of P bridge} = 1.35 V_L \cos \alpha_p \tag{8.8}$$

$$\text{output of N bridge} = 1.35 V_L \cos (\pi - \alpha_p) = -1.35 V_L \cos \alpha_p \tag{8.9}$$

We see that both outputs have the same magnitude of the DC voltage, but mutually reversed polarity. As far as the DC output is concerned, the two converters can operate in parallel, but with reversed polarity for one, which is what we mean by "antiparallel" connection. The range of α is ideally from 0 to 180°. In this range, any one converter can operate with both positive and negative voltages. But each converter can conduct current only in one direction. Therefore we still require two converters to enable load current flow in both directions—one converter for each direction of current. The P converter handles the positive load current and the N converter the negative load current. In this way, the parallel combination can provide an alternating current output.

However, we cannot solidly connect the two converters in parallel as shown in Fig. 8.12 and operate both converters simultaneously, because we looked only at the DC component of the converter voltages. Besides the DC component, each converter has ripple voltages, which will not be the same for both. The ripple voltages will not balance in the same way as the DC voltages, because the firing angles for the two converters are different and the terminal polarities are reversed. If a solid parallel connection is used and the two converters are simultaneously operated, there will be a large flow of ripple current in the loop consisting of the two converters, because the instantaneous voltages of the two converters will not balance, even though they are operated in such a way that the DC components balance.

There are two alternative solutions to this problem. One is to insert reactors as shown in Fig. 8.13(a) to limit the circulating current and keep operating both bridges simultaneously. Two reactors are needed for the bridge configuration as shown in the figure, because the circulating current has to be

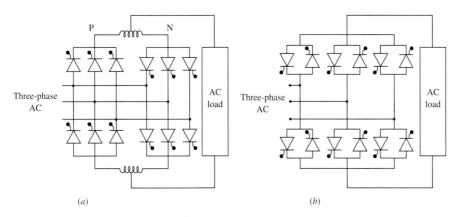

FIGURE 8.13
Cycloconverters: (*a*) circulating current type, (*b*) non-circulating current type.

limited on both ends of the bridge. One reactor will suffice for the midpoint configuration. The circulating current has to be unidirectional, because each bridge can carry current only in one direction. Since the ripple voltage is alternating in nature, this will mean a discontinuous flow of the circulating current. The flow of an AC ripple current can be made possible if this AC current is superimposed on a DC current whose magnitude is larger than the peak value of the AC ripple current. If this is done, both converters will be always in conduction, and the current can shift from one converter to the other, according to its direction, "naturally."

The second solution to the problem eliminates the circulating current altogether. This is done by having only one converter in operation at a time. Only one converter need be active at any given instant, to supply the load current. Which one is to be active will depend on the direction of the load current at any given instant. In this scheme, the thyristor gate pulses will be provided only to the converter that needs to conduct the load current, and the other converter will be made inoperative by blocking the gate pulses to its thyristors. This means that when supplying an alternating load current, the switching between the bridges will have to be made after every half-cycle of the load current. The control circuit has to detect the instant when the current becomes zero, and decide if a current reversal is needed, and then implement the channeling of the gate pulses as required. The necessary logic has to be incorporated in the controller. This makes the controller somewhat more complex. Nevertheless, it is this type of operation, without the flow of circulating current, that is more popular at present. This is because there is no need for the heavy circulating current reactors. Also, the extra power dissipation due to the circulating current in the idle converter is avoided. The power circuit assembly of the dual bridge without circulating current reactors is also simpler, as shown in Fig. 8.13(*b*). The dual-bridge assembly is practically the same as a single-bridge assembly, except for the addition of an inverse parallel thyristor to each one in the single bridge. Both thyristors can share the same snubber circuit and overcurrent fuse, if any.

If, let us say, the bridge P is operated with a firing angle of 0°, its output DC voltage will be maximum and equal to $1.35V_L$, (8.8), because $\cos 0° = 1$. If we now slowly increase the firing angle, the output voltage will decrease in magnitude until α_p becomes 90°, according to the cosine law (8.8). Further increase of α_p will cause the voltage to reverse according to the cosine law, and the voltage reaches the negative maximum at $\alpha_p = 180°$ under the ideal assumption that 180° firing delay is possible. In this way, by slowly varying the firing angle between 0° and 180° and back again to 0°, it is possible to vary the DC voltage through a complete AC cycle, between the positive maximum and the negative maximum, and back again to the positive maximum. This, in principle, is the manner in which the cycloconverter provides a low frequency AC output. This AC, according to our description, is a slowly varying DC, within the positive and negative maximum limits of the two individual converters. We emphasize the word "slow" regarding the variation of the output, because the expression for the DC voltage is based on the typical commutation sequence between the successive phases, and the variation of the firing angle should be such that this sequence of commutation is not altered. The frequency of the AC output will depend on the rate at which the firing angle is changed—the number of complete cycles of change in one second. This frequency has to be considerably less than the AC input frequency of the converter.

Figure 8.14 shows how the output voltage varies over a complete cycle of variation of the firing angle. The DC magnitude corresponding to the prevailing firing angle at any instant is shown by the broken line. The thick lines show the instantaneous output voltage. Overlap between phases during commutation is neglected. We can achieve a sinusoidal waveform for the mean voltage if the firing angle is controlled in the manner described below:

$$V_d = 1.35V_L \cos \alpha_p$$

We may vary α_p as a function of time t in such a way that

$$\cos \alpha_p = k \cos \beta t$$

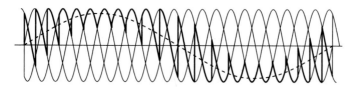

FIGURE 8.14
Instantaneous and mean voltage waveforms of a cycloconverter output.

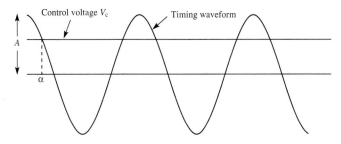

FIGURE 8.15
Cosine wave intersection method for firing angle control of a cycloconverter.

where k is a number less than unity. This will give the mean output voltage at any instant as

$$V_d = 1.35V_L k \cos \beta t \qquad (8.10)$$

This is a sinusoidal AC voltage whose amplitude is $1.35kV_L$ and whose angular frequency is β (rad/s).

By varying the firing angle with respect to time in this manner, the cycloconverter can be made to provide a mean voltage output that is sinusoidal. The amplitude of the sine wave can be adjusted by variation of k, and the output frequency can be independently adjusted by variation of β.

A practical way of implementing this is to use what is known as the cosine wave intersection method, which may be explained by reference to Fig. 8.15. This scheme employs a timing waveform of fixed amplitude A, which has a cosine waveform as shown in the figure. This waveform is compared with a control voltage labeled V_c in the figure. The firing instants corresponding to a firing angle α is obtained by the intersection instant of the control voltage and the cosine waveform, as shown in the figures. At the intersection,

$$V_c = A \cos \alpha$$

from which

$$\cos \alpha = \frac{V_c}{A}$$

Substituting this into the expression for the DC (mean) output voltage, we get

$$V_d = \frac{1.35V_L}{A} V_c \qquad (8.11)$$

We notice from the above equation that the output voltage is proportional to the control signal voltage V_c. Therefore, by making the control voltage a low frequency sinusoid,

$$V_c = Ak \cos \beta t$$

we can implement the required output waveform according to (8.10). For a six-pulse bridge, there are six thyristors to be fired at 60° intervals of the output frequency. The gate pulse timings can be obtained by using six cosine timing waves at 60° intervals and the same control voltage for each.

Ripple voltage considerations. The statements made above concern the waveform of the mean output voltage. The actual waveform, as may be seen from Fig. 8.14 (which is typical, but neglects overlap effects), contains a significant content of ripple. For a low output frequency, the ripple frequency will be high relative to the AC fundamental frequency, and the filtering of the ripple will be easier. As the output frequency is increased, the ripple frequency comes closer to the output frequency, adversely affecting the motor performance in drive applications. In practice, the upper frequency limit for satisfactory use of the cycloconverter is considered to be about one-third of the AC input frequency. Therefore, for a cycloconverter working from a 60 Hz supply, the limit of the upper frequency for satisfactory working is about 20 Hz.

> **Illustrative example 8.7.** The input to a cycloconverter is from a three-phase 208 V, 60 Hz supply. Determine
> (a) the maximum possible amplitude of the sinusoidal component of the output voltage;
> (b) the number of voltage pulse ripples in one period of the AC output.
> The circuit topology of the cycloconverter is a dual three-phase bridge. The output frequency is 10 Hz.
>
> **Solution.** The maximum possible amplitude of the "mean" voltage will be given by (8.10) as
> $$V_d = 1.35 \times 208 = 280.8 \text{ V}$$
>
> Since the configuration is a three-phase bridge, there are six pulses per AC input period. At an output frequency of 10 Hz, there will be six periods of the input AC in one period of the output AC. Therefore there will be 36 voltage ripple pulses in one period of the AC output. We may state here that these ripple pulses are not identical to each other nor will their time durations be exactly the same.

Current and power flow considerations. Our attention so far has been focused on the voltage waveform. If we neglect the ripple content, we may assume that the output waveform is sinusoidal. But the output current waveform will depend on the nature of the load. This current will not necessarily be in phase with the output voltage. The cycloconverter, being a four-quadrant converter, can handle any type of load, because any combination of voltage polarity and current direction is permissible on the output side. This may be illustrated by

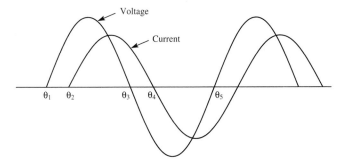

FIGURE 8.16
Output voltage and output current, neglecting ripple.

reference to Fig. 8.16, where we show the voltage and current waveforms for a typical inductive load, in which the current will be lagging with respect to the voltage. The ripple pulses have been ignored in drawing Fig. 8.16.

The active converter and its operating mode, for each interval of the waveform, may be identified from the direction of the current and the polarity of the voltage in the interval concerned. These are listed in Table 8.1. In the table, the inversion mode implies the transfer of power from the load side to the input side. This happens in an AC network because of the reactive nature of the load. A reactive type of load is one that is either inductive or capacitive. With such a load in an AC circuit, there is an oscillatory flow of power between the load and the source, which is what we mean by reactive power. The cycloconverter has the capability to handle reactive power flow. This capability results from the ability of the cycloconverter for reverse power flow, that is, from the load side to the input side. The cycloconverter therefore has an inherent capability to handle the power recouped by regenerative braking of a motor. No special arrangements, such as the use of a braking resistor, are needed to absorb the power returned by braking.

Polyphase output. For three-phase motor control applications, the cycloconverter has to provide three-phase AC output. The cycloconverter that we have

TABLE 8.1

Interval (see Fig. 8.16)	Voltage polarity	Current direction	Active bridge	Operating mode
θ_1 to θ_2	Positive	Negative	N	Inversion
θ_2 to θ_3	Positive	Positive	P	Rectification
θ_3 to θ_4	Negative	Positive	P	Inversion
θ_4 to θ_5	Negative	Negative	N	Rectification

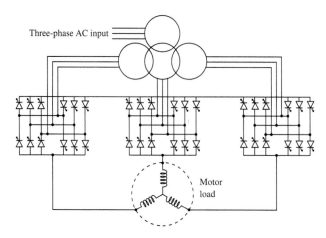

FIGURE 8.17
Cycloconverter with three-phase output.

considered so far provides only one AC output phase. In a cycloconverter, the number of output phases is not dictated by the number of input phases. For example, our dual converter with three-phase input has only a single-phase AC output. Three-phase output can be obtained by having three such cycloconverter blocks, and controlling them in such a way that their outputs have the 120° phase difference between them. Figure 8.17 shows such an arrangement.

In the scheme shown in Fig. 8.17, there are three dual converter bridge pairs, each bridge pair functioning as one phase of the AC output. The load is shown Y-connected. In this arrangement, it is necessary to provide electrical isolation between the inputs to the three cycloconverter phases. This is achieved in Fig. 8.17 by the use of a three-phase transformer with three isolated secondaries, which provide three separate three-phase supplies, with electrical isolation between them. Each three-phase secondary is used as the input to one cycloconverter. If this isolation is not provided, inspection of Fig. 8.17 shows that there will be circulating current between the phases, without any means for limiting it. An alternative means of meeting this isolation requirement in the case of a three-phase motor control application is to isolate the individual motor phases and supply each phase separately from a cycloconverter. This is shown in Fig. 8.18. In other words, we use neither the Y nor the Δ connection, but treat each phase of the motor as a separate load fed from a separate cycloconverter, making sure that the output phases of the cycloconverters have the necessary 120° phase difference needed for three-phase operation of the motor.

In practice, cycloconverters are used for motor drives of very large capacity. The cycloconverter basically uses the dual phase controlled converter technology, which has become well established over the years for large DC motor drives. The cycloconverter needs a larger number of thyristors. For

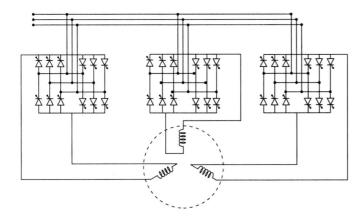

FIGURE 8.18
Three-phase cycloconverter with isolated load phase circuits.

example, the illustrations in Figs. 8.17 and 8.18 show six three-phase bridges which need a total of 36 thyristors. But these thyristors only need to be of the line commutated category, which are less expensive than fast switching inverter-grade thyristors.

8.5 THE LOAD COMMUTATED INVERTER

The main application of this type of inverter is for the control of three-phase synchronous motors. This is not an inverter with a new circuit topology or new switching control technique. It is actually the same as the phase controlled line commutated thyristor converter operating in the inversion mode. Such converters were studied in Chapter 2. Figure 8.19 shows a typical arrangement for a synchronous motor drive from a load commutated inverter.

In Fig. 8.19, the DC link voltage is provided by a line commutated thyristor rectifier. The inverter bridge is shown on the right. The AC terminals of the inverter are connected to the three-phase synchronous motor. When the motor spins, it generates a three-phase AC like a synchronous generator, and

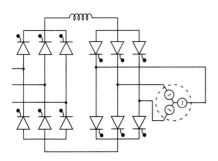

FIGURE 8.19
Synchronous motor drive using a load commutated inverter.

this three-phase AC voltage appears across the AC lines of the inverter. This AC voltage serves to commutate the thyristors of the inverter in the same manner as the line commutation in a line commutated inverter.

In our analysis of the phase controlled inverter in Chapter 2, we saw that for successful operation in the inversion mode, there are limitations on the firing angle. The angle of advance β, which we defined as $180° - \alpha$, where α is the firing delay angle, should have a finite minimum value for successful commutation, because of overlap and the need to provide adequate turn OFF time for the outgoing thyristor. This criterion must be met in gating the thyristors of the inverter. We also saw that for a firing angle α in a phase controlled rectifier, the fundamental AC current lags by an angle equal to α, if we look upon the AC side as the source supplying current into the converter. If we view the AC side as a load, as is the situation in the present case, where the motor is a load on the AC side, we should reverse our reference direction for the positive load current, and it would be more appropriate to say that the AC current is leading by an angle $180° - \alpha$, that is, β. Therefore, for the load commutation process to be successful, the motor should be working at a leading power factor. The theory of synchronous motors tells us that the motor will operate at leading power factor if its field current is made larger than a threshold value. This leads to the conclusion that for a synchronous motor controlled by a load commutated inverter, adequate field current should be provided for the machine to make it operate at a leading power factor. In the line commutated inverter, the timing reference for firing angle control is provided by timing transformers that provide a replica of the voltage waveform on the AC side. This is also possible in the load commutated inverter. But, in practice, it is more convenient to obtain the reference timing signals from a shaft position sensor, which may be a digital encoder. This is possible because the phase voltage waveform is the result of the rotation of the rotor, which carries the field of the motor. Therefore the induced AC voltage due to the rotation of the rotor has a fixed phase orientation with respect to the rotor position. In other words, the zero-crossings of the motor phase voltages are dependent on the angular position of the rotor shaft, relative to the fixed phase winding on the stator. We may look upon the function of the converter thyristors as basically to switch the currents into each phase of the motor at the correct instants of time, that is, when the rotor has the proper angular location with respect to the stator winding. If this is done by actually sensing the rotor position, the switchings will be correct irrespective of the speed of the shaft. The inverter bridge may then be looked upon as functioning in a somewhat similar way to the commutator and brush assembly in a DC motor. The latter serve to switch the currents into the different coils on the rotating armature at the correct instants to provide the unidirectional torque. In the synchronous motor, the armature windings are fixed on the stator. When the switchings are controlled by shaft position signals, our inverter is analogous to an inverted commutator–brush assembly, with a stationary commutator and a rotating brush assembly basically achieving a similar purpose. For this reason, a synchronous motor whose stator is supplied from an inverter with the switchings controlled by a shaft position sensor is called a brushless DC motor.

When a synchronous motor fed from a load commutated inverter is to be started from rest, the initial voltage of the synchronous machine will be too low for successful commutation. The practical way in which this problem is overcome in the scheme shown in Fig. 8.19 is as follows. Initially, the proper thyristors are gated in the inverter according to the shaft position. As the machine begins to rotate and turns through 60° electrical, the time has arrived for the commutation of a thyristor. This follows from the fact that commutations in a three-phase bridge take place at 60° intervals. At this time, there is insufficient phase voltage to turn OFF the thyristor that has to be turned OFF. Therefore, to turn OFF this thyristor, we block the firing pulses to the front end rectifier bridge, effectively starving current to the inverter bridge. In this way, all the thyristors turn OFF, both in the inverter bridge and the rectifier bridge. After providing sufficient turn OFF time to the thyristors, the firing pulses are released to the rectifier bridge, and the next pair of inverter thyristors of the inverter are gated, to provide current to the appropriate phase of the motor. In this manner, the current can be channeled successively, in the proper direction, to the different phases of the motor, and will cause the motor to pick up speed. Normal working of the inverter bridge can commence when the motor speed has reached about 10% of its normal rated speed.

In practice, the load commutated inverter has been used for very large power drives. It basically uses the technology of the phase controlled converters, which is well established at very large power levels. An interesting example of its use has been in a pumped storage power generating system. Here we have a synchronous generator that is normally used as a hydro electric generator driven by a water turbine. During periods when excess power is available on the electrical network, this machine is used as a synchronous motor, and it is driven in the reverse direction. The water turbines are thus made to work as water pumps to pump up water into a storage reservoir, to be later used for generating electricity. The load commutated inverter is used to start and run the synchronous machine as a motor in the reverse direction in the manner described. After it has reached the required speed, it is connected to the AC network, and the inverter is turned OFF. The inverter is used only for starting and running up to the required speed, to enable the machine to be connected directly to the AC network.

8.6 ADJUSTABLE SPEED DRIVES USING THE CAGE-TYPE INDUCTION MOTOR

We have now completed our description of the static converters that are used for AC drives, and we now proceed to describe the drive systems themselves. Each drive system consists of one or more converters, the motor and the controller. We shall first consider drive systems using the squirrel cage induction motor, to be followed by drives using the wound rotor induction motor and the synchronous motor.

8.6.1 Induction Motor Basics

The cage-type induction motor has electrical circuits on both the stator and the rotor. But it is a "singly fed machine" in the sense that all external electrical input is into the stator only. There is no external electrical input into the rotor circuit. The currents in the rotor circuit are all induced from the stator side, by means of the magnetic field in the air gap between the stator and the rotor. This is a great advantage of the machine in that there are no rubbing contacts needed to make electrical connections from the stationary environment to the rotating rotor.

8.6.1.1 THE STATOR. The stator windings are housed in axial slots. For a three-phase machine there are three separate windings—one for each phase. Each phase winding is a distributed winding, consisting of several coils distributed in slots. The objective that should be ideally met by distributing the phase coils in several slots, is to make the magneto motive force (m.m.f.) due to a current in the winding conform to a cosine law, with the maximum along the direction about which the winding is centered. Because the slots are located at discrete intervals along the circumference of the stator, it is impossible to achieve the ideal cosine m.m.f. distribution. The best that can be expected is to obtain a stepped wave distribution that approximates to the cosine law. Such a spatial distribution will have a fundamental component that follows the ideal cosine law. We shall consider only this spatial cosine component and ignore the spatial harmonics for the present. There are actually three sets of identical windings—one for each phase—with a spatial displacement of 120° electrical between them. We shall label the phases A, B and C. For specifying angular measurements along the circumference of the stator, we shall take the reference direction as the radial direction through the center of the A-phase winding. Therefore, when the A-phase winding carries a current i, the magnetic flux density created in the air gap of the machine at a location whose angular coordinate is θ from the reference direction may be expressed as

$$B_a = B_1 \cos \theta$$

where B_1 is the maximum flux density, which occurs at the center of the A-phase winding, that is, at $\theta = 0$. B_1 will be proportional to i_a, where i_a is the current in the A-phase winding. Therefore

$$B_a = Ki_a \cos \theta \qquad (8.12)$$

Similarly, the flux density resulting from a current i_b in the B-phase coils will be maximum in the direction of the center of the B-phase winding. The

B-phase winding itself is located at a spatial angle of $\frac{2}{3}\pi$ from the reference direction. Therefore the flux density due to current i_b in the B-phase winding at a location θ from the reference axis will be

$$B_b = Ki_b \cos{(\theta - \tfrac{2}{3}\pi)} \tag{8.13}$$

Extending this to the C-phase winding, which has a spatial displacement of $\frac{4}{3}\pi$ from the reference direction, the flux density due to a current i_c in the C-phase winding at the location θ from the reference direction will be

$$B_c = Ki_c \cos{(\theta - \tfrac{4}{3}\pi)} \tag{8.14}$$

Resultant flux density due to balanced three-phase currents. If now the three currents i_a, i_b and i_c are a balanced set of three-phase currents, each may be expressed as

$$i_a = I_m \cos{\omega t} \tag{8.15}$$

$$i_b = I_m \cos{(\omega t - \tfrac{2}{3}\pi)} \tag{8.16}$$

$$i_c = I_m \cos{(\omega t - \tfrac{4}{3}\pi)} \tag{8.17}$$

Substitution of these expressions for the currents into the expressions (8.12)–(8.14) for the flux densities gives

$$B_a = KI_m \cos{\omega t} \cos{\theta} = \tfrac{1}{2}KI_m[\cos{(\omega t - \theta)} + \cos{(\omega t + \theta)}] \tag{8.18}$$

$$B_b = KI_m \cos{(\omega t - \tfrac{2}{3}\pi)} \cos{(\theta - \tfrac{2}{3}\pi)} = \tfrac{1}{2}KI_m[\cos{(\omega t - \theta)} + \cos{(\omega t + \theta - \tfrac{4}{3}\pi)}] \tag{8.19}$$

$$B_c = KI_m \cos{(\omega t - \tfrac{4}{3}\pi)} \cos{(\theta - \tfrac{4}{3}\pi)} = \tfrac{1}{2}KI_m[\cos{(\omega t - \theta)} + \cos{(\omega t + \theta - \tfrac{8}{3}\pi)}] \tag{8.20}$$

The resultant flux density at the location θ from the reference direction will be obtained by addition of (8.18)–(8.20) as

$$B = \tfrac{3}{2}KI_m \cos{(\omega t - \theta)} \tag{8.21}$$

because the second terms in the expression on the right-hand side will all add up to zero.

The resultant flux density as given by (8.21), at an angular location θ, shows that the flux density in the air gap has a constant amplitude and rotates at an angular velocity of ω electrical rad/s. At any instant of time t, the flux density follows the cosine law of variation spatially with respect to the spatial angle. If an observer moves along the circumference at an angular velocity such that the angular location of the observer $\theta = \omega t$, the field observed will be constant and unchanging.

Equation (8.21) is the mathematical statement of the concept of the rotating magnetic field, which is basic to the working of three-phase induction and synchronous machines. It means that, in effect, the stator, when carrying the three currents (which have a time phase displacement of one-third of an AC period) in the windings (which have a spatial displacement of one-third of a spatial electrical period), behaves like a magnet that is rotating. The magnetic flux density in the air gap is as if only one of the phase windings is carrying a DC current equal to $\frac{3}{2}$ times the peak phase current, and the entire stator is rotating at an angular velocity of ω electrical rad/s.

The relationship between electrical angle and mechanical angle is $\theta_e = p\theta_m$, where p is the number of pole pairs for which the machine is wound. This is because one complete mechanical angular zone of 360° will have p electrical spatial periods in a machine with p pairs of poles, one electrical spatial period being the angular zone covered by one pole pair.

Therefore the actual mechanical speed of rotation of the rotating magnetic field will depend on the number of pole pairs, and will be given by $1/p$ times ω.

The speed of rotation of the rotating magnetic field is called the synchronous speed. The synchronous speed may be expressed as

$$\omega_{synch} = 2\pi f \quad \text{electrical rad/s}$$

$$= \frac{2\pi f}{p} \quad \text{mechanical rad/s}$$

where f is the frequency of the AC (in Hz).

The synchronous speed may also be expressed as f/p rotations per second $= 60f/p$ rev/min, where p is the number of pole *pairs*.

The above analysis shows that when the stator of the machine is energized from a balanced three-phase supply, the resulting flow of current produces the rotating magnetomotive force (m.m.f.) in the airgap of the machine. The m.m.f. and the resulting flux density have sinusoidal distributions spatially. The rotation at a constant angular velocity causes the flux linkage of each coil to change sinusoidally, causing a sinusoidal voltage to be induced in each stator coil. What is taking place in the stator coils is similar to what takes place in a three-phase transformer. In each phase of a three-phase transformer, a sinusoidal time-varying magnetic flux induces a sinusoidal time-varying e.m.f. The equation for the induced e.m.f. in a transformer is

$$E = 4.44fN\phi \quad \text{r.m.s.} \tag{8.22}$$

where ϕ is the maximum value of the flux through the coils (in Wb).

The same equation also applies for the induced e.m.f. in a motor phase winding, although the sinusoidal variation of flux through the coils is the result of the spatial rotation of the flux in the air gap. We may write

$$E = 4.44N_{eff}f\phi \tag{8.23}$$

where N_{eff} is the effective total number of turns. The effective number of turns is different from the actual number of turns, because the winding is distributed spatially, and the e.m.f.s induced by the rotating flux have a phase difference between successive coils, which have a spatial angular displacement. Therefore the e.m.f.s in the coils should add up as phasors, and the total will be less compared with the transformer, where there is no phase difference between the e.m.f.s in different turns. Also, if the coils are not full "pitched," that is, if each coil is not wide enough to cover one complete pole pitch, there will be a reduction in the e.m.f. in a coil because it does not link with the maximum flux. We may define N_{eff} as

$$N_{eff} = K_w N$$

where N is the total number of turns and K_w is called the winding factor and is defined as $K_w = K_p K_d$ with K_p being the pitch factor, which takes into account any departure of the actual pitch (span) of the coils from the ideal pitch, and K_d being the distribution factor, which takes care of the phase difference between successive coils in adjacent slots of the same phase winding. The winding factor is a constant for the machine, and is less than unity but usually very close to it.

Equation (8.23) shows that the peak flux per phase in the air gap of the rotor is given by

$$\phi = \frac{1}{4.44 K_w N} \frac{E}{f} \qquad (8.24)$$

Equation (8.24) is significant for the speed control of the motor, because it tells us the relationship between induced voltage and frequency for maintaining a certain peak flux level in the air gap of the motor.

8.6.1.2 THE ROTOR CIRCUIT. We may now turn our attention to the rotor of the motor. There are two types of induction motors according to the rotor structure:

1. the wound rotor type,
2. the squirrel cage type.

The wound rotor type of induction motor has three distributed phase windings on the rotor, which are similar to the windings on the stator. These phase windings are usually Y-connected, and the three phase terminals are connected to slip rings mounted on the rotor. There is a stationary brush assembly with a brush for each slip ring. Through the brushes, it is possible to connect the rotor circuit to a stationary electric circuit. In normal use, the slip rings and brushes are for the purpose of inserting additional resistance into each rotor phase winding. The insertion of additional resistance can be shown to increase the starting torque of the motor. Therefore the conventional use of slip rings had been during the starting of the motor, to obtain a high starting torque and to limit the starting current. After starting, the slip rings are short-circuited during normal running. Speed control by inserting resistance in the rotor circuit is inefficient, but is sometimes used for speed control applications in drives such as crane and hoist drives, where continuous operations at low speeds may not be needed. For the present, we shall assume that the rotor phase windings are short-circuited during the normal running of the motor.

In the squirrel cage type of motor, the rotor has conductors in the form of bars in the rotor slots, and all these bars are shorted together at each end of the rotor by an "end ring," to which the conductors are attached. The bars and the two end rings at the ends would make the electrical circuit look like a cage. Hence these motors are called squirrel cage type, or simply cage type. The cage is basically a shorted electrical circuit, and in the normal working of the motor, the cage type rotor behaves in a very similar way to the wound rotor motor with the windings short-circuited. For this reason, for analysis purposes, it is more convenient to replace the cage by an equivalent wound phase, which is short-circuited. This is the approach usually followed in describing the operation of the squirrel cage type of motor also, and we shall do the same.

8.6.1.3 MOTOR OPERATION. We have seen that when the stator is energized from a three-phase supply, a rotating magnetic field is created in the air gap of the motor. The rotating magnetic flux will induce voltages in both the stator coils and the rotor coils, in the same way as the alternating flux in a transformer core induces voltage in both the primary and the secondary coils. If the rotor is held fast and prevented from rotating, the set-up is basically a three-phase transformer with the secondary windings short-circuited. Therefore there will be induced voltages in the rotor phase coils and the resulting currents. The electromagnetic forces resulting from the interaction of the current in the rotor conductors and the air-gap flux will result in a torque, which will cause the rotor to rotate. The basic cause of the creation of the induced voltage, the current and the torque is the relative velocity between the air-gap flux and the rotor conductors. Therefore the direction of rotation will be such as to minimize this relative velocity. In other words, the rotor will rotate in the same direction as the rotating magnetic field. But the rotor speed will not quite reach the synchronous speed, because, as the rotor speed increases, the relative speed between the rotor conductors and the field becomes less and less. Therefore the induced e.m.f., the current and the resulting torque also become less. There will be no induced e.m.f. if the rotor spins at the synchronous speed, because there will then be no relative motion between the field and the rotor conductors.

The difference between the rotor speed and the synchronous speed is called the "slip speed." The slip speed ω_s will be given by

$$\omega_s = \omega_{synch} - \omega_r$$

where ω_r is the actual rotor speed. The fractional slip, usually denoted by s, is the slip speed expressed as a fraction of the synchronous speed:

$$s = \frac{\omega_s}{\omega_{synch}} = \frac{\omega_{synch} - \omega_r}{\omega_{synch}}$$

8.6.1.4 ROTOR CURRENTS. If the rotor is held stationary, the machine is like a transformer whose primary is the stator winding and whose secondary is the rotor. Let E_r be the induced e.m.f. per phase when the rotor is stationary and the stator energized by the rated voltage. Let R_2 be the resistance and X_2 the leakage reactance per phase of the short-circuited rotor. The rotor phase current when stationary will be given by the induced phase voltage divided by the phase impedance:

$$I_2 = \frac{E_2}{R_2 + jX_2}$$

Since the machine is a transformer under standstill conditions, we can use the conventional equivalent circuit of the transformer for current calculations at standstill. This is shown in Fig. 8.20.

When the rotor is rotating with slip s, the relative speed between the rotor conductors and the field will be only $s\omega_{\text{synch}}$. Therefore the AC frequency of the induced rotor voltage will be only sf Hz where f is the AC frequency of the stator input. The induced e.m.f. per phase of the rotor will also decrease in magnitude, because of the decrease in the relative speed with respect to the rotating field. At slip s, the induced phase e.m.f. will be

$$E_s = sE_2$$

where E_2 is the e.m.f. at standstill. The inductive reactance of the rotor phase, which is proportional to the frequency, will also decrease. The reactance at slip s will be

$$X_s = sX_2$$

where X_2 is the rotor phase reactance at standstill. Therefore the rotor phase current at slip s will be given by

$$I_2 = \frac{sE_2}{R_2 + jsX_2}$$

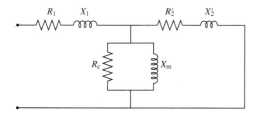

R_1 = stator resistance, X_1 = stator leakage reactance
R_2' = rotor resistance referred to stator side, X_2' = rotor leakage reactance referred to stator side
R_c = equivalent core loss resistance, X_m = magnetic reactance

FIGURE 8.20
Equivalent circuit for one phase of the motor at standstill.

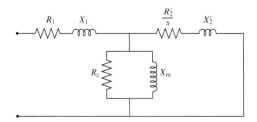

FIGURE 8.21
Equivalent circuit when rotating with slip s.

Dividing both numerator and denominator by s, we may write I_2 as

$$I_2 = \frac{E_2}{R_2/s + jX_2}$$

This shows that we can calculate the rotor phase current at any slip by assuming that the induced rotor e.m.f. and the phase reactance are both unchanged from their standstill values the change being only in the rotor phase resistance from R_2 to R_2/s. Therefore we need only change the resistance value from R_2' to R_2'/s in the equivalent circuit of Fig. 8.20. This modification is shown in Fig. 8.21. The equivalent circuit of Fig. 8.21 can be used at any value of the slip.

When the rotor windings carry current, these currents also create a rotating field with respect to the rotor. Since the rotor frequency is only sf, this field rotates at a speed of $s\omega_{synch}$ *with respect to the rotor.* But the rotor itself is rotating at a speed of $(1-s)\omega_{synch}$ with respect to the stator. From this, we find that the speed of the rotating field due to the rotor currents is also equal to $s\omega_{synch} + (1-s)\omega_{synch} = \omega_{synch}$ rad/s with respect to the stator. From this, we see that the rotating fields due the stator and the rotor are both rotating at the same speed, and there is no relative motion between the two fields. In fact, we can say that the rotating magnetic poles due to stator currents are actually getting locked with the rotating magnetic poles due to the rotor currents, and the two fields are rotating together at the synchronous speed.

The effect of the rotor currents is felt by the stator, through the magnetic field created by the rotor. Since there is no relative motion between the stator and rotor fields, the stator and the rotor are continuing to function as a transformer even when the rotor is rotating, irrespective of the speed of rotation. In this equivalent circuit, the significance of the different parameters are as follows (the circuit itself is for one phase of the motor).

R_1 represents the resistance of one phase of the stator.

X_1 represents the leakage reactance of one phase of the stator circuit. The flux that this represents is the flux linkage of the stator due to the stator phase current, which does not link with the rotor and is therefore called the *leakage* reactance.

R_c is the equivalent resistance for representing the core losses in the magnetic core due to hysteresis and eddy currents.

X_m represents the mutual flux linkage common to both stator and rotor

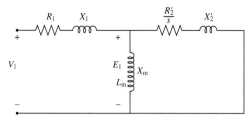

FIGURE 8.22
Commonly used equivalent circuit for one phase of the induction motor.

due to the magnetic flux linking with both the stator and the rotor winding (also called the air-gap flux).

X_2' is the rotor leakage reactance, and represents the flux linkage of the rotor due to rotor currents, which does not link with the stator, because of leakage. The prime on X_2' indicates that it is the value "referred to the stator" in the same way as secondary side parameters are referred to the primary side in a transformer, taking the turns ratio into consideration.

R_2'/s is the fictitious resistance (referred to the stator side) that gives the correct rotor current when the rotor is rotating. When the rotor rotates at slip s, the stator effectively sees the rotor resistance to increase from the actual value of R_2' to a higher value R_2'/s. The increase in effective resistance at slip s is actually $R_2'/s - R_2'$. The power absorbed by this fictitious increase in resistance is the power converted into mechanical power at the shaft of the machine.

In practice, it is usual to make an approximation in the equivalent circuit of Fig. 8.21. The modified equivalent circuit is shown in Fig. 8.22. The simplification consists in removing R_c, which represents the core loss. When the resistance R_c is eliminated from the equivalent circuit, the core loss that it represents is included in the rotational power loss due to friction and wind resistance. The actual power converted to mechanical power is the power in the fictitious increase in resistance $(R_2'/s - R_2')$. We have to subtract the rotational losses including the core losses from this to obtain the useful mechanical power output. It is this equivalent circuit (Fig. 8.22) that is most commonly used in induction motor calculations.

Illustrative example 8.8. A four-pole Y-connected induction motor is working from a three-phase 60 Hz supply of line voltage 208 V. The slip is 4%. The equivalent circuit parameters for one phase of the motor are $R_1 = 0.3\,\Omega$, $X_1 = 1.2\,\Omega$, $X_m = 20\,\Omega$, $R_2' = 0.4\,\Omega$ and $X_2' = 1.5\,\Omega$. Determine
(a) the motor speed;
(b) the input power;
(c) the input power factor.

Solution.
(a) The number of pole pairs for the motor is 2. The synchronous speed at 60 Hz is $60/2\ \text{rev/s} = 1800\ \text{rev/min}$. At a slip s, the motor speed $= (1-s)N_{\text{synch}} = 0.96 \times 1800 = 1728\ \text{rev/min}$.

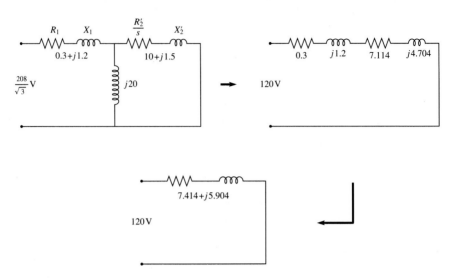

FIGURE 8.23
Circuit reduction in Example 8.8.

Based on the given data, the equivalent circuit of the motor for one phase is drawn in Fig. 8.23. The circuit is reduced to a simple R–L equivalent by the step-by-step reduction shown in Fig. 8.23, and we get the equivalent phase impedance as

$$Z = 7.414 + j5.904$$

$$= 9.478\angle 38.53°$$

The resulting phase current, which is the same as the line current for the Y connection, will be $120/9.478 = 12.66$ A.

(b) Total input power $= \sqrt{3} \times 208 \times 12.66 \times 0.78 = 3557.6$ Watts.

(c) The power factor is $\cos 38.53° = 0.78$ lagging.

8.6.1.5 MECHANICAL POWER OUTPUT AND TORQUE. The torque and output power can also be determined using the equivalent circuit. Since the reactances do not consume any average power, the total power absorbed by the rotor may be expressed from the equivalent circuit of Fig. 8.23, as

$$\text{power} = \frac{I_2'^2 R_2'}{s}$$

But, since the resistance of the rotor circuit is only R_2, the actual power dissipated in the rotor circuit is only $I_2'^2 R_2'$. The rest of the power that goes into the rotor gets converted to mechanical power. Therefore the converted power per phase will be

$$P_{con} \text{ per phase} = I_2^2\left(\frac{R_2'}{s} - R_2'\right) = I_2^2 \frac{R_2'}{s}(1 - s)$$

The total power due to all the three phases will be

$$P_{con} = 3I_2^2 \frac{R_2'}{s}(1 - s)$$

All of the converted power is not available for driving the load coupled to the motor, because some will be lost in overcoming friction and wind resistance. The power lost in this way is called "rotational losses." Therefore the useful power output will be

$$P = P_{con} - \text{rotational losses} = 3I_2^2 \frac{R_2'}{s}(1 - s) - P_{rot}$$

We have to include in the rotational losses the core losses in the magnetic circuit, because it is on this basis that we omitted the resistor R_c that represented the core loss in the circuit of Fig. 8.21 when we changed over to Fig. 8.22.

The induced torque in N m will be given by the converted power in W divided by the rotational speed in rad/s. Therefore the induced torque will be

$$T = 3I_2^2 \frac{R_2'}{s}(1 - s) \Big/ \omega_{synch}(1 - s) = 3I_2^2 \frac{R_2'}{s} \Big/ \omega_{synch}$$

All of this torque will not be available for driving the load, because some of it will be used up in overcoming friction and windage, that is rotational losses. The useful torque output will be

$$T = 3I_2^2 \frac{R_2'}{s} \Big/ \omega_{synch} - T_f$$

where T_f is the torque needed to overcome rotational losses.

Illustrative example 8.9. For the motor in Example 8.8, determine the induced torque and the output power. If the total rotational power loss, inclusive of core losses, is 400 W, determine the useful power output and the efficiency.

Solution. The rotor current I_2' may be determined as follows (see Fig. 8.23):

$$I_1 \frac{j20}{(10 + j1.5) + j20} = 12.66 \frac{j20}{10 + j21.5}$$

$$I_2' = 9.68 + j4.50$$

$$|I_2'| = 10.678 \text{ A}$$

The converted power per phase is $10.678^2 \times 10 \times (1 - 0.4) = 1094.6$ W. The total converted power is $3 \times 1094.6 = 3283.8$ W. So

$$\text{useful power} = \text{converted power} - \text{rotational losses} = 3283.8 - 400 = 2883.8 \text{ W}$$

The total input power into the stator is $3VI \cos \phi = 3 \times 120 \times 12.66 \times 0.78 = 3554.9$ W. So

$$\text{efficiency} = \frac{2883.8}{3554.9} = 0.811 = 81.1\%$$

The synchronous speed is $30 \text{ rev/s} = 188.5 \text{ rad/s}$. The motor speed is $(1 - 0.04) \times 188.5 = 180.96 \text{ rad/s}$.

$$\text{Induced torque} = \frac{\text{converted power}}{\text{speed}} = \frac{3283.8}{180.96} = 18.15 \text{ N m}$$

$$\text{Useful torque} = \frac{\text{useful power}}{\text{speed}} = \frac{2883.8}{180.96} = 15.94 \text{ N m}$$

8.6.1.5 THE PHASOR DIAGRAM FOR ONE PHASE OF THE MOTOR.

The phasor diagram based on the equivalent circuit of Fig. 8.22 is drawn in Fig. 8.24. In this, we have taken the mutual flux labeled ϕ, which links with both stator and rotor along the reference (horizontal) direction. The magnetizing current that is responsible for this mutual flux is labeled I_m and is drawn in phase with the mutual flux ϕ. The induced e.m.f. in the stator phase due to the mutual flux is labled E_1, and leads the flux phasor by 90°. This is called the air-gap voltage, and is the voltage across the terminals of the reactor X_m in the equivalent circuit of Fig. 8.22. The rotor current I_2' lags this voltage by the phase angle of the impedance $R_2'/s + jX_2'$, that is, $\tan^{-1}(sX_2'/R_2')$. The stator phase current is the phasor sum of I_2' and I_m. All the rotor quantities shown on the phasor diagram are quantities "referred" to the stator, in the same way as the secondary quantities are referred to the primary circuit in the case of a transformer equivalent circuit.

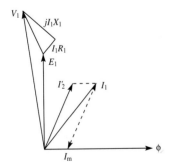

FIGURE 8.24
The phasor diagram for one phase of the motor.

8.6.2 Open-Loop Speed Control of an Induction Motor Fed from a Voltage Source Inverter or Cycloconverter

For simple applications where high speed accuracy is not required, an open-loop control may be sufficient. Figure 8.25(*a*) shows a scheme for doing this using a voltage source converter. The converter may be the six-step voltage source inverter described in Section 8.2.1 or the SPWM inverter described in Section 8.2.2. It could also be the cycloconverter described in Section 8.4. It is assumed that the converter used has the facility for independent adjustment of the output voltage and output frequency.

In the scheme shown in Fig. 8.25(*a*), the adjustment of speed is made by simply varying the frequency of the inverter, thereby varying the synchronous speed of the machine. There will be a small variation of the speed under loaded conditions, because of the variation of slip. But this will usually be small, and seldom more than about 4%. Such a speed variation can also be manually corrected if necessary. However, the simple variation of the inverter frequency alone is unsatisfactory and rarely used, for the following reason.

If reference is made to the expression for the motor voltage given in (8.23) or the expression for the peak flux given by (8.24), we find that the motor flux will change if the frequency alone is adjusted without adjustment of the voltage. If the frequency is increased, the flux will decrease. The weakened flux will adversely affect the ability of the machine to develop torque. If the frequency is decreased, the flux will increase, and may drive the magnetic circuit into saturation. We notice from these equations that the flux is determined by the ratio of the voltage to frequency. Therefore, to maintain

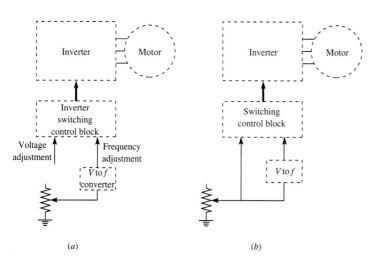

(*a*) (*b*)

FIGURE 8.25
Open-loop speed adjustment.

the correct flux level in the magnetic circuit of the machine, it is necessary to vary the voltage also proportionally when the frequency is changed. Since our converter has the facility for independent adjustment of both voltage and frequency, simultaneous adjustment of both can be made by the arrangement shown in Fig. 8.25(b).

Voltage boost or voltage function generator for low speed operation. The arrangement shown in Fig. 8.25(b) may be satisfactory at high speeds when the voltage required is high. But it will not provide satisfactory performance at low frequencies for the following reason. Referring to the equivalent circuit of Fig. 8.22, the mutual flux in the air gap determines the voltage labeled E_1 across the mutual reactance X_m. What we are adjusting in Fig. 8.25(b) is the terminal voltage V_1 at the input terminal. There is a difference between V_1 and E_1. V_1 is obtained by the phasor addition of E_1 and the voltage drop across the stator phase resistance R_1 and the leakage reactance X_1, as shown by the phasor diagram of Fig. 8.24. At high speeds, when both V_1 and E_1 are large, the drop across the stator phase impedance $R_1 + jX_1$ will be relatively small, and the relative difference between the terminal voltage and the air-gap voltage will be small. But when we go to low speeds, and therefore low voltages, if the adjustment is made on the terminal voltage, the air-gap voltage and therefore the flux level will be too small, adversely affecting the torque performance. What we need to do is to adjust the air-gap voltage when we adjust the speed. It is E_1/f that has to be kept constant, and not V_1/f.

We can improve the operation to some extent by adding a small fixed voltage to the variable voltage that we use for adjusting the converter voltage, as shown in Fig. 8.26(a). We can design the block labeled V_{adj} in such a way that its output voltage follows the line labeled 1 in Fig. 8.26(b) rather than the

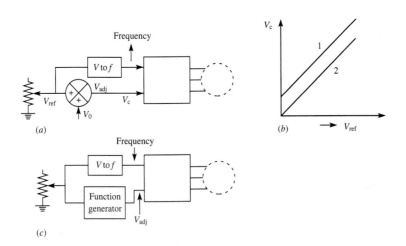

FIGURE 8.26
Two improved methods of terminal volts per hertz adjustment.

line 2. In this way, we can make E_1 conform more closely to the required value when the frequency is varied.

For more accurate adjustment of the voltage, we can use a function generator, which will provide an output voltage that will give a more accurate terminal voltage reference, which is needed to give correct air-gap voltage E_1 for the constant flux operation of the motor when the frequency is varied. The output of the function generator should be shaped on the basis of calculations of the stator impedance voltage drop, using the motor parameters. The arrangement is shown in Fig. 8.26(c).

8.6.3 Closed-Loop Control using a Voltage Source Converter

Simple closed-loop versions of the two open-loop speed control schemes shown in Fig. 8.26 can be implemented by including a speed error detector and error amplifier. In the open-loop schemes, we adjust a voltage that will directly vary the clock frequency of the inverter and also vary the voltage output of the inverter in one of the two ways shown in Fig. 8.26(a) and (c). The closed-loop versions of these schemes work in a similar way. In place of the adjustable voltage, we use the output of an error amplifier that amplifies the error between a speed reference and a speed feedback signal from a tacho generator or some other form of speed sensing device. The wanted speed is given to the controller as the reference voltage labeled ω_{ref} in Fig. 8.27. The actual speed sensed by means of a speed sensor or tachogenerator is fed back to the controller as the feedback signal ω_f. The error detector detects the difference,

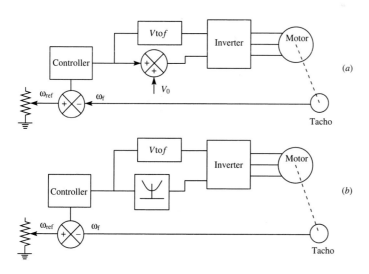

FIGURE 8.27
Closed-loop control versions of the speed control schemes of Fig. 8.26(a) and (b).

and the speed controller amplifies this difference. The output of the controller serves to adjust the clock frequency of the converter through a voltage controlled oscillator. The same voltage also serves to make the terminal voltage adjustment. The method of terminal voltage adjustment in Fig. 8.27(a) is the same as in Fig. 8.26(a), that is, by adding a fixed low voltage V_0. The scheme shown in Fig. 8.27(b) is the same as shown in Fig. 8.26(c), which is the use of a function generator that generates the corrected voltage profile for volts per hertz control. Except for this, the two schemes in Figs. 8.27(a) and (b) are the same.

Slip controlled operation. In Section 8.6.1, in our treatment of induction motor basics, we showed that the torque induced in an induction motor can be calculated by using the equivalent circuit of Fig. 8.22. We showed how this is done in Example 8.9. In this example, the value of the fractional slip s that we used was 4%, which is typical. We can, however, calculate the induced torque in the same way for different values of s. If we do this, obtain values of the induced torque for different values of slip, and plot a graph showing this relationship, the shape of the graph for a typical motor will be as shown in Fig. 8.28. This is actually a torque–speed graph. Both the slip and the speed are shown on the horizontal axis. The slip is zero at the synchronous speed and unity at zero speed. The graph is plotted so that the slip has to be read backwards from the zero value, which occurs at the synchronous speed.

Figure 8.28 is drawn in three separate parts (a), (b) and (c) for the sake of clarity. (a) shows the torque for only positive values of slip from 0 to 1. This corresponds to the motoring mode of operation of the machine, from

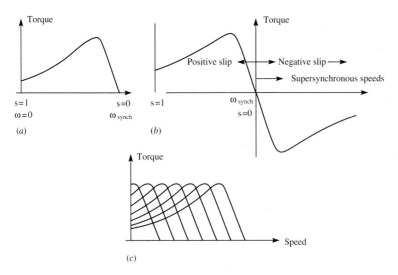

FIGURE 8.28
Torque variation with speed and slip in an induction machine.

synchronous speed down to zero speed. We can use the same equivalent circuit to calculate the torque for negative values of slip. Negative slip means that the machine is spinning at a speed higher than the synchronous speed. This will give negative values for the torque, implying the generating mode of operation. This mode of operation occurs when the machine is regeneratively braked. Regenerative braking is easily implemented in an induction motor drive by simply lowering the inverter frequency, so that the synchronous speed becomes lower than the machine speed, thereby making the slip negative. In Fig. 8.28(*b*), we show the torque for negative values of slip also, corresponding to the braking operation. Figure 8.28(*c*) shows a family of curves for different values of the synchronous speed, that is, different values of the inverter frequency, because the synchronous speed is determined by the inverter frequency.

We see from Fig. 8.28 that the torque is zero at zero slip, that is, at synchronous speed, and increases approximately linearly with slip until the maximum torque is reached. This is true for both the motoring and the generating (negative-slip) modes of operation. If the synchronous speed is varied by varying the inverter frequency, the characteristics are shifted horizontally as shown in (*c*). These curves suggest that torque can be controlled by controlling the slip. It is possible to use a control strategy that has a torque control loop inside a speed control loop, in the same way as we have for the speed control of a separately excited DC motor.

Figure 8.29 shows how torque control can be introduced by means of a controlled slip operation. In the scheme shown, the required slip is provided to the controller as the signal labeled ω_s. This is added to the sensed signal to give the synchronous speed. The actual speed signal is labeled ω_f in the figure. The summing block adds the slip reference voltage to ω_f to give the signal voltage that determines the inverter frequency that will give the required slip. The signal voltage that determines the inverter frequency may be used with a function generator in the same manner as shown in Figs. 8.26 and 8.27 for determining the inverter output voltage.

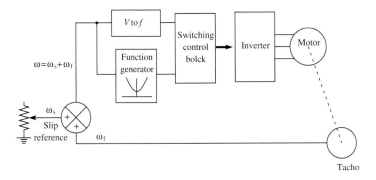

FIGURE 8.29
Slip controlled operation with a voltage source inverter.

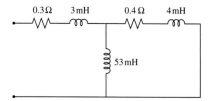

FIGURE 8.30
Phase equivalent circuit of the motors of Example 8.10.

Illustrative example 8.10. The equivalent circuit data for one phase of a three-phase cage-type induction motor are given in Fig. 8.30. The air-gap voltage at its rated frequency of 60 Hz is 108 V. The motor is controlled by a voltage source inverter with controlled slip operation as shown in Fig. 8.29. The commanded frequency in Fig. 8.29 is f. Determine the terminal voltage command to be provided by the function generator for (a) $f = 40$ Hz and (b) $f = 20$ Hz. Assume that the slip command is 4% in both cases.

Solution. To maintain the rated air-gap flux, the airgap voltage at 40 Hz will have to be

$$\frac{40}{60} \times 108 = 72 \text{ V}$$

At 40 Hz, the motor equivalent circuit impedances will be as shown in Fig. 8.31(a) for $s = 4\%$. For $E = (72 + j0)$ V, the currents can be determined as

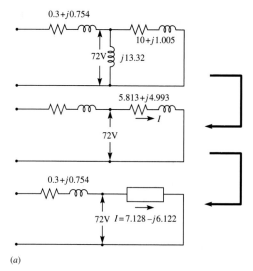

(a)

FIGURE 8.31
(a) Reduction of equivalent circuit for 40 Hz.

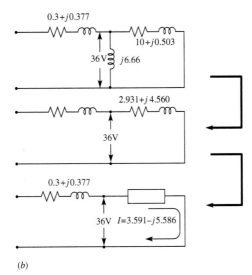

FIGURE 8.31

(b) Reduction of equivalent circuit for 20 Hz.

(b)

follows:

$$(10 + j1.005) \parallel j13.32 = 5.813 + j4.993$$

$$I = \frac{72 + j0}{5.813 + j4.993} = 7.128 - j6.122$$

$$V_1 = (72 + j0) + (7.128 - j6.122)(0.3 + j0.754) = 78.754 + j3.538$$

$$|V_1| = 78.8 \text{ V}$$

At 20 Hz, the motor equivalent impedances will be as shown in Fig. 8.31(b). To maintain the rated airgap flux at this frequency the airgap voltage will have to be $E = 36$ V:

$$(10 + j0.503) \parallel j6.66 = 2.931 + j4.560$$

$$I = \frac{36 + j0}{2.931 + j4.560} = 3.591 - j5.586$$

$$V_1 = (36 + j0) + (3.591 - j5.586)(0.3 + j0.377) = 39.183 - j0.322$$

$$|V_1| = 39.2 \text{ V}.$$

A complete control strategy incorporating the slip controller and an outer speed controller is shown in Fig. 8.32. In this scheme, the slip reference ω_s of Fig. 8.29 is obtained from a speed control amplifier that amplifies the error between the speed reference signal ω_{ref} and the speed signal obtained through

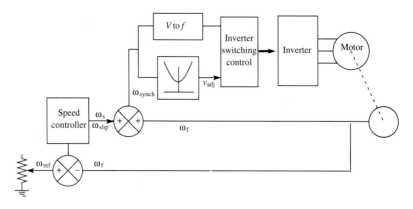

FIGURE 8.32
Speed control scheme using an inner slip control loop and an outer speed control loop.

a tachometer or other speed sensing device. The sensed speed is labled ω_f. The sensed speed is also fed to the summing module, which adds the slip reference ω_s to it and provides the voltage signal that determines the inverter frequency. The air-gap flux control is achieved by using the inverter frequency signal itself, through a function generator to adjust the inverter terminal voltage in the manner explained earlier. This scheme therefore conforms to the classic speed control pattern of motor control, which has an outer speed control loop and an inner torque control loop. The torque control loop here is actually the slip control loop—torque being practically proportional to slip in the control range involved. This scheme will provide regenerative braking also. For braking of the motor, the slip reference signal ω_s will be negative in sign because the speed reference will be less than the actual speed. The negative slip reference will cause the inverter frequency to be lower, causing the machine to function as a generator.

> **Illustrative example 8.11.** A four-pole Y-connected induction motor is fed from an inverter and is spinning at 1810 rev/min. The motor is regeneratively braked by lowering the inverter frequency to 58 Hz. The line voltage at this frequency is 208 V. The equivalent circuit parameters of the motor for one phase are as follows at $f = 58$ Hz; $R_1 = 0.3\,\Omega$, $X_1 = 1.2\,\Omega$, $X_m = 20\,\Omega$, $R_2' = 0.4\,\Omega$ and $X_2' = 1.5\,\Omega$. Determine the power fed back from the machine by braking.
>
> **Solution.** This example can be worked out in the same manner as Example 8.8, except that the slip is negative. The synchronous speed corresponding to 58 Hz is $(58/2)60 = 1740$ rev/min. The fractional slip is $(1740 - 1810)/1740 = -0.04 = -4\%$. The equivalent circuit for one phase for this value of the slip is sketched in Fig. 8.33.
>   By circuit reduction in steps, as shown in the figure, the equivalent impedance of the circuit becomes
>
> $$-6.814 + j5.904$$

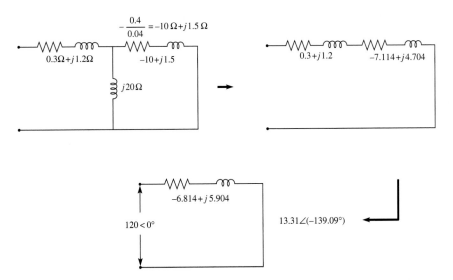

FIGURE 8.33
Circuit reduction in Example 8.11.

The corresponding phase current for the phase voltage of 120 V will be

$$I = \frac{120 + j0}{-6.814 + j5.904} = -10.059 - j8.716$$

$$= 13.31\angle(-139.09°)$$

The power per phase is

$$120 \times 13.31 \cos(-139.09) = -1207 \text{ W}$$

The total power of all three phases is $3 \times -1207 = -3621$ W, the negative sign indicating that the power is being fed back from the machine.

8.6.4 Induction Motor Speed Control Using Controlled Current Converters

When a current controlled inverter is used for the speed control of the induction motor, there is no possibility of an overload current, even if the motor is stalled owing to excessive load torque. This is because the current is automatically limited by the inverter. We have described three types of current controlled inverters:

1. the auto sequentially commutated current source inverter;
2. the hysteresis type current regulated inverter;
3. the pulse width modulated current regulated inverter.

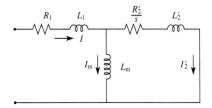

FIGURE 8.34

Equivalent circuit to determine total phase current for a given magnetizing current

In all of these types, the current and frequency of the inverter are independently adjustable. In the auto sequentially commutated current source inverter, the current adjustment is done on the front end rectifier, which is a closed-loop phase controlled thyristor bridge. The frequency adjustment is done by varying the clock frequency of the switching control circuit of the inverter section. In the other two types of inverter, the current amplitude is controlled by varying the amplitude of the reference waveform and the frequency is adjusted by variation of the frequency of the reference sine waveform.

When the induction motor is fed from one of these types of controlled current inverter, the speed is controlled by the frequency adjustment. But, as we have seen, there is need to maintain the air-gap flux at the correct level. The manner in which this may be done is explained below.

Referring to the equivalent circuit, the air-gap flux is determined by the magnetizing current I_m through the mutual inductance I_m (see Fig. 8.34). The stator phase current I divides into the two parallel branches in the inverse ratio of their impedances. On this basis, the expression for I_m will be

$$I_m = \left[\frac{R_2'/s + j\omega L_2'}{R_2'/s + j\omega(L_2' + L_m)}\right]I$$

Multiplication of both numerator and denominator by s gives

$$I_m = \left[\frac{R_2' + js\omega L_2'}{R_2' + js\omega(L_2' + L_m)}\right]I$$

$$= \left[\frac{R_2' + j\omega_2 L_2'}{R_2' + j\omega_2(L_2' + L_m)}\right]I \tag{8.25}$$

Equation 8.25 tells us that with known values of the motor parameters, for any value of the motor frequency, the required value of I can be calculated from the following relationship:

$$I = I_m \frac{R_2' + j\omega_2(L_2' + L_m)}{R_2' + j\omega_2 L_2'} \tag{8.26}$$

where I_m is the required value of the magnetizing current for the correct value of the airgap flux. Equation (8.26) tells us that the value of I is independent of

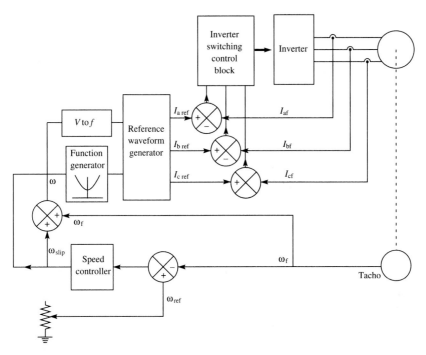

FIGURE 8.35
Closed-loop speed controller using a controlled current inverter.

the stator resistance and leakage reactance. For given values of the motor constants R_2', L_m and L_2', the factor on the right-hand side of (8.26) depends only on the rotor frequency, that is, the slip frequency ω_2. The value of I_m needed can be determined from tests under rated conditions on the motor, with sufficient accuracy. Therefore it is possible to determine the magnitude of the stator current for different values of the slip frequency by means of (8.26). A function generator that gives a signal for the required current for any value of the slip frequency can be assembled, and this can be used in the controller as shown in Fig. 8.35.

The scheme in Fig. 8.35 shows a current regulated inverter whose frequency and output current are independently adjustable by means of the frequency and amplitude of a reference waveform. Instead, we may also use an auto sequentially commutated current source inverter in which the frequency is adjusted by the clock frequency of the inverter switching controller and the current is independently adjusted by means of the current reference to the front end phase controlled rectifier. The speed control amplifier amplifies the speed error, which is the difference between the speed reference ω_{ref} and the speed feedback ω_f. The output of the speed error amplifier is the slip reference ω_{slip}. This is added to the speed ω_f to determine the synchronous speed and therefore the inverter frequency. The output of this summing module serves as

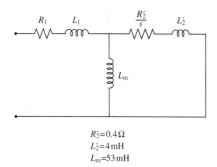

$R_2'=0.4\,\Omega$
$L_2'=4\,\text{mH}$
$L_m=53\,\text{mH}$

FIGURE 8.36
Motor phase equivalent circuit in Example 8.12.

the frequency control of the inverter. The output of the speed error ampliper, which is ω_2 of (8.26), is given to the function generator, which gives an output according to (8.26). This serves as the magnitude adjustment for the current. The switching control block compares the reference current with the actual sensed value of the current, and makes the actual current conform to the reference in each phase.

Illustrative example 8.12. The equivalent circuit data per phase of a three-phase induction motor are given in Fig. 8.36. The magnetizing current needed for the rated airgap flux is 5 A. The motor works from a current controlled inverter, and the closed-loop controller employed is the one shown in Fig. 8.35. The slip frequency command ω_2 corresponds to a frequency of 2 Hz. Determine the corresponding value of the stator current reference to be provided by the function generator for the inverter: (*a*) if this is for driving; (*b*) if it is for braking.

Solution. For driving, the slip is positive, and therefore the rotor frequency $\omega_2 = s\omega$ should be treated as +2 Hz. The corresponding angular frequency $\omega_2 = 2 \times \pi \times 2 = 12.566$ rad/s. Substituting this value of ω_2 and the other given data into (8.26), we get

$$I = I_m \frac{0.4 + j12.566 \times 57 \times 10^{-3}}{0.4 + j12.566 \times 4 \times 10^{-3}}$$

This gives $I = 10.175$ A.

For braking, the slip is negative. This means that the rotor frequency $\omega_2 = s\omega_1$ should be treated as negative. Reference to (8.26) shows that this does not change the magnitudes of the numerator or denominator in the complex impedances. Therefore the magnitude of the current reference will be the same as was calculated for the driving mode, namely 10.167 A. From this, we may also conclude that the function generator waveform in Fig. 8.35 has symmetry about the vertical axis.

We have now completed our treatment of drive systems using the cage-type induction motor. We shall now proceed to take up drive systems using the wound rotor induction motor.

8.7 ADJUSTABLE SPEED DRIVE SYSTEMS USING THE WOUND ROTOR INDUCTION MOTOR

The difference between the wound rotor induction motor and the squirrel cage induction motor is only in the rotor part of the machine. The wound rotor machine has windings on the rotor that are similar to the windings on the stator, although the effective number of turns per phase may not be the same. These rotor windings are typically connected in Y, and the terminals of the Y connection are brought to three slip rings. Access to these windings is possible through brushes that make contact with the slip rings. For this reason, this type of motor is also called a slip ring induction motor. The windings, the slip rings and the brush assembly make this type of motor mechanically less robust than the squirrel cage type of induction motor. The brushes and slip rings are subject to wear, and increase the maintenance requirements of the machine. But, because the rotor circuit is accessible for external connections, it becomes possible to utilize this feature to exercise control over the speed and torque of the motor. Traditionally, the slip rings have been used for inserting resistance in the rotor phases for the purpose of achieving a high torque at starting. During normal running, the rotor circuit is short-circuited. In some applications, such as in crane and hoist drives, where long periods of operation at low speeds are not needed, the resistances are introduced into the rotor circuit not only for starting but also for bringing down the speed. Such schemes involving the introduction of resistances into the rotor circuit are wasteful from the point of view of energy. With developments in power electronics, it is now possible to use static converters to interact with the rotor circuit for purposes of achieving speed control, electronically by static means, without the need for mechanical switching of resistances, or without the wastage of energy associated with resistance control. This section will be devoted to the description of such schemes.

8.7.1 Wound Rotor Induction Motor Basics

The basic theory of the induction motor that we developed in Section 8.6.1 is also applicable to the wound rotor machine. In fact, in Section 8.6.1 we replaced the actual squirrel cage rotor by an equivalent wound rotor that will give the same electrical performance when viewed from the stator side. In the case of the wound rotor machine, there is no need to make this assumption of an equivalent wound rotor, because this is the real situation. The relationships relevant for our present purpose can be obtained from the equivalent circuit

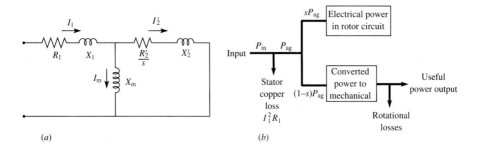

(a)   (b)

FIGURE 8.37
Equivalent circuit for one phase of the induction motor.

of the machine, which we arrived at in Section 8.6.1 and was shown by Fig. 8.22. This equivalent circuit is reproduced in Fig. 8.37(a) for ready reference.

Figure 8.37(b) shows a power flow diagram, conforming to the equivalent circuit (a). In (b), the power input into the stator is labeled P_{in}. The power lost in the stator circuit is the copper loss $I_1^2 R_1$. The rest of the power is called the "air-gap power" P_{ag}. This is the power crossing into the rotor side from the stator side. Since R_2'/s is the only element in the equivalent circuit that consumes power on the rotor side,

$$P_{ag} = \frac{I_2'^2 R_2'}{s}$$

The resistance R_2'/s is actually the sum of two parts the real resistance R_2' and the fictitious resistance $R_2'/s - R_2'$. Together, they make up the total rotor side resistance in the equivalent circuit:

$$\frac{R_2'}{s} = R_2' + \frac{R_2'(1-s)}{s}$$

The first term on the right-hand side represents the actual rotor resistance. The second term is the fictitious resistance, which represents the power converted into mechanical power.

The total power that crosses the air gap from the stator side divides into two parts: the electrical power and the mechanical power. We may divide the air-gap power as follows:

$$P_{ag} = \frac{I_2'^2 R_2'^2}{s} = I_2'^2 R_2' + \frac{I_2'^2 R_2'(1-s)}{s}$$

$$P_{ag} = sP_{ag} + (1-s)P_{ag}$$

(8.27)

The first term on the right-hand side gives the electrical power in the rotor while the second term gives the power converted to mechanical power at the rotor shaft. The power flow diagram is shown in Fig. 8.37(b).

The speed of the rotor is

$$\omega = \omega_{\text{synch}}(1 - s) \quad \text{rad/s}$$

Therefore the induced torque, which is the converted power in W, divided by the speed in rad/s, will be

$$T = \frac{(1 - s)P_{\text{ag}}}{(1 - s)\omega_{\text{synch}}} = \frac{P_{\text{ag}}}{\omega_{\text{synch}}} \tag{8.28}$$

Since the synchronous speed for a machine depends only on the inverter frequency, for any given operating frequency, (8.28) shows that the induced torque is constant for a constant air-gap power. Equation (8.28) may be written as

$$P_{\text{ag}} = T\omega_{\text{synch}} \tag{8.29}$$

Therefore, from (8.27), we get

$$\text{electrical rotor power} = sP_{\text{ag}}$$
$$s = \frac{\text{rotor electrical power}}{T\omega_{\text{synch}}} \tag{8.30}$$

Equation (8.30) shows that the slip is proportional to the rotor electrical power for any given operating torque at a given operating frequency. Therefore it is possible to operate the motor at a fixed frequency, such as the power system frequency of 50 or 60 Hz, and vary the slip, thereby varying the speed of the motor by adjustment of the electrical power in the rotor. This brings us to the two speed control schemes that we shall describe below. The first is simple, and involves the variation of the effective resistance in the rotor circuit. We can increase the rotor power dissipation by inserting additional resistance in the rotor circuit and thereby increase the slip and bring down the speed. We have already stated that this is a wasteful technique from the point of view of energy. But this is a simple method, and is sometimes used where continuous operation at low speeds is not needed. Before the advent of power electronic converters, for resistance control it was necessary to introduce a three-phase adjustable rheostat in series with the slip rings. In the first scheme described below, only a single resistor is used, and its effective value is adjusted by variation of the duty cycle of a static switch that is placed in parallel with it.

The poor efficiency performance of the resistance control technique may be explained as follows. For any value of the air-gap power, increasing the slip means a corresponding decrease of the mechanical power output. For example, at 50% slip, the converted power $(1 - s)P_{\text{ag}}$ is equal to the electrical rotor power sP_{ag}, and therefore the total power loss including stator copper loss and rotational losses is more than the converted power, so that the overall efficiency is less than 50%. At higher slips, the efficiency will be still less. Therefore dissipating more power in the rotor circuit is an inefficient way of controlling the speed. However, the first scheme described below in fact does this. It is described here because it gives a simple, though inefficient, way of controlling the speed electronically, in applications where continuous running at low speeds is not necessary.

8.7.2 Speed Control by Means of a Chopper Controlled Resistance in the Rotor Circuit

In this scheme, shown in Fig. 8.38, the diode bridge rectifies the induced rotor voltage. The DC power is dissipated in the resistor R. The static switch S is placed in parallel with the resistor. This is made to function as a chopper by operating it at a fixed frequency with adjustable duty cycle.

If the smoothing inductance L is assumed to be large, and if the switching frequency is high, the current may be assumed to be unchanged during a switching period of the chopper. The power in the resistor during the ON period of the switch is zero, and the power during the OFF period of the switch is $I_2'^2 R$. During the ON period,

$$\text{energy dissipated in } R = 0$$

During the OFF period,

$$\text{energy dissipated in } R = I_2'^2 R T_{\text{OFF}}$$

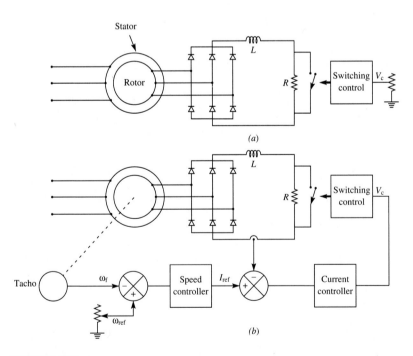

(a)

(b)

FIGURE 8.38
Speed control by static variation of effective rotor circuit resistance.

The average power dissipated in the resistor is

$$\frac{I_2'^2 R T_{OFF}}{T}$$

where T is the chopper period. Therefore the effective value of the resistance with the chopper in operation is

$$R_{\text{eff}} = \frac{T_{OFF}}{T} R = (1 - D)R \tag{8.31}$$

where D is the duty cycle of the chopper. This shows that the effective value of the external resistor can be varied from R to zero by varying the duty cycle of the chopper from 0 to unity. The total electrical power in the rotor includes the power dissipated in the resistance of the rotor windings and in the power semiconductor devices. In the scheme shown in Fig. 8.38(a), the speed adjustment is made manually by varying the control voltage V_c, which varies the duty cycle of the chopper. A closed-loop version of the same scheme is shown in Fig. 8.38(b). In this scheme, there is an outer speed control loop that compares a speed reference signal ω_{ref} against the speed feedback signal ω_f. The speed control amplifier amplifies the speed error and provides the reference labeled I_{ref} to an inner current control loop. The current controller amplifies the current error and provides the control voltage V_c, which adjusts the duty cycle D of the chopper to cancel the error. In this scheme, the stator is supplied from the fixed frequency power system bus.

8.7.3. Speed Variation by Controlled Slip Power Recovery

In the speed control scheme shown in Fig. 8.38, the slip frequency power from the rotor is converted into DC and dissipated in the chopper controlled resistance. The wastage of power in the resistor R makes the scheme very inefficient from the point of view of power utilization. The scheme can be made more efficient if, instead of dissipating the rectified power in the resistance R, this power is returned to the AC system bus by means of a phase controlled inverter. Such a scheme is shown in Fig. 8.39. In this arrangement, there are two converters in cascade, usually called the "subsynchronous cascade" because they provide speed control below the synchronous speed of the machine. The first converter is the diode rectifier bridge, which as in the previous method, rectifies the slip frequency voltage of the rotor. The output of this rectifier is connected in series with a smoothing inductor L to the DC terminals of the second converter, which is a phase controlled thyristor bridge, operating in the inversion mode. The AC side of this bridge is connected to the three-phase AC power system bus. The inverter serves to feed back power from the DC side to the AC power bus by phase controlled inversion. The DC bus voltage may be expected to be small, because the DC is obtained by rectification of the slip frequency voltage from the rotor. Therefore a large

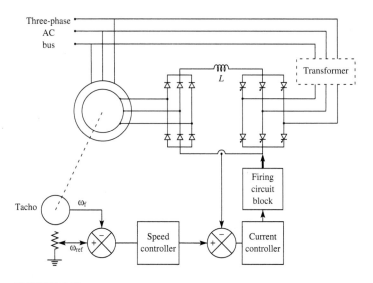

FIGURE 8.39
Speed control using the subsynchronous converter cascade.

angle of advance β will be needed for the inverter to feedback power. This large value of β will adversely affect the power factor. Therefore it will be advisable to use a transformer between the AC system bus and the inverter to match the voltages. This transformer is indicated by the box of broken lines in Fig. 8.39. The power fed back can be adjusted by adjustment of the firing angle of advance β of the inverter.

The scheme shown in Fig. 8.39 is a closed-loop version of the slip power recovery scheme described above. There is an outer speed control loop from which the speed controller provides the reference for an inner current control loop. The inner loop senses the DC current and automatically adjusts the firing angle of the inverter to maintain the DC current at the reference value.

8.8 ADJUSTABLE SPEED DRIVES USING THE SYNCHRONOUS MOTOR

8.8.1 Synchronous Motor Categories

The stator of the synchronous motor is similar to that of the induction motor. In fact, by simply changing the rotor, a three-phase induction motor can be changed into a three-phase synchronous machine. There are three categories of synchronous motors according to the type of rotor used. They are

1. the wound field synchronous motor;
2. the permanent magnet synchronous motor;
3. the synchronous reluctance motor;

The synchronous machine used as a generator is called an alternator. Three-phase alternators are universally used for the generation of electric power in generating stations around the world. Structurally, there is no difference between the synchronous motor and the synchronous generator. These two operating modes are interchangeable in a drive system when changing between driving and braking.

8.8.1.1 THE WOUND FIELD MOTOR. In this type of machine, the rotor has fixed magnetic poles, which are created by causing a DC current to flow in the "field windings" that surround the poles. The DC field current is also frequently called the "excitation" or the "exciting current." The excitation has to be supplied from a source of DC. For this, several arrangements are possible. The source of DC may be a separate DC generator, called the exciter, coupled to the same shaft as the main motor, and the DC may be fed to the rotor windings by means of brushes that make sliding contact with two slip rings mounted on the shaft of the machine, to which the field terminals are connected. Instead of an exciter machine, a rectifier may also be used in an alternative arrangement by means of which DC may be fed to the slip rings from the AC power bus. Brushless excitation systems are also in use. In a typical brushless excitation system, the exciter machine is an AC generator coupled to the shaft of the main machine. The AC is generated in the rotor windings of this machine. This AC is rectified by means of diodes, which are also mounted on the shaft, and the rectified DC is directly fed to the field windings of the main synchronous machine without the need for any brushes and slip rings. In all the arrangements that we have described, there will be means to adjust the field current independently, and thereby adjust the flux created by the fixed magnetic poles of the rotor. When an exciter machine is used, this can be done by adjusting the field current of the exciter. If a stationary rectifier is used, the adjustment can be done using phase control of the thyristor rectifier.

Cylindrical and salient pole types of rotors. Since the rotor windings carry only DC current, there is no necessity for a laminated rotor assembly. The rotor can be made of solid magnetic steel, in which slots are cut longitudinally to house the windings. But for these slots, the rotor may be cylindrical in shape, which will make the air gap uniform. With a cylindrical rotor, the reluctance of the magnetic circuit of a stator coil will not change as the rotor rotates. The alternative arrangement is the "salient pole" structure. In this, the rotor has projecting magnetic poles, alternately north and south. In a salient pole machine, as the rotor rotates, the reluctance of the magnetic circuit at a given location changes. The reluctance of the magnetic path through the center of the poles will be less because of the air gap. This is often called the direct axis of the rotor. The quadrature axis is along the direction through the middle of the inter polar zone. The quadrature axis reluctance will be high, because of the longer air gap along this axis.

When the armature carries three-phase currents, there will be a rotating m.m.f. created by them. The field poles will lock with the rotating magnetic field created by the armature currents, and the rotor will rotate synchronously with the armature rotating field. The flux resulting from the latter field will depend on the reluctance of the magnetic path. This in turn will depend on the angular orientation of the armature rotating m.m.f. with respect to the rotating salient poles. For analysis purposes, we may treat the rotating m.m.f. due to the armature currents as the resultant of two components—one aligned with the direct axis of the rotor and the other with the quadrature axis. In this way, each component of the armature rotating m.m.f. will be seeing constant but separate reluctances.

8.8.1.2 THE PERMANENT MAGNET ROTOR. Several types of high quality permanent magnetic materials have been developed in recent years, and practical permanent magnet machines in the power range of several tens of horsepower have become available. The earlier machines used alloys of nickel, cobalt and iron, which go by the name "alnico," of which several grades are available. Later, ceramic permanent magnets came to be developed. Ceramic magnets are manufactured by a sintering process, and consist of ferrite oxides of barium or strontium. These ferrite permanent magnet materials possess higher coercivity than alnico. The coercive force is a measure of the magnet's ability to withstand demagnetization due to demagnetizing fields that may occur in a machine. The most recent developments in permanent magnet materials are the rare earth magnets, and these represent a big step in permanent magnet technology. The most important materials are the samarium–cobalt magnets and the neodymium–iron–boron magnets. Samarium–cobalt magnets have coercive forces three to five times that of ceramic magnets, and are highly suited for electric motor applications. But their main disadvantage is their cost: at present, neodymium–iron–boron magnets have the highest coercive force available. This material may become attractive for motor applications in terms of cost.

Rotor structure. There are several geometrical ways in which the magnet may be mounted on the rotor. The two main classifications are

1. surface mounted magnets;
2. interior magnets.

High quality permanent magnet materials are not mechanically strong, and have to be suitably supported on the rotor. Strong epoxies or bolts may be used for the purpose. Shrink-fitted cylinders may be used in interior mounted magnet structures in high speed machines.

Permanent magnet synchronous machines are basically salient pole machines. But the saliency may not always be visible from the outside of the rotor. The direct axis and quadrature axis reluctances are different. The permanent magnet materials have low relative permeability. For this reason, the effective air gap of a permanent magnet motor is actually higher than the actual air gap. As a consequence, the effect of armature m.m.f. will be comparatively less in a permanent magnet machine. Also, depending on the magnetic structure of the rotor, the quadrature axis reluctance can be lower than the direct axis reluctance. This is in contrast to the conventional salient pole machine with wound field structure. Since the field flux is constant in a permanent magnet machine, for analysis purposes it may be treated as a wound field machine with a constant current source for the field excitation.

Damper windings. Synchronous motors often have an additional winding on the rotor, which is short-circuited. This is called the damper winding. The winding functions like the shorted rotor of an induction motor. The synchronous machine develops torque only at the synchronous speed, and for this reason it is not a self-starting motor when energized from a fixed frequency supply. The induction motor action of the damper windings helps to accelerate and come close to the synchronous speed, at which the stator rotating field is able to lock in with the rotor poles and enable the machine to run at synchronous speed. The damper winding also serves to damp out oscillations that are superimposed on the synchronous speed of the machine for any reason. At the synchronous speed, there is no relative velocity between the air-gap flux and the damper windings. But if, for any reason, a change occurs in the speed of the rotor, or the rotor tends to oscillate about its mean speed, then currents are induced in the damper windings, as in an induction motor. The resulting torque opposes the speed changes from synchronism, and in this way damps any oscillations that may occur.

8.8.1.3 THE SYNCHRONOUS RELUCTANCE MOTOR. The synchronous reluctance motor has a salient pole structure for its rotor. But there are no windings on the rotor. When the stator is energized by three-phase currents, a rotating field is created, in the same way as in the induction motor, which rotates at the synchronous speed. The physical effect on the rotor is such as to make it move in such a way as to minimize the reluctance seen by the field. The rotor moves in much the same way as a piece of iron will move when a magnet is moved in its proximity. The torque that causes the rotor to rotate is called the reluctance torque. We shall derive an expression for the reluctance torque when we analyze the salient pole synchronous machine subsequently.

8.8.2 Wound Field Synchronous Motor— Mathematical Relationships

The physical description of the working of the synchronous motor is as follows. When the stator is energized from a three-phase supply and carries three-phase currents, a rotating magnetic field is created in the air gap by the stator

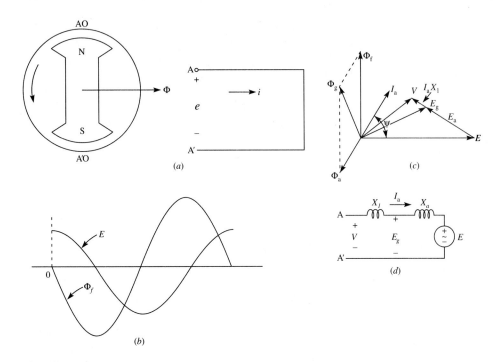

FIGURE 8.40
Synchnonous machine: assignment of voltage reference polarities, current and flux reference directions, phasor diagram and equivalent circuit.

currents. The stator, although stationary, as far as its magnetic field is concerned behaves like a rotating magnet with alternate north and south poles, rotating at the synchronous speed. The rotor has fixed magnetic poles created by the DC field current. The poles of the stator rotating field lock with the fixed magnetic poles on the rotor and cause the rotor to rotate together with the stator rotating flux. As the flux in the air gap rotates, an AC voltage is induced in the stator windings by the resultant air-gap flux. We shall focus our attention on one of the three phases of the machine. We may symbolically represent this phase circuit by the coil labeled A–A' in Fig. 8.40(a). In this figure, we show the rotor as a rotating magnet whose position at time $t = 0$ is at right-angles to the axis of the coil A–A', with the north pole pointing upwards as shown. The rotation is assumed to be counterclockwise, as indicated in the figure. We shall assume the rotor flux distribution to be sinusoidal. As the rotor rotates, a sinusoidal e.m.f. will be induced in the stator coil. At the instant shown in the figure, which is our reference time $t = 0$, the induced voltage will be seen to be maximum, and is about to start decreasing. By the application of the right-hand rule for the induced voltages, we can see that the polarity at $t = 0$ of the induced e.m.f. will be such as to make terminal A positive with respect to terminal A'. We shall clearly define the reference polarities for voltages, the reference direction for the current in the coil and the physical reference direction for positive flux through the coil so that there will be no ambiguity as regards the interpretation of the phasor diagram, the waveforms and the physical phenomena in the circuit itself. The induced voltage will be treated as *positive* when this voltage will make *terminal A positive with respect to terminal A'*. Current will be treated as *positive* when it is flowing *from A to A' inside the coil*. This is as marked by the arrow in Fig. 8.40. Flux linkage of the coil will be treated as *positive* when its direction is *opposite to the flux that will be created by a positive current through the coil*. These assigned reference directions and polarities are clearly indicated in Fig. 8.40. On the basis of the above *assignments of reference polarities and directions,* we can verify that at the instant $t = 0$, the flux linking the coil is zero and is starting to become negative. The induced e.m.f. will be at the maximum positive value, and will be starting to decrease. On this basis, we have sketched the waveforms of the rotor flux through the coil and the induced e.m.f. in Fig. 8.40(*b*). We notice that the e.m.f. will be lagging the flux by 90°. The phasor diagram is also sketched in Fig. 8.40(*c*) in conformity with this, showing E horizontally along the reference direction, and the rotor field flux labeled ϕ_f vertically at 90° ahead of E. The actual current in the coil will depend on the phase and magnitude of the external voltage connected to the terminals A–A' of the coil, and the impedance of the coil itself. We shall assume that these quantities are such that the current phasor is as shown in the phasor diagram. The power flowing into the coil per phase will be positive, being equal to

$$EI_a \cos \psi$$

where ψ is the phase angle between the armature phase current I_a and E. Here E is the phase voltage induced by the rotor flux ϕ_f alone. This is positive, and therefore consistent with the motoring operation, because, for motoring, power has to be pushed into the phase circuit, in contrast to the generating mode, where power will be coming out of the phase coil.

When the armature carries currents, which will be three-phase for a balanced three-phase machine, there will be a rotating m.m.f. due to the armature currents. This is called the armature reaction m.m.f., or simply armature reaction. The reference directions that we chose for the armature current and the flux are not consistent, because, according to our assignment of reference directions, a positive current will produce a flux in the negative direction through the coil. Therefore we have to draw the armature reaction flux phasor in opposition to the current phasor. This is done in the phasor diagram of Fig. 8.40(c). The resultant air-gap flux phasor will be the phasor sum of the field flux and the armature reaction flux. This resultant is labeled ϕ_g in the phasor diagram. The total induced voltage due to the resultant rotating field is the phasor sum of the induced voltage due to the armature reaction flux and the rotor flux. This is the air-gap voltage, and is labeled E_g in the phasor diagram. To obtain E_g, we have to add the e.m.f. due to armature reaction flux. This e.m.f. is lagging the armature reaction m.m.f. and the armature reaction flux by 90°, that is, leading the armature current phasor by 90°. The induced e.m.f. due to the armature reaction flux is labeled E_a in the phasor diagram. To get the terminal voltage of the phase, we again have to add the drop in the phase leakage reactance and the resistance to the air-gap voltage E_g. Usually the reactances are large compared with the resistances at normal speeds and frequencies, and we have therefore neglected the resistance in comparison with the reactance drops. We have added this phase leakage reactance drop I_aX_1 at right-angles to the current and ignored the resistance drop, to obtain the terminal voltage, which is labeled V in the phasor diagram.

We notice that the voltages due to armature reaction and the armature leakage inductance that have to be added to the rotor induced phase voltage E have the same phase, and both are proportional to the current. Therefore, for voltage calculations, we can include the effect of armature reaction by adding an extra reactance X_a to the armature leakage reactance, and the equivalent phase circuit will be as shown in Fig. 8.40(d). Since X_1 and X_a are in series, they can be added together and treated as one reactance

$$X_s = X_1 + X_a$$

This sum X_s is called the synchronous reactance, and is used as a single reactance for calculations of voltage regulation.

The equivalent circuit for calculation of the terminal voltage phasor is shown in Fig. 8.41(a) and the corresponding phasor diagram is shown in Fig. 8.41(b). In this figure, the terminal phase voltage V is drawn horizontally as

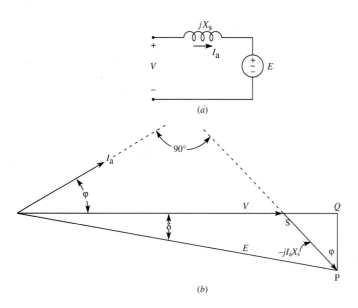

FIGURE 8.41
Phasor diagram of phase voltages.

the reference phasor, and the phase current is shown to be leading the terminal voltage by the angle φ. We may express E as

$$E = V - jI_aX_s$$

The phasor diagram in (b) reflects this equation.

Power flow considerations. The power consumed by the motor per phase will be given by:

$$P = VI_a \cos \varphi$$

where φ is the phase angle between the phase voltage and the phase current.

In the phasor diagram of Fig. 8.41(b), we have made a simple geometrical construction by drawing a vertical line PQ upwards from the tip of the E phasor to the V phasor. The tip of the V phasor is labeled S. Inspection shows that the angle SPQ is equal to the phase angle φ, which is the angle by which the phase current leads the terminal phase voltage phasor. Also, the height PQ will be given by

$$PQ = I_aX_s \cos \varphi$$

This gives

$$I_a \cos \varphi = \frac{PQ}{X_s}$$

The power per phase becomes equal to

$$\frac{V(PQ)}{X_s}$$

Therefore, with constant power operation at constant voltage, the height PQ will be constant. If the field current is varied under these conditions, the locus of variation of the point P, that is, the tip of the phasor E, will be a horizontal line. If, say, the field current is decreased, so that the rotor flux and therefore the induced voltage E due to the rotor flux become less, the tip of the E phasor, that is, the point P, moves to the left. This makes the power factor angle SPQ less. If the field current is progressively decreased, the power factor will become less and less leading, and will become unity when the angle becomes zero. Further decrease in the field will cause the angle to lag, and the power factor angle will become more and more in the lagging direction. Therefore, in a synchronous motor, the power factor is adjustable by variation of the field current. This is a useful property, because at unity power factor, the input current has the lowest value for a given power.

The phase angle between the terminal voltage V and the induced voltage E due to the rotor flux is labeled δ in the phasor diagram in Fig. 8.41. This is called the load angle. The power can be expressed in terms of the load angle in the following manner. The height PQ may be written as

$$PQ = I_aX_s \cos \varphi = E \sin \delta$$

From this,

$$I_a \cos \varphi = \frac{E \sin \delta}{X_s}$$

Therefore the power per phase is

$$VI_a \cos \varphi = \frac{VE}{X_s} \sin \delta$$

The total power due to all the three phases will be

$$P = \frac{3VE}{X_s} \sin \delta \tag{8.32}$$

Torque expression. The speed of the motor is the same as the synchronous speed, which, for a frequency f, will be $2\pi f$ electrical rad/s, and so

$$\omega = \frac{2\pi f}{p} \quad \text{mechanical rad/s}$$

where p is the number of pole pairs. Under our assumption of negligible power loss in the rotor resistance, the power converted to mechanical power will be the same as the input power. With this assumption, the induced torque (in N m) will be

$$T = \frac{p}{2\pi f} \frac{3VE}{X_s} \sin \delta \tag{8.33}$$

Thus we find that the torque is proportional to the sine of the load angle δ.

If the motor is spinning at no load, with zero load torque, neglecting friction and wind resistance, the load angle will be zero. This means that the induced voltage E due to the rotor flux will be in phase with the terminal voltage applied to the motor. If a load torque is now applied to the shaft, the machine will momentarily slow down, causing the induced voltage E to lag, resulting in a finite load angle and the creation of torque. The load angle will adjust to match the applied load torque, and the machine will continue to spin at the synchronous speed. The slowing down will be only momentary, to create the necessary load torque. If the load torque is steadily increased, the load angle will steadily increase further to balance the load torque. The machine will develop the maximum torque when the load angle δ is 90°. Any further increase in load torque will be beyond what the machine can balance, and the machine will go out of synchronism and stall. The maximum torque will be given by putting $\sin \delta = 1$ in (8.33):

$$T_{max} = \frac{p}{2\pi f} \frac{3VE}{X_s} \tag{8.34}$$

Similar statements can be made for the electric braking operation, during which the machine will function as a generator. Braking is initated by slowly reducing the AC frequency. If, as stated above, the machine is spinning under no load with zero load torque, the load angle δ will be zero. If the synchronous speed is now reduced by a small decrease in the AC frequency, this will cause the V phasor to slow down, causing the E phasor to advance, since it is already spinning at the previous synchronous speed. The load angle δ will become negative, causing a braking torque to be developed, which will slow down the rotor, so that it will spin at the new lower synchronous speed. There is also a maximum braking torque, which will correspond to $\delta = -90°$.

Power and torque considerations in salient pole machines. In a salient pole machine, the armature reaction flux is dependent on the angular orientation of the armature reaction m.m.f. with respect to the rotor. Both the armature reaction m.m.f. and the rotor are rotating at synchronous speed, and there is no relative motion between the two. In this case, for analysis purposes, we treat the armature current as the resultant of two components. One component, I_d, produces the m.m.f. along the direct axis of the rotor, while the other, I_q, produces the m.m.f. along the quadrature axis of the rotor. Figure 8.42 shows the phasor diagram of the terminal voltage V and the induced voltage E due to the rotor flux. The reluctance along the direct axis is less because of the smaller air gap. Therefore the armature reaction flux per ampere due to the direct axis component I_d will be relatively greater. We have to use a higher value of reactance to represent the armature reaction e.m.f. for the direct axis component. We shall label the direct axis synchronous reactance as X_d and the quadrature axis synchronous reactance as X_q. Using separate values of reactances, we have to modify the relationship between V and E as follows:

$$V = E + jI_d X_d + jI_q X_q$$

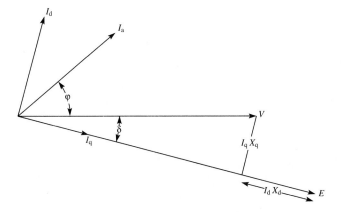

FIGURE 8.42
Phasor diagram for a salient pole motor.

The phasor diagram of Fig. 8.42 is drawn on this basis.

The load angle δ, defined as the phase angle between the terminal voltage and the induced voltage due to the rotor flux, is indicated in the phasor diagram. Referring to the latter,

$$V \sin \delta = I_q X_q \tag{8.35}$$

$$V \cos \delta = E - I_d X_d \tag{8.36}$$

Also,

$$I_a \cos \varphi = I_q \cos \delta + I_d \sin \delta$$

The power per phase is

$$V I_a \cos \varphi = V I_q \cos \delta + V I_d \sin \delta$$

We may replace I_d and I_q by the values obtained from (8.35) and (8.36), to give

$$
\begin{aligned}
V I_a \cos \varphi &= V \frac{V \sin \delta}{X_q} \cos \delta + \frac{V(E - V \cos \delta)}{X_d} \sin \delta \\
&= \frac{VE \sin \delta}{X_d} + \frac{V^2 \sin 2\delta}{2} \left(\frac{1}{X_q} - \frac{1}{X_d} \right) \\
&= \frac{VE \sin \delta}{X_d} + \frac{V^2 \sin 2\delta}{2} \frac{X_d - X_q}{X_d X_q}
\end{aligned}
\tag{8.37}
$$

The expression for the air-gap torque may be obtained by dividing the total power of all three phases by the synchronous speed:

$$T = \frac{3VE \sin \delta}{(\omega_{synch}/p) X_d} + \frac{3V^2 \sin 2\delta}{\omega_{synch}/p} \frac{X_d - X_q}{2 X_d X_q} \tag{8.38}$$

Here ω_{synch} is the synchronous speed in electrical rad/s and ω_{synch}/p is the speed in mechanical rad/s. The expression for torque in (8.38) has two terms. The second results from the fact that the direct and quadrature axis reluctances are different. This term is called the reluctance torque.

In a synchronous reluctance motor, there is no field winding on the rotor. Therefore the field-induced voltage $E = 0$, and the first term in (8.38) vanishes. The torque in this machine is the reluctance torque given by the second term:

$$T = \frac{3pV^2 \sin 2\delta}{\omega_{synch}} \frac{X_d - X_q}{2 X_d X_q}$$

where $\omega_{synch} = 2\pi f$.

8.8.3 Speed Control of Synchronous Motors Fed from Voltage Source Inverters—Open-Loop Control

Since a synchronous motor spins at the synchronous speed at all loads, an open-loop speed control scheme may suffice for many applications. Since the

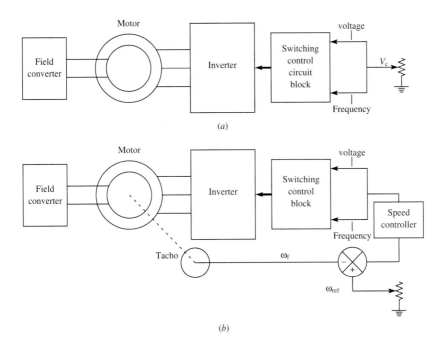

(a)

(b)

FIGURE 8.43
Speed control of a synchronous motor fed from a voltage source inverter.

number of pole pairs of a given machine is fixed, the synchronous speed for the machine will be determined by the AC frequency alone. The motor runs at the synchronous speed, and the speed does not change with load or the voltage. Therefore an open-loop speed controller might suffice for single or multiple motor drives where several motors have to spin at the same speed. Although the speed depends only on the frequency, the induced voltage is proportional to the frequency and speed, and it will be necessary to adjust the voltage also for satisfactory current and torque. A simple open-loop control scheme on these lines is shown in Fig. 8.43(a).

In the scheme shown, the motor is fed from a three-phase voltage source inverter. The switching control circuit of the inverter has independent adjustment for the frequency and voltage. The frequency adjustment is by means of a voltage-to-frequency converter module inside the switching control circuit block, which determines the clock frequency of the inverter switching circuit. The field of the motor may be supplied from another converter, and may be manually adjusted to give a power factor close to unity. If the motor is of the permanent magnet type, there will be no field circuit, and no adjustment of the field is possible. The speed adjustment is made by varying the control voltage V_c, which is adjusted manually. This varies the clock frequency of the inverter switching control circuit and therefore the synchronous speed. The same voltage V_c also varies the AC output voltage of the inverter. This may be by varying the DC input voltage of the inverter or by varying the modulation index, depending on the type of inverter.

Because the same control voltage V_c varies both the frequency and the voltage of the inverter, the voltage control is automatically implemented when the frequency is varied.

8.8.4 Closed-Loop Version of the Controller

A closed-loop version of the same scheme is indicated in Fig. 8.43(b). In this scheme, the control voltage V_c that adjusts the frequency and voltage of the inverter may be obtained from an error amplifier that amplifies the error between the actual speed and the reference speed. The commanded speed is set by means of the speed reference signal ω_{ref}, and the actual speed feedback signal ω_f is obtained from a tacho.

8.8.5 Automatic Closed-Loop Adjustment of Power Factor

The scheme described above does not enable automatic adjustment of the field current to give the optimum power factor. Automatic adjustment of the power factor can be incorporated by a scheme such as the one shown in Fig. 8.44. The primary means of adjustment of power factor is the variation of the field current. This is possible in a wound field machine. If the motor is operated at a power factor of unity, the current drawn by it will have the lowest magniutde for a given power input, and therefore the lowest internal copper losses. In the scheme shown in Fig. 8.44, the motor voltage and current are sensed and a functional circuit module computes the phase angle between the two and therefore the power factor. The computed value of the power factor is compared against the commanded value. The error is amplified by the error amplifier, and its output varies the field current of the motor so as to make the power factor conform to the commanded value.

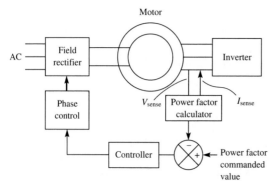

FIGURE 8.44

Automatic closed-loop adjustment of power factor.

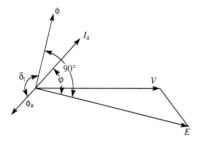

FIGURE 8.45
Phasor diagram for operation with current control.

8.8.6 Synchronous Motor Operation from a Current Controlled Inverter

Operation of a synchronous motor from a current controlled converter is simpler and more convenient. The torque becomes independent of the stator leakage inductance and resistance. The expression for the torque may be obtained by reference to the phasor diagram of Fig. 8.45. In Fig. 8.45, φ represents the phase angle between the induced e.m.f. E due to the rotor flux and the phase current I_a. The total converted power per phase is

$$P = EI_a \cos \varphi$$

Now E is proportional to the rotor flux ϕ and the speed, which is the synchronous speed ω. Therefore we may write the above expression as

$$P = K\omega\phi I_a \cos \varphi$$

where K is a constant of proportionality. The torque will be given by the power divided by the angular velocity. Thus

$$T = K\phi I_a \cos \varphi$$

The angle between the rotor flux and the armature reaction flux is called the torque angle, and is labeled δ_t in the phasor diagram. We can see by inspection that

$$\delta_t = \tfrac{1}{2}\pi + \varphi$$

Therefore the expression for the torque becomes

$$T = K\phi I_a \sin \delta_t \tag{8.40}$$

Equation (8.40) shows that the torque is independent of the speed, and is determined by the field flux, the armature current and the torque angle. The field flux is constant for a permanent magnet machine. It is adjustable by means of the field current in a wound rotor machine.

8.8.7 Self-Controlled Operation

In the synchronous motor, the rotation is synchronous with the AC frequency output of the inverter. The inverter frequency is determined by a clock oscillator in the switching control circuit block. It is possible to operate the switching control circuit in synchronism with timing signals received from a position sensor coupled to the shaft of the motor. In this way, the motor will always be in synchronism with the inverter, and there will be no question of the machine going out of synchronism. If the machine slows down owing to load torque, the inverter frquency also decreases correspondingly, because the frequency is determined by the shaft speed. Therefore the machine continues to be in synchronism with the inverter. Such an arrangement is analogous to the operation of a commutator and brush assembly in a DC motor. What the commutator and brush assembly does in a DC machine is to switch the current into each armature coil in such a way that the current direction in each coil is always such as to maintain the same direction of torque. When a conductor moves from the north pole zone to the south pole zone, the current direction in that conductor is automatically reversed, so that the induced torque does not change sign. The commutator and brushes keep doing this for every coil as the rotor rotates. The switching circuit of our inverter, when controlled by signals from the shaft position sensor, functions as an electronic commutator that switches the current in the proper direction in the stator phase as the rotor poles rotate and the relative orientation of the coil with respect to the poles changes. In a DC machine, there are several coils, and correspondingly a large number of commutator segments. In the three-phase machine, there are only the three phase circuits on the stator, and the role of our electronic commutator is limited to switching current in these three circuits only. But, basically, a synchronous motor operating in a self-controlled manner in this way is like a DC motor. It is synchronous only in name, but the speed control is similar to that of a DC motor. A permanent magnet synchronous motor, with self-controlled switching, using a shaft position sensor, is often called a brushless DC motor.

Figure 8.46 shows the arrangement for self-controlled switching of the inverter using shaft position signals. By means of the appropriate position signal, it is possible to switch current in a particular phase of the motor when

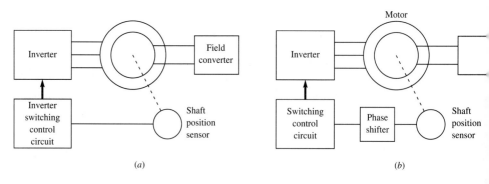

(a) *(b)*

FIGURE 8.46
Self-controlled synchronous motor drive with facility for torque angle adjustment.

this phase has the desired orientation with respect to a pole on the rotor. In this figure, the inverter is of the constant current type. Therefore, with this arrangement, the current phasor will give a certain orientation of the armature m.m.f. with respect to the rotor flux phasor. In other words, the torque angle will have a fixed value. It will be possible to adjust the torque angle by introducing a phase shifting circuit, which will phase-shift the shaft position signals before being fed to the inverter switching control circuit. This modification is shown in Fig. 8.46(b). With this, the torque angle becomes adjustable.

8.8.8 Self-Controlled Synchronous Motor Fed from a Load Commutated Inverter

The load commutated inverter was described in Section 8.5. Figure 8.47 shows a control scheme for a self-controlled synchronous motor drive using a load commutated inverter. In this, the DC link is provided by a thyristor bridge, which is phase-controlled in such a way as to give a controlled current operation. The inverter is also a thyristor bridge. The turn ON signals of the inverter thyristors are derived from the shaft position signals. Therefore each thyristor is turned ON at the preselected instant when the rotor has the required orientation with respect to the associated phase winding, to give motoring torque. Each phase current is turned ON in this manner as the rotor rotates. When an incoming thyristor is turned ON, the outgoing thyristor turns OFF by line commutation. The commutating line voltage is the induced voltage in the motor phase. For successful commutation, it is necessary to provide a minimum angle of advance $\beta_{\min}$ in a line commutated inverter. Because of

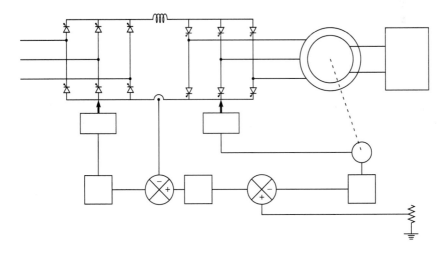

FIGURE 8.47
Self-controlled synchronous motor drive using a load commutated inverter.

this, the current in a phase has to be leading with respect to the phase voltage; that is to say, the motor must be operated at a leading power factor. For this, the field current should be kept sufficiently high to give leading power factor operation. There will be difficulty for commutation at starting and at low speeds, as explained earlier in Section 8.5, because of insufficient induced voltage for successful commutation. We have explained that this difficulty can be overcome by momentarily blocking the gate pulses to the rectifier bridge thyristors and thereby starving the DC link of current, to make the inverter thyristor go OFF.

The scheme is useful for high power applications, and has the advantage that no force commutation circuits are needed for the inverter thyristors

PROBLEMS

8.1 The DC link voltage of a three-phase voltage source inverter is provided by a single-phase diode bridge rectifier fed from a 120 V single-phase supply. The capacitor on the input side has a value of 2000 μF. The motor fed from the inverter is being regeneratively braked, causing a reverse DC current of 20 A. Determine

(a) the rate at which the DC link voltage will rise due to the braking;

(b) the value of the braking resistor needed on the DC side.

8.2 A Y-connected three-phase synchronous AC motor drive is supplied from a three-phase six-step inverter. The DC rails of the inverter are fed from a single-phase 240 V, 60 Hz supply through a diode bridge rectifier. Determine

(a) the r.m.s. phase voltage of the motor;

(b) the r.m.s. fundamental sinusoidal component of the motor phase voltage;

(c) the r.m.s. values of the two lowest order harmonics in the motor phase voltage.

8.3 A three-phase sinusoidally pulse width modulated inverter is supplied from a DC link voltage of 250 V. The load is Y-connected. Determine

(a) the r.m.s. total voltage of one phase;

(b) the fundamental frequency component of the phase voltage of the load;

(c) the total harmonic content in the phase voltage and the total harmonic distortion factor.

Assume that the modulation index is 100% and that the pulse repetition frequency is arbitrarily large.

8.4. The number of pulses per half-cycle of a sinusoidally pulse width modulated voltage waveform is 9. Determine the r.m.s. value of the SPWM waveform for 100% modulation index. Also determine, by Fourier analysis, the fundamental frequency component of the voltage (r.m.s. value) and the total harmonic distortion factor.

8.5. Repeat Problem 8.4 for a pulse number of 5 per half-cycle for

(a) 100% modulation index;

(b) 70% modulation index.

8.6. The DC input current to a three-phase auto sequentially commutated current source inverter is 50 A. The load on the AC side of the inverter consists of three identical impedances, which are Y-connected, each phase consisting of a resistance of 1 Ω in series with an inductance of 2 mH. The AC output frequency of the inverter is 30 Hz. Determine the r.m.s. value of the fundamental frequency component of the load phase voltage.

8.7. The three-phase load on the output side of a three-phase auto sequentially commutated current source inverter is Y-connected. The DC current input to the inverter is 20 A. For a load phase determine
(*a*) the r.m.s. total phase current;
(*b*) the r.m.s. fundamental sinusoidal component of the phase current;
(*c*) the total harmonic distortion factor.

8.8. A three-phase auto sequentially commutated current source inverter feeds a Δ-connected load. The DC current input to the inverter is 20 A. Determine
(*a*) the r.m.s. total phase current;
(*b*) the r.m.s. fundamental component of the phase current;
(*c*) the r.m.s. harmonic current and the total harmonic distortion factor.

8.9. The three-phase load on an auto sequentially commutated current source inverter consists of three identical impedances that are Δ-connected. Each phase impedance consists of a resistance of 3 Ω in series with an inductance of 20 mH. The AC frequency of the inverter is 40 Hz. Determine the fundamental frequency component of the load phase voltage. The DC current input to the inverter is 5 A.

8.10. A three-phase load consists of three identical impedances, each consisting of a resistance of 2 Ω in series with an inductance of 10 mH. These are Y-connected and fed from a three-phase current source inverter of the auto sequentially commutated type. The DC current input to the inverter is 10 A and the AC frequency of the inverter is 40 Hz. Determine the r.m.s. values of
(*a*) the fundamental frequency components of the phase current and the phase voltage;
(*b*) the frequency of the two lowest order harmonics in the phase current, and the r.m.s. value of each;
(*c*) the r.m.s. values of the two lowest order harmonics in the phase voltage.

8.11. A current controlled inverter employs the PWM type of current control. The triangular carrier used has a frequency of 5 kHz. The peak-to-peak amplitude of the triangular carrier waveform is 10 V. The inverter frequency is 20 Hz. Determine
(*a*) the number of positive and negative voltage pulses at a load terminal in one inverter AC period;
(*b*) the maxiimum width of a pulse and the value of the control voltage at which this occurs;
(*c*) the widths of the positive and negative pulses at the load terminal for a control voltage of +4 V.

8.12. A three-phase four-pole Δ-connected induction motor is working from a 208 V, 60 Hz supply. It is driving a load spinning at 1728 rev/min and drawing a line current of 15 A. The input power factor is 0.8. The resistance of a stator phase winding is 0.5 Ω. The rotational losses, inclusive of core losses, are 400 W total. Determine the torque available for driving the load.

8.13. A three-phase Y-connected induction motor has the following equivalent circuit parameters at a frequency of 50 Hz: $R_1 = 0.9\,\Omega$, $X_1 = 1.2\,\Omega$, $R_2' = 0.6\,\Omega$,

$X_2' = 0.8\,\Omega$ and $X_m = 20\,\Omega$. The motor is being regeneratively braked, and at a certain point during the braking operation the inverter frequency is 50 Hz and the slip is -3%. Determine the power being fed back by the motor. The voltage of the DC rails of the inverter is 250 V. Determine, ignoring losses in the inverter, the reversed DC current. What will be the proper value of the braking resistor needed at this instant? The inverter is of the SPWM type with an arbitrarily large carrier frequency.

8.14. A six-pole induction motor driving a battery powered vehicle is spinning at 840 rev/min. It is to be regeneratively braked with slip of 4%. Determine the frequency of the inverter needed at

(a) the commencement of braking;

(b) when the speed has dropped to 600 rev/min.

8.15. The equivalent circuit parameters of a squirrel cage induction motor are $R_1 = 0.8\,\Omega$, $L_1 = 4\,\text{mH}$, $R_2' = 0.6\,\Omega$, $L_2' = 3\,\text{mH}$ and $L_m = 66\,\text{mH}$. The magnetizing phase current I_m at the rated conditions of the motor is 8 A. The motor is being supplied from a current regulated inverter, and the slip frequency command is 2 Hz. What should be the stator current for maintaining the proper value of the magnetizing current if the motor is driving a load?

8.16. Rework Problem 8.15 to determine the inverter output current magnitude if the motor is being regeneratively braked with a negative slip of the same magnitude.

8.17. The controlled current inverter shown in Fig. P8.17 uses the hysteresis type of controller. In an experiment, the reference current given to the inverter is fixed at 20 A, and the hysteresis limits are set at 19 A and 21 A. The hysteresis controller turns OFF the static switches S_1 and S_4 and turns ON S_2 and S_3 at the instant when the current rises to the upper hysteresis limit, and reverses the above operations when the current drops to the lower hysteresis limit. If the inverter is allowed to repetitively operate with these reference input and hysteresis limits, determine the ON time and the OFF time of each switch, and the repetitive switching frequency.

8.18. A phase controlled cycloconverter consists of two three-phase thyristor bridges in antiparallel fed from a 400 volt 60 Hz three-phase supply. The firing instants of the thyristors are obtained on the basis of the cosine wave intersection scheme, in which a cosine timing waveform of amplitude 10 V is employed for each thyristor.

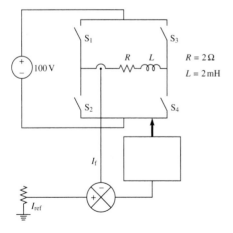

FIGURE P8.17

Intersection of this timing wave with a control voltage V_c determines the firing instants. The control voltage used may be expressed as

$$V_c = 8 \sin (2\pi \times 15t)$$

Determine the frequency and the peak value of the AC voltage output of the cycloconverter.

8.19. In a speed control scheme that uses the chopper controlled resistor in the rotor circuit of a wound rotor induction motor, a resistor of 6 Ω is used, in parallel with a chopper operated at a high switching frequency. Determine the chopper duty cycle to give the following values of the effective resistances:
(a) 4.0 Ω;
(b) 2.0 Ω;
(c) 1.5 Ω.

8.20. The speed of a wound rotor induction motor is controlled using the sub-synchronous cascade of converters in the rotor circuit. The motor is a four-pole machine working from a 60 Hz AC bus. For a given load torque, the air-gap power is 7500 W. The total electric power going into the rotor is adjusted to 2500 W by adjusting the power fed back through the phase controlled inverter of the converter cascade. Determine the speed of the motor in rev/min.

8.21. A four pole three-phase Y-connected synchronous motor is supplied from an inverter at a line voltage of 208 V and a frequency of 40 Hz. The induced phase voltage due to the rotor flux is 150 V. The synchronous rectance per phase is 1.2 Ω and the resistance is negligible. Determine the torque developed at a load angle of 20°. Determine the maximum torque that the motor can develop before it stalls.

CHAPTER
9

VECTOR CONTROL OF AC MOTOR DRIVES

9.1 INTRODUCTION

The steady state accuracy that can be achieved using inverter-fed AC motors is as good as what is possible using the separately excited DC motor. However, the "dynamic performance" attainable in AC motor drives is generally not as good. Dynamic performance is the measure of the fastness with which the motor can respond to changes in the commanded speed or torque. Therefore, in spite of the development of highly reliable inverter-fed AC motor drives, the DC motor continued to maintain its superiority for drives that demanded high dynamic performance, such as in reversible sheet rolling mills in the metallurgical industry and many machine tool drives. Recent years have seen the evolution of a new control strategy for AC motors, called "vector control," which has made a fundamental change in this picture of AC motor drives in regard to dynamic performance. Vector control makes it possible to control an AC motor in a manner similar to the control of a separately excited DC motor, and achieve the same quality of dynamic performance. Vector control recognizes the fact that the inferior dynamic performance of AC motor drives is not because of a basic limitation of the AC motor itself, but because of the manner in which power is fed to the motor and the way this is controlled. For this reason, vector control is the most significant development in the area of adjustable speed electric motor drives in recent years.

The high quality of dynamic performance of the separately excited DC motor is a consequence of the fact that its armature circuit and the field circuit are magnetically decoupled. In a DC motor, the m.m.f. produced by the field current and the m.m.f. produced by the armature current are spatially in quadrature. Therefore there is no magnetic coupling between the field circuit and the armature circuit. Because of the repetitive switching action of the commutator on the rotor coils as the rotor rotates, this decoupling continues to exist irrespective of the angular position or speed of the rotor. This makes it possible to effect fast current changes in the armature circuit, without being hampered in this by the large inductance of the field circuit. Since the armature current can change rapidly, the machine can develop torque and accelerate or decelerate very quickly when speed changes are called for, and attain the demanded speed in the fastest manner possible. As in the DC motors, in AC motors also, the torque production is the result of the interaction of a current and a flux. But in the AC induction motor, in which the power is fed on the stator side only, the current responsible for the torque and the current responsible for producing the flux are not easily separable. The underlying principle of vector control is to separate out the component of the motor current responsible for producing the torque and the component responsible for producing the flux in such a way that they are magnetically decoupled, and then control each independently, in the same way as is done in a separately excited DC motor.

Traditionally, we use the equivalent circuit of a motor to explain its performance and to derive quantitative relationships. Generally, equivalent circuits are based on steady state operating conditions. The primary objective of vector control is to achieve good performance when speed and torque conditions change. Therefore simple equivalent circuits of AC machines are not very convenient to explain the technique of vector control. To do this in a direct manner, the concept of "space vectors" is most convenient, and we shall use that approach. In this chapter, we shall first introduce the concept of space vectors, and then make use of this concept to explain the techniques of vector control in three-phase induction and synchronous motors. The implementation of vector control is comparatively easier in synchronous motors, because the field circuit that is on the rotor is fed separately and is independently adjustable. The orientation of the field flux with respect to the rotor is fixed and does not change with rotation. In the cage type induction motor, which is a singly fed machine fed from the stator side only, it is more difficult to separate the torque component and the field component of the stator input current. Also, the rotor carries AC currents and the rotor field is not stationary with respect to the rotor itself. This makes the implementation of vector control less simple than in the synchronous motor. But the squirrel cage induction motor is the more widely preferred machine for drive applications, because it is cheaper, mechanically sturdier and has relatively very little maintenance requirements. Therefore our primary interest is in the vector control of induction motors, and, unless otherwise stated, we shall have the induction motor in focus, during the major part of our treatment in this chapter.

9.2 SPACE VECTORS

The magnetic circuit of a three-phase induction motor consists of a stator, a rotor and an air gap of uniform thickness between the stator and the rotor. The stator winding consists of coils located in slots on the stator. The span or pitch of a coil will depend on the number of poles for which the machine is wound. The angular zone of one pole is 180° electrical. Therefore, in a two-pole winding, when we traverse the air gap once circumferentially, we will have traversed 360° electrical, which therefore will be the same as the actual number of mechanical degrees that we have traversed. Therefore the actual angles will be the same in both electrical and mechanical measures in a two-pole machine. If, however, the machine is wound for four poles, the angular zone of one pole will be only 90° mechanical, whereas by our definition of electrical angle, it will be 180° electrical. Therefore, in a four-pole machine, one mechanical degree is equivalent to two electrical degrees. In general, we can state that in a machine wound for p pole pairs, one mechanical degree will be the same as p electrical degrees. To make our description easier to visualize, in this chapter, we shall keep in focus a two-pole machine, in which the electrical and mechanical angles are the same. This does not affect the generality of our treatment, because, when dealing with a machine with more than one pole pair, we can always bring in the integral ratio between the electrical and mechanical angles. Figure 9.1 shows the stator, air gap and rotor of the machine. Also shown on the stator is a coil whose sides are labeled A and A'. This is a full-pitch coil, by which we mean that the angular span of the coil extends over one complete zone of a pole, or 180° electrical. Let us assume that the coil has n turns and that it carries a current i A. This coil will therefore create a m.m.f. of ni ampere-turns. The magnetic circuit around which this m.m.f. will be acting consists of the stator, the rotor and two air gaps. This is indicated by the large circular paths drawn around A and A' in Fig. 9.1(b). The stator and rotor are made of laminated magnetic steel. To make our treatment simple, we shall assume that the magnetic steel is perfect, which means that the permeability is infinite, and therefore the reluctance is zero. Therefore the entire m.m.f. appears across the two air gaps in the magnetic path. Let g m be the radial length of the air gap. Therefore the magnetic field intensity across each air gap will be given by

$$H = \frac{ni}{2g} \text{ ampere-turns}$$

and the resulting flux density B will be given by

$$B = \mu_0 H = 4\pi \times 10^{-7} \times \frac{ni}{2g} \text{ T} \tag{9.1}$$

where $\mu_0 = 4\pi \times 10^{-7}$ H/m is the permeability of free space.

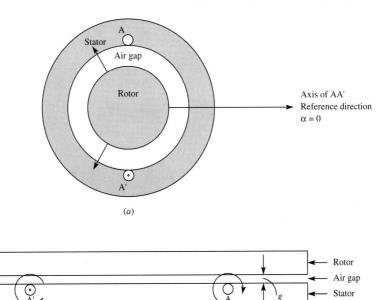

(a)

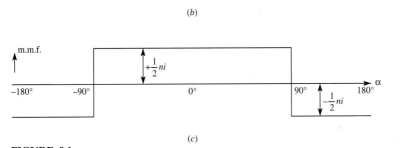

(b)

(c)

FIGURE 9.1
AC machine with uniform air gap.

Figure 9.1(*b*) is a "developed" or a "cut and opened out" diagram of the magnetic circuit, where the circumferential air gap is shown horizontally. The stator is shown below and the rotor above. The m.m.f., and therefore the flux density distribution around the air gap, is plotted in Fig. 9.1(*c*). In this, we have treated a radially inward flux density (from the stator towards the rotor) as positive. We notice from Fig. 9.1(*c*) that the spatial distributions of the m.m.f. and therefore the flux density B are rectangular. If we take the reference direction for measuring angles as the axis of the coil, towards the right, as indicated in (*a*), the flux density B has a constant positive value given by (9.1) from $\alpha = -90°$ to $\alpha = +90°$, and the negative of this value in the interval $\alpha = +90°$ to $\alpha = +270°$ (or $-90°$).

In induction and synchronous motors, what we want is not a rectangular flux density distribution as shown in Fig. 9.1. The ideal distribution is a sinusoidal spatial distribution. The flux density should actually peak along the

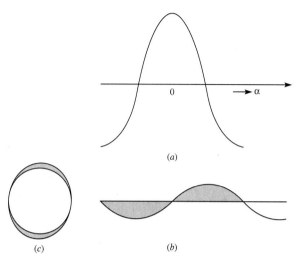

FIGURE 9.2
Ideal distributions of flux density
and turns density.

axis of the winding, and then decrease as a cosine function of the angle measured from the axis.

Figure 9.2(a) shows the ideal distribution of flux density around the air gap. For a meaningful definition of the "space vector," it is important that the spatial flux density distribution follow the cosine law, as sketched in Fig. 9.2 We can actually approach this ideal flux density distribution, to some extent, by having a distributed winding with several coils distributed around the periphery. To get a smooth spatial distribution of B, it is necessary to have a continuous distribution of the coil winding, which is impossible because of the need to house the turns in discrete slots. All the same, to emphasize the importance of a sinusoidal spatial distribution for the definition of space vectors, we shall investigate the type of distribution of the turns of the winding that can result in a spatially sinusoidal m.m.f. distribution. For this purpose, we shall define a turn density function D as

$$D = \frac{dN}{d\alpha}$$

where dN is the number of conductors in an angular interval $d\alpha$. We shall arbitrarily try the following function for D:

$$D = K \sin \alpha$$

Since the total number of turns, N, is equal to the number of conductors in the zone from $\alpha = 0$ to $\alpha = \pi$, it will be given by

$$N = \int_0^\pi K \sin \alpha \, d\alpha = 2K$$

This gives $K = \frac{1}{2}N$, and the expression for the theoretical distribution becomes

$$D = \tfrac{1}{2}N \sin \alpha \qquad (9.2)$$

With a conductor density distribution given by (9.2), if the winding carries a current of I A, the m.m.f. distribution will be found to conform to the ideal spatial cosine wave. This can be seen from the following analysis. Referring to Fig. 9.2, the m.m.f. acting along the reference zero direction will be the ampere-turns of all the coils that have a side in the zone from $\alpha = 0$ to $\alpha = \pi$:

$$\int_0^\pi \tfrac{1}{2}NI \sin \alpha \, d\alpha = NI \quad \text{ampere-turns}$$

The m.m.f. will be a maximum along $\alpha = 0$. At any other location, making an angle θ with the reference axis, the m.m.f. will be that due to all the coils that have a side in the zone from $\alpha = \theta$ to $\alpha = \pi - \theta$:

$$\int_\theta^{\pi - \theta} \tfrac{1}{2}NI \sin \alpha \, d\alpha = NI \cos \theta \qquad (9.3)$$

This is the ideal cosine distribution that we need.

Therefore, if we can have a winding that is continuously distributed according to (9.2), it will be possible to achieve the ideal cosine distribution of the m.m.f. and the flux density. The continuous conductor density of $D = \frac{1}{2}N \sin \alpha$ is shown symbolically in Fig. 9.2(b) horizontally, and depicted circumferentially in Fig. 9.2(c). This distribution of conductor density is effectively a current density distribution, which may be expressed as J A/rad:

$$J = \tfrac{1}{2}NI \sin \alpha$$

The current density distribution and the m.m.f. distribution can both be seen as sinusoidal with a spatial phase displacement of 90°. The magnitude of the m.m.f. distribution is proportional to the integral of the current density distribution. In a real machine, it is impossible to achieve the ideal sinusoidal m.m.f. distribution, because the windings are housed in slots, which are displaced by discrete angles. The best that can be expected is to achieve a stepped m.m.f. waveform, which approximates to a sine wave. Such a practical stepped waveform will have a fundamental sinusoidal Fourier component and several spatial harmonics. Each harmonic component gives an additional multiple-pole structure. Therefore, our practical machine, with its coils housed in discrete slots, has a superposition of the fundamental pole pair and several integral multiple-pole pairs. In our present analysis, we are looking only at the fundamental pole pair. This is really the practical implication of our assumption that the machine is ideally wound with the ideal turn density distribution given by (9.2). Our analysis can, in principle, be refined for a practical machine by extending it to the individual harmonic pole pair structures and superposing the effects. We shall, however, confine our attention solely to the fundamental spatial flux density distribution here.

9.2.1 The Stator Current Space Vector

Let us assume that a DC current I_{as} A is flowing in our winding. This results in a spatially sinusoidal distribution of the flux density B, which has its peak magnitude along the axis of the coil, which we are considering as our reference direction $\alpha = 0$. We associate this directional property of I_{as} by representing this current by a "current space vector" $\mathbf{I_{as}}$. The direction of the space vector $\mathbf{I_{as}}$ will be along the axis of the winding, which in our example is the direction $\alpha = 0$. The magnitude of the space vector will be I_{as}. Its contribution will be equal to $I_{as} \cos \alpha$ in a radial direction making an angle α with respect to the direction of the space vector. In this respect, the space vector $\mathbf{I_{as}}$ is similar to a physical vector such as a velocity or a force. The component of such a physical vector along a direction that makes an angle α with it is also given by multiplying it by $\cos \alpha$. This makes it possible to use vectorial addition of two or more space vectors to get their resultant, in exactly the same manner as we add two or more physical vectors to find their resultant by the parallelogram law of addition.

In a three-phase induction or synchronous motor, there are three windings on the stator—one for each phase. The windings are identical. The only difference is that the spatial orientations of the windings have a mutual displacement of 120° between them. This is shown symbolically in Fig. 9.3. In this figure, the three windings are labeled A_s, B_s and C_s, where the letters A, B and C denote the three phases and the subscript s denotes the fact that these windings are on the stator. Because of the orientations of the axes of these windings, the current space vector for a phase current I_a in phase A will be in the direction of the axis of the A-phase winding, that is, $\alpha = 0$, since we are

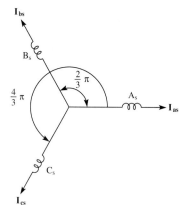

FIGURE 9.3
The orientations of the three stator phase windings.

taking this as our reference direction for measuring angles. The current space vector for a current I_b in phase B will be in the direction of the axis of the B-phase winding, that is, $\alpha = 120°$ ($\frac{2}{3}\pi$ in radian measure), and the current space vector for a current I_c in the C-phase winding will be in the direction of the axis of the C-phase winding, that is, $\alpha = 240°$ or $-120°$ ($\frac{4}{3}\pi$ or $-\frac{2}{3}\pi$ in radian measure).

If each one of the three phase windings carry DC currents I_a, I_b, and I_c respectively, the current space vectors will be vectors in the directions indicated. The resultant stator current space vector can be obtained by vector addition of the three individual current phase vectors. This is illustrated by the following example.

Illustrative example 9.1. Take the axis of the A-phase winding as the reference axis.

(a) Draw the current space vector for the individual phase currents.

(b) Obtain the resultant stator current space vector by vector addition for the following values of the phase currents:

(i) $I_a = 1.0$ A, $I_b = -0.5$ A and $I_c = -0.5$ A;

(ii) $I_a = 0.866$ A, $I_b = 0$ A and $I_c = -0.866$ A;

(iii) $I_a = 0.5$ A, $I_b = 0.5$ A and $i_c = -1.0$.

Solution. For each of the three sets of values of the phase currents, the individual current space vectors are first drawn and indicated by appropriate labels on the left-hand side in Fig. 9.4. Each current space vector is in the direction of the axis of the corresponding phase winding. However, whenever the magnitude given has a negative sign, the direction of that space vector is shown reversed.

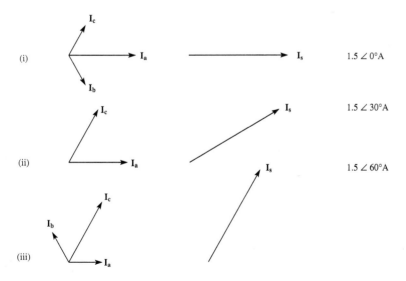

FIGURE 9.4
Current space vectors in Example 9.1.

For each set, the resultant current space vector, obtained by vectorial addition of the individual phase current space vectors, is drawn on the right-hand side, and labeled $\mathbf{I}_s$, the stator current space vector.

In Example 9.1 we did not indicate any time variation of the phase currents. Whether or not the individual currents vary with respect to time, the values are at a given instant of time. We actually chose the values in this example to correspond to instantaneous current values in a balanced three-phase system of currents, according to the following equations:

$$i_a = 1.0 \cos \omega t$$

$$i_b = 1.0 \cos \left(\omega t - \tfrac{2}{3}\pi \right)$$

$$i_c = 1.0 \cos \left(\omega t - \tfrac{4}{3}\pi \right)$$

The three sets of values in Example 9.1 correspond to $\omega t = 0$, $\omega t = \tfrac{1}{6}\pi$ and $\omega t = \tfrac{1}{3}\pi$. The resultant stator current space vector has the same magnitude of 1.5 A for each of the three given sets of values of the currents. The differences are only in the directions. As we change ωt in steps of 30°, the direction of the resultant stator current space vector also changed in steps of 30°. This will be found to be in conformity with what we shall derive later regarding the behavior of the stator current space vector of a three-phase machine excited by a balanced system of three-phase currents.

In solving Example 9.1, we graphically drew each individual phase current space vector and found the resultant by means of vector addition using the parallelogram law of addition of vectors. There is an alternative way of adding space vectors, which is often more convenient. This second method is to resolve each individual vector in terms of its components along two mutually perpendicular axes and then add the components along each axis. This procedure will also give the components of the resultant space vector along these directions, which we can again combine to get the resultant. This method has direct relevance to the practical implementation of vector control, and therefore we shall discuss it further. We shall choose the reference direction along the stator a-phase winding A_s, as one of the two directions and call this the "direct axis." We shall choose the second direction at an angle $\tfrac{1}{2}\pi$ with respect to the direct axis and call this the "quadrature axis." Every space vector may be expressed as a complex number, with the real part as its component along the direct axis and the imaginary part as its component along the quadrature axis. We shall state, for the sake of definiteness, that the angles are measured in the counterclockwise direction as viewed from a preselected observer side of the motor shaft. We shall label the winding that makes an angle of 120° with respect to the direct axis as the b-phase winding and its axis as the B_s axis. Therefore the C_s axis will be in the direction $\tfrac{4}{3}\pi$ (or $-\tfrac{2}{3}\pi$) with respect to the A_s axis.

To illustrate, let us find the resultant stator current space vector when the phase currents are I_{as}, I_{bs} and I_{cs} respectively. The expression for each phase current space vector in complex form will be

$$\mathbf{I_{as}} = I_{as} + j0 = I_{as}e^{j0} \tag{9.4}$$

$$\mathbf{I_{bs}} = I_{bs}(\cos \tfrac{2}{3}\pi + j \sin \tfrac{2}{3}\pi)$$
$$= I_{bs}(-\tfrac{1}{2} + j\tfrac{1}{2}\sqrt{3}) = I_{bs}e^{j(2\pi/3)} \tag{9.5}$$

$$\mathbf{I_{cs}} = I_{cs}(\cos \tfrac{4}{3}\pi + j \sin \tfrac{4}{3}\pi)$$
$$= I_{cs}(-\tfrac{1}{2} - j\tfrac{1}{2}\sqrt{3}) = I_{cs}e^{j(4\pi/3)} \tag{9.6}$$

The resultant stator current space vector will be given by

$$\mathbf{I_s} = I_{as} + I_{bs}e^{j(2\pi/3)} + I_{cs}e^{j(4\pi/3)} \tag{9.7}$$

9.2.2 The Time Variation of Current Space Vectors

In our three-phase machine, each phase current space vector has a fixed direction in space, along the axis of the respective winding. The magnitude of each phase current space vector is the magnitude of the current, which may or may not vary with time. We have not so far brought in time variation. The space vector concept represents a spatially sinusoidal variation, and should not be confused with sinusoidal variation with respect to time. In AC circuit theory the "phasor" concept is used to indicate sinusoidal variation of a current with respect to time. The term "space phasor" has been suggested, and sometimes used, instead of the term "space vector" to indicate the sinusoidal variation with respect to spatial angle. But we have chosen to use the term space vector, because it is the one more widely used in literature, and also because here we are concerned with the directional aspect in space, and a vector is in fact the name associated with a physical quantity that has a spatial direction. In circuit theory, when dealing with sinusoidal alternating quantities, it is usual to express time differences in angular measure, taking one time period of the AC as 2π radians. Therefore, when expressing angles, it is important to be very clear as to whether the angle concerned is a spatial angle, or a time phase difference expressed as an angle.

If we now assume the current in each phase winding to be time-varying, the magnitude of each individual phase current space vector will also be varying with respect to time, although the direction for each phase current space vector will have a fixed orientation, which is along the axis of the winding. But when we do the vector addition to find the resultant of the three current space vectors, the instantaneous resultant may be time-varying, both in magnitude and spatial direction. To indicate a variation with respect to time explicitly, we may use the following symbolism:

$$\mathbf{i_s}(t) = i_s(t)e^{j\alpha(t)}$$

where $\mathbf{i}_s(t)$ is the current space vector whose magnitude at the instant t is $i_s(t)$ and whose direction is $\alpha(t)$.

9.2.3 Resultant Current Space Vector when the Individual Phase Currents are Sinusoidally Time-Varying and Constitute a Balanced Three-Phase System

We shall determine the resultant when the individual phase currents constitute a balanced system of three-phase currents and are expressed as time functions in the following manner:

$$i_{as} = \sqrt{2}\, I_s \cos{(\omega t + \epsilon)} \tag{9.8}$$

$$i_{bs} = \sqrt{2}\, I_s \cos{(\omega t + \epsilon - \tfrac{2}{3}\pi)} \tag{9.9}$$

$$i_{cs} = \sqrt{2}\, I_s \cos{(\omega t + \epsilon - \tfrac{4}{3}\pi)} \tag{9.10}$$

By assigning the proper spatial directions, each of the above currents may be expressed as a space vector as follows:

$$\mathbf{I}_{as} = \sqrt{2}\, I_s \cos{(\omega t + \epsilon)}e^{j0} \tag{9.11}$$

$$\mathbf{I}_{bs} = \sqrt{2}\, I_s \cos{(\omega t + \epsilon - \tfrac{2}{3}\pi)}e^{j(2\pi/3)} \tag{9.12}$$

$$\mathbf{I}_{cs} = \sqrt{2}\, I_s \cos{(\omega t + \epsilon - \tfrac{4}{3}\pi)}e^{j(4\pi/3)} \tag{9.13}$$

Addition of the above three gives the resultant current space vector, which we shall call the "stator current space vector" and label as $\mathbf{I}_s(t)$:

$$\mathbf{I}_s(t) = \mathbf{I}_{as}(t) + \mathbf{I}_{bs}(t) + \mathbf{I}_{cs}(t) \tag{9.14}$$

Addition and simplification using usual trigonometric relationships will give

$$\mathbf{I}_s(t) = \tfrac{3}{2}\sqrt{2}\, I_s e^{j(\omega t + \epsilon)} \tag{9.15}$$

The physical interpretation of (9.15) is as follows.

1. The stator current space vector has a constant magnitude equal to $I_s = \tfrac{3}{2}\sqrt{2}\, I_s$.
2. The stator current space vector rotates with a constant angular velocity equal to ω rad/s. Therefore the air-gap flux density distribution B will have a spatially sinusoidal distribution at any given instant of time t. This distribution will be rotating at a constant angular velocity of ω rad/s. This is the familiar phenomenon of the rotating magnetic field, described in basic induction motor theory. In vector control, under transient conditions, the three-phase currents may not be balanced. Therefore the magnitude and/or

the angular velocity of the stator current space vector may not be constant under such conditions.

9.2.4 Transformation from the Actual Three-Phase Machine to an Equivalent Two-Phase Machine and Back

In our three-phase machine, at any instant of time, the component of the resultant stator current space vector along the "direct axis" (which we specified earlier as the A_s axis), is equal to the sum of the components of the three individual phase current space vectors along the same axis. Similarly, the component along the quadrature axis is the sum of the individual components along this axis. From this, we can conceive an equivalent two-phase machine that will have exactly the same stator current space vector. This equivalent two-phase machine will have one stator phase winding D_s centered along the direct axis and another identical stator phase winding Q_s centered along the quadrature axis. The required direct axis winding current $i_{s(ds)}$ and the quadrature axis winding current $i_{s(qs)}$ are easily determined by resolving the individual current space vectors along the two directions. These will therefore be given by

$$I_{s(ds)} = I_{as} + I_{bs} \cos \tfrac{2}{3}\pi + I_{cs} \cos \tfrac{4}{3}\pi$$

$$= I_{as} - \tfrac{1}{2}I_{bs} - \tfrac{1}{2}I_{cs} \tag{9.16}$$

$$I_{s(qs)} = I_{bs} \cos (\tfrac{2}{3}\pi - \tfrac{1}{2}\pi) + I_{cs} \cos (\tfrac{4}{3}\pi - \tfrac{1}{2}\pi)$$

$$= \tfrac{1}{2}\sqrt{3}I_{bs} - \tfrac{1}{2}\sqrt{3}I_{cs} \tag{9.17}$$

Equations (9.16) and (9.17) give us a means for transforming the three-phase currents into equivalent two-phase currents.

The reverse transformation also becomes unique if we make the following assumption, which is generally true for a three-phase system of currents. This assumption is that, the sum of the phase currents in the three-phase motor is zero. This will be always true if the windings are Y-connected with isolated neutral. Therefore, there are only two independently variable currents in the three-phase motor, because

$$I_{as} + I_{bs} + I_{cs} = 0 \tag{9.18}$$

Now, when we want to make the reverse transformation from the equivalent two-phase machine to the actual three-phase machine, we can treat (9.16)–(9.18) as three simultaneous equations in I_{as}, I_{bs} and I_{cs}, and solve for the latter quantities. This procedure gives

$$I_{as} = \tfrac{2}{3}I_{s(ds)} \tag{9.19}$$

$$I_{bs} = -\tfrac{1}{3}I_{s(ds)} + \sqrt{\tfrac{1}{3}} I_{s(qs)} \tag{9.20}$$

$$I_{cs} = -\tfrac{1}{3}I_{s(ds)} - \sqrt{\tfrac{1}{3}} I_{s(qs)} \tag{9.21}$$

The transformation of three-phase currents into equivalent two-phase currents according to (9.16) and (9.17) is very convenient for analysis. This is because there is magnetic decoupling between the currents in a two-phase machine, since the two-phase windings are oriented in quadrature spatially. The direct axis current does not contribute to a flux component linking with the quadrature axis, and vice versa. The reverse transformation from the equivalent two-phase to three-phase currents according to (9.19)–(9.21) are also necessary, because the real machine is a three-phase machine with three-phase currents. In the control strategy to be described later, the actual three-phase currents are first transformed into the equivalent two-phase values, to simplify the computational steps. The required currents for the equivalent two-phase machine are computed by the controller on the basis of the vector control strategy. These are then transformed back to the three-phase equivalents. The controller commands the inverter to implement these three-phase currents. Therefore one transformation from three- to two-phase and a second transformation from two- to three-phase are generally required to be implemented by the control circuit. The controller has to incorporate functional circuit blocks that will perform these and the other computations necessary to implement the vector control strategy. Figure 9.5(a) is a representation of a computational circuit block, which is designed to transform input signal voltages proportional to the three-phase currents into output voltage signals proportional to the equivalent two-phase currents. The internal circuit of this block is designed to multiply each signal by the appropriate numerical coefficients and do the arithmetic according to (9.16) and (9.17), and thereby provide the transformed output signals. Figure 9.5(b) is a representation of a similar computational circuit block that does the reverse transformation on input signal voltages proportional to two-phase currents, performing the arithmetic according to (9.19)–(9.21), and provides output signal voltages proportional to the three-phase currents. We shall later use these computational blocks to describe the implementation of vector control of an AC motor. We do not propose to go into the internal details of these and the other

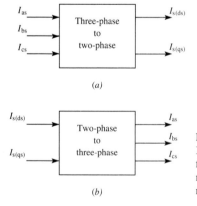

(a)

(b)

FIGURE 9.5
Representation of computational circuit blocks for transformation (a) from three-phase to equivalent two-phase machine; (b) from two-phase to equivalent three-phase machine.

computational circuit blocks that we shall need later for implementing vector control. The need to perform a significant amount of on-line computations was initially looked upon as the disadvantage of vector control. This difficulty has disappeared in recent years, with the availability of powerful microprocessor chips that can perform all the required computations with adequate speed.

9.2.5 Stator Voltage Space Vector and Stator Flux Linkage Space Vector

We defined each stator phase current space vector by making its magnitude equal to the phase current magnitude and the direction along the axis of the phase winding. We can define a "stator phase voltage space vector" in exactly the same way by assigning its direction along the axis of the phase winding, the magnitude being equal to the same phase voltage magnitude. We can combine the three individual stator phase voltage space vectors and obtain the resultant by vector addition, and call this the "stator voltage space vector." A stator "flux-linkage space vector" can also be defined following the same procedure for flux linkages of the phase windings. The following equations define the stator voltage space vector and the stator flux-linkage space vector in terms of the individual stator voltage and flux linkage space vectors for each phase:

$$\mathbf{V}_s(t) = V_{as}(t) + V_{bs}(t)e^{j2\pi/3} + V_{cs}(t)e^{j4\pi/3} \tag{9.22}$$

$$\mathbf{\Psi}_s(t) = \Psi_{as}(t) + \Psi_{bs}(t)e^{j2\pi/3} + \Psi_{cs}(t)e^{j4\pi/3} \tag{9.23}$$

Notice that (9.22) and (9.23) are on the same lines as (9.7) for the resultant stator current space vector in terms of the individual phase current space vectors. In the above equations, each term on the right-hand side is the space vector corresponding to one phase. This is expressed in terms of the magnitude and the angle, that is, in polar form. We shall use bold letters as shown on the left-hand side in the above equations to represent space vectors as such, implying both magnitude and directional angle. A space vector can be expressed either in polar form in terms of the magnitude and angle (direction) or in rectangular form as a complex number in terms of its components along two mutually perpendicular reference axes. The two components in the rectangular complex number expression of the *resultant* space vector are the components in the equivalent two-phase machine—the real part corresponding to the direct axis and the imaginary part to the quadrature axis. The expressions for the stator voltage space vector and the stator flux linkage space vector in rectangular form are

$$\mathbf{V}_s(t) = V_{s(ds)}(t) + jV_{s(qs)}(t) \tag{9.24}$$

$$\mathbf{\Psi}_s(t) = \Psi_{s(ds)}(t) + j\Psi_{s(qs)}(t) \tag{9.25}$$

9.2.6 Self and Mutual Inductances of the Stator Windings

Equation (9.25) gives the flux linkage in weber-turns at an instant t for each phase of the equivalent two-phase machine. The first term on the right-hand side gives the flux linkage of the direct axis winding while the second term gives the flux linkage of the quadrature axis winding. Each stator winding of the equivalent two-phase machine will have a flux linkage when the winding carries a current. In the two-phase machine, there will be no mutual inductance between the two phase windings because the windings are oriented at 90° relative to each other. There will be no flux linkage in the quadrature axis winding due to current in the direct axis winding, and vice versa. Each phase winding of the two-phase machine has a self-inductance, which will be the same for both the windings, because they can be seen to be identical from considerations of symmetry. We shall represent this self-inductance of each winding by L_s. The self-inductance of a winding is the flux linkage per ampere resulting from the current in the same winding. We shall assume that the magnetic circuit is linear without any saturation. This means that L_s can be taken as constant. Therefore we can write (9.25) in terms of the currents and the self-inductance as follows:

$$\Psi_s(t) = L_s I_{s(ds)}(t) + j L_s I_{s(qs)}(t)$$

or as

$$\Psi_s(t) = L_s \mathbf{I}_s(t) \tag{9.26}$$

Equation (9.26) gives the stator flux linkage contribution due to the stator currents only. When currents flow in the rotor circuits, there will be additional contributions to the stator flux linkage space vector, because of the mutual flux created by the rotor circuit currents. To take this into account, we shall now turn our attention to the rotor.

9.2.7 Rotor Windings

Rotors of induction motors may be either of cage type or wound type. In the cage type, the rotor electric circuit consists of bars of copper or aluminum located in the rotor slots. These bars are short-circuited by "end rings" on either end. The wound rotor type has three phase windings, similar to the three phase windings on the stator. In normal operation, these windings are short-circuited. Both the cage-type rotor and the wound rotor function in a similar manner, and therefore, for analysis purposes, it is a common practice to replace the cage rotor by an equvalent wound rotor. We shall follow the same practice whether or not the rotor is cage type. We shall picture our rotor as having three phase distributed windings, which are Y-connected with isolated neutral, similar to the stator. Further, to avoid having to bring the turns ratio into our equations, we shall assume the same number of turns for the stator and rotor phases. These assumptions do not create any practical difficulties, because all electrical measurements and inputs are on the stator side. Therefore the actual values of quantities such as voltages, currents and impedances, which are of interest to us on the rotor, are the values "referred to the stator side" in the same way as we use values "referred" to one of the sides, say the primary side, in the analysis of a transformer. In our case, the primary side is the stator. The rotor is the secondary side.

We shall use subscript "r" to identify rotor windings and other rotor quantities. To express the rotor circuit space vectors, we shall use a coordinate reference frame considered as fixed to the rotor, with its reference zero direction along the axis of the rotor A-phase winding A_r. In this rotor reference frame, the rotor currents, voltages and flux linkages will be treated as space vectors in exactly the same way as we previously described for stator currents, voltages and flux linkages, in the stationary reference frame fixed to the stator. Further, to simplify our analysis, we shall replace the three-phase rotor windings by the equivalent two-phase windings, with the direct axis windings aligned along the axis of the A-phase rotor winding A_r. These steps are exactly what we followed for the transformation of the stator three-phase winding to the equivalent two-phase winding in the stator (stationary) reference frame.

Also, on similar lines to (9.26) for the stator, we shall define L_r, the self-inductance of the rotor, as the ratio

$$L_r = \frac{\text{rotor flux linkage space vector due to rotor currents only}}{\text{rotor current space vector}}$$

9.2.8 Transformation of Space Vector Coordinates from One Reference Frame to Another

To study the effects of rotor based space vectors such as currents and flux linkages on the stator side, we have to look at them from the reference frame fixed to the stator. For example, the same rotor current space vector will have a different set of coordinates in the stator reference frame from those in the rotor reference frame if the rotor is turned through an angle θ_r. We shall now proceed to find the equations that relate the coordinates of the same space vector in two different spatial reference frames between which there is a spatial angular displacement. Let us, for example, consider the resultant rotor current space vector, which may be expressed in polar and rectangular forms respectively by the following two equations:

$$\mathbf{I_r} = I_r e^{j\alpha} \tag{9.27}$$

$$= I_{r(dr)} + j I_{r(qr)} \tag{9.28}$$

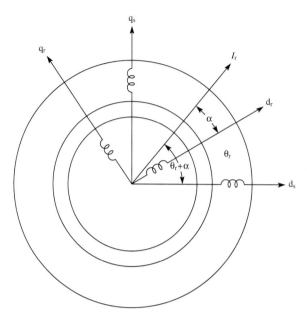

FIGURE 9.6
The change in angular coordinates of a space vector when changing from the reference frame fixed to the rotor to the reference frame fixed to the stator.

If the rotor reference frame and the stator reference frame are in alignment, the spatial coordinates of $\mathbf{I}_r$ or any space vector will be the same in the stator reference frame also. If now we rotate the rotor, and therefore the rotor reference frame, by an angle θ_r, the space vector will have rotated through the same angle θ_r when viewed from the stator reference frame. This is shown in Fig. 9.6. Therefore the new expression for $\mathbf{I}_r$ in the stator reference frame will be

$$\mathbf{I}_r = I_r e^{j\alpha} e^{j\theta_r} \tag{9.29}$$

We multiplied the original expression by $e^{j\theta_r}$ to effect the spatial rotation through θ_r rad. For this reason, the factor $e^{j\theta_r}$ is called a vector rotator. To obtain the rectangular coordinates in the stator reference frame, we shall rewrite the terms of each of the factors on the right-hand side in (9.29) in rectangular form, and complete the multiplication. This will give, in the stator frame,

$$\mathbf{I}_{r(s)} = (I_{r(dr)} + jI_{r(qr)})(\cos \theta_r + j \sin \theta_r)$$

$$= I_{r(dr)} \cos \theta_r - I_{r(qr)} \sin \theta_r + j(I_{r(dr)} \sin \theta_r + I_{r(qr)} \cos \theta_r) \tag{9.30}$$

In writing the above expressions, we have continued to follow the policy of indicating the reference axis by means of the letters within brackets in the subscripts. For example (dr) in the subscript means that the quantity concerned is along the direct axis of the rotor reference frame, (qr) means the quadrature axis of the rotor reference frame, etc. Using a similar symbolism for the rotor current vector in the stator reference frame, the real and imaginary terms on the right-hand side of (9.30), give us separately the following expressions for the rotor current space vector in the stator reference frame:

$$I_{r(ds)} = I_{r(dr)} \cos \theta_r - I_{r(qr)} \sin \theta_r \tag{9.31}$$

$$I_{r(qs)} = I_{r(dr)} \sin \theta_r + I_{r(qr)} \cos \theta_r \tag{9.32}$$

In each of (9.31) and (9.32), the term on the left-hand side is the rectangular coordinate of the rotor current space vector as seen from the stator reference frame. On the right-hand side are the rectangular coordinates of the same space vector in the reference frame fixed to the rotor. Therefore these two equations are the transformation equations for transforming the rectangular coordinates of a space vector in the rotor reference frame into their rectangular coordinates in the stator reference frame when the rotor, and therefore the reference frame fixed to the rotor, is turned through an angle θ_r with respect to the stator frame. These equations basically tell us how the rectangular coordinates change when a space vector is operated on by the vector rotator, as in (9.29).

Equations (9.31) and (9.32), as also (9.29), are very general, and can be used for transforming any space vector from one reference frame to another when there is an angular displacement between the two reference frames. By prefixing a negative sign to the angle, and using the appropriate symbolism, we can use the same equations for writing down the equations for a reverse transformation from the stator frame to the rotor frame. It should be noted that when we change the reference frame from which we view the space vector in this manner, the modulus (magnitude) of the space vector does not change, but only the angle. The angle will be a function of time if one frame is rotating with respect to the other. If one reference frame is rotating with respect to the other, the coefficients on the right-hand sides of (9.31) and (9.32) will not be constant, but will be functions of time.

This type of transformation between reference frames is useful in both the analysis and the practical implementation of vector control. Figure 9.7 is a symbolic representation that we shall use to represent a practical computational circuit block for implementing the transformation from, say, the rotor to

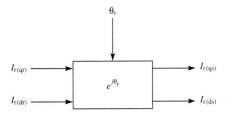

FIGURE 9.7
Symbolic representation of vector rotator for transformation of coordinates from the rotor to the stator reference frame.

the stator reference frame. The inputs to this circuit block, which performs the vector rotator operation, are the voltage signals proportional to the rotor coordinates and the angular displacement θ_r of the rotor reference frame with respect to the stator reference frame. The internal circuit of the vector rotator is designed to perform the multiplications and additions according to (9.31) and (9.32), and provide the output signal voltages proportional to the transformed coordinates.

When transformations are made between two reference frames, one of which is rotating with respect to the other, the angle input θ_r will be a function of time. Therefore there could be differences in the nature of the input and output signal voltages of the vector rotator circuit block of Fig. 9.7 as regards their variation with respect to time. This can be illustrated by a striking example. Consider for instance the case of a three-phase motor when the stator windings carry balanced three-phase AC currents. We have shown earlier that, under this condition, the stator current space vector has a constant amplitude and is rotating at the constant synchronous speed. The signal representing this space vector in a synchronously rotating reference frame will therefore have a constant DC value, because this space vector will be stationary and will have a fixed magnitude and direction in such a reference frame. When transformed into a stationary reference frame, this signal will become sinusoidal AC.

9.2.9   Mutual Inductance between Stator and Rotor Circuits

The effect of the rotor on the electrical circuit of the stator is felt through the mutual magnetic flux linking the stator and the rotor windings. This mutual magnetic flux linkage is the medium through which the input power into the stator is transferred to the rotor. It is therefore fundamental to the functioning of the induction motor. For a quantitative mathematical treatment of this mutual flux linkage using the concept of space vectors, we shall define a constant mutual inductance parameter M, whose significance can be explained as follows.

Let us first assume that the rotor is stationary with zero angular displacement between the stator and the rotor reference axes. Figure 9.8(a)

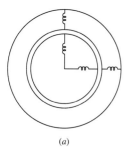

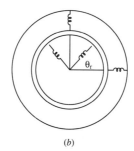

(a)   (b)

FIGURE 9.8
Orientations for mutually induced flux linkage calculation due to a current space vector in a different reference frame.

shows the orientations of the direct and quadrature axis windings on both the stator and the rotor in this position, in which the direct axis winding of the rotor will be aligned with the direct axis winding of the stator. The quadrature axes also will be similarly aligned. To avoid any possible ambiguity about the algebraic sign of the mutual inductance, we shall choose the positive reference direction of the rotor winding current as the one that will create flux in the same direction as a positive current in the stator winding. In this position, the mutual inductance between the stator direct axis winding and the rotor direct axis winding will have its maximum value, whose magnitude we shall denote by M. The mutual inductance between the quadrature windings will also have the maximum value M in this position, because the direct and quadrature windings are identical on the stator and also on the rotor. Under our assumed conditions of ideal sinusoidal flux density distribution and continuous distribution of turns as stated earlier in (9.2), it can be shown that the mutual inductance varies as a cosine function of the relative orientation of the windings. This means that the mutually induced flux linkage will vary as the cosine of the angle θ_r if the rotor is turned through this angle.

Let us now send a current $i_{r(dr)}$ through the rotor direct winding only, with the stator and the rotor in alignment. The rotor current space vector will be oriented along the direct axis. This will cause a flux linkage $L_r i_{r(dr)}$ for the rotor and a flux linkage $M i_{r(dr)}$ for the stator, where L_r is the self-inductance of the rotor winding. Treated as space vectors, these two flux linkage space vectors will be in the direction of the direct axis, which we have aligned to be the same for both the stator and the rotor. Because of the presence of the air gap between the stator and the rotor, there will be some leakage flux, and therefore the flux linking with the stator will be less than the flux linking with the rotor. Since we have assumed the same number of turns for the stator and the rotor windings, M will be less than L_r. We shall define a "rotor leakage factor" σ_r based on the following relationship:

$$L_r = (1 + \sigma_r)M \qquad (9.33)$$

We shall now send a current $i_{r(qr)}$ through the rotor quadrature axis also, still keeping the rotor stationary and in alignment with the stator. This will cause a self-induced flux linkage equal to $L_r i_{r(qr)}$ in the rotor quadrature winding, and a mutually induced flux linkage equal to $M i_{r(qr)}$ in the stator quadrature winding, in exactly the same manner as we described for the direct axis windings.

We shall now turn the rotor through an angle θ_r, keeping the currents unchanged. The orientations of the windings will now be as shown in Fig. 9.8(b). The mutually induced flux linkage of the direct axis winding of the stator will now be due to two currents: the direct axis current $I_{r(dr)}$ of the rotor and the quadrature axis rotor current $I_{r(qr)}$. These contribute respectively

$$I_{r(dr)}M \cos \theta_r \quad \text{and} \quad I_{r(qr)}M \cos (\theta_r + \tfrac{1}{2}\pi)$$

Therefore the total direct axis component of the mutually induced flux linkage of the stator, due to both components of the rotor current is the sum of these contributions:

$$\Psi_{s(ds)} = M(I_{r(dr)} \cos \theta_r - I_{r(qr)} \sin \theta_r) \tag{9.34}$$

An identical procedure to that described above will give the total flux linkage of the quadrature axis stator coil due to both the direct axis and quadrature axis rotor currents as

$$\Psi_{s(qs)} = M(I_{r(dr)} \sin \theta_r + I_{r(qr)} \cos \theta_r) \tag{9.35}$$

The expressions in parentheses on the right-hand sides of (9.34) and (9.35) are the same as those on the right-hand sides of (9.31) and (9.32). Therefore, by substitution from those equations, we may write

$$\Psi_{s(ds)} = MI_{r(ds)}$$

$$\Psi_{s(qs)} = MI_{r(qs)}$$

Therefore the total flux linkage space vector mutually induced by the rotor current may be written using space vectors in the stator reference frame as

$$\Psi_{s(s)} = M(I_{r(ds)} + jI_{r(qs)}) = M\mathbf{I}_{r(s)} = M\mathbf{I}_{r(r)}e^{j\theta_r} \tag{9.36}$$

Equation (9.36) tells that the mutually induced flux linkage space vector of the stator due to the rotor current can be obtained by transforming the rotor current into stator coordinates and multiplying by M. Equation (9.36) gives only the mutually induced stator flux linkage due to rotor currents. If the stator also is carrying current, the total flux linkage space vector of the stator will be the sum of the flux linkage due to self-inductance by this current plus the mutually induced flux linkage due to the rotor current. Therefore we can write the expression for the total stator flux linkage space vector in the stator reference frame as follows:

$$\Psi_{s(s)} = L_s\mathbf{I}_{s(s)} + M\mathbf{I}_{r(s)}$$

$$= L_s\mathbf{I}_{s(s)} + M\mathbf{I}_{r(r)}e^{j\theta_r} \tag{9.37}$$

Equation (9.37) shows how we can write the expression for the total flux linkage space vector in one reference frame when the total flux linkage is due to self-inductance by the current space vector in the same natural reference frame plus the flux linkage due to mutual induction by a current space vector in a different reference frame that has an angular displacement. The same expression is valid if the second reference frame is rotating with respect to the first, in which case the angle will be time-varying.

We can use this approach and also write the total flux linkage space vector of the rotor when both stator and rotor are carrying currents, as follows:

$$\mathbf{\Psi}_{r(r)} = L_r\mathbf{I}_{r(r)} + M\mathbf{I}_{s(s)}e^{-j\theta_r} \tag{9.38}$$

9.3 VOLTAGE EQUATIONS FOR AN INDUCTION MOTOR USING SPACE VECTORS

Using space vectors, we can write a single equation each for the stator and rotor circuits instead of different equations for the individual phases. The stator voltage equation in terms of space vectors will be as follows, in the reference frame fixed to the stator:

$$\mathbf{U}_{s(s)} = R_s\mathbf{I}_{s(s)} + \frac{d\mathbf{\Psi}_{s(s)}}{dt} \tag{9.39}$$

In this equation, R_s is the resistance per phase of each of the stator direct and quadrature winding and $\mathbf{\Psi}_s$ is the total flux linkage space vector of the stator, in stator coordinates, as given by (9.37).

Since the rotor windings are short-circuited in an induction motor, the rotor terminal voltage is zero. Therefore the rotor voltage equation may be expressed in rotor coordinates as

$$\mathbf{U}_r = 0 = R_r\mathbf{I}_{r(r)} + \frac{d\mathbf{\Psi}_{r(r)}}{dt} \tag{9.40}$$

Here R_r is the resistance per rotor phase of each of the direct and quadrature axis windings and $\mathbf{\Psi}_{r(r)}$ is the total flux linkage space vector of the rotor in rotor coordinates. $\mathbf{\Psi}_{r(r)}$ is given by (9.38). By substituting the expression given in (9.38) and differentiating, we may write the rotor voltage equation as follows:

$$0 = R_r\mathbf{I}_{r(r)} + \frac{d}{dt}(L_r\mathbf{I}_{r(r)} + M\mathbf{I}_{s(s)}e^{-j\theta_r}) \tag{9.41}$$

9.4 EQUATION FOR THE ELECTROMAGNETIC TORQUE IN AN INDUCTION MACHINE USING SPACE VECTORS

Torque is created in an electric motor by the interaction of a magnetic field and a current. The simplest example is the DC motor. In this machine, the field current produces the field flux. The armature current creates the armature m.m.f. (called the armature reaction m.m.f.) and flux, which in effect makes the armature into an electromagnet whose poles are oriented at 90° with respect to the main field poles. The torque developed by the motor results from the mutual force between the stator and the rotor magnets tending to align themselves. In a DC motor whose brushes are placed in the correct angular position, the two magnets are at right-angles, and the switching action of the commutator and brushes maintains this 90° orientation at all times as the rotor rotates. In an ideal induction motor also, the stator and the rotor behave as magnets, and the torque is developed by their tendency to align themselves. But the angle may not be 90° as in the DC motor. In the induction motor with the ideal assumption of sinusoidal flux density distribution, the torque is proportional to the sine of the angle between the stator and rotor poles. This will be a maximum when the angle is maintained at 90° as in a DC motor.

The production of the force on the individual conductor of the rotor, and therefore the production of the torque of the machine, is basically due to the interaction of a current and a flux. If the current and the flux linkage are expressed as space vectors, it is possible to express the torque as the "cross-product" of these two space vectors. The cross-product is obtained by the rules of vector multiplication of vector algebra. For this, we multiply the magnitudes of the two vectors and the sine of the angle between them. The expression for the torque may be written as

$$\mathbf{T} = K\mathbf{\Psi}_{r(r)} \times \mathbf{I}_{r(r)} \qquad (9.42)$$

Here K is a proportionality constant, $\times$ indicates vector multiplication, and $\mathbf{\Psi}_{r(r)}$ and $\mathbf{I}_{r(r)}$ are respectively the flux linkage space vector and the rotor current space vector, both expressed in the rotor reference frame. By the rules of vector multiplication, this can be expressed as

$$T = K\Psi_r I_r \sin \beta \qquad (9.43)$$

where β is the angle between the two space vectors. This angle is measured in the direction from the first vector to the second. The product itself should be treated as another vector, in our case the torque vector. The direction of the vector product is perpendicular to plane of the two vectors, and is given by the right-hand rule. Its direction is the same as the direction in which a right-handed screw will move when we turn in the direction from the first vector to the second. Therefore the order in which we multiply the vectors is important in vector multiplication. If we reverse the order, the direction of the cross-product vector will be reversed.

The "reaction torque" experienced by the stator will have the same *magnitude,* and may be expressed in a similar manner as (9.42):

$$\mathbf{T} = K\mathbf{\Psi}_{s(s)} \times \mathbf{I}_{s(s)} \qquad (9.44)$$

The torque expressions are valid irrespective of the reference frame in which the two space vectors are expressed, as long as both are expressed in the same reference frame. This is evident because if both space vectors are changed to a new reference frame, their mutual angle does not change, and neither do their magnitudes.

9.5 VECTOR CONTROL STRATEGY FOR AN INDUCTION MOTOR

The induction motor is a singly fed machine, receiving electrical input from the stator side only. The input current into the stator is ultimately responsible for creating both the flux in the machine and the current, which together produce the torque. The objective in vector control of an induction motor is to separate out the stator current component responsible for creating the flux and the one responsible for creating the torque, and control the two independently, in a decoupled manner, as in a separately excited DC motor.

There are different ways of implementing the vector control strategy according to the choice of the reference frames for the space vectors concerned. A typical choice is the reference frame fixed to the stator flux linkage space vector and oriented in the direction of this space vector as the reference direction. Alternatively, it could be a reference frame fixed to the rotor flux linkage space vector and oriented in its direction. Another alternative is the magnetizing current space vector. The "magnetizing current" may be defined by reference to the equivalent circuit of the induction motor, which is reproduced in Fig. 9.9. In this equivalent circuit, all the quantities are referred quantities, referred to the stator. There are three currents, namely the stator current I_1, the rotor current referred to the stator I_2' and the magnetizing current labeled I_m in the figure. As can be seen, this current I_m, is the stator equivalent of the current responsible for the flux linkage that links with both the rotor and the stator, that is, the mutual flux linkage.

The most convenient and commonly used choice of reference frame for the vector control of an induction motor is the reference frame fixed to the space vector representing the total flux linkage of the rotor, which includes the mutual flux linkage due to I_m in Fig. 9.9 and the rotor leakage flux represented by X_2' in the same figure.

The total flux linkage space vector of the rotor will be the sum of the self-induced flux linkages due to the rotor current space vector and the

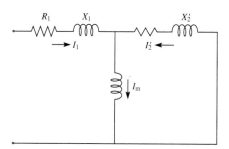

FIGURE 9.9
Currents responsible for the different flux linkages.

mutually induced flux linkage due to the stator current space vector. On this basis, the total rotor flux linkage space vector may be written in the stator reference frame as

$$\mathbf{\Psi}_{r(s)} = L_r \mathbf{I}_{r(r)} e^{j\theta_r} + M\mathbf{I}_{s(s)} \tag{9.45}$$

Here L_r is the self-inductance of the rotor, assumed constant (no magnetic saturation). The factor $e^{j\theta_r}$ is the vector rotator to transform to the stator coordinates.

We shall now define a rotor magnetizing current space vector $\mathbf{I}_{mr}$ in stator coordinates, which will give the total rotor flux linkage (9.45) when multiplied by the mutual inductance M, that is

$$M\mathbf{I}_{mr(s)} = \mathbf{\Psi}_{r(s)} = L_r \mathbf{I}_{r(r)} e^{j\theta_r} + M\mathbf{I}_{s(s)} \tag{9.46}$$

On division by M, this gives the rotor magnetizing current space vector $\mathbf{I}_{mr}$ as

$$\mathbf{I}_{mr(s)} = \frac{L_r}{M} \mathbf{I}_{r(r)} e^{j\theta_r} + \mathbf{I}_{s(s)}$$

$$= \mathbf{I}_{s(s)} + (1 + \sigma_r)\mathbf{I}_{r(r)} e^{j\theta_r} \tag{9.47}$$

Here σ_r is the rotor leakage factor defined by

$$1 + \sigma_r = \frac{L_r}{M} \tag{9.48}$$

We may write the following expression for the rotor current space vector in the stator reference frame using (9.47):

$$\mathbf{I}_{r(s)} = \frac{\mathbf{I}_{mr(s)} - \mathbf{I}_{s(s)}}{1 + \sigma_r} \tag{9.49}$$

Torque equation in terms of I_{mr}. We may now substitute for the flux linkage space vector (9.46) and the rotor current space vector (9.49) in the expression for torque as given by (9.42), *but in the stator coordinates*;

$$T = KM\mathbf{I}_{mr(s)} \times \frac{\mathbf{I}_{mr(s)} - \mathbf{I}_{s(s)}}{1 + \sigma_r} \tag{9.50}$$

The cross product of $\mathbf{I}_{mr(s)}$ with itself yields zero, because both are the same vectors and the cross-product of a vector with itself is zero. Therefore the expression (9.51) for the torque may be written as

$$T = KM\mathbf{I}_{mr(s)} \times \frac{-\mathbf{I}_{s(s)}}{1 + \sigma_r} \tag{9.51}$$

By reversing the order of multiplication, and using a new constant of proportionality, this expression for the torque may be written as

$$\mathbf{T} = K_1(\mathbf{I}_{s(s)} \times \mathbf{I}_{mr(s)}) \tag{9.52}$$

By the rules of vector multiplication this may be written as

$$T = K_1 I_s I_{mr} \sin \delta \tag{9.53}$$

in terms of the magnitudes and the mutual angle of the vectors. Equation (9.53) is the fundamental relationship at the basis of vector control. We may express its meaning as follows. The torque generated by the motor is proportional to the product of the magnitudes of the stator current space vector $\mathbf{I}_s$ and the rotor field current space vector $\mathbf{I}_{mr}$, and the sine of the angle between them. The space vectors may be expressed in any reference frame, as long as both are expressed in the same one, because the mutual angle and magnitudes do not change when both are transformed to a new reference frame.

9.6 THE FIELD-ORIENTED REFERENCE FRAME

For the implementation of vector control using (9.52) or (9.53), it is convenient to use a reference frame fixed to the rotor magnetizing current space vector $\mathbf{I}_{mr}$ and oriented in the direction of $\mathbf{I}_{mr}$. This means that the reference zero direction will be that of the maximum rotor flux linkage, including the mutual flux linkage and the leakage flux linkage of the rotor. This is a rotating reference frame, rotating with the total flux linkage space vector of the rotor. It is called the "field frame," and the coordinates of a space vector in this reference frame are called the "field coordinates." Vector control implemented using the field frame is also known as "field oriented control."

In this reference frame, the rotor magnetizing current space vector will be oriented in the direction of zero angle, that is, the reference direction. This means that, in this reference frame, the space vector representation of the magnetizing current I_{mr} will be simply its magnitude with the angle coordinate zero. Let us denote by ρ the angular orientation of this field reference frame with respect to the stationary (stator) reference frame. This angle will be varying with time as the field frame rotates. Therefore, if $\mathbf{I}_{mr}$ is the rotor magnetizing current space vector as expressed in the stator frame, it will be equal to $\mathbf{I}_{mr}e^{-j\rho}$ in the field frame. The stator current space vector $\mathbf{I}_{s(s)}$ in the stator reference frame may be expressed in the field reference frame as $\mathbf{I}_{s(s)}e^{-j\rho}$. We shall denote by δ the angular coordinate of the stator current space vector in the field frame. Therefore, in the field frame, the stator current space vector may be expressed as

$$\mathbf{I}_{s(f)} = I_s e^{j\delta} \tag{9.54}$$

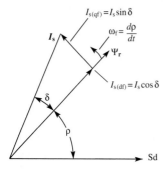

FIGURE 9.10
Torque and field components of the stator current space vector in the field frame.

Here I_s is the magnitude and δ is the phase angle with respect to the magnetizing current space vector $\mathbf{I_{mr(f)}}$ in the field frame. The stator current space vector may be expressed in rectangular coordinates in the field frame as

$$\mathbf{I_{s(f)}} = I_{s(df)} + jI_{s(qf)} \tag{9.55}$$

Here the two components are the direct axis component $I_{s(df)}$ and the quadrature axis component $I_{s(qf)}$ in the field frame. These are respectively

$$I_{s(df)} = I_s \cos \delta \tag{9.56}$$

$$I_{s(qf)} = I_s \sin \delta \tag{9.57}$$

Figure 9.10 shows the angular orientations of the stator current space vector in the field frame and the angular orientation ρ of the field frame relative to the stator frame at an instant of time.

Using (9.57), the expression for the torque (9.53) may be written as

$$T = K_1 I_{mr} I_{s(qf)} \tag{9.58}$$

This equation shows that the quadrature component $I_{s(qf)}$ of the stator current space vector in field coordinates is the torque-producing component. The direct component $I_{s(df)}$ in field coordinates is the field component responsible for the field flux. This component can be seen in Fig. 9.10 to be in alignment with the rotor field flux linkage space vector $\mathbf{\Psi_r}$.

The methodology of vector control of an induction motor using the field frame. Vector control is implemented by controlling the torque component and the field component separately in the same way as in the closed-loop control of a separately excited DC motor. The torque component $I_{s(qf)}$ is controlled in a manner analogous to the control of the armature current in a DC motor. The field component $I_{s(df)}$ is controlled in the same manner as the control of the field current in a DC motor controller. The two controls can be effected in a decoupled manner, and can result in the same level of high dynamic performance as in the DC motor.

We have stated that the field frame is oriented at an angle ρ with respect

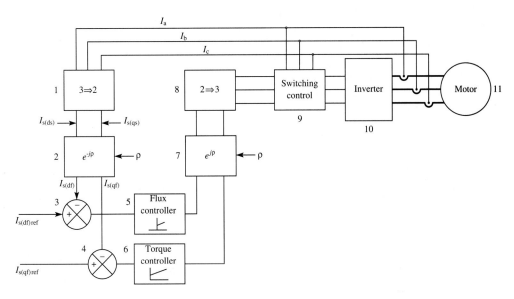

FIGURE 9.11
Methodology of vector control when values of the references for the direct and quadrature axis stator current components in the field frame as well as the angular coordinate ρ of the field frame are available.

to the stationary reference frame. If this information ρ is available in the form of a signal voltage proportional to it then we may explain the methodology of vector control as follows, by reference to Fig. 9.11. Later, we shall describe how this angle signal voltage can be acquired.

The block schematic of Fig. 9.11 is limited to the description of how the direct and the quadrature axis components of the stator current space vector in the field frame are controlled. It is assumed that the controller provides the reference values for these components $I_{s(df)}$ and $I_{s(qf)}$ through other circuit blocks that are not shown in this figure, but which are part of the complete controller and will be described later. It is also assumed that the signal representing the angle ρ of the field frame is available. The method of obtaining it will also be explained later.

In Fig. 9.11, the sensed values of the three motor phase currents I_a, I_b and I_c are input to the circuit block labeled 1. This is a computational block for transformation from three-phase to equivalent two-phase. The outputs of block 1 are the direct axis component $I_{s(ds)}$ and the quadrature axis component $I_{s(qs)}$ of the actual sensed values of the motor current in the stator reference frame. These are transformed into the field reference frame by the vector rotator block, which is labeled 2. For performing this vector rotation, block 2 must have the signal voltage ρ, which we have assumed is available from other circuit blocks not shown in this figure. The outputs of block 2 are the direct and quadrature axis components of the stator current space vector, $I_{s(df)}$ and

$I_{s(qf)}$, in the field reference frame. The error detector modules labeled 3 and 4 compare these outputs of block 2 against the respective references, which are provided by other circuit blocks not shown in the figure. The output of block 3 is the field current component error $I_{s(df)ref} - I_{s(df)}$. The output of block 4 is the torque current component error $I_{s(qf)ref} - I_{s(qf)}$. These two error voltages are amplified by the respective control amplifiers labeled 5 and 6. The former is the flux controller and the latter is the torque controller. The outputs of the two controllers are in the field coordinates. They are transformed into the stator coordinates by the vector rotator block labeled 7. The outputs of the vector rotator block 7 are two-phase values. They are transformed back to equivalent three-phase values by the $2 \Rightarrow 3$ transformation block labeled 8. The outputs of block 8 are given to the switching control block, labeled 9, of the inverter. The inverter block is labeled 10, and is a current regulated three-phase inverter. Current regulation may be by means of the hysteresis type of switching or the PWM type of switching. The switching control block implements the switchings of the inverter in such a way as to make the individual currents conform to the reference values. The output of the inverter are fed to the three-phase induction motor labeled 11 in the figure.

9.7 ACQUISITION OF THE ROTOR FLUX LINKAGE VECTOR

From the above description of the vector control methodology, it is evident that the acquisition of the rotor flux linkage space vector is crucial to the implementation of vector control. There are different alternative ways to acquire this signal. Depending on the manner of acquisition of the flux linkage signal, vector control implementation methods are classified as direct and indirect methods. In a direct method, the flux linkage signal is acquired directly either by having special sensors on the machine that sense the field, or by determining the flux linkage space vector directly by a so-called flux model. In an indirect method, the speed of the flux linkage space vector with respect to the rotor (the so called slip speed) is determined and integrated to determine the angle of movement of the field with respect to the rotor. This is added to the measured angle moved by the rotor to obtain ρ. In this chapter, we shall limit our treatment to a direct method. A direct method using flux sensors will mean that the motor should be specially equipped with flux sensors, which may be coils or Hall effect sensors. In any case, this will need specially built machines, and is therefore not applicable generally to all induction motors. Therefore the direct method that we shall describe will use a "flux model."

To derive a flux model, let us first recall (9.40), which is the rotor voltage equation of the machine in space vector form in the rotor reference frame:

$$0 = R_r \mathbf{I_r} + \frac{d\mathbf{\Psi_r}}{dt} \tag{9.40}$$

From (9.46), the rotor flux linkage in stator coordinates is given by

$$\boldsymbol{\Psi}_{\mathbf{r(s)}} = M\mathbf{I}_{\mathbf{mr(s)}} \tag{9.46}$$

In the field reference frame, the magnetizing current space vector has zero phase angle and has magnitude I_{mr}. Therefore this space vector in the stator frame may be expressed as

$$\mathbf{I}_{\mathbf{mr(s)}} = I_{mr}e^{j\rho}$$

where ρ is the angular displacement of the field frame with respect to the stator frame. Therefore we may write (9.46)

$$\boldsymbol{\Psi}_{\mathbf{r(s)}} = MI_{mr}e^{j\rho} \tag{9.59}$$

In the rotor frame, this will become

$$\boldsymbol{\Psi}_{\mathbf{r(r)}} = MI_{mr}e^{j(\rho - \theta_r)} \tag{9.60}$$

Therefore

$$\frac{d\boldsymbol{\Psi}_{\mathbf{r(r)}}}{dt} = Me^{j(\rho - \theta_r)}\frac{dI_{mr}}{dt} + j\left(\frac{d\rho}{dt} - \frac{d\theta_r}{dt}\right)MI_{mr}e^{j(\rho - \theta_r)} \tag{9.61}$$

Here $d\rho/dt$ is the angular velocity of the field frame, which we shall denote by ω_f, and $d\theta_r/dt$ is the angular velocity of the rotor, which we shall denote by ω_r. Therefore we can write (9.61) as

$$\frac{d\boldsymbol{\Psi}_{\mathbf{r(r)}}}{dt} = Me^{j(\rho - \theta_r)}dI_{mr} + j(\omega_f - \omega_r)MI_{mr}e^{j(\rho - \theta_r)} \tag{9.62}$$

The rotor current space vector $\mathbf{I}_{\mathbf{r(s)}}$ in the stator frame is given by (9.49) as

$$\mathbf{I}_{\mathbf{r(s)}} = \frac{\mathbf{I}_{\mathbf{mr(s)}} - \mathbf{I}_{\mathbf{s(s)}}}{1 + \sigma_r}$$

This may be expressed in the rotor frame as

$$\mathbf{I}_{\mathbf{r(r)}} = \frac{\mathbf{I}_{\mathbf{mr(s)}} - \mathbf{I}_{\mathbf{s(s)}}}{1 + \sigma_r}e^{-j\theta_r} \tag{9.63}$$

We can now substitute (9.62) and (9.63) into the rotor voltage equation (9.40), which is recalled above. This gives

$$0 = R_r\frac{\mathbf{I}_{\mathbf{mr(s)}} - \mathbf{I}_{\mathbf{s(s)}}}{1 + \sigma_r}e^{-j\theta_r} + Me^{j(\rho - \theta_r)}\frac{dI_{mr}}{dt} + j(\omega_f - \omega_r)MI_{mr}e^{j(\rho - \theta_r)} \tag{9.64}$$

We can simplify (9.64) by multiplying throughout by $e^{-j(\rho-\theta_r)}$. This procedure gives

$$0 = R_r \frac{\mathbf{I}_{mr(s)} - \mathbf{I}_{s(s)}}{1 + \sigma_t} e^{-j\rho} + M \frac{dI_{mr}}{dt} + j(\omega_f - \omega_r)MI_{mr} \tag{9.64'}$$

We shall further multiply throughout by $(1 + \sigma_r)/R_r$, to give

$$0 = (\mathbf{I}_{mr(s)} - \mathbf{I}_{s(s)})e^{-j\rho} + \frac{L_r}{R_r} \frac{dI_{mr}}{dt} + j\frac{L_r}{R_r}(\omega_f - \omega_r)I_{mr} \tag{9.65}$$

In writing (9.65) we have replaced $(1 + \sigma_r)M$ by L_r.

The ratio L_r/R_r is the inductive time constant of the rotor circuit. We shall denote it by τ_r. Also, the space vector $\mathbf{I}_{mr}$ in (9.65) is in the stator frame. Since this frame makes an angle $-\rho$ with respect to the field frame and the angle coordinate of the same space vector is zero in the field frame, we may write this space vector in polar coordinates in the stator frame as $I_{mr}e^{j\rho}$. With these substitutions, (9.65) becomes

$$\mathbf{I}_{s(s)}e^{-j\rho} = I_{mr} + \tau_r \frac{dI_{mr}}{dt} + j\tau_r(\omega_f - \omega_r)I_{mr} \tag{9.66}$$

Here the term on the left-hand side is the stator current space vector in the field frame expressed in polar form. We may express it in the rectangular form and rewrite the equation as follows:

$$I_{s(df)} + jI_{s(qf)} = I_{mr} + \tau_r \frac{dI_{mr}}{dt} + j\tau_r(\omega_f - \omega_r)I_{mr} \tag{9.67}$$

We can equate the real and imaginary parts in (9.67) separately, and write the following two equations:

$$I_{s(df)} = I_{mr} + \tau_r \frac{dI_{mr}}{dt} \tag{9.68}$$

$$I_{s(qf)} = \tau_r(\omega_f - \omega_r)I_{mr} \tag{9.69}$$

Equation (9.68) is a differential equation, which can be symbolically written, using the operator p for differentiation with respect to time, as

$$I_{s(df)} = (1 + \tau_r p)I_{mr} \tag{9.70}$$

The solution of the differential equation (9.70) gives I_{mr}. The solution may be written symbolically as

$$I_{mr} = \frac{1}{1 + \tau_r p} I_{s(df)} \tag{9.71}$$

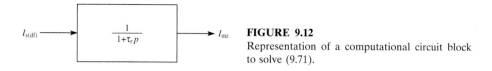

FIGURE 9.12
Representation of a computational circuit block to solve (9.71).

A computational circuit block designed to solve the above differential equation may be symbolically represented as in Fig. 9.12.

Using (9.69), the angular velocity ω_f of the field frame may be expressed as follows:

$$\omega_f = \frac{I_{s(qf)}}{\tau_r I_{mr}} + \omega_r \qquad (9.72)$$

Using (9.71) and (9.72), we can obtain a flux model for computing I_{mr} and ρ as shown in Fig. 9.13. To explain the flux model of Fig. 9.13, we may start from the signals labeled I_a, I_b and I_c, which are the sensed values of the instantaneous stator input current in the three phases at any instant. These signals are transformed into equivalent two-phase currents by the $3 \Rightarrow 2$ phase transformation block, which was shown earlier in Fig. 9.5(a). The signal outputs of the $3 \Rightarrow 2$ transformation block are the direct and quadrature axis components of the stator current space vector in the stationary frame of reference, which are labeled $I_{s(ds)}$ and $I_{s(qs)}$ respectively. These two signals constitute the inputs into the vector rotator block labeled $e^{-j\rho}$. The vector rotator block needs the angle signal ρ to do the vector rotation. We shall show that this can be obtained from a later stage in the flux model. The vector rotator block transforms the inputs into the field coordinates. The outputs of the vector rotator block are the direct axis and quadrature axis components of the stator current space vector in field coordinates, which are $I_{s(df)}$ and $I_{s(qf)}$, respectively. The direct axis component $I_{s(df)}$ is used as the input into the circuit block that we showed earlier in Fig. 9.12. This circuit block outputs the rotor magnetizing current space vector I_{mr} in the field frame, that is, the modulus or magnitude of this space vector. For doing this, the circuit block also needs the data on rotor time constant τ_r. We may mention here that the exact value of the time constant varies somewhat, because the rotor resistance

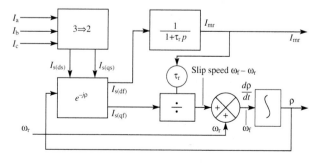

FIGURE 9.13
Flux model for computing I_{mr} and ρ.

changes to some extent with temperature. Therefore, as the motor delivers the load, the rotor temperature rises, and so there is a decrease in the time constant τ_r. The variation of inductance due to the nonlinearity of the magnetic circuit, which may result from the onset of magnetic saturation, is another reason that may introduce inaccuracy in the value of τ_r. These are causes of inaccuracy in the working of the controller. Methods of overcoming these errors are beyond the scope of our treatment and will not be discussed here. The signal I_{mr} that is obtained from this block is multiplied by τ_r by the multiplier block shown as a circle. The output of the multiplier block is $\tau_r I_{mr}$, and this is given as the input to the divider block indicated by the division symbol in the figure. The other input to the divider block is the signal representing the quadrature axis stator current component signal $I_{s(qf)}$ obtained from the vector rotator. The output of the divider block is $I_{s(qf)}/\tau_r I_{mr}$, which is the first term on the right-hand side in (9.72). Equation (9.72) states that this is equal to $\omega_f - \omega_r$, which is the angular speed of the field vector with respect to the rotor, and is known as the "slip speed." The speed of the field frame ω_f is obtained by adding to the slip speed the actual sensed rotor speed ω_r obtained from a speed sensor. The addition is done by the summing block shown. The output of the summing block, which is $d\rho/dt$, is integrated by the integrator block, shown by the integration signal. The output signal of the integrator is the angle signal ρ. This is one of the two outputs of the flux model, the other being I_{mr}. The output ρ is also internally fed back to the vector rotator inside the flux model, as we stated earlier, and indicated in the figure.

9.8 EXAMPLE OF A COMPLETE VECTOR CONTROL SCHEME FOR AN INDUCTION MOTOR

We have now prepared the ground to describe a complete vector control scheme for an induction motor using the field-oriented reference frame of the total rotor flux linkage space vector. We have chosen a direct implementation method using a flux model, which we derived earlier and showed in Fig. 9.13, for the acquisition of the flux linkage space vector. Vector control can be implemented by impressed voltages in the stator using a voltage source inverter or impressed currents in the stator by means of a current regulated inverter. Implementation using a current regulated inverter is simpler, and therefore the scheme we have chosen to describe is based on a controlled current inverter, which may be of hysteresis or controlled current PWM switching type. The latter type has the advantage that the maximum switching frequency is predetermined.

We have stated earlier that vector control functions similar to the control of a separately excited DC motor, where the armature current and the field current are separately controlled in a decoupled manner. In such a DC motor controller, there is typically an outer speed control loop and an inner current (torque) control loop for the armature current control. The same policy is used in the vector control scheme described below, for the induction motor. For describing the features of the scheme, we have identified the different functional circuit blocks, which totally make up the controller, in Fig. 9.14. These are basically computational blocks that we have individually described earlier.

This scheme has separate control loops for the control of the torque by adjusting the torque component $I_{s(qf)}$ and for adjusting the field component $I_{s(df)}$ of the stator current space vector in the field frame. The torque control loop is implemented by an outer speed control loop. In these respects, the control policy is identical to that for a separately excited DC motor. As in the case of the DC motor, the field control policy is as follows. Below what is known as the "base speed," the field current reference is maintained at its fixed maximum value, corresponding to the field flux under rated conditions of the motor. Therefore, in the lower speed range, which corresponds to the speeds below the base speed, there are no changes in the demanded field component. Speed changes in this range are implemented by the torque control loop. Therefore, in this range, the machine can develop up to the same maximum torque at all speeds. This speed range is therefore called the constant-torque range. Above the base speed, the controller reduces the field current reference, to enable the speed to go up. But an increase in the maximum value of the torque current reference is not possible, because the inverter current cannot be made to exceed the rated value. Therefore there is a proportionate reduction in the torque of the machine as the field is weakened. Therefore the maximum power that the machine can develop will remain approximately constant above the base speed. Therefore the speed range above base speed, which is the field-weakening range, is called the constant-horsepower region. In these respects also, the speed control strategy is identical to the case of the separately excited DC motor.

The operation of the control scheme of Fig. 9.14 may be described specifically as follows. The commanded speed is given as the speed reference signal, labeled ω_{ref}, to the error detector block labeled 1. The second input to block 1 is the actual speed, labeled ω_r, fed back from a tacho, labeled 16, or other speed sensing device. The output of the block 1 is the speed error $\omega_{ref} - \omega_r$. This is the input to the speed control amplifier, which may be a PI controller. It is labeled 2 in the figure. The output of the speed controller serves as the reference input to the inner torque control loop. This reference input is labeled $I_{s(qf)ref}$ and is fed to the torque error detector, labeled 3. The second input to this error detector is the prevailing value of the torque component, labeled $I_{s(qf)}$. This is computed by the flux model block, which is the same as the flux model described earlier in Fig. 9.13. The flux model block is labeled 10 in the figure. Its inputs are the sensed values of the motor stator current and the speed, and its internal functional organization was shown in Fig. 9.13. $I_{s(qf)}$ is the signal output by the vector rotator block in Fig. 9.13. The error detector determines the torque component error $I_{s(qf)ref} - I_{s(qf)}$. This is the input to the second error amplifier, labeled 4, which is the torque

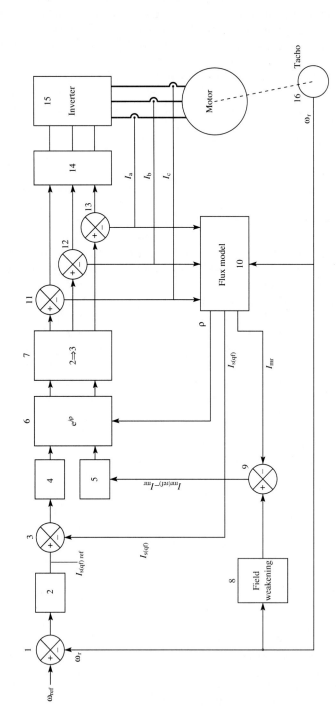

FIGURE 9.14
Rotor flux-oriented vector control scheme using a current controlled inverter.

error amplifier. The output of the torque controller is the input reference for implementing the commanded value of the quadrature component $I_{s(qf)}$ of the stator current in the field reference frame. This is one of the inputs to the circuit block labeled 6. The second input to 6 is the input reference for implementing the commanded value of the direct component $I_{s(df)}$ of the stator current space vector in the field frame, which is obtained from the block labeled 5, the field error amplifier. The input to the field error amplifier is the field current error $I_{mr(ref)} - I_{mr}$ obtained from block 9, which is the field error detector. The inputs to the field error detector are the field current reference and the computed value of the actual prevailing field current obtained from the flux model. The field current reference is provided by the field-weakening block, labeled 8. The field-weakening module gives a constant output for all speeds below the base speed, but starts to decrease the output when the sensed speed exceeds the base speed.

The circuit block 6 is a vector rotator that transforms the inputs from the field frame coordinates to the stator frame coordinates. The angle signal ρ needed for this is obtained from the flux model (block 10). The output of the vector rotator is in equivalent two-phase values. These are transformed into the equivalent three-phase values by the $2 \Rightarrow 3$ transformation block, labeled 7. The outputs of block 7 control the switching of the inverter to make the individual line currents conform to the reference values. The comparison blocks, labeled 11, 12 and 13, and the switching control block, labeled 14, serve to achieve the purpose of making the line current outputs from the inverter conform to the reference values, with negligible error. The inverter, labeled 15, is a current controlled inverter in which the current control may be by the hysteresis type of switching or by PWM switching. In either case, the output currents are made to conform to the computed reference value in each individual line of the three-phase supply to the motor.

It is evident from Fig. 9.14 that vector control uses a number of computational blocks that have to perform on-line computations for the satisfactory working of the controller. These computations can be done by means of analog or digital circuits. With analog methods, it is generally more difficult to achieve the required accuracy. By using a dedicated microprocessor, it is possible to implement the required computations reliably and with adequate speed and accuracy. It is this facility for reliable practical implementation using digital signal processing that has made it possible to realize high dynamic performance in AC drives using the vector control principle. This is contributing to the increasing use of AC motor drives with vector control at the present time.

9.9 VECTOR CONTROL OF SYNCHRONOUS MOTOR DRIVES

In a synchronous motor, the field poles are fixed to the rotor frame. Therefore it is possible to obtain the angle coordinate signal of the field directly using a shaft position sensor, which may be a digital encoder. Because of this, the implementation of vector control is somewhat easier in synchronous motor drive systems than in induction motor drives. A synchronous motor may be of the wound field type or of the permanent magnet type. Permanent magnet synchronous motors with vector control are highly suited for many types of servo motor drives, such as in machine tool feed drives, industrial robots etc. There are basically two categories of permanent magnet synchronous motors, depending on how the magnets are mounted on the rotor. These are

1. motors with surface-mounted magnets on the rotor;
2. motors with interior permanent magnet rotors.

In interior permanent magnet machines, the magnets are usually held inside a steel magnetic core on the rotor. As a result, the outer rotor surface is uniform. Such machines are mechanically sturdier and are more suitable for high speed applications.

The incremental relative permeability of the permanent magnet material is very close to unity. Therefore the region occupied by the permanent magnet is magnetically closely equivalent to an air gap. Rotors with interior permanent magnets have typically a lower reluctance and therefore higher reactance along the quadrature axis as compared with the direct axis. This is in contrast to the field wound salient pole synchronous motor, which typically has a lower reactance along the quadrature axis, because of the larger air gap and therefore larger reluctance along this axis. Also, in the interior permanent magnet machines, eddy currents are liable to be created in the rotor casing, which therefore makes the casing behave similarly to damper windings. For an accurate dynamic modeling of the flux linkage, this effect as well as the "reverse saliency," meaning the lower quadrature axis reluctance as compared with the direct axis reluctance, should be taken into account.

In the permanent magnet machine with surface-mounted magnets, the space between the magnetic poles is filled with nonmagnetic material. Because the material of the permanent magnet has a relative incremental permeability close to unity, the effective air gap is large and almost uniform. Also, the resistivity of the permanent magnet material is high. Therefore it has effectively no damper winding effect. (The armature reaction effects are less because of the larger effective air gap.) Therefore damper winding effects and saliency can be ignored in modeling the flux. These effects generally will have to be considered also in a wound field salient pole machine with damper windings.

For a simplified description of vector control using a synchronous motor, we have therefore chosen a drive employing a synchronous motor with surface-mounted permanent magnets. In such a machine, the torque component of the stator current vector is the quadrature axis component in the field reference frame. Since this is a permanent magnet machine, the field flux is constant. But it is possible to achieve field-weakening above the base speed by introducing a negative direct axis component in the armature current space vector in the field frame. But, because the effective air gap is large, and therefore the effective reluctance seen by the armature reaction m.m.f. is large, the extent of field-weakening possible in this manner is not large. Figure 9.15 shows the vector control scheme for a synchronous motor with surface-mounted permanent magnets. This scheme does employ field-weakening above the base speed.

This scheme employs vector control in the rotor field-oriented reference frame. The angle signal θ_r for the rotor field is obtained using a shaft position sensor, which may be a digital encoder. From the same signal, the angular velocity of the rotor is also computed, by the differentiating circuit block labeled 1. The commanded speed is given to the controller by means of the speed reference signal labeled ω_{ref}. The error detector block, labeled 2, compares the reference speed against the actual speed ω and outputs the speed error signal $\omega_{ref} - \omega$. The error amplifier, which may be a PI controller, labeled 3, is the speed controller, which amplifies the speed error. The output of the speed controller is the torque reference, which is the reference for the quadrature axis component of the stator current space vector in the field frame, which we have indicated as $I_{s(qf)ref}$. The error detector block, labeled 4, compares the torque component reference with the computed value of the prevailing torque component, labeled $I_{s(qf)}$, obtained from the vector rotator block, labeled 15. The vector rotator block (15) outputs both the direct axis component $I_{s(df)}$ and the quadrature axis component $I_{s(qf)}$ of the prevailing value of the stator current space vector, by transforming the same quantities from the stator frame into the field frame. The inputs to the vector rotator block 15 are obtained from the $3 \Rightarrow 2$ transformation block, labeled 14. The inputs to block 14 are the sensed values of the three line currents of the motor, obtained through the fast current sensing devices shown symbolically in the figure. The output of block 4 is the error in the torque component $I_{s(qf)}$ in the field frame. The field-weakening block is labeled 16 in the figure. This provides the reference for the direct axis component of the stator current space vector in the field frame. This reference is zero for all speeds below the base speed and negative for speeds above the base speed. The block labeled 17 is the error detector for the direct axis component of the stator current space vector in the field frame, which we have indicated by $I_{s(df)}$. The output of block 17 is given to the field component controller, which is the error amplifier block lableld 6. The input of 6 is the error in the direct axis component in the field frame. The outputs of the controllers 5 and 6 are the reference value commands for the stator current space vector. These are in the field frame. These are transformed into the stationary frame by the vector rotator block, labeled 7. These are again transformed into the corresponding three-phase values by the $2 \Rightarrow 3$ transformation block, labeled 8. The outputs of this block serve to implement the switching of the inverter in such a way as to make the output current

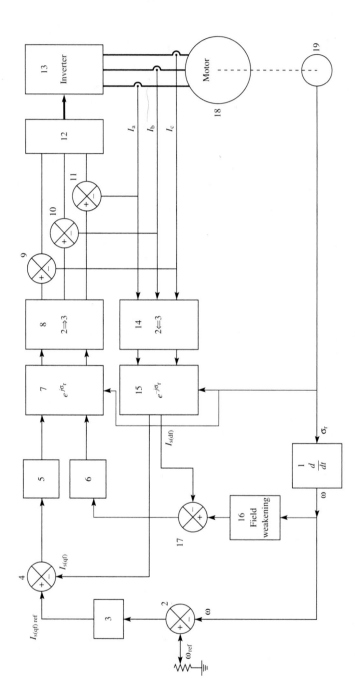

FIGURE 9.15
Vector control scheme for a permanent magnet synchronous motor with surface-mounted magnets.

conform to the commanded reference values of the stator currents. The comparison blocks, labeled 9, 10 and 11, and the switching control block, labeled 12, serve to implement the switching of the inverter to achieve this objective. The inverter itself is a controlled current type. It could be using the hysteresis or the PWM type of current control. Both types are designed to achieve the same purpose, which is to make the output current conform to the reference value at every instant. For regenerative braking, the sign of the torque component will be reversed.

9.10 CHAPTER 9 SUMMARY

In this chapter, we introduced the main features of the technique of vector control, applicable to induction and synchronous types of three-phase AC motors. The sequence of presentation may be summarized as follows.

The objective of vector control is to achieve high dynamic performance in AC motor drive systems, that is, to make the drive system respond in a fast manner to changes in the demanded speed or torque. The separately excited DC motor drive is inherently capable of high dynamic performance, because the field and armature currents are magnetically decoupled and can be controlled independently. Vector control achieves the same level of dynamic performance in a similar manner, by separating out the components of the stator current that are responsible for torque production and for field production, and controlling each component independently in a decoupled manner, in the same way as in a DC motor.

The conventional equivalent circuit approach is not very well suited for the mathematical treatment of the technique of vector control. This is best done by using the concept of space vectors. Therefore we first introduced the space vectors and used them in our description. A sinusoidal m.m.f. and flux density distribution in the air gap of the motor is necessary for the representation by means of space vectors. For the purpose of highlighting this fact, we showed that, in principle, it is possible to achieve a spatial sinusoidal m.m.f. distribution in the air gap of the machine if it can be wound with a sinusoidal conductor density distribution. This, however, is impractical when the coils have to be located in slots with discrete spacing. However, the space vector approach is valid when we separately treat the fundamental sinusoidal component of the spatial m.m.f. distribution.

With the space vector approach, we can combine the individual stator phase currents of the three-phase machine into one single-stator current space vector. The same can be done for voltages and flux linkages. These single space vectors of the three-phase machine can be expressed in rectangular coordinates as a complex number with the real part along the spatial direct axis and the imaginary part along the quadrature axis. In this way, we can transform our three-phase machine into an equivalent two-phase machine in which the windings are spatially in quadrature and therefore have no mutual magnetic coupling. We derived the transformation equations for transformation in either direction, that is from three-phase to two-phase and in reverse. We repeated the procedure for the windings of the machine on the rotor also, which we assumed can be replaced by windings equivalent to those on the stator.

We looked at a space vector of a rotor quantity, and showed how the coordinates change when seen from the reference frame fixed to the stator, when the rotor moves by an angle. We derived the general transformation equations for transforming the coordinates of a space vector between reference frames when there is a relative angular displacement or velocity. We used such transformations to take into account the mutually induced flux linkages between the stator circuit and the rotor circuits, for which we defined a constant mutual inductance parameter M.

Using the concept of space vectors, we could write a single voltage equation for the stator, using space vectors instead of the conventional method of writing separate equations for each phase. The same could be done for the rotor also. We also wrote a single equation for the electromagnetic torque, as the vector product of two space vectors—a current space vector and a flux linkage space vector. We then defined a new reference frame, oriented in the direction of the total flux linkage space vector of the rotor and rotating with it. It was shown that the torque component of the stator current space vector is the quadrature axis component of the stator current space vector in this field reference frame. The field component of the stator current space vector is the direct axis component of the stator current space vector in the same reference frame. We described the methodology of vector control, as independently controlling the direct axis component and the quadrature axis component of the stator current space vector, in the field-oriented reference frame.

To implement vector control in this manner, we need to transform the stator current space vector into the field-oriented reference frame. For this, we needed the angle signal of the field frame with respect to the stationary frame. We derived a dynamic flux model from which the magnitude and angle signals of the rotor magnetizing current can be obtained. This flux model was derived using the rotor voltage equation of the motor. The flux model requires inputs of the stator current signals and the rotor speed signal. It also needs the data relating to the inductive time constant of the rotor. We pointed out that the inductive time constant of the rotor will vary to some extent during operation, owing to the variation of rotor resistance with temperature and the variation of rotor inductance with magnetic saturation. This is a source of inaccuracy for the controller.

Using the flux model, we described a complete vector control scheme, based on the reference frame oriented along the the total flux linkage space vector of the rotor, for a squirrel cage induction motor using impressed stator currents. The stator currents are provided by a current regulated three-phase inverter, which may be of hysteresis or PWM type. We separately showed the different functional blocks that constitute the controller. These are basically computational blocks. The facility to implement these computational blocks using a dedicated microprocessor chip has made it possible to achieve the benefits of vector control in practical induction motor drive systems.

In synchronous motor drive systems, the magnetic poles on the rotor have fixed orientation with respect to the rotor. Therefore rotor position sensors can give the angle signal corresponding to the spatial orientation of the rotor field. This makes it easier to implement vector control in a synchronous machine. However, if the machine has damper windings on the rotor, the currents induced in these will affect the magnitude and orientation of the rotor flux. This and the saliency effect should be taken into consideration in the dynamic modeling of the rotor flux linkage. For a simplified description of vector control, we chose a permanent magnet synchronous motor with surface-mounted magnets, in which the saliency and the damper winding effects may be neglected. We concluded our treatment of vector control with a description of such a scheme using impressed stator currents. The scheme employs field-weakening also above base speed. Field-weakening in this permanent magnet machine is implemented by creating a negative direct axis component in the stator current space vector.

PROBLEMS

9.1. The phase winding of an AC induction motor may be assumed to be ideal and continuously distributed according to (9.2). The total number of turns is 400. The air gap between the stator and the rotor is 3 mm. Assume the magnetic core to have infinite permeability. For a constant current of 8 A through the winding, determine

(a) the flux density along the axis of the winding;

(b) the flux density in a direction making an angle of 60° with the axis of the winding.

9.2. The stator of a two-pole motor has a single full-pitch coil of 50 turns and is carrying a constant current of 12 A. The air gap is 2.5 mm. The mean diameter of the air gap is 24 cm and the effective axial length 30 cm. Assume that the magnetic core material is perfect with infinite permeability.

(a) Sketch the spatial distribution of m.m.f. and the air-gap flux density.

(b) Determine the amplitude of the fundamental Fourier component of the spatial flux density distribution.

(c) Determine the magnitude and direction of the total flux linkage space vector of the coil.

9.3. The currents in the three individual phase windings of a three-phase induction motor are $I_a = 12.94$ A, $I_b = -48.3$ A, and $I_c = 35.36$ A. Determine the magnitude and direction of the resultant stator current space vector. Use the axis of the A-phase winding as the reference axis for expressing angles.

9.4. The instantaneous phase currents in a three-phase synchronous motor are $I_a = -40$ A, $I_b = 80$ A and $I_c = -40$ A. Sketch the phase current space vectors individually and find the resultant stator current space vector. Express the stator current space vector in polar form.

9.5. Table P9.5 gives the instantaneous currents of a three phase machine at intervals of one-eighth of an AC period, starting from the instant of maximum current in the A phase. The fourth row of the table is for entering the magnitude of the resultant stator current space vector. The fifth row is for entering the spatial angle of the resultant stator current space vector. Complete the table by completing the fourth and fifth rows. The latter is to be in degrees (electrical)

TABLE P9.5

i_a	100.00	70.71	0	−70.71	−100	−70.71	0	70.71
i_b	−50	25.88	−86.60	96.59	50	−25.88	−86.60	−96.59
i_c	−50	96.59	−86.60	−25.88	50	96.59	86.60	25.88
I_s (magnitude)								
I_s (direction)								

9.6. In Table P9.5, suitably modify if necessary the angle entries in the fifth row to mechanical angles if the machine is (a) a two-pole machine; (b) a six-pole machine.

9.7. The instantaneous phase voltages of a three-phase motor are $V_a = 84.85$ V, $V_b = 31.06$ V and $V_c = -115.91$ V. Determine the magnitude and direction of the stator voltage space vector.

9.8. The resultant stator voltage space vector at a particular instant of time in a three phase machine may be expressed as $V_s = 150e^{j\pi}$. The b- and c-phase voltages are each 50 V. Determine the a-phase voltage at this instant.

9.9 The individual flux linkages of the three phase windings of a three-phase machine at a particular instant are $\Psi_a = 0.1414$ Wb-turns, $\Psi_b = 0.0518$ Wb-turns and $\Psi_c = -0.1932$ Wb-turns. Determine the resultant stator flux linkage space vector.

9.10. A Δ-connected three-phase induction motor is fed from an auto sequentially commutated thyristor current source inverter. The individual phase current waveforms are shown in Fig. P9.10 (these are in conformity with Fig. 8.6). According to these waveforms, the phase currents stay constant during 60° intervals and undergo changes at the end of each 60° interval. Compute the magnitude and direction of the stator current space vector in each 60° interval, using the phase current values during the interval indicated by the waveforms. Comment on the movement of the stator current space vector.

9.11. The three stator phase currents of a three-phase motor may be expressed as follows:

$$i_a = 20 \cos 377t$$

$$i_b = 20 \cos (377t - \tfrac{2}{3}\pi)$$

$$i_c = 20 \cos (377t - \tfrac{4}{3}\pi)$$

Determine the magnitude and the angular velocity of the stator current space vector in electrical rad/s and in rev/min if the number of poles is (a) 2; (b) 8.

9.12. The instantaneous phase currents of a three phase motor are $i_a = 10$ A, $i_b = -5$ A and $i_c = -5$ A. Determine the direct axis and quadrature axis currents of an equivalent two-phase machine to give the same magnitude and direction for the stator current space vector.

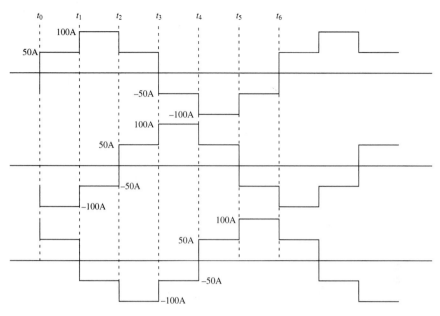

FIGURE P9.10

9.13. Repeat Problem 9.12 for the following values of the instantaneous phase currents: $i_a = 7.07$ A, $i_b = 2.59$ A and $i_c = -9.66$ A.

9.14. The direct and quadrature axis currents in a two-phase machine at a particular instant are $i_d = 5$ A and $i_q = 8.66$ A. Determine the equivalent phase currents of a three-phase machine whose windings are Y-connected with isolated neutral.

9.15. Repeat Problem 9.14 for the following values of the direct and quadrature axis currents:

(a) $i_d = 7.07$ A and $i_q = 7.07$ A;
(b) $i_d = -7.07$ A and $i_q = 7.07$ A.

9.16. The rotor current space vector in rotor coordinates may be expressed at a particular instant of time in rectangular coordinates as

$$\mathbf{I}_{r(r)} = 7.07 + j7.07 \quad \text{A}$$

Transform this into stator coordinates for the following values of the rotor angle with respect to the stator: (a) $\theta_r = 0$; (b) $\theta_r = 45°$; (c) $\theta_r = -45°$.

9.17. The rotor voltage space vector of a three-phase motor may be expressed in rectangular coordinates in the rotor reference frame as

$$\mathbf{V}_{r(r)} = 86.6 + j50 \quad \text{V}$$

Transform this into stator coordinates for the following values of the rotor angle with respect to the stator: (a) $\theta_r = 0°$; (b) $\theta_r = 90°$; (c) $\theta_r = 150°$.

9.18. The phase currents on the stator of a three-phase machine are $i_a = 70.71$ A, $i_b = 25.88$ A and $i_c = -96.59$ A. Determine the resultant stator current space vector and express it in rectangular coordinates in

(a) the stator reference frame;

(b) the rotor reference frame if the rotor makes an angle of 90° with the stator frame.

9.19. The stator current space vector may be expressed in rectangular coordinates at a particular instant in the stator frame as

$$\mathbf{I}_{s(s)} = -50 - j86.6 \quad \text{A}$$

Transform this into rotor coordinates if the rotor is making an angle of 180° with respect to the stator.

9.20. The stator flux linkage space vector may be expressed in stator coordinates as

$$\mathbf{\Psi}_{s(s)} = 0.5 + j0.866 \quad \text{Wb-turns}$$

Transform this into rotor coordinates when the rotor is making an angle of 60° with respect to the stator.

9.21. A three-phase induction motor is operating under steady conditions. The stator is receiving the input from a 60 Hz supply, and therefore the stator current space vector and flux linkage space vector are both rotating with a steady speed of 377 ($= 2\pi \times 60$) electrical rad/s. The motor slip is 4%, and therefore the rotor is rotating at 0.96×377 electrical rad/s. The rotor frequency is 0.04×60, corresponding to the 4% slip. Determine the angular velocity of the rotor current space vector relative to a reference frame fixed to the stator flux linkage space vector.

INDEX